Intermediate Dynamics

Edward J. Haug

Center for Computer Aided Design
and
Department of Mechanical Engineering
College of Engineering
The University of Iowa

PRENTICE HALL Englewood Cliffs, New Jersey 07632

Library of Congress Cataloging-in-Publication Data

Haug, Edward J.
Intermediate dynamics / Edward J. Haug.
p. cm.
Includes bibliographical references and index.
ISBN 0-13-478926-1
1. Machinery, Dynamics of. 2. Dynamics. 3. Matrices.
I. Title.
TJ170.H386 1992
621.8'11--dc20 91-9976
CIP

Acquisitions Editor: Doug Humphrey
Production Editor: Bayani Mendoza de Leon
Cover Designer: Joe Di Domenico
Prepress Buyer: Linda Behrens
Manufacturing Buyers: Patrice Fraccio and Dave Dickey
Supplements Editor: Alice Dworkin

Printed in the United States of America
10 9 8 7 6 5 4 3 2 1

ISBN 0-13-478926-1

9 780134 789262 90000>

ISBN 0-13-478926-1

Prentice-Hall International (UK) Limited, *London*
Prentice-Hall of Australia Pty. Limited, *Sydney*
Prentice-Hall Canada Inc., *Toronto*
Prentice-Hall Hispanoamericana, S.A., *Mexico*
Prentice-Hall of India Private Limited, *New Delhi*
Prentice-Hall of Japan, Inc., *Tokyo*
Simon & Schuster Asia Pte, Ltd., *Singapore*
Editora Prentice-Hall do Brasil, Ltda., *Rio de Janeiro*

Dedicated to

WILLIAM G. NUSSER

Bright, Multifaceted, and Uncompromising as a Fine Diamond

Contents

Preface

The goal of this book is a systematic development of the basic theory and methods of mechanical system dynamic analysis, at a somewhat higher level than is introduced in beginning engineering dynamics courses. The book is intended for use in a course at the junior or senior undergraduate and beginning graduate level, or for self-study by practicing engineers. The development builds on the foundations of basic engineering mechanics, mathematics, and physics and is more rigorous and general than the beginning engineering dynamics course, preparing the student for analysis of the dynamics of broad classes of engineering systems.

Although there are numerous modern textbooks at the beginning and advanced levels, books at the intermediate level are typically several decades old and present precomputer analytical methods. Owing to the complexity and nonlinearity of differential and algebraic equations of motion, conventional practice in teaching dynamics is the presentation of a large number of "principles" and special techniques that yield either closed-form solutions or analytical information regarding solutions for simple systems. The complexity of the equation of dynamics has led to an extensive "bag of tricks" that have historically been used by dynamics specialists to solve problems. In contrast, methods that are suitable for formulation and solution of equations of dynamics with the aid of modern high speed computers place a premium on being systematic in formulation.

A unified virtual work, or variational, approach for formulation of the equations of dynamics of particles and rigid bodies is adopted in this book. This variational theory is consistent with modern methods of structural mechanics that have led to general and widely applicable finite element structural analysis methods. A matrix form of vector analysis is used thoughout the book, leading to systematic formulations of governing equations of motion that are well suited for digital computer numerical solution. Energy and momentum concepts are treated as derived relations rather than as independent principles. The goal in adopting this approach is to develop a unified theory of dynamics that can be clearly understood and broadly applied.

Variational methods for derivation of governing differential equations of motion and systems of differential-algebraic equations for constrained dynamic systems are developed for particles and rigid bodies. Many examples are presented, to illustrate basic ideas and use of the methods presented. Care is taken to develop and illustrate properties of relationships between generalized coordinates and global (or Cartesian) coordinates of particles and bodies in space. Examples of both proper and improper selection of generalized coordinates are presented, to guide the reader in formulation of valid dynamic models and derivation of equations of motion. Counterexamples are presented to illustrate pitfalls that lead to both analytical and computational difficulties.

Chapter 1 contains a review of Newtonian dynamics of particles. Newton's laws of motion for a single particle are reviewed and applied to prototype examples. The basic physical laws of dynamics of systems of particles are presented and used to model the dynamics machines and predict reaction forces between components. Finally, impulse–momentum and kinetic energy–work relations are derived for particles and their use is illustrated.

Chapter 2 introduces concepts of kinematics of particles, using both Cartesian and generalized coordinates to locate particles in space. Essential mathematical properties of relationships between generalized coordinates and Cartesian coordinates are developed and illustrated, including examples of pitfalls that must be avoided. Concepts of holonomic and nonholonomic constraints are introduced. Relationships between virtual displacements and generalized coordinate variations are developed based on concepts of differential calculus.

Chapter 3 develops the variational equations of motion for particles, beginning with Newton's laws of motion for a single particle. Lagrange multipliers are introduced to define constraint reaction forces, which are then used in formulating constrained equations of motion. A general matrix formulation of constrained equations of motion for systems of particles is presented and mathematical properties of constraint equations are identified that guarantee existence and uniqueness of a solution. Pitfalls associated with improper models and formulations of problems are illustrated. Lagrange's equations of motion are derived and their use in applications is studied. Numerical results are presented for selected problems, using methods presented in Appendices B, C, and D. Although the book is not focused on numerical methods, it is important that the student gain confidence that if care is taken in formulating equations, reliable digital computer numerical methods can be used to obtain solutions and insights.

Chapter 4 develops the theory of orientation of a rigid body in space, using basic properties of direction cosine orientation transformation matrices. Angular velocity of a rigid body is introduced from both mathematical and physical points of view. Virtual displacements and virtual rotations of bodies are similarly introduced and their properties are developed, in anticipation of their use in the virtual work formulation of the equations of motion of a rigid body. Finally, Euler parameters and Euler angles are defined as orientation generalized coordinates and their properties are developed.

Chapter 5 presents a self-contained derivation of the equations of motion of a rigid body, modeled as a collection of differential masses that are subject to distance constraints, using only Newton's laws for individual particles. Centroidal body-fixed reference frames are introduced to simplify the equations of motion. The inertia tensor is obtained as a direct result of the derivation. Variational Newton–Euler equations of motion are derived and reduced to the Newton–Euler differential equations of motion for unconstrained bodies in space. Examples that illustrate derivation of the equations of motion of a single constrained rigid body are presented, using the variational approach.

The sixth and final chapter is an introduction to the theory of multibody dynamics. The systematic Cartesian coordinate formulation for constrained systems is developed and illustrated for both planar and spatial systems. The power of combining the variational approach and relative coordinates between constrained systems of bodies is illustrated through application to example problems. Finally, emerging recursive methods for dynamic analysis of multibody systems are introduced.

Four appendices are provided, to make the book as nearly self-contained as possible. Appendix A summarizes vector and matrix algebra notation and basic relations needed throughout. Appendix B presents theoretical and numerical methods for solving linear and nonlinear equations that arise in dynamics. Appendix C summarizes basic results from the theory of ordinary differential equations and differential-algebraic equations that arise in constrained mechanical system dynamics. Finally, Appendix D presents basic numerical methods for solution of differential and differential-algebraic equations.

Examples are presented throughout to assist the reader in understanding how the methods developed are used. The differential and differential-algebraic equations of motion derived in most examples are highly nonlinear, hence they cannot be solved analytically. Numerical results are presented for selected examples, using numerical methods of Appendix D, to provide the reader with confidence that numerical results can be systematically obtained and analyzed to understand the dynamic performance of machines. A computer program that will allow readers to carry out their own numerical calculations is included in the Solutions Manual. Instructors and individual readers may wish to obtain this computer program, or to write their own, to develop experience in modeling and analysis of numerical results. For more detail on computational methods in dynamics, the reader is referred to Refs. 12 and 13.

This book has been used as part of a three course dynamics program for students in mechanical, civil, and related branches of engineering at The University of Iowa. It serves as a junior or senior elective and as a beginning graduate course. For nonspecialists in dynamics, it provides a firm theoretical foundation and experience with methods of formulating governing equations of mechanical system dynamics. For graduate students specializing in the field of machine dynamics, it serves as the foundation upon which advanced theoretical and computational methods of dynamics are based. A two-course sequence in computational methods for kinematics and dynamics of mechanical systems that focuses on methods of computer generation and solution of equations of motion uses this book as both a theoretical foundation and as an introduction to methods and concepts of formulating equations of mechanical system dynamics.

Thanks are due to many colleagues and students who have made contributions to the development of this book, including J. G. Andrews, K. E. Atkinson, D. S. Bae, S. T. Chern, K. T. Jeong, D. Y. Jo, and S. S. Kim. Finally, the author is indebted to C. K. Mills and S. M. Van Fosson for their dedication in preparation of the manuscript and illustrations.

Edward J. Haug

1

Newtonian Dynamics of Systems Modeled as Particles

Newton's equations of motion and conventional vector analysis methods (see Appendix A) are used in this chapter to obtain useful forms of equations of motion of systems in which moving bodies are modeled as particles. While direct application of Newton's equations of motion is shown to be useful in some applications, limitations are noted in their use for analysis of more realistic mechanical systems.

Ideas presented in this chapter for dynamics of particles are applicable to broad classes of dynamic systems. The reader is encouraged to study the underlying concepts and methods carefully, in the context of particles where kinematic complexity is minimal. Once Newtonian methods are understood in the particle setting, they are readily extended to more complex systems, in which kinematic complexities lead to messy expressions and manipulations.

1.1 NEWTON'S LAWS OF MOTION

The physical foundation on which the theory of dynamics is based is a collection of three laws of motion, which are easily stated and understood for *particles,* or *point masses*. These laws are stated and interpreted in this section. They represent the only fundamental physics required for development of a broadly applicable theory of dynamics of mechanical systems.

Definition of position, velocity, and acceleration of a point in space presupposes the existence of some frame of reference. For the purposes of dynamics, it is important that the frame of reference be an *inertial reference frame,* a frame in which Newton's laws of motion are valid. An inertial reference frame may be thought of as a frame fixed in some ideal space relative to which motion of stars, planets, and so on, can be described. In most engineering applications, a reference frame that is fixed to the surface of the earth is employed. This is technically not an inertial reference frame, due to motion of the earth relative to the sun and other stars. Accuracy associated with an *earth-fixed reference frame* is, however, often acceptable for engineering purposes. In an inertial reference frame, the following are *Newton's laws of motion:*

First Law: A particle continues in its state of rest or constant velocity if the resultant force acting on the particle is zero.
Second Law: The product of the mass of a particle and its acceleration is equal to the resultant force acting on the particle.
Third Law: To every action there is an equal and opposite reaction; that is, the mutual forces of interaction between two particles are equal in magnitude and opposite in direction.

In many applications, the forces of action and reaction between pairs of particles are not only equal in magnitude and opposite in direction, as guaranteed by the third law, but they are *collinear;* that is, they lie along the same straight line. This is true for gravitational forces and forces of contact, for example. Forces of interaction due to charged particles in an electric field need not, however, be collinear. The reader should recall, therefore, that in general, Newton's third law does not guarantee that forces of action and reaction are collinear.

The resultant force in Newton's second law, or *Newton's equations of motion,* is the vector sum of all forces that act on the particle. The acceleration of a particle is the second time derivative of the position vector $\mathbf{r} = [x, y, z]^T$ (see Appendix A for vector and matrix notation) of the particle in an inertial reference frame and the mass of a particle is denoted m. Newton's second law may thus be written in vector form as

$$m\ddot{\mathbf{r}} = \mathbf{f} \tag{1.1.1}$$

where $\mathbf{f}$ is the resultant of all forces that act on the particle.

The relation between mass and weight of a particle is established by the characteristic *gravitational acceleration* g, which has magnitude g and is directed toward the center of the earth. The *weight* w of the particle is a scalar quantity that is related to its mass and the gravitational acceleration by

$$w = mg \tag{1.1.2}$$

Units employed in defining positions of particles, forces acting on particles, and masses of particles must be consistently defined in order that Eqs. 1.1.1 and

TABLE 1.1.1 UNITS OF LENGTH, FORCE, AND MASS AND CONVERSION FACTORS (CF) TO STANDARD INTERNATIONAL (SI) UNITS[a]

System	Length	CF	Force	CF	Mass	CF	g
SI	meter (m)	1.0	newton (N)	1.0	kilogram (kg)	1.0	9.8 m/s^2
CGS	centimeter (cm)	10^{-2}	dyne (dyn)	10^{-5}	gram (g)	10^{-3}	980 cm/s^2
English	foot (ft)	0.3048	pound (lb)	4.448	slug (sl)	14.59	32.2 ft/s^2

[a] Multiply by CF to obtain SI units.

1.1.2 may be employed. *Metric units* or *Standard International* (SI) *units* are preferred. *English gravitational units* and alternative forms of metric units have, however, historically been employed. In all systems of units, the unit of time is seconds. Length, force, and mass, however, are characterized differently in the different systems of units. Table 1.1.1 defines units for several of the standard systems and conversion factors to SI units.

Note that if $\mathbf{f}(t)$ and m are known in Eq. 1.1.1, Newton's equations of motion are second-order differential equations. As is well known from the theory of ordinary differential equations (see Appendix C), *initial conditions* must be specified on $\mathbf{r}$ and $\dot{\mathbf{r}}$ at an initial time t_0 to yield a unique solution for $\mathbf{r}(t)$:

$$\begin{aligned} \mathbf{r}(t_0) &= \mathbf{r}^0 \\ \dot{\mathbf{r}}(t_0) &= \dot{\mathbf{r}}^0 \end{aligned} \tag{1.1.3}$$

Equations 1.1.1 and 1.1.3 comprise an *initial-value problem* that determines the motion of a particle.

Example 1.1.1: A projectile of mass 50 kg is launched in the x-y plane at an angle of θ from the horizontal (the x axis), with an initial speed of 700 m/s, as shown in Fig. 1.1.1. Neglecting drag due to its motion in air, Eqs. 1.1.1 and 1.1.3 are

$$\begin{aligned} 50\ddot{\mathbf{r}} &= [0,\ -490,\ 0]^T \\ \mathbf{r}(0) &= [0,\ 0,\ 0]^T \\ \dot{\mathbf{r}}(0) &= 700[\cos\theta,\ \sin\theta,\ 0]^T \end{aligned}$$

The solution of this initial-value problem is

$$\begin{aligned} z(t) &= 0.0 \text{ m} \\ x(t) &= (700\cos\theta)t \text{ m} \\ y(t) &= (700\sin\theta)t - 4.9t^2 \text{ m} \end{aligned}$$

or

$$\mathbf{r}(t) = \begin{bmatrix} (700\cos\theta)t \\ (700\sin\theta)t - 4.9t^2 \\ 0 \end{bmatrix}$$

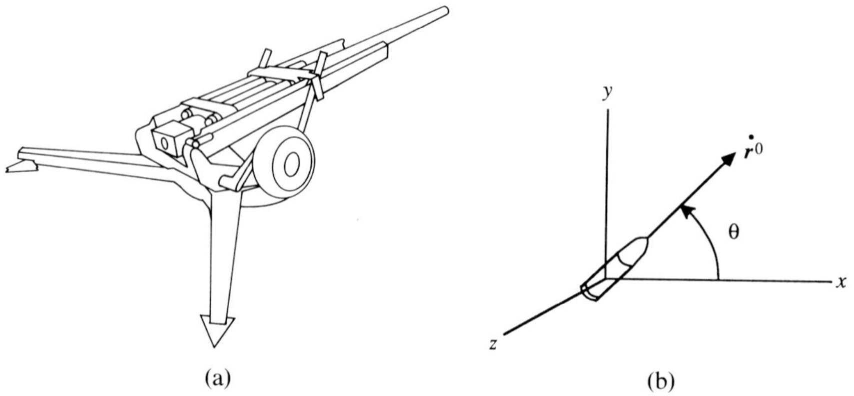

Figure 1.1.1 Projectile launched in x–y plane.

The trajectory followed by the projectile is a parabola in the x-y plane. Trajectories for three values of the angle of elevation θ, $0 < \theta_1 < \theta_2 < \theta_3 < \pi/2$, are shown in Fig. 1.1.2. It is clear from Fig. 1.1.2 that there is some angle of elevation that maximizes the x coordinate of the point where the projectile impacts earth (approximated as the x-z plane); that is, that maximizes the range of the projectile.

To find the angle θ that maximizes range, note first that the projectile strikes earth at a time $t^* > 0$ when $y(t^*) = 0$; that is, $(700 \sin \theta)t^* - 4.9(t^*)^2 = 0$. Thus

$$t^* = \frac{700 \sin \theta}{4.9}$$

Evaluating $x(t^*)$ gives the range R as

$$R = 10^5 \cos \theta \sin \theta$$

The range, as a function of angle of elevation θ, is maximum when $dR/d\theta = 0$; that is,

$$(10^5)(-\sin^2\theta + \cos^2\theta) = 0$$

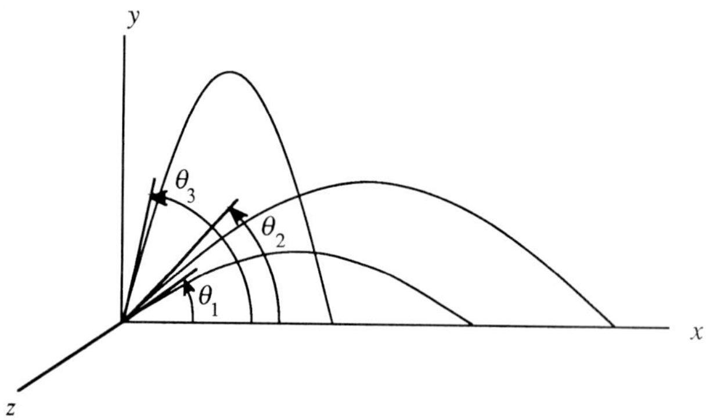

Figure 1.1.2 Trajectories of projectile.

or

$$\theta = \frac{\pi}{4}$$

The maximum range is thus

$$R_{max} = 10^5\left(\frac{\sqrt{2}}{2}\right)^2 = 50{,}000 \text{ m}$$

or 50 km (about 30 miles). Recall that it was assumed that drag on the projectile was zero, so this is an optimistic estimate of the range that can be achieved.

Selection of a convenient inertial reference frame can often simplify and permit easier interpretation of the equations of motion of a system. If an x'-y'-z' Cartesian reference frame is parallel to and moves with constant velocity $\mathbf{v}$ relative to an inertial x-y-z Cartesian reference frame, as shown in Fig. 1.1.3, then

$$\mathbf{r}^P = \mathbf{r}'^P + \mathbf{v}t \tag{1.1.4}$$

where P is any point in space. Since the x'-y'-z' and x-y-z frames remain parallel,

$$\mathbf{f}^P = \mathbf{f}'^P \tag{1.1.5}$$

Newton's equations of motion for a particle of mass m located at point P, written in the x-y-z inertial frame, are

$$m\ddot{\mathbf{r}}^P = \mathbf{f}^P \tag{1.1.6}$$

Differentiating Eq. 1.1.4 twice with respect to time, noting that $\mathbf{v}$ is constant, and substituting this result and Eq. 1.1.5 into Eq. 1.1.6 yields

$$m\ddot{\mathbf{r}}'^P = \mathbf{f}'^P \tag{1.1.7}$$

The equations of motion in an x'-y'-z' reference frame that is translating with constant velocity relative to an inertial frame are thus of the same form as the

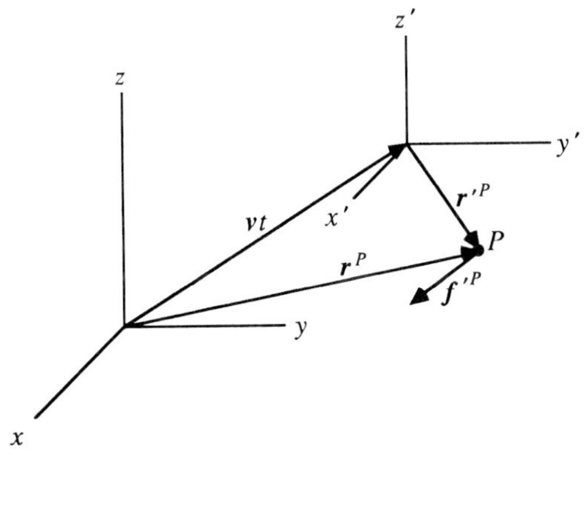

Figure 1.1.3 x'-y'-z' Frame translating with constant velocity $\mathbf{v}$.

equations of motion in the inertial reference frame. Therefore, any reference frame that translates with constant velocity (but no angular velocity) relative to an inertial frame is also an inertial reference frame.

Newton's equations of motion are adequate for analysis of individual particles that move freely in space, as illustrated by Example 1.1.1. Even for constrained particle dynamics, in which the form of kinematic constraints is simple, Newton's equations of motion can sometimes be used to advantage, if they are carefully applied. Techniques that can be used for direct application of Newton's laws of motion for particle dynamics are illustrated in the remainder of this section, primarily with examples.

Example 1.1.2: A particle of mass m moves in a viscous fluid, as shown in Fig. 1.1.4. In addition to gravitational force that acts in the negative y direction, a viscous drag force $\mathbf{F}^d$ acts to retard velocity $\dot{\mathbf{r}}$; that is, $\mathbf{F}^d = -c\dot{\mathbf{r}}$, where c is a coefficient of viscosity.

From Newton's second law, the equations of motion of the particle are

$$\begin{aligned} m\ddot{\mathbf{r}} &= \mathbf{F}^d - mg\mathbf{j} \\ &= -c\dot{\mathbf{r}} - mg\mathbf{j} \end{aligned} \tag{1.1.8}$$

where $\mathbf{j}$ is the unit vector along the positive y axis.

At the initial time $t = 0$, the particle is at the origin of the Cartesian reference frame; that is,

$$\mathbf{r}(0) = \mathbf{0} \tag{1.1.9}$$

and has the velocity

$$\dot{\mathbf{r}}(0) = [v_x, v_y, v_z] \tag{1.1.10}$$

Motion of the particle is uniquely determined by the solution of the initial-value problem of Eqs. 1.1.8 to 1.1.10.

The general solution of the differential equation of Eq. 1.1.8 is

$$\mathbf{r} = \begin{bmatrix} a_x \\ a_y \\ a_z \end{bmatrix} + \begin{bmatrix} b_x \\ b_y \\ b_z \end{bmatrix} e^{-ct/m} - \begin{bmatrix} 0 \\ \dfrac{mg}{c} \\ 0 \end{bmatrix} t \tag{1.1.11}$$

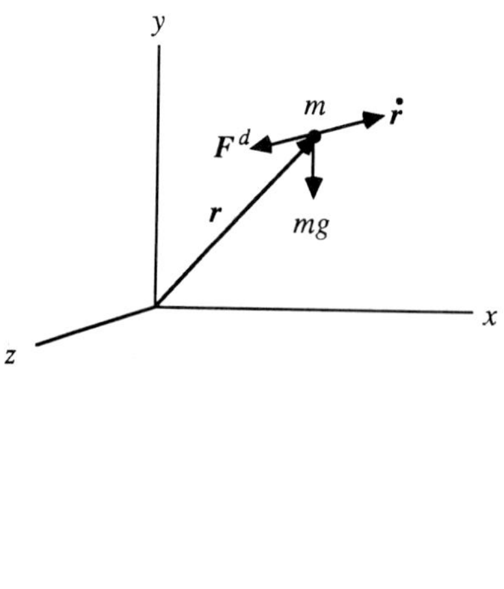

Figure 1.1.4 Particle moving in viscous medium.

where a_x, a_y, a_z, b_x, b_y, and b_z are constants of integration that are to be determined to satisfy the initial conditions of Eqs. 1.1.9 and 1.1.10. Substituting the general solution of Eq. 1.1.11 into the initial condition of Eq. 1.1.10 yields

$$\dot{\mathbf{r}}(0) = -\frac{c}{m}\begin{bmatrix} b_x \\ b_y \\ b_z \end{bmatrix} - \begin{bmatrix} 0 \\ \dfrac{mg}{c} \\ 0 \end{bmatrix} = \begin{bmatrix} v_x \\ v_y \\ v_z \end{bmatrix}$$

which determines b_x, b_y, and b_z. Substituting the resulting form of $\mathbf{r}$ in Eq. 1.1.11 into Eq. 1.1.9 yields

$$\mathbf{r}(0) = \begin{bmatrix} a_x \\ a_y \\ a_z \end{bmatrix} - \frac{m}{c}\begin{bmatrix} v_x \\ v_y \\ v_z \end{bmatrix} - \frac{m^2 g}{c^2}\begin{bmatrix} 0 \\ 1 \\ 0 \end{bmatrix} = \mathbf{0}$$

which determines the remaining constants of integration. The motion of the particle in a viscous medium is thus

$$\mathbf{r} = (1 - e^{-ct/m})\left(\frac{m}{c}\begin{bmatrix} v_x \\ v_y \\ v_z \end{bmatrix} + \frac{m^2 g}{c^2}\begin{bmatrix} 0 \\ 1 \\ 0 \end{bmatrix}\right) - \begin{bmatrix} 0 \\ \dfrac{mg}{c} \\ 0 \end{bmatrix} t \tag{1.1.12}$$

Note the fundamental difference in the form of solution in this example and motion of a particle in a vacuum with no drag in Example 1.1.1. The exponential term in Eq. 1.1.12 represents a decay effect that is typically associated with damping in mechanical systems.

Let $c = 0.1$ to represent drag on the projectile of Example 1.1.1. The time t^* at which the projectile strikes earth is determined from Eq. 1.1.12 by

$$y(t^*) = 0 = (1 - e^{-t^*/500})(35 \times 10^4 \sin\theta + 245 \times 10^4) - 4.9 \times 10^3 t^*$$

Using a second-order Taylor approximation for the exponential yields

$$t^* \approx \frac{3.5 \times 10^8 \sin\theta}{3.5 \times 10^5 \sin\theta + 2.45 \times 10^6}$$

At $\theta = \pi/4$, this is $t^* \approx 91.7$ s. Substituting this result into Eq. 1.1.12,

$$x(t^*) \approx 41{,}160 \text{ m}$$

or about 41 km (25 miles), a 20% decrease in range due to drag.

Example 1.1.3: A particle of mass m is connected to the origin of the x-y-z inertial frame by a linear plastic spring, as shown in Fig. 1.1.5. The force in the spring, with tension as positive, is $f = k(l - l_0)$, where k is the *spring constant*, l is the length of the spring, and $l_0 > 0$ is the *free length* of the spring.

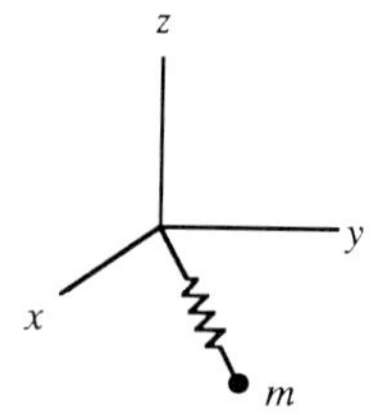

Figure 1.1.5 Particle at end of spring.

Since $l = |\mathbf{r}| = \sqrt{x^2 + y^2 + z^2}$ and the force acting on the particle when the spring is in tension is directed toward the origin of the inertial frame,

$$\mathbf{F}^s = \frac{-f}{|\mathbf{r}|}\mathbf{r} = -\frac{k(l - l_0)}{l}\mathbf{r}$$

With gravitational force acting in the negative z direction, Newton's equations of motion of the particle are

$$m\ddot{\mathbf{r}} = -\frac{k(l - l_0)}{l}\mathbf{r} - mg\mathbf{k}$$

It is interesting to note that even though the force–deflection relation for the spring is linear, the force vector as a function of generalized coordinate $\mathbf{r}$ is nonlinear, due to the factor $(l - l_0)/l$. Thus the equations of motion are nonlinear in $\mathbf{r}$.

To illustrate the use of numerical methods such as those presented in Appendix D to analyze the response predicted for such nonlinear differential equations, consider the following data: $m = 1$ kg, $g = 9.89665$ m/s^2, $k = 100$ N/m, and $l_0 = 0.9019335$ m. The value of l_0 was selected so that the particle would hang at equilibrium with $z = -1$ m (i.e., with the spring slightly deformed). This allows the model to be used to simulate the motion of a pendulum whose length is 1 m.

As a first numerical experiment, initial conditions are selected as $x(0) = \dot{x}(0) = 0$, $y(0) = 0$, $\dot{y}(0) = 0.1$ m/s, $z(0) = -1$, and $\dot{z}(0) = 0$. As expected, on theoretical grounds, numerical integration results yield $x(t) = 0$. The solid curve in Fig. 1.1.6(a) shows numerical results for $y(t)$. The value of $z(t)$ varies imperceptibly from the value -1. To see the effect of a change in the initial velocity, simulations are carried out with initial velocities $\dot{y}(0) = 0.05$ and 0.15 m/s. The results shown in Fig. 1.1.6(a) indicate that the amplitude of swing of the pendulum model changes, but that there is very little change in the period of motion. Thus with small initial velocity, this model simulates the motion of a simple pendulum well.

As a second numerical experiment, numerical calculations are carried out with the following initial conditions: $x(0) = 0.1$ m, $\dot{x}(0) = 0.1$ m/s, $y(0) = 0.1$ m, $\dot{y}(0) = 0$ m/s, $z(0) = -0.98995$ m, and $\dot{z}(0) = 0.010102$ m/s. The initial conditions on z are selected to be consistent with the rigid pendulum model of Example 3.1.4. Plots of $x(t)$ and $y(t)$ obtained from numerical integration are shown in Fig. 1.1.6(b). An alternative plot of results is given in Fig. 1.1.6(c), which shows the path traced by the particle, as viewed from the positive z axis. Figure 1.1.6(b) and (c) show that the particle returns to essentially its initial position and velocity after 2 s (i.e., the small-amplitude motion of the spherical pendulum model is essentially periodic).

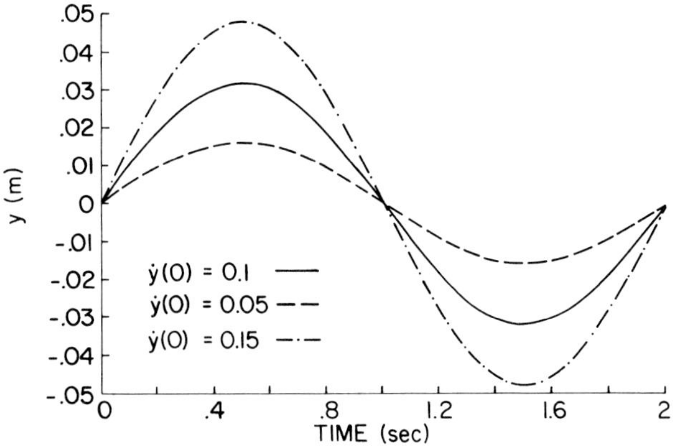

(a) y vs time for planar motion

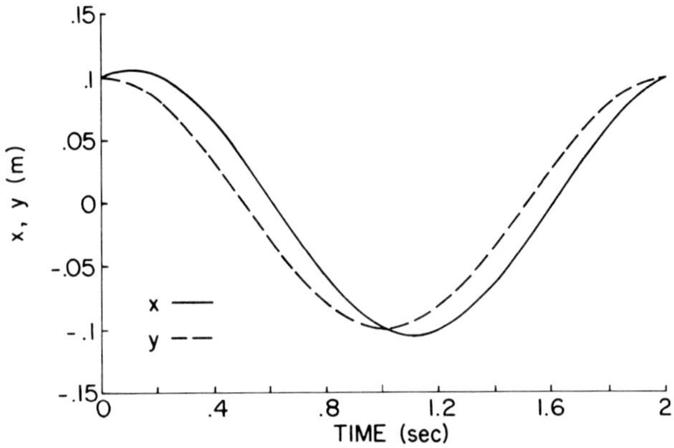

(b) x and y vs time for spatial motion

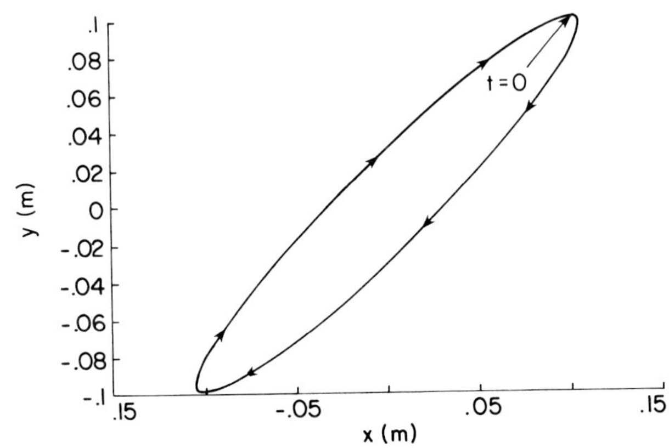

(c) y vs x for spatial motion

Figure 1.1.6 Numerical results for particle on spring.

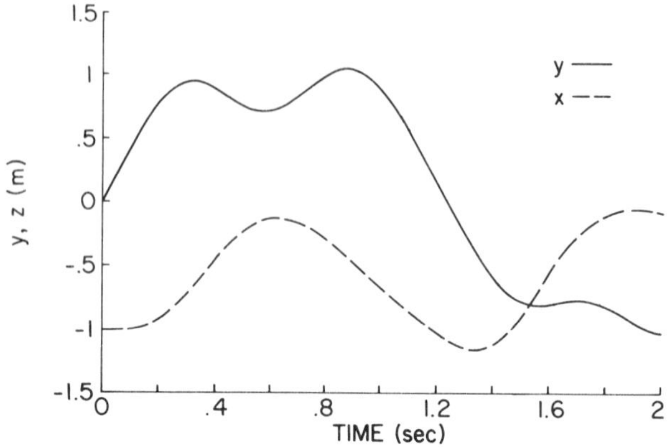

(d) y and z vs time for large amplitude planar motion

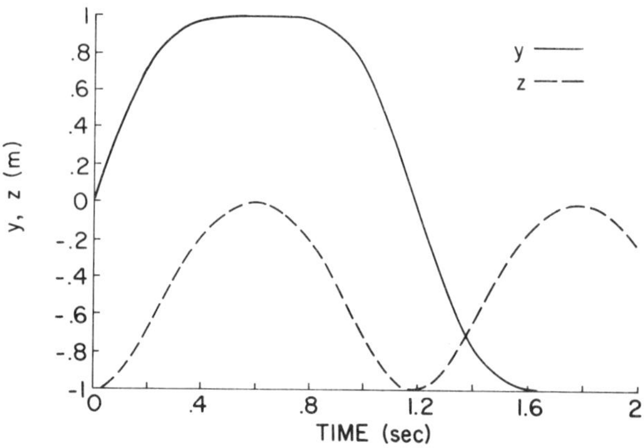

(e) y and z vs time for stiff spring, large amplitude planar motion

Figure 1.1.6 (*continued*)

Consider next a set of initial conditions that leads to large-amplitude motion; for example, $x(0) = \dot{x}(0) = 0$, $y(0) = 0$, $\dot{y}(0) = 4.42896$ m/s, $z(0) = -1$m, and $\dot{z}(0) = 0$. The initial condition on $\dot{y}(0)$ is selected so that an actual simple pendulum will swing up to a horizontal position, stop, and swing back. Numerical results show that $x(t) = 0$, as expected. Plots of $y(t)$ and $z(t)$ in Fig. 1.1.6(d) show that large motion indeed occurs, but that the model fails to behave as a rigid simple pendulum. Since $z\,(1.35) \approx -1.15$, it is clear that the spring is stretched significantly. The oscillation in the $y(t)$ curve also illustrates this behavior. To improve the model, the spring stiffness is increased to $k = 10{,}000$ N/m and the free length set to $l_0 = 0.999019$. Numerical results for this stiff spring model of the simple pendulum are shown in Fig. 1.1.6(e). Note that the mass indeed rises to $z = 0$ at about $t = 0.55$ s, comes to rest, and drops downward. Note also that the spring does not stretch significantly and the motion is essentially periodic. These results may be compared with predictions made in Example 3.1.4 for a rigid pendulum model.

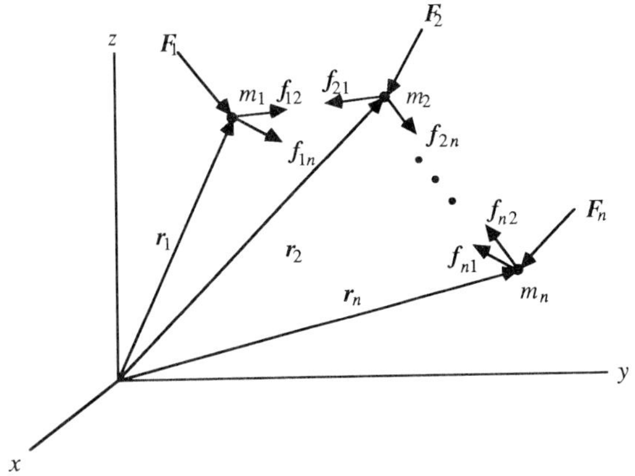

Figure 1.1.7 System of n particles.

Consider the system of n bodies modeled as particles shown schematically in Fig. 1.1.7. Particle i has mass m_i and is located by a position vector $\mathbf{r}_i$ from the origin of an inertial reference frame to the particle. The forces that act on each particle include an external applied force $\mathbf{F}_i$ and forces of interaction between particles $\mathbf{f}_{ij}$, $j \neq i$. For analytical simplicity, define $\mathbf{f}_{ii} = \mathbf{0}$. The total force that acts on the ith particle is the summation of external and internal forces. Thus, for body i, Newton's equations of motion are

$$m_i\ddot{\mathbf{r}}_i = \mathbf{F}_i + \sum_{j=1}^{n} \mathbf{f}_{ij}, \qquad i = 1, \ldots, n \tag{1.1.13}$$

This system of equations thus describes the motion of the system of n particles.

Since the forces of interaction between particles of the system must satisfy Newton's law of action and reaction, the force on particle i due to particle j must be equal to the negative of the force that acts on particle j due to particle i. In vector form, this is

$$\mathbf{f}_{ij} = -\mathbf{f}_{ji}, \qquad i,j = 1, \ldots, n \tag{1.1.14}$$

Note that since $\mathbf{f}_{ii} = \mathbf{0}$, Eq. 1.1.14 holds even for $i = j$.

If the external forces that act on each particle are known and the nature of the force that acts between particles i and j is known, the system of equations in Eq. 1.1.13 may be explicitly written. However, the forces of interaction between particles will generally depend on the position of the particles; that is,

$$\mathbf{f}_{ij} = \mathbf{f}_{ij}(\mathbf{r}_i, \mathbf{r}_j) \tag{1.1.15}$$

where the function on the right side of Eq. 1.1.15 depends on position vectors of the two particles involved. If particles i and j are connected by a linear spring with spring rate k and free length l_0, as in Example 1.1.3, the force of interaction is

$$\mathbf{f}_{ij} = \frac{k(|\mathbf{r}_j - \mathbf{r}_i| - l_0)}{|\mathbf{r}_j - \mathbf{r}_i|}(\mathbf{r}_j - \mathbf{r}_i) \tag{1.1.16}$$

Note that for even this simple connection between masses, the dependence of the forces of interaction on the position coordinates of the particles is quite complex. Thus the equations of motion of Eq. 1.1.13 are usually highly nonlinear.

Since Eq. 1.1.13 must hold for each particle in the system, these equations can be summed to obtain

$$\sum_{i=1}^{n} m_i \ddot{\mathbf{r}}_i = \sum_{i=1}^{n} \mathbf{F}_i + \sum_{i=1}^{n} \sum_{j=1}^{n} \mathbf{f}_{ij} \tag{1.1.17}$$

which is valid for the entire system of particles. Using Eq. 1.1.14, the double sum of Eq. 1.1.17 may be written in the form

$$\sum_{i=1}^{n} \sum_{j=1}^{n} \mathbf{f}_{ij} = \frac{1}{2} \sum_{i=1}^{n} \sum_{j=1}^{n} (\mathbf{f}_{ij} + \mathbf{f}_{ji}) = \mathbf{0} \tag{1.1.18}$$

Thus Eq. 1.1.17 reduces to

$$\sum_{i=1}^{n} m_i \ddot{\mathbf{r}}_i = \sum_{i=1}^{n} \mathbf{F}_i \tag{1.1.19}$$

Define the total mass m of the system as the sum of the individual masses,

$$m = \sum_{i=1}^{n} m_i \tag{1.1.20}$$

The *center of mass* or *centroid* of the system of particles is defined as

$$\mathbf{r}^c = \frac{1}{m} \sum_{i=1}^{n} m_i \mathbf{r}_i \tag{1.1.21}$$

where $\mathbf{r}^c$ is the vector from the origin of the inertial reference frame to the center of mass. Since the total mass and each of the particle masses are constant, Eq. 1.1.21 can be differentiated twice with respect to time to obtain

$$m\ddot{\mathbf{r}}^c = \sum_{i=1}^{n} m_i \ddot{\mathbf{r}}_i \tag{1.1.22}$$

Finally, defining the total external force that acts on the system of particles as

$$\mathbf{F} = \sum_{i=1}^{n} \mathbf{F}_i \tag{1.1.23}$$

Eq. 1.1.19 may be written as

$$m\ddot{\mathbf{r}}^c = \mathbf{F} \tag{1.1.24}$$

This result shows that the motion of the center of mass of a system of particles moves as if the system were a concentrated mass m that is acted on by the vector sum $\mathbf{F}$ of all external forces that act on the system of particles.

♦ **Example 1.1.4:** To illustrate the behavior of the centroid of a system of particles, the center of mass of pellets of a shotgun shell moves as if the collection were a single projectile. Assume that there are negligible differences in the drag forces that act on different pellets. This is a good approximation, since the initial velocity of each of the pellets is essentially the same. The value of this observation in calculation of impact patterns of pellets on the target is significant. A very good approximation of the center of mass of the impact pattern may be obtained simply by integrating the equations of motion of Eq. 1.1.24 as if the total system were a single projectile, as in Examples 1.1.1 and 1.1.2. If the relative velocities of the pellets at the muzzle of the shotgun is known, the distribution of particles in the impact pattern, relative to the center of mass, can readily be calculated.

While extensive treatments of special problems, such as orbital dynamics of spacecraft, could be studied at this point, essentially all direct applications of Newton's equations of motion for unconstrained systems that are modeled as particles reduce to initial-value problems such as those encountered in this section. The study of such problems generally reduces to special techniques and tricks for solving nonlinear differential equations and analysis of qualitative properties of solutions. Problems 1.1.1 to 1.1.3 are recommended for the reader who wishes to gain practice in formulating and solving equations of motion using Newton's laws.

1.2 DYNAMICS OF PARTICLES CONSTRAINED TO MOVE ON CURVES

While the dynamics of bodies that can be modeled as particles that move in space under the action of known forces is of interest in the study of ballistics of projectiles, trajectories of satellites, astrophysics, and other areas of application, broad classes of applications are excluded. For example, in the study of dynamics of machines such as vehicles, manufacturing equipment, robots, aircraft landing gear, and engines, bodies that make up a machine are constrained to move relative to one another in well-established patterns. In fact, the fundamental purpose of a machine is to control motion and transmit load. Thus, relative motion that is permitted between bodies in a machine is often constrained by connections between bodies. For example, the wheel of an automobile is restrained to rotate about a spindle or axle, much as the crankshaft of an engine is constrained to rotate about an axis that is fixed in the engine housing. The main attention in this book is on the dynamics of constrained systems.

The purpose of this section is to introduce concepts of constrained system dynamics by considering examples of bodies that can be modeled as particles that are constrained to move along established paths in space. Constraint reaction forces are introduced and the form of the governing equations of motion is shown to be a mixed system of differential-algebraic equations. A fundamental technique for reducing differential-algebraic equations of motion to differential equations is illustrated through simple examples, which serve to guide development of more general methods later in the book. Finally, the concept of Coulomb friction, which depends on the normal constraint reaction force between a body and the surface on which it slides, is introduced and illustrated.

Example 1.2.1: A vehicle of mass m is constrained to move without friction along a straight track, as shown in Fig. 1.2.1. Define the directed distance q of the particle that represents the vehicle from the point on the track defined by $\mathbf{r}_0$. With the unit vector $\mathbf{u}$ along the track, the vector $\mathbf{r}$ that locates the particle is

$$\mathbf{r} = \mathbf{r}_0 + q\mathbf{u} \tag{1.2.1}$$

Since $\mathbf{r}_0$ and $\mathbf{u}$ are constant and only q in Eq 1.2.1 depends on time, differentiating Eq. 1.2.1 yields the velocity and acceleration of the particle as

$$\dot{\mathbf{r}} = \dot{q}\mathbf{u} \tag{1.2.2}$$

$$\ddot{\mathbf{r}} = \ddot{q}\mathbf{u} \tag{1.2.3}$$

Since no friction acts between the particle and the track along which it moves, the *constraint reaction force* $\mathbf{F}^C$ must act perpendicular to the track, as shown in Fig. 1.2.2. Gravity acts in the negative z direction. Newton's equations of motion for the particle are thus

$$m\ddot{\mathbf{r}} = \mathbf{F}^C - mg\mathbf{k} \tag{1.2.4}$$

If the particle is located at $\mathbf{r}_0$ and has unit velocity in the direction $\mathbf{u}$ at $t = 0$, the initial conditions are

$$\begin{aligned} \mathbf{r}(0) &= \mathbf{r}_0 \\ \dot{\mathbf{r}}(0) &= \mathbf{u} \end{aligned} \tag{1.2.5}$$

It is important to note that these initial conditions are consistent with the constraints that the particle must move along the straight track. If an initial position is specified that is not on the track or if an initial velocity is specified that is not parallel to the track, no solution can exist. While definition of initial position and velocity that are consistent with the constraint in this elementary example is obvious, subtle difficulties can arise when more complex systems are treated.

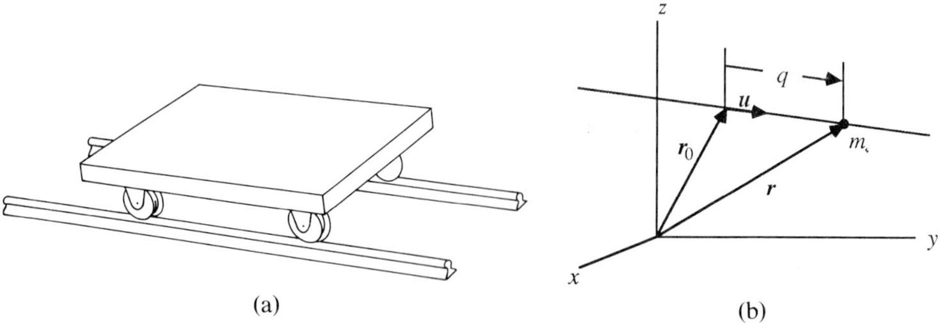

Figure 1.2.1 Vehicle on straight track.

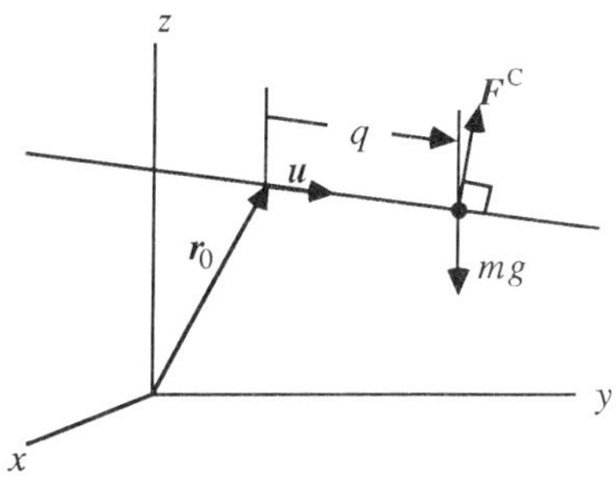

Figure 1.2.2 Forces on particle.

Substituting $\ddot{\mathbf{r}}$ from Eq. 1.2.3 into Eq. 1.2.4 yields

$$m\ddot{q}\mathbf{u} = \mathbf{F}^C - mg\mathbf{k} \tag{1.2.6}$$

Note that while this is a system of three equations, only derivatives of the single variable q appear. The unknown constraint reaction force $\mathbf{F}^C$, which satisfies $\mathbf{u}^T\mathbf{F}^C = 0$, appears algebraically as an unknown in the equations of motion. Thus Eq. 1.2.6 is not a classical differential equation. Rather, it is a mixed system of differential and algebraic equations, called *differential-algebraic equations* (DAE). While such systems arise often in the study of mechanics, the mathematical theory of differential-algebraic equations is remarkably underdeveloped. A brief account of the theory is presented in Appendix D.

In applications with particles that are constrained to move along curves or surfaces without friction, the constraint reaction force must be perpendicular to the curve or surface along which the particle is constrained to move. This suggests that the constraint reaction force may be suppressed by taking the scalar product of both sides of the equations of motion with vectors that are tangent to the curve or surface.

Example 1.2.2: Taking the scalar product of both sides of Eq. 1.2.6 with the unit vector **u** yields the scalar equation

$$m\ddot{q} = -mg\mathbf{u}^T\mathbf{k} \tag{1.2.7}$$

since $\mathbf{u}^T\mathbf{F}^C = 0$. This is a single second order differential equation in the variable q. Using the relations of Eq. 1.2.3 and the initial conditions of Eq. 1.2.5, initial conditions on q are obtained as

$$\begin{aligned} q(0) &= 0 \\ \ddot{q}(0) &= 1 \end{aligned} \tag{1.2.8}$$

The initial-value problem of Eqs. 1.2.7 and 1.2.8 may be viewed as the equations of motion of the particle tangent to the track, hence eliminating explicit appearance of the constraint reaction force. The solution of Eqs. 1.2.7 and 1.2.8 is

$$q = t - \frac{g\mathbf{u}^T\mathbf{k}t^2}{2} \tag{1.2.9}$$

Substituting this result into Eq. 1.2.3 yields

$$\mathbf{r} = \mathbf{r}_0 + \left(t - \frac{g\mathbf{u}^T\mathbf{k}t^2}{2}\right)\mathbf{u} \tag{1.2.10}$$

This result may be used in Eq. 1.2.4 of Example 1.2.1 to determine the constraint reaction force $\mathbf{F}^C = mg[\mathbf{K} - (\mathbf{u}^T\mathbf{k})\mathbf{u}]$

In general, for a particle that is constrained to move along a curve in space, the form of velocity and acceleration relations is much more complicated.

Example 1.2.3: Consider a vehicle that is driven up a spiral parking ramp, modeled as a particle that moves along a spiral, as shown in Fig. 1.2.3.

In terms of the angle ϕ,

$$\mathbf{r} = \begin{bmatrix} R\cos\phi \\ R\sin\phi \\ a\dfrac{\phi}{2\pi} \end{bmatrix} \tag{1.2.11}$$

where R is the radius of the ramp and a is the height advance due to one revolution of the spiral path (the distance between floors of the parking garage). Differentiating Eq. 1.2.11 yields the velocity

$$\dot{\mathbf{r}} = \begin{bmatrix} -R\sin\phi \\ R\cos\phi \\ \dfrac{a}{2\pi} \end{bmatrix}\dot{\phi} \equiv \boldsymbol{\tau}\dot{\phi} \tag{1.2.12}$$

where $\boldsymbol{\tau}$ is tangent to the helix. Note that

$$\begin{aligned} \boldsymbol{\tau}^T\boldsymbol{\tau} &= R^2(\sin^2\phi + \cos^2\phi) + \frac{a^2}{4\pi^2} \\ &= R^2 + \frac{a^2}{4\pi^2} \end{aligned}$$

so $\boldsymbol{\tau}$ has constant length. Differentiating Eq. 1.2.12 yields the acceleration,

$$\ddot{\mathbf{r}} = \boldsymbol{\tau}\ddot{\phi} + \begin{bmatrix} -R\cos\phi \\ -R\sin\phi \\ 0 \end{bmatrix}\dot{\phi}^2 \tag{1.2.13}$$

If $\mathbf{F}^A$ is the net forward thrust on the vehicle due to torque applied to the driving wheels, $\mathbf{F}^A$ is parallel to $\boldsymbol{\tau}$. Let the driving thrust be T. Then

$$\mathbf{F}^A = \frac{T}{\sqrt{R^2 + a^2/(4\pi^2)}}\boldsymbol{\tau} \tag{1.2.14}$$

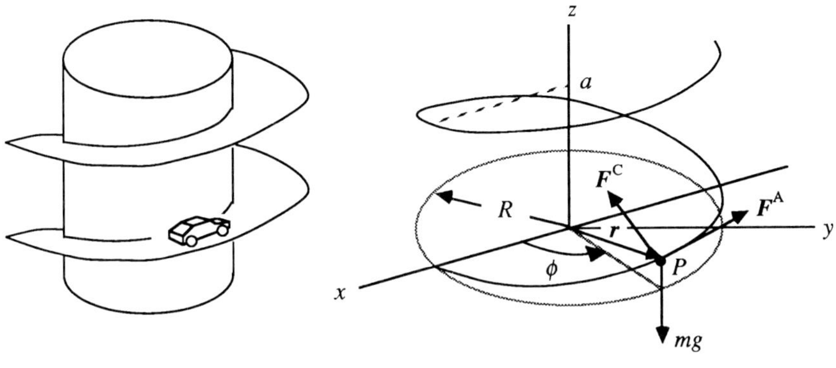

Figure 1.2.3 Vehicle on spiral ramp.

Also note that the constraint force $\mathbf{F}^C$ on the vehicle is perpendicular to the path (i.e., $\boldsymbol{\tau}^T\mathbf{F}^C = 0$) and the gravitational force is in the negative z direction.

Using the forces acting on the vehicle and the acceleration of Eq. 1.2.13, Newton's equations of motion are

$$m\boldsymbol{\tau}\ddot{\phi} - mR\begin{bmatrix}\cos\phi \\ \sin\phi \\ 0\end{bmatrix}\dot{\phi}^2 = \frac{T}{\sqrt{R^2 + a^2/(4\pi^2)}}\boldsymbol{\tau} + \mathbf{F}^C - mg\mathbf{k} \qquad (1.2.15)$$

These equations involve the unknown reaction force components that appear algebraically and nonlinear terms in ϕ. This is typical of the equations encountered in constrained system dynamics.

While analytical solution of Eq. 1.2.15 appears to be out of the question, specialized results can be obtained. First, recalling that $\boldsymbol{\tau}^T\mathbf{F}^C = 0$, both sides of Eq. 1.2.15 may be multiplied by $\boldsymbol{\tau}^T$ to obtain

$$m\left(R^2 + \frac{a^2}{4\pi^2}\right)\ddot{\phi} = \frac{T}{\sqrt{R^2 + \dfrac{a^2}{4\pi^2}}} - \frac{mga}{2\pi} \qquad (1.2.16)$$

In this very special case, the term in Eq. 1.2.15 associated with $\dot{\phi}^2$ is suppressed, along with $\mathbf{F}^c$, and a very simple equation in $\ddot{\phi}$ is obtained.

Note that in order to accelerate up the ramp, the thrust T delivered by the drivetrain of the vehicle must satisfy

$$T \geq \frac{mga}{\sqrt{4\pi^2R^2 + a^2}} \qquad (1.2.17)$$

so that $\ddot{\phi} \geq 0$. If the vehicle starts at the bottom of the ramp at rest, after picking up a parking stub, $\phi(0) = \dot{\phi}(0) = 0$. With a constant driving thrust that is greater than the value on the right of Eq. 1.2.17, the solution of the initial-value problem is

$$\begin{aligned}\phi(t) &= \frac{T_0\sqrt{4\pi^2R^2 + a^2} - mga}{2m(4\pi^2R^2 + a^2)}t^2 \\ &\equiv \alpha t^2\end{aligned} \tag{1.2.18}$$

The results of Examples 1.2.1 to 1.2.3 were obtained deceptively easily. Most constrained system applications will lead to more complex equations that are not easily solved. The method used in these examples is, however, general. Information regarding the path that is to be traversed is used to derive expressions for velocity and acceleration of the body, in terms of a variable that defines the position of the body on the path. If the body can be acceptably modeled as a particle, Newton's equations of motion are written, accounting for applied forces and constraint reaction forces. Multiplying both sides of the equations of motion by the transpose of the tangent to the curve (i.e., taking the scalar product of both sides with the tangent), the constraint reaction force is suppressed and a reduced equation of motion is obtained. If this equation can be solved, motion of the body is determined. The known motion can then be substituted into Newton's equations of motion and the constraint reaction force can be recovered.

➧ **Example 1.2.4:** Returning to the vehicle driving up the parking ramp of Example 1.2.3, the constraint reaction force $\mathbf{F}^C$ is now determined. For the solution $\phi(t)$ of Eq. 1.2.18, $\dot{\phi} = 2\alpha t$ and $\ddot{\phi} = 2\alpha$. Substituting these results and $\boldsymbol{\tau}$ of Eq. 1.2.12 into Eq. 1.2.15, the constraint reaction force is obtained as

$$\begin{aligned}\mathbf{F}^C &= 2m\alpha\begin{bmatrix}-R\sin\alpha t^2 \\ R\cos\alpha t^2 \\ \dfrac{a}{2\pi}\end{bmatrix} - 4\alpha^2 mR\begin{bmatrix}\cos\alpha t^2 \\ \sin\alpha t^2 \\ 0\end{bmatrix}t^2 \\ &\quad - \frac{T_0}{\sqrt{R^2 + a^2/(4\pi^2)}}\begin{bmatrix}-R\sin\alpha t^2 \\ R\cos\alpha t^2 \\ \dfrac{a}{2\pi}\end{bmatrix} + mg\begin{bmatrix}0 \\ 0 \\ 1\end{bmatrix}\end{aligned} \tag{1.2.19}$$

➧ **Example 1.2.5:** As an illustration of a use of the equations of motion that is called *inverse dynamics*, which is to determine actuator forces required to cause prescribed motion, consider again the vehicle on the spiral parking ramp of Example 1.2.3. The drivetrain thrust T_c is to be determined to drive the vehicle up the ramp with $\dot{\phi} = c$ (i.e., $\phi = ct$). From Eq. 1.2.16,

$$T_c = \frac{mga}{\sqrt{4\pi^2R^2 + a^2}}$$

will sustain any desired constant speed, since $\ddot{\phi} = 0$. The constraint reaction force $\mathbf{F}^C$, which is the tire traction force that must be sustained for safe operation, is, however, dependent on c. From Eq. 1.2.15,

$$\mathbf{F}^C = -mRc^2\begin{bmatrix}\cos ct\\ \sin ct\\ 0\end{bmatrix} - \frac{T_c}{\sqrt{R^2 + a^2/(4\pi^2)}}\begin{bmatrix}-R\sin ct\\ R\cos ct\\ \dfrac{a}{2\pi}\end{bmatrix} + mg\begin{bmatrix}0\\0\\1\end{bmatrix}$$

A good approximation of the lateral tire force required is

$$[-\cos ct,\ -\sin ct,\ 0]\mathbf{F}^C = -mRc^2$$

Applications thus far in this section have dealt with constraints that are workless (i.e., forces $\mathbf{F}^C$ of constraint do no work). If friction exists between a particle and the curve or surface on which it moves, the situation is quite different. Consider specifically the *Coulomb friction* model, in which the friction force $\mathbf{F}^f$ that is tangent to sliding surfaces opposes motion and is proportional to the magnitude N of the normal force between the surfaces when sliding occurs; that is,

$$|\mathbf{F}^f| = \mu_d N \tag{1.2.20}$$

during relative motion, where μ_d is the *dynamic coefficient of friction*, or just the *coefficient of friction*.

The Coulomb friction model, which is often a good representation of reality, leads to a further complication beyond the fact that the friction force depends on the normal force. If there is no relative motion between surfaces, the magnitude of the friction force is bounded by the limit

$$|\mathbf{F}^f| \leqslant \mu_s N \tag{1.2.21}$$

where $\mu_s \geqslant \mu_d$ is the *static coefficient of friction*. In reality, a *friction lock* exists between the surfaces when there is no relative motion and the inequality of Eq. 1.2.21 is strict. This is equivalent to the existence of a kinematic constraint on relative tangential motion along the surfaces, as long as the inequality in Eq. 1.2.21 is strict. The magnitude of the resulting friction force, sometimes called the *stiction force*, is determined by the constraint condition.

➧ **Example 1.2.6:** Consider again the vehicle of Example 1.2.1 that moves along a straight track, but with friction between the vehicle and the track. If the vehicle is in motion, the friction force is

$$\mathbf{F}^f = -\mu_d F^C \operatorname{sgn} \dot{q}\mathbf{u} \tag{1.2.22}$$

where the notation of Example 1.2.1 is used and

$$\operatorname{sgn} \alpha = \begin{cases} 1, & \text{if } \alpha > 0 \\ -1, & \text{if } \alpha < 0 \end{cases} \tag{1.2.23}$$

Note that the value of sgn α is undefined for $\alpha = 0$. Newton's equations of motion when $\dot{q} \neq 0$ are thus

$$m\ddot{q}\mathbf{u} = \mathbf{F}^C - mg\mathbf{k} - \mu_d F^C \operatorname{sgn} \dot{q}\mathbf{u} \tag{1.2.24}$$

If $\mathbf{v}$ and $\mathbf{w}$ are two orthogonal unit vectors that are also orthogonal to $\mathbf{u}$, then taking the scalar product of both sides of Eq. 1.2.24 with $\mathbf{u}$, $\mathbf{v}$, and $\mathbf{w}$ and using the fact that $\mathbf{u}^T\mathbf{F}^C = 0$ yields

$$\begin{aligned} m\ddot{q} &= -mg\mathbf{u}^T\mathbf{k} - \mu_d F^C \operatorname{sgn} \dot{q} \\ 0 &= F_{\mathbf{v}}^C - mg\mathbf{v}^T\mathbf{k} \\ 0 &= F_{\mathbf{w}}^C - mg\mathbf{w}^T\mathbf{k} \end{aligned} \tag{1.2.25}$$

where $F_{\mathbf{v}}^C = \mathbf{v}^T\mathbf{F}^C$ and $F_{\mathbf{w}}^C = \mathbf{w}^T\mathbf{F}^C$. Since $F^C = \sqrt{(F_{\mathbf{v}}^C)^2 + (F_{\mathbf{w}}^C)^2}$, the first of Eqs. 1.2.25 reduces to

$$m\ddot{q} = -mg\mathbf{u}^T\mathbf{k} - \mu_d mg \operatorname{sgn} \dot{q} \sqrt{(\mathbf{v}^T\mathbf{k})^2 + (\mathbf{w}^T\mathbf{k})^2} \tag{1.2.26}$$

As a particular case, let $\mathbf{u} = \mathbf{i}$ (i.e., the particle slides along the x axis). Choosing $\mathbf{v} = \mathbf{j}$ and $\mathbf{w} = \mathbf{k}$, Eq. 1.2.26 becomes

$$m\ddot{q} = -\mu_d mg \operatorname{sgn} \dot{q}$$

If $\dot{q}(0) = 1$ and $q(0) = 0$, then as long as $\dot{q} > 0$, sgn $\dot{q} = 1$. The solution is thus

$$q(t) = t - \frac{\mu_d g t^2}{2}$$

$$\dot{q}(t) = -\mu_d g t + 1$$

The particle comes to rest at $t = 1/(\mu_d g)$, at which time $q = 1/(2\mu_d g)$.

Example 1.2.7: Consider next the vehicle of Examples 1.2.3 and 1.2.4 that drives up a helical ramp, but with a dynamic coefficient of friction μ_d that accounts for rolling resistance and $R = a = 1.0$ m. Denote the length of the tangent vector $\boldsymbol{\tau}$ defined in Example 1.2.3 as $|\boldsymbol{\tau}| = \sqrt{4\pi^2 + 1/(2\pi)} \equiv 1/\beta$. Thus the friction force acting on the vehicle is

$$\mathbf{F}^f = -\beta\mu_d F^C \operatorname{sgn} \dot{\phi}\boldsymbol{\tau}$$

Since $\mathbf{F}^A = -mg\mathbf{k} + \mathbf{F}^f + \beta T\boldsymbol{\tau}$, the equations of motion of the vehicle are, from Eq. 1.2.15,

$$m\left(\boldsymbol{\tau}\ddot{\phi} + \begin{bmatrix} -\cos\phi \\ -\sin\phi \\ 0 \end{bmatrix}\dot{\phi}^2\right) = -mg\mathbf{k} - \beta(\mu_d F^C \operatorname{sgn} \dot{\phi} + T)\boldsymbol{\tau} + \mathbf{F}^C \tag{1.2.27}$$

Note that the unit vector

$$\mathbf{v} = [-\cos\phi, \ -\sin\phi, \ 0]^T$$

is orthogonal to $\boldsymbol{\tau}$. Finally, using the vector product notation of Appendix A.3, the unit vector

$$\mathbf{w} = \frac{1}{|\tilde{\boldsymbol{\tau}}\mathbf{v}|}\tilde{\boldsymbol{\tau}}\mathbf{v} = \frac{1}{2\pi}[\sin\phi,\ -\cos\phi,\ 2\pi]^T$$

is orthogonal to both $\boldsymbol{\tau}$ and $\mathbf{v}$. Taking the scalar product of both sides of Eq. 1.2.27 with $\boldsymbol{\tau}$, $\mathbf{v}$, and $\mathbf{w}$ yields

$$\begin{aligned} \frac{m}{\beta^2}\ddot{\phi} &= -\frac{mg}{2\pi} - \frac{\mu_d F^C \operatorname{sgn}\dot{\phi} + T}{\beta} \\ m\dot{\phi}^2 &= F_{\mathbf{v}}^C \\ 0 &= -mg + F_{\mathbf{w}}^C \end{aligned} \tag{1.2.28}$$

where $F_{\mathbf{v}}^C = \mathbf{v}^T\mathbf{F}^C$ and $F_{\mathbf{w}}^C = \mathbf{w}^T\mathbf{F}^C$. Thus

$$F^C = \sqrt{(F_{\mathbf{v}}^C)^2 + (F_{\mathbf{w}}^C)^2} = m\sqrt{\dot{\phi}^4 + g^2}$$

Substituting this result into the first of Eqs. 1.2.28 yields the differential equation of motion

$$\ddot{\phi} = -\frac{\beta^2 g}{2\pi} - \beta(\mu_d \operatorname{sgn}\dot{\phi}\sqrt{\dot{\phi}^4 + g^2} + T) \tag{1.2.29}$$

which is not trivially solvable in closed form, even if sgn $\dot{\phi}$ is known.

➧ **Example 1.2.8:** Consider the pair of blocks, modeled as particles, shown in Fig. 1.2.4 that slide on a rough surface, with equal static and dynamic coefficients of friction $\mu_s = \mu_d \equiv \mu$ between each block and the surface. The blocks are connected by a spring with the force–deflection relation $F^S = k(x_2 - x_1 - 1)$, tension positive. Initial conditions are $x_1(0) = 1$, $x_2(0) = 2$, $\dot{x}_1(0) = v_0$, and $\dot{x}_2(0) = 0$. For numerical calculations, let $m = k = 1$.

Block 2 will remain at rest until the force in the spring reaches magnitude μg. Let t_1 be the time at which F^S first reaches the magnitude μg, if it ever does. Thus for $t \le t_1$, $x_2 = 2$. As long as $\dot{x}_1 > 0$, the equation of motion of block 1 is just

$$\ddot{x}_1 = -\mu g + (1 - x_1)$$

The general solution is

$$x_1 = (1 - \mu g) + C_1 \cos t + C_2 \sin t$$

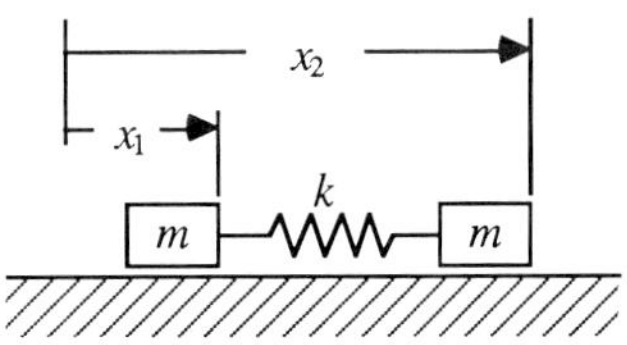

Figure 1.2.4 Blocks on a rough surface.

The initial conditions imply that $C_1 = \mu g$ and $C_2 = v_0$. Thus

$$\dot{x}_1 = \mu g \sin t + v_0 \cos t$$

Presuming that the second block does not slip, which is an assumption that is to be checked, the first block comes to rest at t_1^* when

$$\dot{x}(t_1^*) = -\mu g \sin t_1^* + v_0 \cos t_1^* = 0$$

so $t_1^* < \pi/2$. Using the trigonometric identity $\cos^2 t_1^* + \sin^2 t_1^* = 1$,

$$\cos t_1^* = \frac{\mu g}{\sqrt{(\mu g)^2 + v_0^2}}$$

$$\sin t_1^* = \frac{v_0}{\sqrt{(\mu g)^2 + v_0^2}}$$

Thus

$$\begin{aligned} x_1(t_1^*) &= (1 - \mu g) + \frac{(\mu g)^2}{\sqrt{(\mu g)^2 + v_0^2}} + \frac{v_0^2}{\sqrt{(\mu g)^2 + v_0^2}} \\ &= (1 - \mu g) + \sqrt{(\mu g)^2 + v_0^2} \end{aligned}$$

Since the second block has been assumed not to slip, the force in the spring is

$$\begin{aligned} F^S(t_1^*) &= 1 - x_1(t_1^*) \\ &= \mu g - \sqrt{(\mu g)^2 + v_0^2} \end{aligned}$$

If $|F^S(t_1^*)| > \mu g$, the second block would have slipped before t_1^*. To find the critical value v_0^C of v_0, below which no slipping of the second block will occur, set

$$| F^S(t_1^*) | = \sqrt{(\mu g)^2 + (v_0^C)^2} - \mu g = \mu g$$

Thus

$$v_0^C = \sqrt{3}\, \mu g$$

For $v_0 \leq v_0^C$, block 1 slides to the right and stops at $x_1(t_1^*)$ and block 2 never moves. No additional motion occurs.

If $v_0 > v_0^C$, block 2 begins to slide at the time $t_1^\dagger$ and, since $\dot{x}_1(t_1^\dagger) > 0$, both blocks slide to the right for some time interval $t_1^\dagger < t < t_2^\dagger$. At time $t_2^\dagger$, either block 1 or block 2 comes to rest. To determine which comes to rest, the simultaneous equations of motion

$$\ddot{x}_1 = -\mu g + (x_2 - x_1 - 1)$$

$$\ddot{x}_2 = -\mu g - (x_2 - x_1 - 1)$$

must be solved, with initial conditions

$$x_1(t_1^\dagger) = (1 - \mu g) + \sqrt{(\mu g)^2 + v_0^2}$$

$$\dot{x}_1(t_1^\dagger) = -\mu g \sin t_1^\dagger + v_0 \cos t_1^\dagger$$

$$x_2(t_1^\dagger) = 2$$

$$\dot{x}_2(t_1^\dagger) = 0$$

The conditions $\dot{x}_1(t_2^\dagger) = 0$ and $\dot{x}_2(t_2^\dagger) = 0$ yield two values of $t_2^\dagger$, the smaller of which defines the time at which one of the blocks comes to rest, or sticks. A new equation of motion for the other block results. This process is continued until both blocks stick.

These examples illustrate the complexity of equations of motion of even a single constrained particle with Coulomb friction. The complexity dictates use of numerical methods of solution and logic associated with friction locks, or stiction. This requires the logic and speed of a digital computer.

1.3 DYNAMICS OF CONSTRAINED SYSTEMS OF PARTICLES

Dynamic models of realistic mechanical systems can occasionally be created using several particles that are constrained to move relative to one another to model the behavior of the actual mechanical system. In such cases, direct application of Newtonian mechanics can yield governing equations of motion. Examples are studied in this section to illustrate this direct application of Newton's equations of motion. The most important conclusion to be drawn from this exercise, however, is that the direct application of Newtonian dynamics of particles is impractical for dynamic analysis of most mechanical systems. The concepts illustrated do provide the foundation for methods that are developed later in the book, so they are recommended for careful study here.

Example 1.3.1: The *dumbbell* shown in Fig. 1.3.1(a) is modeled as two particles of equal mass m that are connected by a massless bar of unit length. The free body diagram of Fig. 1.3.1(b) shows both the applied forces on the particles and the constraint reaction forces due to the bar, $\mathbf{F}_1^C$ and $\mathbf{F}_2^C$. By Newton's third law, $\mathbf{F}_2^C = -\mathbf{F}_1^C$. Furthermore, since the massless bar behaves as a truss member, the reaction forces $\mathbf{F}_1^C$ and $-\mathbf{F}_2^C$ must be along the bar (i.e., they are collinear). Defining F^C to be the magnitude of the constraint reaction force, with tension positive,

$$\begin{aligned} \mathbf{F}_1^C &= F^C(\mathbf{r}_2 - \mathbf{r}_1) \\ \mathbf{F}_2^C &= -F^C(\mathbf{r}_2 - \mathbf{r}_1) \end{aligned} \tag{1.3.1}$$

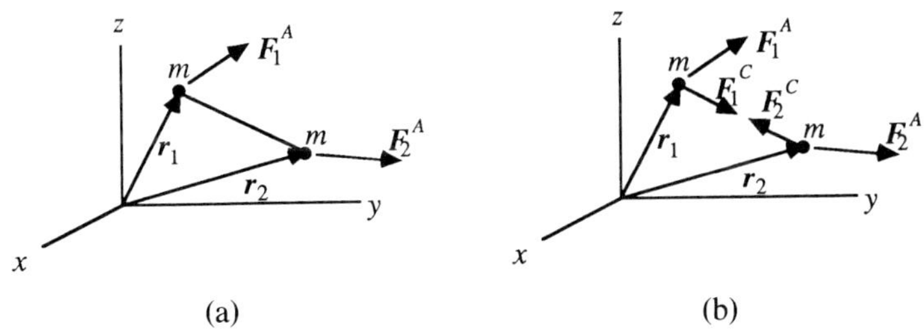

Figure 1.3.1 Dumbbell modeled as constrained particles.

Using Eqs. 1.3.1, Newton's equations of motion of the particles are

$$\begin{aligned} m\ddot{\mathbf{r}}_1 &= \mathbf{F}_1^A + F^C(\mathbf{r}_2 - \mathbf{r}_1) \\ m\ddot{\mathbf{r}}_2 &= \mathbf{F}_2^A - F^C(\mathbf{r}_2 - \mathbf{r}_1) \end{aligned} \tag{1.3.2}$$

This may be viewed as a system of six equations in the seven variables $\mathbf{r}_1$, $\mathbf{r}_2$, and F^C. Note that no derivatives of the constraint reaction force F^C appear. The additional condition that determines F^C is the algebraic constraint condition that the masses remain a unit distance apart:

$$(\mathbf{r}_2 - \mathbf{r}_1)^T(\mathbf{r}_2 - \mathbf{r}_1) - 1 = 0 \tag{1.3.3}$$

Equations 1.3.2 and 1.3.3 comprise a system of seven equations in seven unknowns, presuming the applied forces are known. Since the unknown F^C appears algebraically and Eq. 1.3.3 is an algebraic equation, this is a system of differential-algebraic equations. Note, furthermore, that both Eqs. 1.3.2 and 1.3.3 are nonlinear in the unknowns. Thus the prospect of obtaining an analytic solution of these equations of motion is dim. For an introduction to the theory and numerical methods of solving DAE, the reader is referred to Appendices C and D.

➧ **Example 1.3.2:** As a very simplistic model of the engine *slider–crank mechanism,* consider the constrained particle model shown in Fig. 1.3.2. The particle of mass m_1 represents the mass of the crank and a fraction of the mass of the connecting rod, lumped at the joint between the crank and the connecting rod. Mass m_2 contains the remaining fraction of the mass of the connecting rod and the mass of the piston. Mass m_1 is constrained to remain a distance $l_1 < 1$ from the center of the crank, m_2 must slide long the x axis of the x-y plane, and the distance between m_1 and m_2 is unity. With $\mathbf{r}_1 = [x_1, y_1]^T$ and $\mathbf{r}_2 = [x_2, y_2]^T$, these constraint conditions are

$$\begin{aligned} \mathbf{r}_1^T\mathbf{r}_1 - l_1^2 &= 0 \\ \mathbf{r}_2^T\mathbf{j} &= 0 \\ (\mathbf{r}_2 - \mathbf{r}_1)^T(\mathbf{r}_2 - \mathbf{r}_1) - 1 &= 0 \end{aligned} \tag{1.3.4}$$

The constraint reaction force on m_1 due to the crank is

$$\mathbf{F}_1^{C_1} = -F^{C_1}\frac{1}{l_1}\mathbf{r}_1 \tag{1.3.5}$$

The constraint reaction forces on m_1 and m_2 due to the connecting rod are, as in the

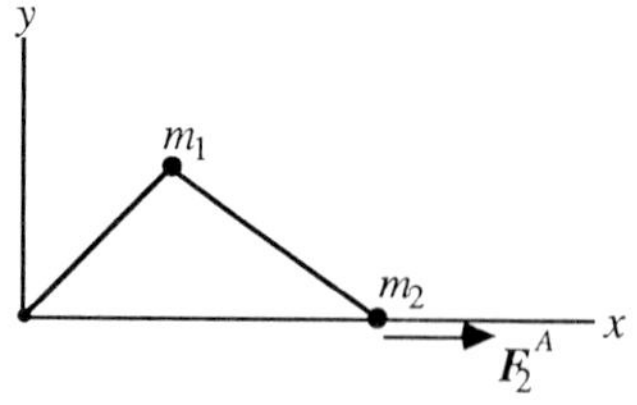

Figure 1.3.2 Simplistic particle model of slider–crank mechanism.

dumbbell of Example 1.3.1,

$$
\begin{aligned}
\mathbf{F}_1^{C_2} &= F^{C_2}(\mathbf{r}_2 - \mathbf{r}_1) \\
\mathbf{F}_2^{C_2} &= -F^{C_2}(\mathbf{r}_2 - \mathbf{r}_1)
\end{aligned}
\tag{1.3.6}
$$

Finally, the constraint reaction force on m_2 due to sliding without friction along the x axis is

$$
\mathbf{F}_2^{C_3} = F^{C_3}\mathbf{j} \tag{1.3.7}
$$

Accounting for the force F_2^A on the piston due to combustion of fuel in the cylinder of an engine and ignoring gravitational effects, Newton's equations of motion are

$$
\begin{aligned}
m_1\ddot{\mathbf{r}}_1 &= -\frac{F^{C_1}}{l_1}\mathbf{r}_1 + F^{C_2}(\mathbf{r}_2 - \mathbf{r}_1) \\
m_2\ddot{\mathbf{r}}_2 &= -F^{C_2}(\mathbf{r}_2 - \mathbf{r}_1) + F^{C_3}{}_{\mathrm{j}} + F_2^A\mathbf{i}
\end{aligned}
\tag{1.3.8}
$$

Equations 1.3.4 and 1.3.8 comprise a system of seven equations in the seven variables $\mathbf{r}_1$, $\mathbf{r}_2$, F^{C_1}, F^{C_2}, and F^{C_3}. It is a nonlinear system of DAE, with little prospect for finding an analytical solution. A numerical solution can, however, be obtained using methods of Appendix D.

If masses m_1 and m_2 are judiciously selected, the slider–crank model of Example 1.3.2 may yield reasonable and useful predictions of gross motion of an engine, under the influence of fuel combustion forces. This model is, however, completely unrealistic for prediction of bearing forces that act in an engine. In the actual engine, the crank is carefully balanced so that its center of mass is at the center of the pivot axis, not at the connection between the crank and the connecting rod. Thus the constraint reaction force F^{C_1} in Example 1.3.2 is physically meaningless. Constraint reaction force F^{C_2} may yield a reasonable estimate of the axial force in the connecting rod of the engine. The third constraint of Eqs. 1.3.4 realistically represents the kinematic function of the connecting rod (i.e., maintaining a fixed distance between the piston and crank pivot); hence the axial force transmitted by the connecting rod may be realistic. Finally, the lateral force of constraint F^{C_3} on the piston, which is of value in designing piston bearing surfaces, is suspect. Since the mass of the connecting rod is modeled as concentrated at its ends, the substantial lateral component of acceleration of the connecting rod, which influences the reaction force F^{C_3} between the piston and cylinder wall, is not accounted for in this model.

Examples 1.3.1 and 1.3.2 illustrate that direct application of Newton's equations of motion to model the dynamics of machines is at best approximate in nature. Whereas particle models may be systematically defined and constraint reaction forces accounted for, the resulting differential-algebraic equations of motion are far too complicated to permit an analytical solution. Thus numerical solution techniques such as those outlined in Appendix D are required to obtain quantitative information from even particle models of most machines. Finally, to obtain adequate realism, machine components must be modeled using many particles to represent the actual distribution of mass in machine components. This yields very high dimensional models that are impractical for most applications.

It may be concluded that while direct applications of Newton's equations of motion for particles can yield insights into the nature of machine dynamics, more refined methods are required for realistic dynamic analysis of machines. The foundation for such methods is presented in Chapters 2 and 3, where particle models are used to develop basic ideas and analytical techniques. The full implementation of methods that are suitable for broad classes of applications is presented in Chapters 4 through 6, where rigid-body models of machine components and systems are used to more precisely account for the effect of distribution of mass in components and systems.

1.4 MOMENTUM AND KINETIC ENERGY

Even though direct application of Newton's equations of motion for dynamics of systems modeled as particles is impractical in general, conservation relations that can be derived from Newton's equations of motion often yield valuable insight and information, even without constructing a complete solution of the equations of motion. Two fundamental derived relations that follow from Newton's equations of motion are introduced in this section. The first relation involves momentum and impulse, which yields a valuable tool for analysis of impact. The second relation involves kinetic energy and work, which yields valuable insights for systems in which external forces are conservative.

Momentum and Impulse

The *momentum of a particle* is defined as the product of the mass of the particle and its velocity:

$$\mathbf{P} = m\dot{\mathbf{r}} \tag{1.4.1}$$

For a particle with constant mass, taking the time derivative of Eq. 1.4.1 and using the first of Eqs. 1.1.1, Newton's equation of motion of Eq. 1.1.1 may be written as

$$\dot{\mathbf{P}} = m\ddot{\mathbf{r}} = \mathbf{f} \tag{1.4.2}$$

Note that if the resultant force **f** on the particle is zero, its momentum remains constant. More generally, integrating both sides of Eq. 1.4.2 with respect to time from t_1 to t_2 yields

$$\int_{t_1}^{t_2} \dot{\mathbf{P}}\, dt = \int_{t_1}^{t_2} \mathbf{f}\, dt \equiv \bar{\mathbf{f}} \tag{1.4.3}$$

where $\bar{\mathbf{f}}$ is defined as the *impulse* of the force **f** from time t_1 to time t_2. From the fundamental theorem of calculus [3], the integral on the left of Eq. 1.4.3 is simply the change in linear momentum, leading to the *impulse–momentum relation*

$$\mathbf{P}(t_2) - \mathbf{P}(t_1) = \bar{\mathbf{f}} \tag{1.4.4}$$

Thus if the time integral of the resultant force that acts on a body over a period of time is used to calculate its impulse, the change in momentum of the body may be determined immediately.

Example 1.4.1: Concepts of impulse and momentum are valuable in studying impact between bodies. Consider the classical problem of impact of spherical billiard balls of masses m_1 and m_2 that lie in the x-y plane of an inertial frame, as shown in Fig. 1.4.1. The initial velocity of mass m_1 is presumed known and leads to contact with mass m_2, which is initially at rest. Let $\mathbf{u}$ be the common normal between the balls at impact time t^*, directed from m_1 to m_2, as shown in Fig. 1.4.1.

The velocities prior to impact are

$$\begin{aligned} \dot{\mathbf{r}}_1(t^*-) &= [1, 1, 0]^T \\ \dot{\mathbf{r}}_2(t^*-) &= [0, 0, 0]^T \end{aligned} \tag{1.4.5}$$

Applying Newton's third law of action and reaction, the forces $\mathbf{f}_{ij}$, $i,j = 1,2$, of interaction between the balls during the contact event may be written in the form

$$\mathbf{f}_{12} = -f(t)\mathbf{u} = -\mathbf{f}_{21} \tag{1.4.6}$$

where $f(t) \geqslant$ and contact between the balls is frictionless, so the force along the normal $\mathbf{u}$ to the contacting surfaces is zero, except for the very short interval of time in which balls are in contact. Define the beginning and end of contact as t_1 and t_2, such that $t_1 < t^* < t_2$. Presuming that the impact force acts essentially instantaneously, at time t^*, $\mathbf{u}$ may be treated as constant and the impulse of force $\mathbf{f}_{12}$ is

$$\begin{aligned} \bar{\mathbf{f}}_{12} &= \lim_{|t_2 - t_1| \to 0} \left[-\int_{t_1}^{t_2} f(t)\, dt \right] \mathbf{u} \\ &= -\bar{f}\,\mathbf{u} \end{aligned} \tag{1.4.7}$$

where $\bar{f}$ is the magnitude of the impulse of the contact force. Thus, from Eq. 1.4.6,

$$\bar{\mathbf{f}}_{21} = \bar{f}\mathbf{u} \tag{1.4.8}$$

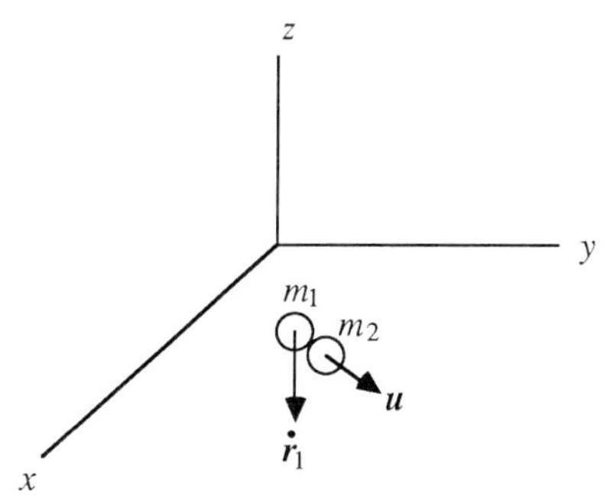

Figure 1.4.1 Billiard balls in oblique impact.

Applying the impulse–momentum relation of Eq. 1.4.4 to both balls, using Eqs. 1.4.7 and 1.4.8, gives

$$\begin{aligned} m_1[\dot{\mathbf{r}}_1(t^*+) - [1, 1, 0]^T] &= -\bar{f}\mathbf{u} \\ m_2\dot{\mathbf{r}}_2(t^*+) &= \bar{f}\mathbf{u} \end{aligned} \tag{1.4.9}$$

One additional relationship is needed to determine the impulse $\bar{f}$, hence determining the velocities $\dot{\mathbf{r}}_1(t^*+)$ and $\dot{\mathbf{r}}_2(t^*+)$ from Eq. 1.4.9.

Presume that experiments have been performed on impact between the balls and that a *coefficient of restitution* e, $0 \leq e \leq 1$, has been determined such that the normal component of relative velocity after impact is equal to the negative of the coefficient of restitution times the normal component of relative velocity before impact:

$$\mathbf{u}^T[\dot{\mathbf{r}}_1(t^*+) - \dot{\mathbf{r}}_2(t^*+)] = -e\mathbf{u}^T[1, 1, 0]^T$$

Solving for velocities after impact from Eq. 1.4.9 as functions of $\bar{f}$ and substituting them into this relation yields

$$\mathbf{u}^T\left\{[1, 1,0]^T - \frac{\bar{f}}{m_1}\mathbf{u} - \frac{\bar{f}}{m_2}\mathbf{u}\right\} = -e[1, 1, 0]\mathbf{u}$$

This equation determines the impulse of the contact force as

$$\bar{f} = \frac{m_1m_2}{m_1 + m_2}(1 + e)[1, 1, 0]\mathbf{u} \tag{1.4.10}$$

Substituting this result into Eq. 1.4.9 yields the desired velocities after impact as

$$\begin{aligned} \dot{\mathbf{r}}_1(t^*+) &= [1, 1, 0]^T - \frac{m_2}{m_1 + m_2}(1 + e)([1, 1, 0]\mathbf{u})\mathbf{u} \\ \dot{\mathbf{r}}_2(t^*+) &= \frac{m_1}{m_1 + m_2}(1 + e)([1, 1, 0]\mathbf{u})\mathbf{u} \end{aligned} \tag{1.4.11}$$

Note that if the coefficient of restitution were $e = 1$, if $\mathbf{u} = [1, 1, 0]^T$, and if $m_1 = m_2$, the first ball comes to rest after impact and the second ball proceeds with the velocity of the first ball after impact.

The momentum relation of Eq. 1.4.2 may be regarded as equivalent to Newton's equation of motion of Eq. 1.1.1. Indeed, momentum concepts have conventionally been used in the literature for development of the equations of motion of particles [6–11]. The more direct approach adopted in this book uses Newton's equations of motion for particles, without the need for reliance on derived "principles" involving momentum.

Kinetic Energy and Work

Much as Newton's equations of motion, written in the form of Eq. 1.4.2, are integrated over a time interval to derive impulse–momentum relations, the scalar product of both sides of Newton's equations of motion of Eq. 1.1.1 with a differential displacement $d\mathbf{r}$ may be formed and integrated over a time interval $t_1 > t > t_2$, to obtain

$$\int_{\mathbf{r}_1}^{\mathbf{r}_2} m\ddot{\mathbf{r}}^T \, d\mathbf{r} = \int_{\mathbf{r}_1}^{\mathbf{r}_2} \mathbf{f}^T \, d\mathbf{r} \tag{1.4.12}$$

where $\mathbf{r}_i \equiv \mathbf{r}(t_i)$, i = 1,2. Both sides of Eq. 1.4.12 must be regarded as *line integrals*, taken over the path followed by the particle that satisfies Newton's equations of motion from time t_1 to time t_2. Since $\mathbf{f}^T \, d\mathbf{r}$ is the *differential work* done by force $\mathbf{f}$ over the differential displacement $d\mathbf{r}$, the integral on the right of Eq. 1.4.12 is a work expression. Hence the integral on the left of Eq. 1.4.12 must also have the units and significance of work.

Consider first the integral on the right of Eq. 1.4.12. If the resultant force $\mathbf{f}$ that acts on the particle is a function of position [i.e., if $\mathbf{f} = \mathbf{f}(\mathbf{r})$] and if the *differential form* $\mathbf{f}^T \, d\mathbf{r}$ is the differential of a *work function* $W(\mathbf{r})$; that is, if

$$\begin{aligned} dW &= \frac{\partial W}{\partial x} dx + \frac{\partial W}{\partial y} dy + \frac{\partial W}{\partial z} dz \\ &\equiv \frac{\partial W}{\partial \mathbf{r}} d\mathbf{r} = \mathbf{f}^T \, d\mathbf{r} \end{aligned} \tag{1.4.13}$$

where differential calculus notation of Appendix A is used, the right side of Eq. 1.4.12 becomes

$$\int_{\mathbf{r}_1}^{\mathbf{r}_2} \mathbf{f}^T \, d\mathbf{r} = \int_{\mathbf{r}_1}^{\mathbf{r}_2} dW = W(\mathbf{r}_2) - W(\mathbf{r}_1) \tag{1.4.14}$$

This is the work done by the applied force $\mathbf{f}$ as the particle traverses the trajectory defined as the solution of Newton's equations of motion between $\mathbf{r}_1$ and $\mathbf{r}_2$.

Consider next the left side of Eq. 1.4.12. Using the chain rule of differentiation, $d\mathbf{r} = \dot{\mathbf{r}} \, dt$. Thus the integrand on the left of Eq. 1.4.12 may be written as

$$m\ddot{\mathbf{r}}^T \, d\mathbf{r} = m\ddot{\mathbf{r}}^T\dot{\mathbf{r}} \, dt = m(\ddot{x}\dot{x} + \ddot{y}\dot{y} + \ddot{z}\dot{z}) \, dt \tag{1.4.15}$$

Using the product rule of differentiation, this is

$$\begin{aligned} m\ddot{\mathbf{r}}^T \, d\mathbf{r} &= \frac{d}{dt} [\tfrac{1}{2}m(\dot{x}^2 + \dot{y}^2 + \dot{z}^2)]dt \\ &= \frac{d}{dt} [\tfrac{1}{2}m \, \dot{\mathbf{r}}^T\dot{\mathbf{r}}] \, dt \end{aligned} \tag{1.4.16}$$

Denoting the *kinetic energy* as T,

$$T \equiv \tfrac{1}{2} \, m\dot{\mathbf{r}}^T\dot{\mathbf{r}} \tag{1.4.17}$$

Eq. 1.4.16 is simply

$$m\ddot{\mathbf{r}}^T \, d\mathbf{r} = \frac{d}{dt} T \, dt = \dot{T} \, dt \tag{1.4.18}$$

Using the fundamental theorem of calculus [3], the integral on the left side of Eq. 1.4.12 is thus

$$\int_{\mathbf{r}_1}^{\mathbf{r}_2} m\ddot{\mathbf{r}}^T\, d\mathbf{r} = \int_{t_1}^{t_2} \dot{T}\, dt = T(t_2) - T(t_1) \tag{1.4.19}$$

Substituting Eqs. 1.4.14 and 1.4.19 into Eq. 1.4.12 yields the *kinetic energy–work relation*

$$T(t_2) - T(t_1) = W(\mathbf{r}_2) - W(\mathbf{r}_1) \tag{1.4.20}$$

This result simply states that the change in kinetic energy of a particle over a period of time is equal to the work done by the resultant external force on the particle over the same period of time. Note that this relation is valid only if the force **f** that acts on the particle is derivable from a work function, in the sense of Eq. 1.4.14. For more details regarding such *conservative forces,* see Chapter 3. In particular, it is well known that gravitational forces and forces that act in elastic springs are conservative.

Example 1.4.2: Let a particle (thought of as a bead) slide without friction along a path (thought of as a wire) whose shape is described by the equation $y = h(x)$ in the x-y plane, as shown in Fig. 1.4.2. Thus the position of the particle is

$$\mathbf{r} = \begin{bmatrix} y \\ h(x) \end{bmatrix} \tag{1.4.21}$$

With gravity acting in the positive y direction and declaring the work function at $y = 0$ to be zero, Eq. 1.4.14 defines W as

$$\begin{aligned} W(\mathbf{r}(t_2)) &= \int_{y(t_1)}^{y(t_2)} mg\, dy \\ &= mg[y(t_2) - y(t_1)] \\ &= mgh(x(t_2)) \end{aligned} \tag{1.4.22}$$

where the particle is presumed to be at the origin of the x-y frame at t_1. If the particle is released from rest at time t_1, then $T(t_1) = 0$ and Eq. 1.4.20 results in the relation

$$T(t_2) = mgh(x(t_2)) \tag{1.4.23}$$

Using the notation $h' = dh/dx$, the chain rule of differentiation may be used with

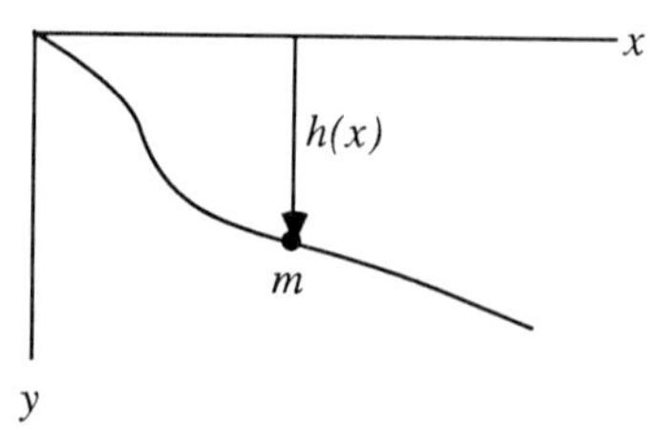

Figure 1.4.2 Particle sliding on wire.

Eq. 1.4.21 to write Eq. 1.4.23 in the form

$$\tfrac{1}{2}m[\dot{x}^2 + h'(x)^2\dot{x}^2] = mgh(x) \qquad (1.4.24)$$

Presuming that $\dot{x}$ remains positive, this equation may be solved to obtain

$$\dot{x} = \sqrt{\frac{1 + 2gh(x)}{1 + h'(x)^2}} \qquad (1.4.25)$$

Thus, in this example, the kinetic energy–work relation of Eq. 1.4.20 yields a first-order differential equation of motion. As will be seen in Chapter 3, this is a *first integral* of the equations of motion that follows from conservation of energy.

To further illustrate use of first integrals, Eq. 1.4.25 may be written in differential form as

$$dt = \sqrt{\frac{1 + h'(x)^2}{2gh(x)}}\, dx \qquad (1.4.26)$$

Integrating both sides of this equation yields

$$t_2 = t_1 \int_0^{x(t_2)} \sqrt{\frac{1 + h'(x)^2}{2gh(x)}}\, dx \qquad (1.4.27)$$

The time taken to traverse the curve from $x = 0$ to a specified value of $x(t_2) = \bar{x}$ may be found by simply evaluating the integral on the right of Eq. 1.4.27. This result leads to interesting optimization questions such as: What shape of the wire [i.e., what function $h(x)$] leads to the shortest transit time from $x = 0$ to $x = 1$? This is the classical *brachistochrone problem* [9] of the calculus of variations, which is typical of broad classes of optimal control and optimal design applications.

PROBLEMS

Section 1.1

1.1.1 The linear spring shown in Fig. P1.1.1 has unit free length and spring rate k and the damper has damping coefficient $c > 0$ (i.e., damping force $= -c\dot{l}$, where l is the length of the damper). Gravity acts in the negative z direction and there is no other

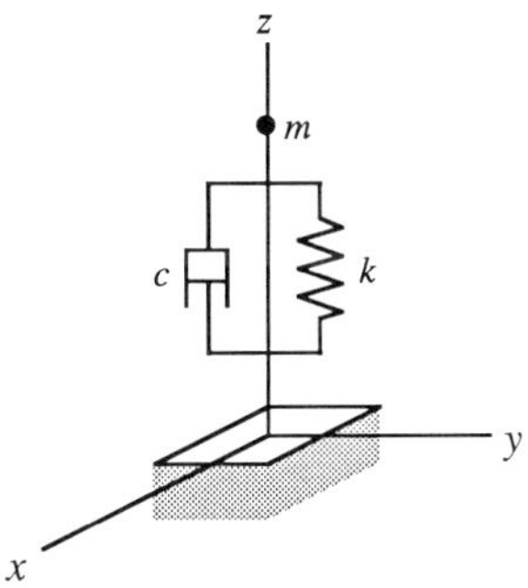

Figure P1.1.1

external force. Write and solve the equations of motion of the particle, with initial conditions $\mathbf{r}(0) = [0, 0, 1]^T$ and $\dot{\mathbf{r}}(0) = [0, 0, 1]^T$. Show that $x(t) = y(t) = 0$, so the spring–damper force acts only in the z direction.

1.1.2 A particle of mass m is launched vertically upward from the origin of an inertial reference frame with velocity v_0. If the drag due to motion in air is cv^2, opposing the velocity, write and solve the equations of motion.

1.1.3 Let a Cartesian reference frame with origin at the center of the earth be treated as an inertial reference frame. If the inverse square law $F = mM/r^2$, with M as the mass of the earth and r as the distance from the center of the earth, defines the force acting on a satellite of mass m (directed toward the center of the earth), write the equations of motion in space for the satellite.

Section 1.2

1.2.1 The simple pendulum shown in Fig. P1.2.1 is used in many mechanisms as a timing device (e.g., a clock). Modeling the pendulum with mass m concentrated as a particle at the end and with gravity acting in the negative y direction:

(a) Write the equations of motion for the particle, accounting for the axial constraint force in the bar. Define the resulting DAE.

(b) Use the fact that $x = \sin\theta$ and $y = -\cos\theta$ to write the equations of motion in terms of θ, $\dot{\theta}$, and $\ddot{\theta}$.

(c) Solve the equations derived in part (b) for $\ddot{\theta}$ and the constraint force in the bar, as functions of θ and $\dot{\theta}$.

(d) Using the small-angle approximations $\sin\theta \approx \theta$ and $\cos\theta \approx 1$, simplify the equations of motion derived in part (b) and repeat part (c).

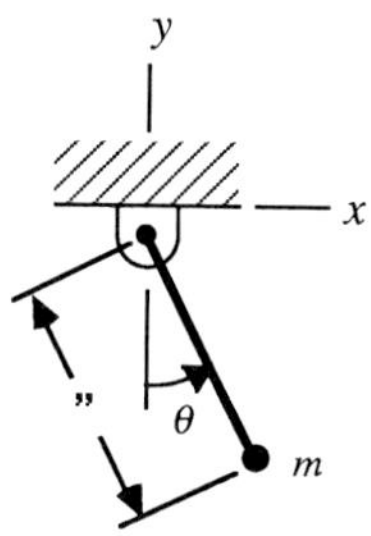

Figure P1.2.1

1.2.2 Consider the particle of unit mass shown in Fig. P1.2.2 that slides without friction along the parabola $y = x^2$ in the x-y plane, where gravity acts in the negative y

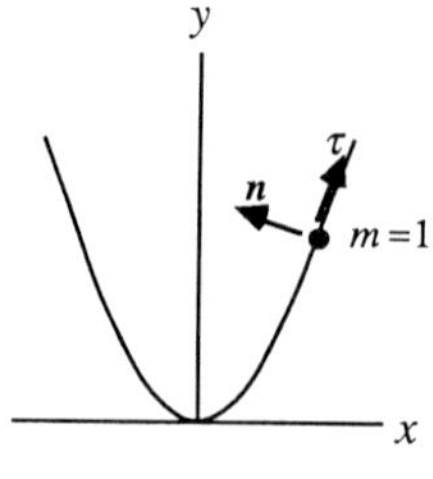

Figure P1.2.2

direction. If the particle is at rest with $x = y = 1$ at $t = 0$, write the constrained equations of motion. (*Hint:* Use the fact that $\boldsymbol{\tau} = [1, 2x]^T$ is tangent to the curve, hence $\mathbf{n} = [-2x, 1]^T$ is normal to the curve, but it is not a unit vector.)

1.2.3 Use the facts that $\dot{y} = 2x\dot{x}$ and $\ddot{y} = 2x\ddot{x} + 2\dot{x}^2$ to write the equations of motion in Problem 1.2.2 in terms of only x, $\dot{x}$, $\ddot{x}$, and the constraint force. Solve the reduced equations of motion for $\ddot{x}$ and the constraint force as functions of x and $\dot{x}$.

1.2.4 Apply a thrust force T to the particle of Problem 1.2.2, tangent to the curve, and repeat the derivation of equations in Problems 1.2.2 and 1.2.3. Determine the thrust $T = T(t)$ and the associated constraint force that would yield the prescribed x coordinate of the particle, $x = \sin \omega t$.

1.2.5 Consider the block of unit mass shown in Fig. P1.2.5, modeled as a particle that slides along an inclined plane with slope $\frac{3}{4}$. The coefficient of friction between the surfaces is $\mu = 0.3$ and gravity acts in the negative y direction. If the block is initially at the origin of the x-y frame and has a velocity of 1 m/s up the incline, determine how far the block will slide before it comes to rest.

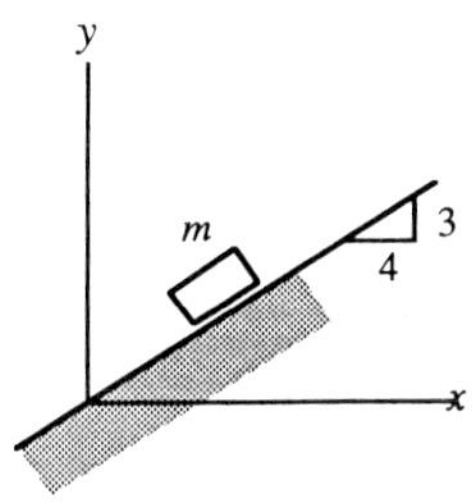

Figure P1.2.5

Section 1.3

1.3.1 Write the equations of motion for the two-particle model of the double pendulum shown in Fig. P1.3.1, where the lengths of the bars are unity, gravity acts in the negative y direction, and the constraint forces act along the bars. Define the resulting DAE.

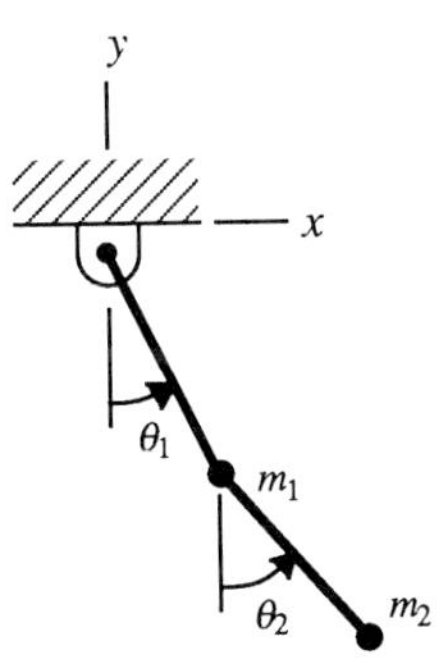

Figure P1.3.1

1.3.2 Use the facts that $x_1 = \sin \theta_1$, $y = -\cos \theta_1$, $x_2 = \sin \theta_1 + \sin \theta_2$, and $y_2 = -\cos \theta_1 - \cos \theta_2$ to write the equations of motion of Problem 1.3.1 in terms of θ_i, $\dot{\theta}_i$, and $\ddot{\theta}_i$, $i = 1, 2$. Solve the resulting equations for $\ddot{\theta}_1$, $\ddot{\theta}_2$, and the two unknown constraint forces as functions of θ_i and $\dot{\theta}_i$, $i = 1, 2$.

1.3.3 Consider the two-particle model of the spatial four-bar mechanism shown in Fig. P1.3.3, with bars of unit length and gravity acting in the negative z direction. Use the fact that constraint forces act along the three bars and that the pivots at points A and B support constraint forces on the particles in the x and y directions, respectively, to write the equations of motion of the two-particle system and define the resulting DAE.

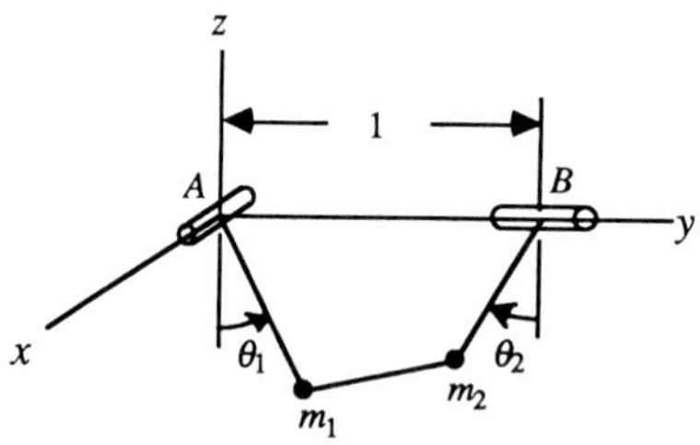

Figure P1.3.3

1.3.4 Use the facts that $x_1 = 0$, $y_1 = \sin \theta_1$, $z_1 = -\cos \theta_1$, $x_2 = \sin \theta_2$, $y_2 = 1$, and $z_2 = -\cos \theta_2$ to write the equations of motion of Problem 1.3.3 in terms of θ_i, $\dot{\theta}_i$, and $\ddot{\theta}_i$, $i = 1, 2$. Show that all constraint forces except the force in the bar between the masses can be eliminated. Define the reduced DAE.

Section 1.4

1.4.1 Analyze the motion of the impacting balls of Example 1.4.1 if the coefficient of restitution is $e = 0$.

1.4.2 Use the kinetic energy–work relation for the conservative system of Problem 1.2.1 to solve for $\dot{\theta}$ as a function of θ. Show that differentiating this relation with respect to time yields the second order differential equation obtained in part (c) of Problem 1.2.1.

1.4.3 Repeat Problem 1.4.2 for the double pendulum of Problem 1.3.1.

1.4.4 Write the kinetic energy–work relation for the conservative spatial four-bar mechanism of Problem 1.3.3.

SUMMARY OF KEY FORMULAS

Newton's Equations of Motion

$$m\ddot{\mathbf{r}} = \mathbf{f} \tag{1.1.1}$$

$$\mathbf{r}(t_0) = \mathbf{r}^0, \qquad \dot{\mathbf{r}}(t_0) = \dot{\mathbf{r}}^0 \tag{1.1.3}$$

$$m_i\ddot{\mathbf{r}}_i = \mathbf{F}_i + \sum_{j=1}^{n} \mathbf{f}_{ij}, \qquad \mathbf{f}_{ij} = -\mathbf{f}_{ij} \tag{1.1.13, 14}$$

Momentum and Impulse

$$\mathbf{P} = m\dot{\mathbf{r}} \tag{1.4.1}$$

$$\bar{\mathbf{f}} = \int_{t_1}^{t_2} \mathbf{f}\, dt, \qquad \mathbf{P}(t_2) - \mathbf{P}(t_1) = \bar{\mathbf{f}} \tag{1.4.3, 4}$$

Kinetic Energy and Work

$$dW = \mathbf{f}^T\, d\mathbf{r}, \qquad \int_{\mathbf{r}_1}^{\mathbf{r}_2} \mathbf{f}^T\, d\mathbf{r} = W(\mathbf{r}_2) - W(\mathbf{r}_1) \tag{1.4.13, 14}$$

$$T = \tfrac{1}{2}m\dot{\mathbf{r}}^T\dot{\mathbf{r}}, \qquad T(t_2) - T(t_1) = W(\mathbf{r}(t_2)) - W(\mathbf{r}(t_1)) \tag{1.4.17, 20}$$

2

Kinematics of Systems Modeled as Particles

Kinematics is the study of the geometry of motion, independent of forces that may influence the motion. This chapter is devoted to a systematic description of the motion of individual particles and systems of particles that are constrained relative to one another, to extend and overcome some of the shortcomings of Newtonian methods introduced in Chapter 1. Although not all systems can be modeled as particles, some can. For example, the satellite shown in Fig. 2.0.1(b) may be modeled as a particle to study its orbit and the influence of gravitational attraction of the earth and moon, as suggested by the model shown in Fig. 2.0.1(a). This particle model, however, is clearly not adequate for the study of attitude control of the satellite. The concepts introduced in this chapter are applicable in the context of particles. More important, they serve as the foundation for later study of more complex rigid-body kinematics.

2.1 PARTICLES MOVING IN SPACE

The position of a particle in space may be characterized by any of the coordinate systems that are commonly used in engineering analysis. Several such coordinate systems are introduced and analyzed in this section and limitations on their use are identified.

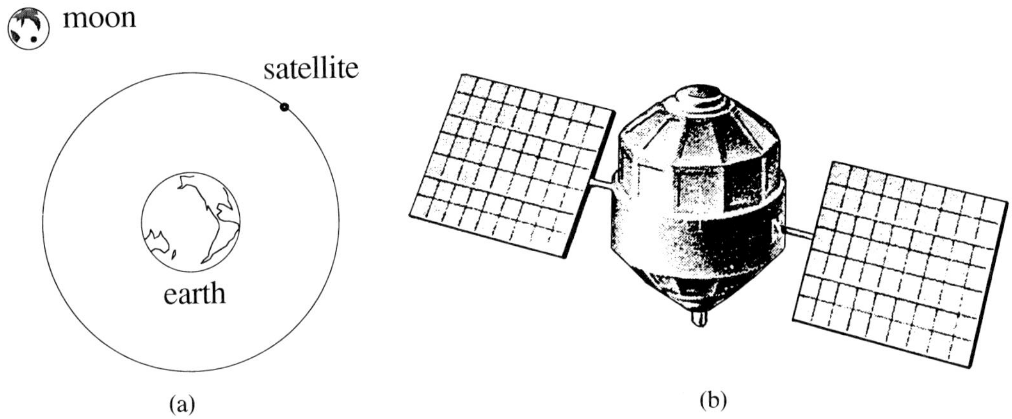

Figure 2.0.1 Particle model of satellite.

Cartesian Coordinates

As explained in Section A.1 of Appendix A, right-hand orthogonal reference frames, called *Cartesian reference frames*, are most commonly used to locate points in space. An x-y-z Cartesian reference frame, shown in Fig. 2.1.1, is used to locate a particle P in three-dimensional space by the vector

$$\mathbf{r} = [x, y, z]^T \tag{2.1.1}$$

of its components in the x-y-z frame. The use of a column matrix of the components of a vector to represent the vector, as in Eq. 2.1.1, is explained and illustrated in detail in Appendix A.

Cartesian coordinates represent the most fundamental and generally applicable set of coordinates to locate points in space. In numerous applications, however, such as orbital motion of a satellite around the earth or motion that is expected to follow some curve or surface in space, a set of coordinates that define position, called *generalized coordinates*, may be used to advantage. For a particle that moves in three-dimensional space, three generalized coordinates $\mathbf{q} = [q_1, q_2, q_3]^T$ are required. The relation between generalized coordinates

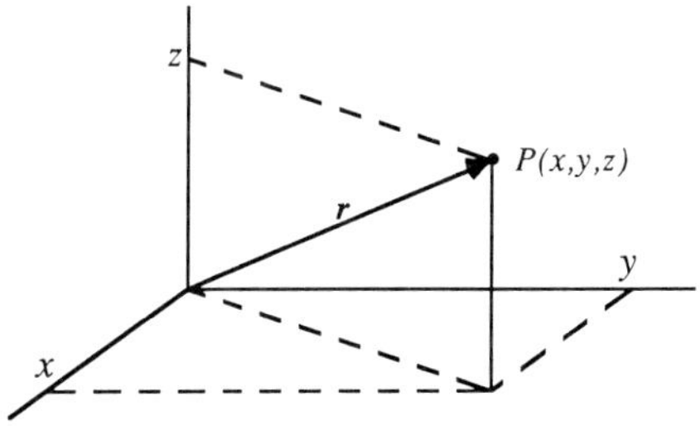

Figure 2.1.1 Cartesian coordinates of point P.

and Cartesian coordinates is defined by a *coordinate transformation*, or *mapping*

$$\mathbf{r} = \mathbf{r}(\mathbf{q}) = \begin{bmatrix} x(\mathbf{q}) \\ y(\mathbf{q}) \\ z(\mathbf{q}) \end{bmatrix} \tag{2.1.2}$$

where the function $\mathbf{r}(\mathbf{q})$ should be *continuous* and in dynamics applications should have two derivatives. The mapping of Eq. 2.1.2 is continuous if the functions $x(\mathbf{q})$, $y(\mathbf{q})$, and $z(\mathbf{q})$ are continuous in $\mathbf{q}$.

Continuity of the mapping of Eq. 2.1.2 is illustrated schematically in Fig. 2.1.2. If V is a small neighborhood of a point $\mathbf{r}^0 = \mathbf{r}(\mathbf{q}^0)$, there must be some small neighborhood U of $\mathbf{q}^0$ so that for all $\mathbf{q}$ in U, $\mathbf{r}(\mathbf{q})$ is in V. If the mapping is not continuous, there will be points $\mathbf{q}^*$ at which arbitrarily small changes in generalized coordinates lead to large changes in the position vector $\mathbf{r}$ of a particle (i.e., $\mathbf{q}$ cannot be used to control the position of the particle).

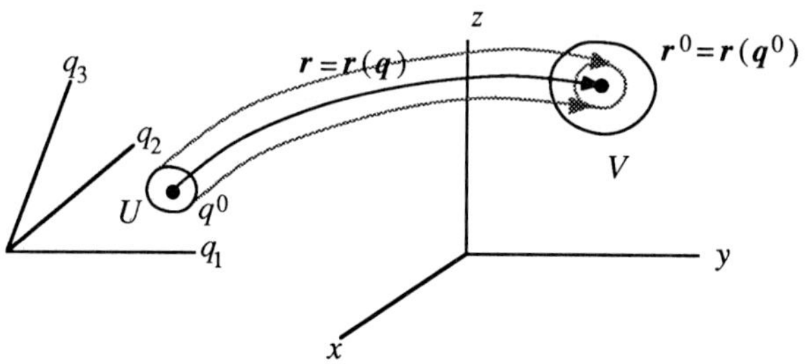

Figure 2.1.2 Continuity of the mapping $\mathbf{r} = \mathbf{r}(\mathbf{q})$.

To be useful in kinematic and dynamic analysis, generalized coordinates should uniquely define the position of a particle in space; that is, if $\mathbf{q}_1$ is fixed and $\mathbf{r}(\mathbf{q}_1) = \mathbf{r}(\mathbf{q}_2)$, the only solution for $\mathbf{q}_2$ in a neighborhood of $\mathbf{q}_1$ is $\mathbf{q}_2 = \mathbf{q}_1$. A mapping for which this property holds is said to be *one-to-one*, or *locally one-to-one*. A point $\mathbf{q}^*$ at which there are neighboring points $\mathbf{q}^\dagger \neq \mathbf{q}^*$, but with $\mathbf{r}(\mathbf{q}^\dagger) = \mathbf{r}(\mathbf{q}^*)$, is called a *singular point* of the mapping, as illustrated in Fig. 2.1.3. The practical difficulty at a singular point $\mathbf{q}^*$ is not that a knowledge of $\mathbf{q}$ fails to yield $\mathbf{r}$, since Eq. 2.1.2 establishes this relation, even continuously. Rather, when $\mathbf{q}$ is used in kinematic analysis, kinematic equations that determine the motion of

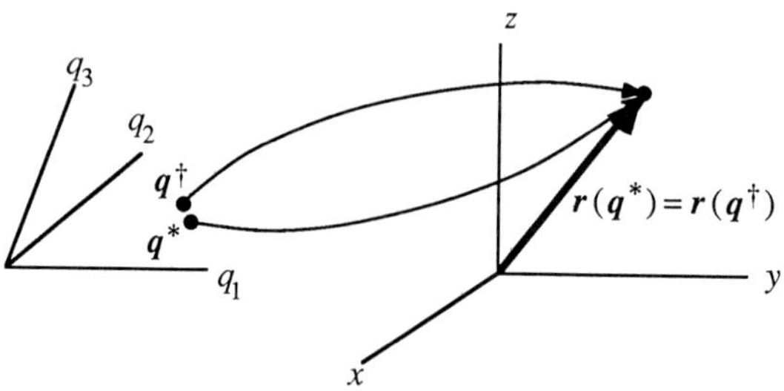

Figure 2.1.3 Singular point of the mapping $\mathbf{r} = \mathbf{r}(\mathbf{q})$.

a particle must be solved for $\mathbf{q}$. Since these equations will have multiple solutions at a singular point $\mathbf{q}^*$, severe difficulties in analysis will be encountered.

Cylindrical Coordinates

If a particle is expected to move on or approximately along the surface of a circular cylinder with its axis of symmetry along the Cartesian z axis, the *cylindrical coordinates* of Fig. 2.1.4 may be used to advantage. Cylindrical generalized coordinates are denoted

$$\mathbf{q} = [s, \phi, z]^T \tag{2.1.3}$$

where $s \geq 0$ and $0 \leq \phi < 2\pi$. The relationship between Cartesian coordinates $\mathbf{r}$ and cylindrical coordinates $\mathbf{q}$ are defined by the mapping

$$\mathbf{r}(\mathbf{q}) = \begin{bmatrix} s \cos \phi \\ s \sin \phi \\ z \end{bmatrix} \tag{2.1.4}$$

The mapping of Eq. 2.1.4 is continuous since each of the functions in the vector on the right of Eq. 2.1.4 is continuous in s, ϕ, and z. For any $\mathbf{r} = [x^0, y^0, z^0]^T$ with $(x^0)^2 + (y^0)^2 \neq 0$, there is just one value $s^0 = \sqrt{(x^0)^2 + (y^0)^2}$ and ϕ^0 in the range $0 \leq \phi < 2\pi$ such that $\cos \phi^0 = x^0/s^0$ and $\sin \phi^0 = y^0/s^0$. Thus the mapping of Eq. 2.1.4 is nonsingular if $s \neq 0$. Difficulties arise, however, for any point P on the z axis (i.e., $\mathbf{r}^* = [0, 0, z^*]^T$). Clearly, $z = z^*$ and $s^* = 0$, but the value of ϕ cannot be determined uniquely. In fact $\mathbf{q} = [0, \phi, z^*]^T$, for all ϕ in the range $0 \leq \phi < 2\pi$, gives $\mathbf{r}^*$, as shown in Fig. 2.1.5. Thus any point on the z axis (i.e., for any $\mathbf{q}^0$ of the form $\mathbf{q}^0 = [0, \phi, z]^T$) is a singular point for the mapping of Eq. 2.1.4.

As will be seen later, difficulties will be encountered in kinematic and dynamic analysis if cylindrical coordinates are used and the path of motion of a particle intersects the z axis. It is important to recognize that the singularity

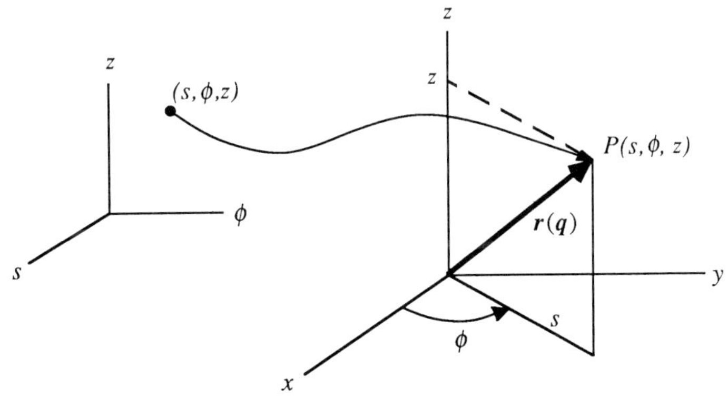

Figure 2.1.4 Cylindrical coordinates.

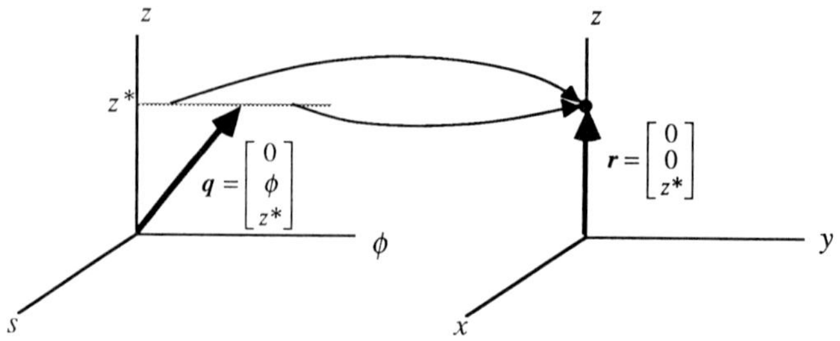

Figure 2.1.5 Singular points for cylindrical coordinates.

encountered is not a result of a bad design or a catastrophic physical situation. Rather, it is due to selection of generalized coordinates that have an inherently singular behavior (i.e., due to a bad mathematical model).

Spherical Coordinates

Consider next the *spherical coordinates* described in Fig. 2.1.6, which locate points in space using the spherical generalized coordinate vector

$$\mathbf{q} = [\rho, \phi, \psi]^T \tag{2.1.5}$$

where $\rho = |\mathbf{r}| \geq 0$, $0 \leq \phi < 2\pi$, and $-\pi/2 \leq \psi \leq \pi/2$. The Cartesian position vector of point P may be written in terms of spherical coordinates as

$$\mathbf{r} = \begin{bmatrix} \rho \cos \psi \cos \phi \\ \rho \cos \psi \sin \phi \\ \rho \sin \psi \end{bmatrix} \tag{2.1.6}$$

It is not difficult to show (Prob. 2.1.1) that for any $\mathbf{r}^0 = [x^0, y^0, z^0]$ with $x^{0\,2} + y^{0\,2} \neq 0$ (i.e., for points not on the z axis), there is just one value of spherical

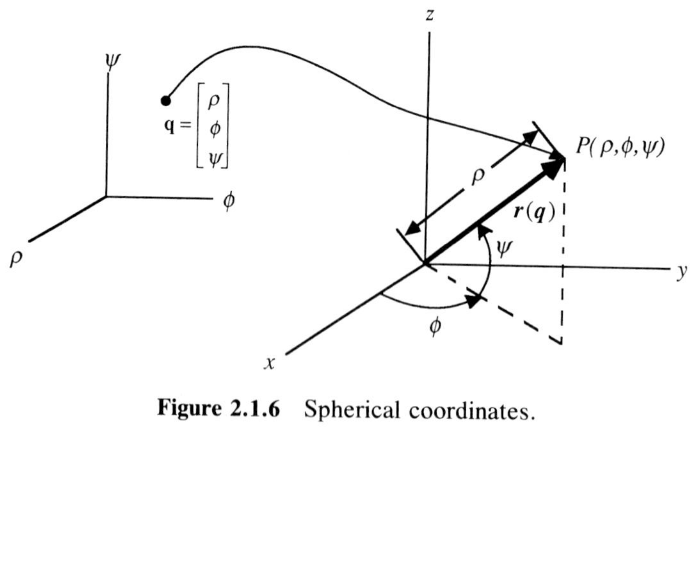

Figure 2.1.6 Spherical coordinates.

coordinates, $\mathbf{q}^0$, such that $\mathbf{r}^0 = \mathbf{r}(\mathbf{q}^0)$. For any point $\mathbf{r}^* = [0, 0, z^*]^T$, however, $\rho^* = |z^*|$ and $\psi = \pi/2 \text{ sgn } z^*$, where sgn z^* is the algebraic sign of z^*, yields $\mathbf{r}^*$, for any value of ϕ. Thus such points are singular points for the mapping of Eq. 2.1.6. The set of points $\mathbf{q}$ that map into $\mathbf{r}^*$, for $z^* > 0$, is shown in Fig. 2.1.7.

The singularity for spherical coordinates at $\mathbf{r}^\dagger = [0, 0, 0]^T$ is even more severe. Any $\mathbf{q}^\dagger = [0, \phi, \psi]^T$ satisfies $\mathbf{r}^\dagger = \mathbf{r}(\mathbf{q}^\dagger)$. As shown in Fig. 2.1.8, any point in the ϕ-ψ plane of ρ-ϕ-ψ space maps into the point $\mathbf{r}^\dagger = [0, 0, 0]^T$.

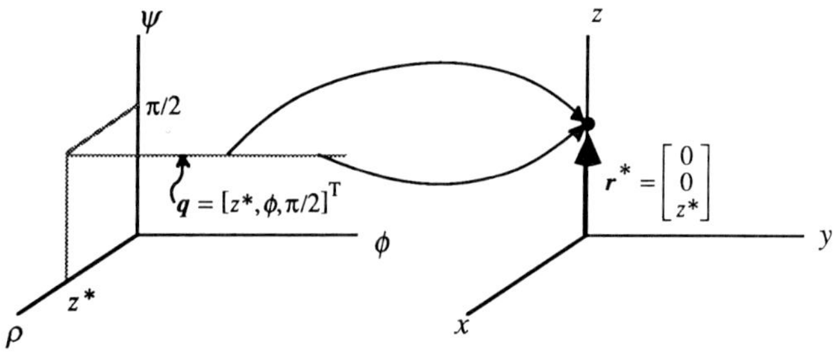

Figure 2.1.7 Singular points for spherical coordinates, $z^* > 0$.

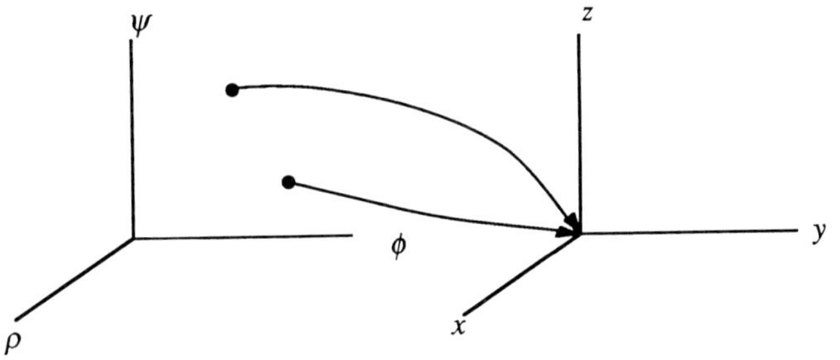

Figure 2.1.8 Singular points for spherical coordinates, $\mathbf{r}^\dagger = \mathbf{0}$.

2.2 VECTOR AND MATRIX DIFFERENTIATION

In kinematics and dynamics of mechanical systems, vectors that represent positions of particles, or points on rigid bodies, are functions of time or some other variables. In analyzing velocities and accelerations of particles, time derivatives and partial derivatives with respect to generalized coordinates are needed. In this section, these derivatives are defined and *matrix calculus notation* that is used throughout the book is introduced.

In analyzing velocities and accelerations, time derivatives of vectors that locate points must be calculated. Consider a vector $\vec{r}(t)$ with components $\mathbf{r} \equiv \mathbf{r}(t) = [x(t), y(t), z(t)]^T$ in a *x-y-z stationary Cartesian reference frame* (i.e., $\vec{i}$, $\vec{j}$, and $\vec{k}$ are of constant length and fixed orientation). The *time derivative of a vector* $\vec{r}$

is the *velocity* of the point that is located by the vector, denoted by $\dot{\vec{r}}$,

$$\dot{\vec{r}} \equiv \frac{d}{dt}\vec{r}(t) = \frac{d}{dt}[x(t)\vec{\imath} + y(t)\vec{\jmath} + z(t)\vec{k}]$$

$$= [\dot{x}(t)]\vec{\imath} + [\dot{y}(t)]\vec{\jmath} + [\dot{z}(t)]\vec{k}$$

where $\dot{} \equiv d/dt$. Note that this is valid only if the vectors $\vec{\imath}$, $\vec{\jmath}$, and $\vec{k}$ are constant in time. In algebraic vector notation, this is simply

$$\dot{\mathbf{r}} \equiv \frac{d}{dt}\mathbf{r}(t) = [\dot{x}(t), \dot{y}(t), \dot{z}(t)]^T \tag{2.2.1}$$

Thus, for vectors that are written in terms of their components in a stationary Cartesian reference frame, the derivative of a vector is obtained by differentiating its components.

Example 2.2.1: A balancing mass, modeled as a particle P, is attached to the periphery of a flywheel that rotates with constant angular speed ω about the z axis of a Cartesian frame, as shown in Fig. 2.2.1(a). Thus particle P moves with constant speed along a circle of radius R in the x-y plane of a stationary Cartesian reference frame, as shown in Fig. 2.2.1. Using cylindrical coordinates with $\phi = \omega t$ and $z = 0$, from Eq. 2.1.4, its position vector is

$$\mathbf{r} = \begin{bmatrix} R\cos\omega t \\ R\sin\omega t \\ 0 \end{bmatrix}$$

where R and ω are constant. Its velocity is thus

$$\dot{\mathbf{r}} = \begin{bmatrix} -R\omega\sin\omega t \\ R\omega\cos\omega t \\ 0 \end{bmatrix}^T$$

The time derivative of the sum of two vectors $\mathbf{a}(t)$ and $\mathbf{b}(t)$ (Prob. 2.2.1) is

$$\frac{d}{dt}(\mathbf{a}(t) + \mathbf{b}(t)) = \dot{\mathbf{a}} + \dot{\mathbf{b}} \tag{2.2.2}$$

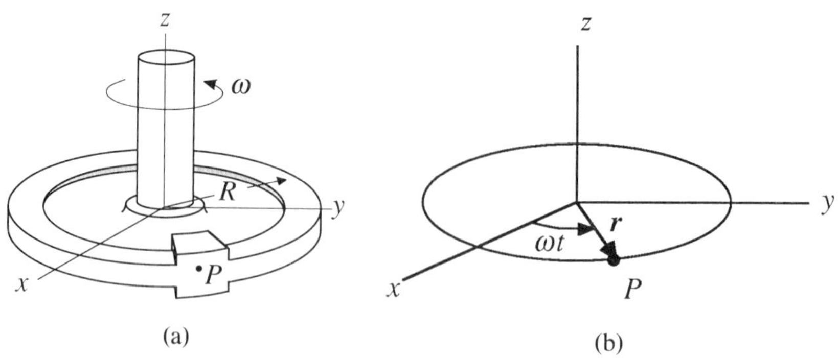

Figure 2.2.1 Balancing mass on flywheel.

which is analogous to the differentiation rule that the derivative of a sum is the sum of the derivatives. The following vector forms of the *product rule of differentiation* are also valid (Prob. 2.2.2):

$$\frac{d}{dt}(\alpha \mathbf{a}) = \dot{\alpha}\mathbf{a} + \alpha\dot{\mathbf{a}} \tag{2.2.3}$$

$$\frac{d}{dt}(\mathbf{a}^T\mathbf{b}) = \dot{\mathbf{a}}^T\mathbf{b} + \mathbf{a}^T\dot{\mathbf{b}} \tag{2.2.4}$$

where $\alpha(t)$ is a scalar function of time.

➧ **Example 2.2.2:** Let the length of a vector $\mathbf{r}(t)$ be fixed (i.e., $\mathbf{r}(t)^T\mathbf{r}(t) = c$). Differentiating both sides of this equation, using Eq. 2.2.4 and observing that $\mathbf{r}^T\dot{\mathbf{r}} = \dot{\mathbf{r}}^T\mathbf{r}$ yields

$$\dot{\mathbf{r}}^T\mathbf{r} = 0 \tag{2.2.5}$$

If $\mathbf{r}$ is a position vector that locates a point relative to the origin of a stationary reference frame, then $\dot{\mathbf{r}}$ is the velocity of that point. Hence Eq. 2.2.5 indicates that the velocity of a point whose distance from the origin is constant is orthogonal (see Appendix A) to the position vector of the point. Note that Eq. 2.2.5 is satisfied by the velocity vector of Example 2.2.1, which is to be expected since there, $\mathbf{r}^T\mathbf{r} = R^2$.

The second time derivative of $\mathbf{r}(t)$ is the *acceleration* of the point that is located by the vector in a stationary Cartesian reference frame, denoted as

$$\ddot{\mathbf{r}} \equiv \frac{d}{dt}(\dot{\mathbf{r}}(t)) = [\ddot{x}(t), \ddot{y}(t), \ddot{z}(t)]^T \tag{2.2.6}$$

Thus, for vectors that are written in terms of their components in a stationary Cartesian reference frame, acceleration may be calculated in terms of second time derivatives of components of the vector.

➧ **Example 2.2.3:** The acceleration of the particle that is located by the vector $\mathbf{r}(t)$ in Example 2.2.1 is

$$\ddot{\mathbf{r}} = \begin{bmatrix} -R\omega^2 \cos \omega t \\ -R\omega^2 \sin \omega t \\ 0 \end{bmatrix} = -\omega^2\mathbf{r}$$

This is the classical *centripetal acceleration* of a point that moves on a circular path, since the direction of $\ddot{\mathbf{r}}$ is opposite the direction of $\mathbf{r}$.

Just as in differentiation of a vector whose components are functions of time, the *derivative of a matrix* whose components depend on t may be defined. Consider a matrix $\mathbf{B}(t) = [b_{ij}(t)]$ (see Appendix A). The derivative of $\mathbf{B}(t)$ with respect to time is defined as

$$\dot{\mathbf{B}} \equiv \frac{d}{dt}\mathbf{B}(t) \equiv [\dot{b}_{ij}(t)] \tag{2.2.7}$$

With this definition and elementary rules of differentiation (Prob. 2.2.3),

$$\frac{d}{dt}(\mathbf{B}(t) + \mathbf{C}(t)) = \dot{\mathbf{B}} + \dot{\mathbf{C}} \tag{2.2.8}$$

$$\frac{d}{dt}(\mathbf{B}(t)\mathbf{C}(t)) = \dot{\mathbf{B}}\mathbf{C} + \mathbf{B}\dot{\mathbf{C}} \tag{2.2.9}$$

$$\frac{d}{dt}(\alpha(t)\mathbf{B}(t)) = \dot{\alpha}\mathbf{B} + \alpha\dot{\mathbf{B}} \tag{2.2.10}$$

where $\alpha(t)$ is a scalar function of time.

In dealing with systems of particles whose locations are described by several variables, a *matrix calculus* notation can be effectively employed. To introduce the notation used here, let $\mathbf{q} = [q_1, \ldots, q_k]^T$ be a k-vector of real variables, $\alpha(\mathbf{q})$ be a scalar-valued differentiable function of $\mathbf{q}$, and $\mathbf{a}(\mathbf{q}) = [a_1(\mathbf{q}), \ldots, a_n(\mathbf{q})]^T$ be an n-vector of differentiable functions of $\mathbf{q}$. Using i and j as row and column indices, the following notations are defined:

$$\alpha_{\mathbf{q}} \equiv \frac{\partial \alpha}{\partial \mathbf{q}} \equiv \left(\frac{\partial \alpha}{\partial q_j}\right)_{1\times k} \tag{2.2.11}$$

$$\mathbf{a}_{\mathbf{q}} \equiv \frac{\partial \mathbf{a}}{\partial \mathbf{q}} \equiv \left(\frac{\partial a_i}{\partial q_j}\right)_{n\times k} \tag{2.2.12}$$

The matrix $\mathbf{a}_{\mathbf{q}}$ is called the *Jacobian matrix* (or simply *Jacobian*) of the vector function $\mathbf{a}(\mathbf{q})$.

Example 2.2.4: A power tool shown in Fig. 2.2.2(a) is modeled as a particle P. The tool is mounted on a control arm that translates along and rotates about the z axis of a Cartesian reference frame. Thus P moves on a cylinder of radius R, as shown in Fig. 2.2.2(b). It may be located by cylindrical coordinates $s = R$, $\phi = q_1$, and $z = q_2$ in Eq. 2.1.6 as

$$\mathbf{r}(\mathbf{q}) = \begin{bmatrix} R\cos q_1 \\ R\sin q_1 \\ q_2 \end{bmatrix}$$

The length of the vector is a scalar-valued function,

$$\begin{aligned} l(\mathbf{q}) &= [\mathbf{r}(\mathbf{q})^T\mathbf{r}(\mathbf{q})]^{1/2} = [R^2\cos^2 q_1 + R^2\sin^2 q_1 + q_2^2]^{1/2} \\ &= [R^2 + q_2^2]^{1/2} \end{aligned}$$

Applying Eq. 2.2.11 to $l(\mathbf{q})$ yields

$$l_{\mathbf{q}} = \left[0, \frac{q_2}{[R^2 + q_2^2]^{1/2}}\right]$$

Applying Eq. 2.2.12 to $\mathbf{r}(\mathbf{q})$ gives

$$\mathbf{r}_{\mathbf{q}} = \begin{bmatrix} -R\sin q_1 & 0 \\ R\cos q_1 & 0 \\ 0 & 1 \end{bmatrix}$$

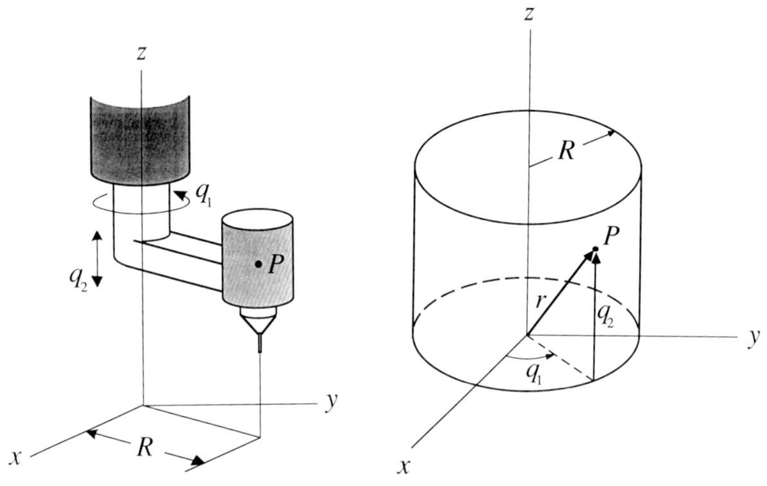

Figure 2.2.2 Power tool modeled as particle on cylinder.

Note that the derivative of a scalar function with respect to a vector variable in Eq. 2.2.11 gives a row matrix. This is one of the few vector symbols in the book that is a row matrix rather than the more common column matrix (see Appendix A). Note also that as defined by Eq. 2.2.12, the derivative of a vector $\mathbf{a}$ whose elements are functions of the vector variable $\mathbf{q}$ is a matrix. The *partial derivative subscript notation* used here is helpful in deriving needed relations without becoming entangled in cumbersome index and partial differentiation notation. To take advantage of this notation, however, it is critically important that the correct vector or matrix definition of derivatives be used.

The partial derivative of the scalar product of two n-vector functions $\mathbf{a}(\mathbf{q}) = [a_1(\mathbf{q}), \ldots, a_n(\mathbf{q})]^T$ and $\mathbf{b}(\mathbf{q}) = [b_1(\mathbf{q}), \ldots, b_n(\mathbf{q})]^T$, by careful manipulation, is

$$\begin{aligned}
\frac{\partial}{\partial \mathbf{q}}(\mathbf{a}^T\mathbf{b}) &= \frac{\partial}{\partial \mathbf{q}}\left(\sum_{l=1}^{n} a_l(\mathbf{q})b_l(\mathbf{q})\right) \\
&= \left[\frac{\partial}{\partial q_j}\left(\sum_{l=1}^{n} a_l(\mathbf{q})b_l(\mathbf{q})\right)\right] \\
&= \left[\sum_{l=1}^{n}\left(\frac{\partial a_l}{\partial q_j} b_l + a_l \frac{\partial b_l}{\partial q_j}\right)\right] \\
&= \left[\sum_{l=1}^{n}\left(b_l \frac{\partial a_l}{\partial q_j} + \sum_{l=1}^{n} a_l \frac{\partial b_l}{\partial q_j}\right)\right] \\
&= \mathbf{b}^T \frac{\partial \mathbf{a}}{\partial \mathbf{q}} + \mathbf{a}^T \frac{\partial \mathbf{b}}{\partial \mathbf{q}} \\
&= \mathbf{b}^T\mathbf{a}_{\mathbf{q}} + \mathbf{a}^T\mathbf{b}_{\mathbf{q}}
\end{aligned} \tag{2.2.13}$$

Note that what might have intuitively appeared to be the appropriate product rule of differentiation is not even defined, much less valid:

$$\frac{\partial}{\partial \mathbf{q}}(\mathbf{a}^T\mathbf{b}) \neq \mathbf{a}_\mathbf{q}^T\mathbf{b} + \mathbf{a}^T\mathbf{b}_\mathbf{q}$$

➧ **Example 2.2.5:** An alternative method of calculating $l_\mathbf{q}$ in Example 2.2.4 is to differentiate

$$l^2(\mathbf{q}) = \mathbf{r}^T(\mathbf{q})\mathbf{r}(\mathbf{q})$$

using Eq. 2.2.13, to obtain

$$2ll_\mathbf{q} = 2\mathbf{r}^T\mathbf{r}_\mathbf{q}$$

Thus

$$l_\mathbf{q} = \frac{1}{l}\mathbf{r}^T\mathbf{r}_\mathbf{q}$$

From Example 2.2.4, this gives

$$l_\mathbf{q} = \frac{1}{[R^2 + q_2^2]^{1/2}}[0, q_2]$$

which agrees with the direct calculation presented there.

If $\mathbf{a}(\mathbf{q}) = [a_1(\mathbf{q}), \ldots, a_m(\mathbf{q})]^T$ and $\mathbf{q}(\mathbf{b}) = [q_1(\mathbf{b}), \ldots, q_n(\mathbf{b})]^T$ are vector functions of vector variables $\mathbf{q} = [q_1, \ldots, q_n]^T$ and $\mathbf{b} = [b_1, \ldots, b_k]^T$, respectively, the *chain rule of differentiation*, in matrix calculus form, is obtained as

$$\begin{aligned}
\mathbf{a}_\mathbf{b} &\equiv \left[\frac{\partial a_i(\mathbf{q}(\mathbf{b}))}{\partial b_j}\right]_{m\times k} \\
&= \left[\sum_{l=1}^{n} \frac{\partial a_i}{\partial q_l}\frac{\partial q_l}{\partial b_j}\right]_{m\times k} \\
&= \left(\frac{\partial \mathbf{a}}{\partial \mathbf{q}}\right)_{m\times n}\left(\frac{\partial \mathbf{q}}{\partial \mathbf{b}}\right)_{n\times k} \\
&= \mathbf{a}_\mathbf{q}\mathbf{q}_\mathbf{b}
\end{aligned} \tag{2.2.14}$$

A major application of the chain rule of differentiation that is central to kinematics and dynamics is relating Cartesian velocities and accelerations to time derivatives of generalized coordinates. If $\mathbf{q} = \mathbf{q}(t)$ is a k-vector, then the mapping

$$\mathbf{r} = \mathbf{r}(\mathbf{q}(t)) \tag{2.2.15}$$

yields $\mathbf{r}$ as a function of time. Using the chain rule of differentiation of Eq. 2.2.14, with $\mathbf{a}(\mathbf{q}) = \mathbf{r}(\mathbf{q})$ and $\mathbf{q} = \mathbf{q}(t)$,

$$\begin{aligned}\dot{\mathbf{r}} &= \mathbf{r_q}\dot{\mathbf{q}} \\ &= \sum_{l=1}^{k} \frac{\partial r_i}{\partial q_l}\dot{q}_l\end{aligned} \tag{2.2.16}$$

This is called the *velocity equation*, which relates the *generalized velocity* $\dot{\mathbf{q}}$ and $\dot{\mathbf{r}}$. It appears to be a linear relation between $\dot{\mathbf{q}}$ and $\dot{\mathbf{r}}$, but the coefficient matrix $\mathbf{r_q}$ depends on $\mathbf{q}$, so it changes as the system moves. This is one of the main sources of nonlinearities that arise in virtually all applications of kinematics and dynamics.

Recalling that $\mathbf{r_q}$ depends on $\mathbf{q}$ and that $\dot{\mathbf{r}}$ and $\dot{\mathbf{q}}$ depend on time, the chain rule of differentiation can be applied to Eq. 2.2.16, to obtain the *acceleration equation*

$$\ddot{\mathbf{r}} = \mathbf{r_q}\ddot{\mathbf{q}} + (\mathbf{r_q}\dot{\mathbf{q}})_\mathbf{q}\dot{\mathbf{q}} \tag{2.2.17}$$

which relates the *generalized acceleration* $\ddot{\mathbf{q}}$ and $\ddot{\mathbf{r}}$. The second term on the right may be written in any of the following alternative forms:

$$\begin{aligned}(\mathbf{r_q}\dot{\mathbf{q}})_\mathbf{q}\dot{\mathbf{q}} &= \frac{\partial}{\partial \mathbf{q}}[(\mathbf{r_q}\dot{\mathbf{q}})]\dot{\mathbf{q}} \\ &= \sum_{j=1}^{k} \mathbf{r}_{\mathbf{q}q_j}\dot{\mathbf{q}}\dot{q}_j \\ &= \frac{\partial}{\partial \mathbf{q}}\left[\sum_{l=1}^{k}\frac{\partial r_i}{\partial q_l}\dot{q}_l\right]\dot{\mathbf{q}} \\ &= \left[\sum_{l=1}^{k}\frac{\partial^2 r_i}{\partial q_l\,\partial q_j}\dot{q}_l\right]\dot{\mathbf{q}} \\ &= \left[\sum_{j=1}^{k}\sum_{l=1}^{k}\frac{\partial^2 r_i}{\partial q_l\,\partial q_j}\dot{q}_l\dot{q}_j\right]_{3\times 1}\end{aligned} \tag{2.2.18}$$

Note that while the first term on the right of Eq. 2.2.17 appears to be linear in $\ddot{\mathbf{q}}$, just as in the velocity equation, $\mathbf{r_q}$ depends on $\mathbf{q}$, so the same kind of nonlinearity is present. The more severe form of nonlinearity that appears on the right of Eq. 2.2.17 is in the second term. As seen by the expansion of this term in Eq. 2.2.18, it is quadratic in the generalized velocities and the terms $\partial^2 r_i/\partial q_l\,\partial q_j$ depend on $\mathbf{q}$. It is the second term on the right of Eq. 2.2.17 that sets the study of dynamics apart from other branches of mechanics.

An observation on notation is in order here. The second term on the right of Eq. 2.2.17 is left with an operation of partial derivative of the vector $\mathbf{r_q}\dot{\mathbf{q}}$ with respect to $\mathbf{q}$ to be carried out. The matrix calculus notation system used here does not permit writing the intuitive expansion

$$(\mathbf{r_q}\dot{\mathbf{q}})\,\mathbf{q} \neq \mathbf{r_{qq}}\dot{\mathbf{q}}$$

because $\mathbf{r_{qq}}$ is not defined. To define it would require introduction of concepts of tensor calculus. To avoid this extension of notation and arithmetic concepts, the operator notation on the right of Eq. 2.2.17 or one of the explicit expansion forms of Eq. 2.2.18 will be used.

➧ **Example 2.2.6:** If $\mathbf{q} = \mathbf{q}(t)$ in Example 2.2.4, then using $\mathbf{r_q}$ derived there and Eq. 2.2.14,

$$\begin{aligned} \dot{\mathbf{r}} &= \mathbf{r_q}\dot{\mathbf{q}} \\ &= \begin{bmatrix} -(R \sin q_1)\dot{q}_1 \\ (R \cos q_1)\dot{q}_1 \\ \dot{q}_2 \end{bmatrix} \end{aligned}$$

Carrying out the expansion on the right of Eq. 2.2.17 yields

$$\ddot{\mathbf{r}} = \begin{bmatrix} -(R \sin q_1)\ddot{q}_1 \\ (R \cos q_1)\ddot{q}_1 \\ \ddot{q}_2 \end{bmatrix} - R\begin{bmatrix} \cos q_1 \\ \sin q_1 \\ 0 \end{bmatrix}(\dot{q}_1)^2 \tag{2.2.19}$$

If m is the mass of the power tool and if the machine is preprogrammed to rotate with a given history $q_1(t)$ and translate with a history $q_2(t)$, the resultant force that must act on the tool is

$$\mathbf{F} = m\ddot{\mathbf{r}}$$

where $\ddot{\mathbf{r}}$ is given by Eq. 2.2.19. The controls engineer can now determine what capacity of actuator is required to create the desired motion. This is an example of *inverse dynamic analysis*, where desired time histories are specified, Cartesian accelerations are calculated, and required forces are determined.

If $\mathbf{B}$ is a constant $m \times n$ matrix and $\mathbf{p}$ and $\mathbf{q}$ are m- and n-vectors of variables, respectively, the following useful relations may be verified (Prob. 2.2.4):

$$\frac{\partial}{\partial \mathbf{q}}(\mathbf{Bq}) = \mathbf{B} \tag{2.2.20}$$

$$\frac{\partial}{\partial \mathbf{p}}(\mathbf{p}^T\mathbf{Bq}) = \mathbf{q}^T\mathbf{B}^T \tag{2.2.21}$$

$$\frac{d}{dt}(\mathbf{p}^T\mathbf{Bq}) = \dot{\mathbf{p}}^T\mathbf{Bq} + \mathbf{p}^T\mathbf{B}\dot{\mathbf{q}} \tag{2.2.22}$$

In particular, if $\mathbf{B}$ is a symmetric $n \times n$ constant matrix,

$$\frac{\partial}{\partial \mathbf{q}}(\mathbf{q}^T\mathbf{Bq}) = 2\mathbf{q}^T\mathbf{B} \tag{2.2.23}$$

$$\frac{d}{dt}(\mathbf{q}^T\mathbf{Bq}) = 2\mathbf{q}^T\mathbf{B}\dot{\mathbf{q}} \tag{2.2.24}$$

2.3 THE IMPLICIT FUNCTION THEOREM

A variety of different variables are used to locate points in space, for purposes of kinematic and dynamic analysis. Constraint equations that involve these variables may restrict relative motion of points and bodies and establish relations among the variables. One of the most important results of calculus is the implicit function theorem, which provides information regarding properties of solutions of equations. This theorem is introduced in this section, using the language of multivariable calculus introduced in Section 2.2. Its use is illustrated in determining independent variables that satisfy constraint equations and in testing for acceptability of coordinate transformations.

Let $\mathbf{u} = [u_1, \ldots, u_n]^T$ and $\mathbf{v} = [v_1, \ldots, v_m]^T$ be vectors of real variables that are to satisfy the system of n equations

$$\mathbf{\Phi}(\mathbf{u}, \mathbf{v}) = \begin{bmatrix} \Phi_1(\mathbf{u}, \mathbf{v}) \\ \vdots \\ \Phi_n(\mathbf{u}, \mathbf{v}) \end{bmatrix} = \mathbf{0} \tag{2.3.1}$$

Equation 2.3.1 may be viewed as n equations in the n variables $\mathbf{u}$, with $\mathbf{v}$ as a set of m parameters. It is important to know when this equation can be solved for $\mathbf{u}$ as a function of $\mathbf{v}$ and, if such a solution exists, when derivatives of $\mathbf{u}$ with respect to $\mathbf{v}$ exist. An important result from multivariable calculus [3] is the following.

Implicit Function Theorem: Let the functions $\Phi_i(\mathbf{u}, \mathbf{v})$ have k continuous derivatives with respect to all components of $\mathbf{u}$ and $\mathbf{v}$. If $\mathbf{u}^0$ and $\mathbf{v}^0$ satisfy Eq. 2.3.1; that is, if

$$\mathbf{\Phi}(\mathbf{u}^0, \mathbf{v}^0) = \mathbf{0} \tag{2.3.2}$$

and if the *Jacobian* $\mathbf{\Phi_u}$ of $\mathbf{\Phi}$ with respect to $\mathbf{u}$ is nonsingular at this point, that is, if

$$|\mathbf{\Phi_u}(\mathbf{u}^0, \mathbf{v}^0)| \neq 0 \tag{2.3.3}$$

then there exists a unique, k times continuously differentiable solution of Eq. 2.3.1 for $\mathbf{u}$ as a function of $\mathbf{v}$,

$$\mathbf{u} = \mathbf{f}(\mathbf{v}) \tag{2.3.4}$$

in a neighborhood of $\mathbf{v}^0$, that is, for all $\mathbf{v}$ satisfying

$$(\mathbf{v} - \mathbf{v}^0)^T(\mathbf{v} - \mathbf{v}^0) < \delta^2 \tag{2.3.5}$$

for some $\delta > 0$.

Example 2.3.1: To see geometrically what the implicit function theorem says for two variables, consider the single equation

$$\Phi(u, v) = \sin u - v = 0$$

The graph of points that satisfy the equation is shown in Fig. 2.3.1. At the point $(u^0, v^0) = (0, 0)$, the equation is satisfied and

$$|\,\Phi_u(u^0, v^0)| = |\cos 0\,| = 1 \neq 0$$

Thus there is the unique solution

$$u = \arcsin v$$

in a neighborhood of $v^0 = 0$. In this case, the length of the interval defined by Eq. 2.3.5 is $2\delta_0 < \pi$.

At $v^1 = 0.8$, $u^1 = 0.927$ rad satisfies the equation and $|\,\Phi_u(u^1, v^1)| = 0.6$, so the same form of unique solution exists, but δ in Eq. 2.3.5 is $\delta_1 < 0.073$.

Finally, at $v^2 = 1$, $u^2 = \pi/2$ satisfies the equation, but $|\,\Phi_u(u^2, v^2)| = 0$. Thus there is no unique solution in any interval about $v^2 = 1$. For any $\epsilon > 0$, for $v = v^2 + \epsilon$, there is no solution and for $v = v^2 - \epsilon$, there are two distinct solutions.

Provided that the hypotheses of the theorem are met, the implicit function theorem provides the following fundamental information:

1. A solution of the form of Eq. 2.3.4 exists.
2. The solution is unique for all $\mathbf{v}$ near $\mathbf{v}^0$.
3. The solution is k times continuously differentiable.

This theorem is local in nature, since the unique solution is guaranteed to exist only in a possibly small neighborhood of the point $\mathbf{v}^0$, given by Eq. 2.3.5. If, however, there exist $\mathbf{u}^0$ and $\mathbf{v}^0$ that satisfy Eq. 2.3.2 and if Eq. 2.3.3 holds everywhere in a simply connected set [3], the conclusion of the theorem can be extended to the entire set.

As will become clear in the study of kinematics and dynamics, the implicit function theorem plays a crucial role in formulation of problems and in determining existence and regularity of solutions. Considering the importance of the theorem, several examples and applications are studied in the remainder of this section.

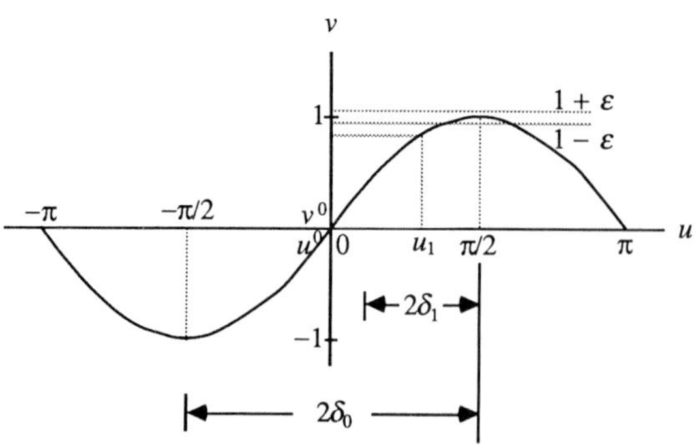

Figure 2.3.1 Solutions of $\sin u - v = 0$.

Example 2.3.2: Let **u** and **v** be 2-vectors and

$$\boldsymbol{\Phi}(\mathbf{u}, \mathbf{v}) = \begin{bmatrix} u_1 - v_2 \\ u_1^2 + u_2 - v_1 \end{bmatrix} = \mathbf{0}$$

At $\mathbf{v}^0 = [1, 1]^T$ and $\mathbf{u}^0 = [1, 0]^T$, $\boldsymbol{\Phi}(\mathbf{u}^0, \mathbf{v}^0) = \mathbf{0}$ and

$$|\boldsymbol{\Phi}_{\mathbf{u}}(\mathbf{u}^0, \mathbf{v}^0)| = \begin{vmatrix} 1 & 0 \\ 2 & 1 \end{vmatrix} = 1 \neq 0$$

The hypotheses of the implicit function theorem are thus satisfied and there is a unique solution for **u** as a function of **v** in some neighborhood of $\mathbf{v}^0$. The unique solution, in this case, is simply

$$\mathbf{u} = \begin{bmatrix} v_2 \\ v_1 - v_2^2 \end{bmatrix} \equiv \mathbf{f}(\mathbf{v})$$

This solution happens to be valid for all **v**.

Example 2.3.3: Let **u** and **v** be 2-vectors and

$$\boldsymbol{\Phi}(\mathbf{u}, \mathbf{v}) = \begin{bmatrix} u_1 - v_2 \\ u_1 + u_2^2 - v_1 \end{bmatrix} = \mathbf{0}$$

At $\mathbf{v}^0 = [1, 1]^T$ and $\mathbf{u}^0 = [1, 0]^T$, $\boldsymbol{\Phi}(\mathbf{u}^0, \mathbf{v}^0) = \mathbf{0}$, but

$$|\boldsymbol{\Phi}_{\mathbf{u}}(\mathbf{u}^0, \mathbf{v}^0)| = \begin{vmatrix} 1 & 0 \\ 1 & 0 \end{vmatrix} = 0$$

The hypotheses of the implicit function theorem are not satisfied. Thus in any neighborhood of $\mathbf{v}^0$ there may be points **v** where either existence or uniqueness of solution for **u** is lost. Consider the point $\mathbf{v} = [1 + \epsilon, 1]^T$, where ϵ is as small as desired, so **v** is as close to $\mathbf{v}^0$ as desired. Then

$$\mathbf{u} = \begin{bmatrix} 1 \\ \pm\sqrt{\epsilon} \end{bmatrix}$$

It is fascinating that if ϵ is chosen positive, two solutions exist. Conversely, if ϵ is chosen as negative, no real solution exists.

Example 2.3.4: Consider applying the implicit function theorem to the equation of Example 2.3.3 at arbitrary points **u** and **v**. The Jacobian test gives

$$|\boldsymbol{\Phi}_{\mathbf{u}}| = \begin{vmatrix} 1 & 0 \\ 1 & 2u_2 \end{vmatrix} = 2u_2$$

Thus, as long as $u_2 \neq 0$, the Jacobian is nonsingular. It might appear that as long as $u_2 \neq 0$, then for any **v** there is a unique solution of $\boldsymbol{\Phi}(\mathbf{u}, \mathbf{v}) = \mathbf{0}$ for **u**. This does not follow from the implicit function theorem, however, because the theorem can only be applied in neighborhoods of points $(\mathbf{u}^0, \mathbf{v}^0)$ for which $\mathbf{\Phi}(\mathbf{u}^0, \mathbf{v}^0) = \mathbf{0}$. For example, if $\mathbf{v}^0 = [0, 1]^T$, then there is no $\mathbf{u}^0$ that will satisfy the equation, and no conclusion can be drawn.

An important application of the implicit function theorem is defining a set of variables that can be determined from constraint equations involving a set of $n + m$ variables $\mathbf{q} = [q_1, \ldots, q_{n+m}]^T$ that satisfy the m equations

$$\mathbf{\Phi}(\mathbf{q}) \equiv \begin{bmatrix} \Phi_1(\mathbf{q}) \\ \vdots \\ \Phi_m(\mathbf{q}) \end{bmatrix} = \mathbf{0} \tag{2.3.6}$$

It is helpful to determine a subset of the variables in $\mathbf{q}$ that can be determined as solutions of Eq. 2.3.6, depending on the remaining components of $\mathbf{q}$. Existence of such a *partitioning* of $\mathbf{q}$ is guaranteed by the following.

Coordinate Partitioning Theorem: Let the function $\mathbf{\Phi}(\mathbf{q})$ be c times continuously differentiable, and let $\mathbf{q}^0$ satisfy Eq. 2.3.6. If the Jacobian $\mathbf{\Phi}_{\mathbf{q}}(\mathbf{q}^0)$ has full row rank (see Appendix A), there exists a subset $\mathbf{u} = [q_{k_1}, \ldots, q_{k_m}]^T$ of $\mathbf{q}$ for which

$$\left| \left[\frac{\partial \Phi_i(\mathbf{q}^0)}{\partial q_{k_j}} \right]_{m \times m} \right| = | \mathbf{\Phi}_{\mathbf{u}}(\mathbf{q}^0)| \neq 0 \tag{2.3.7}$$

Defining $\mathbf{v} = [q_{k_{m+1}}, \ldots, q_{k_{m+n}}]^T$ as the vector that contains the components of $\mathbf{q}$ that are not in $\mathbf{u}$, there exists a unique c times continuously differentiable solution of Eq. 2.3.6 of the form

$$\mathbf{u} = \mathbf{f}(\mathbf{v}) \tag{2.3.8}$$

which is valid in a neighborhood of $\mathbf{v}^0$.

The existence of a set of columns of the Jacobian matrix $\mathbf{\Phi}_{\mathbf{q}}$ for which Eq. 2.3.7 holds follows from the condition that in order for the rank (see Appendix A) of the Jacobian to be m, there must be an $m \times m$ nonsingular submatrix [2]. The remaining conclusions of the coordinate partitioning theorem follow directly from the implicit function theorem. Numerical methods for determining $\mathbf{u}$ are presented in Appendix B.

Example 2.3.5: Let $\mathbf{q}$ be a 4-vector and

$$\mathbf{\Phi}(\mathbf{q}) = \begin{bmatrix} q_1 - q_2 \\ q_1 + q_3^2 - q_4 \end{bmatrix} = \mathbf{0}$$

Note that this is the equation of Example 2.3.3, with $q_1 = u_1$, $q_2 = v_2$, $q_3 = u_2$, and $q_4 = v_1$. At $\mathbf{q}^0 = [1, 1, 0, 1]^T$, $\mathbf{\Phi}(\mathbf{q}^0) = \mathbf{0}$ and

$$\mathbf{\Phi}_q(\mathbf{q}^0) = \begin{bmatrix} 1 & -1 & 0 & 0 \\ 1 & 0 & 0 & -1 \end{bmatrix}$$

Clearly, any 2 × 2 submatrix that contains the third column will be singular, so $q_3 = u_2$ cannot be uniquely determined, as shown in Example 2.3.3. Since the corresponding sub-Jacobian of each pair of the variables q_1, q_2, and q_4 is nonsingular, however, each pair can be uniquely determined as differentiable functions of the remaining variables; for example,

$$\begin{bmatrix} q_2 \\ q_4 \end{bmatrix} = \begin{bmatrix} q_1 \\ q_1 + q_3^2 \end{bmatrix}$$

As a second application of the implicit function theorem, consider a *coordinate transformation*, or *mapping* from $\mathbf{v} = [v_1, \ldots, v_m]^T$ to $\mathbf{u} = [u_1, \ldots, u_n]^T$, where $n \geq m$,

$$\mathbf{u} = \mathbf{g}(\mathbf{v}) \tag{2.3.9}$$

This functional relationship indicates that for each **v**, a value of **u** is determined. It is important to know, however, whether there is just one **v** that determines a given **u**. If different values of **v** lead to the same **u** then trouble will arise if **v** is to be calculated from a knowledge of **u**. More precisely, a well-behaved mapping of the form of Eq. 2.3.9 should be *one-to-one*; that is, if

$$\mathbf{u}^i = \mathbf{g}(\mathbf{v}^i), \qquad i = 1, 2 \tag{2.3.10}$$

then **g** is said to be one-to-one if $\mathbf{u}^1 = \mathbf{u}^2$ implies that $\mathbf{v}^1 = \mathbf{v}^2$ in some neighborhood of $\mathbf{v}^2$.

One-to-One Mapping Theorem: An analytical condition that guarantees the mapping of Eq. 2.3.9 to be one-to-one is that the Jacobian $\mathbf{g_v}$ have *full column rank* (see Appendix A). This is equivalent to existence of a subset of **g** that consists of $g_{k_1}, \ldots, g_{k_m}$ such that

$$\left| \left[\frac{\partial g_{k_i}(\mathbf{v}^0)}{\partial v_j} \right]_{m \times m} \right| \neq 0 \tag{2.3.11}$$

To see that this is true, the implicit function theorem may be applied to the subset of equations

$$\left[g_{k_i}(\mathbf{v}) \right]_{m \times 1} = \left[u_{k_i} \right]_{m \times 1} \tag{2.3.12}$$

Since the Jacobian of these equations is nonsingular, due to Eq. 2.3.11, Eq. 2.3.12 has a unique solution for **v** and the mapping is indeed one-to-one.

As in the special case of Section 2.1, a point $\mathbf{v}^0$ at which the mapping of Eq. 1.6.9 fails to be one-to-one is called a *singular point* of the mapping. The importance of singular points will become more apparent as generalized coordinates are introduced to locate points in space. An elementary example that illustrates difficulties that can be encountered in using an inappropriate set of coordinates involves *spherical coordinates*.

Example 2.3.6: Let point P on the surface of the unit sphere (sphere of radius 1) be located by spherical coordinates ϕ and ψ of Eq. 2.1.6, with $\rho = 1$. As shown in Fig. 2.3.2,

$$\mathbf{r} \equiv \begin{bmatrix} x \\ y \\ z \end{bmatrix} = \begin{bmatrix} \cos\phi\cos\psi \\ \sin\phi\cos\psi \\ \sin\psi \end{bmatrix} \equiv \mathbf{r}(\phi, \psi) \tag{2.3.13}$$

The Jacobian of $\mathbf{r}$ with respect to $\mathbf{q} = [\phi, \psi]^T$ is

$$\mathbf{r_q} = \begin{bmatrix} -\sin\phi\cos\psi & -\cos\phi\sin\psi \\ \cos\phi\cos\psi & -\sin\phi\sin\psi \\ 0 & \cos\psi \end{bmatrix}$$

To determine the column rank of $\mathbf{r_q}$, set a linear combination of its columns equal to zero,

$$\mathbf{r_q}\boldsymbol{\alpha} = \mathbf{0}$$

where $\boldsymbol{\alpha} = [\alpha_1, \alpha_2]^T$. If $\cos\psi \neq 0$, the third of these equations implies that $\alpha_2 = 0$. The first two equations thus reduce to

$$-\sin\phi\cos\psi\,\alpha_1 = 0$$

$$\cos\phi\cos\psi\,\alpha_1 = 0$$

Since $\sin\phi$ and $\cos\phi$ are not both zero for any ϕ, as long as $\cos\psi \neq 0$, $\alpha_1 = 0$. This shows that as long as $\cos\psi \neq 0$, the mapping of Eq. 1.6.13 is one-to-one. However, with $\cos\psi = 0$, the first column of $\mathbf{r_q}$ is zero and $\mathbf{r_q}$ has column rank 1. The mapping of Eq. 2.3.13 may not be one-to-one when $\cos\psi = 0$ (i.e., at the poles of the sphere on the z axis). To see that the mapping is in fact not one-to-one there, note that with $\psi = \pi/2$, $\mathbf{r} = [0\ 0\ 1]^T$, for any value of ϕ. Thus the spherical coordinates of Eq. 2.3.13 fail to uniquely specify points on the surface of the unit sphere in neighborhoods of the poles on the z axis.

Example 2.3.7: For the full set of spherical coordinates in space, the Jacobian of the mapping of Eq. 2.1.6 from spherical to Cartesian coordinates is

$$\mathbf{r_q} = \begin{bmatrix} \cos\psi\cos\phi & -\rho\cos\psi\sin\phi & -\rho\sin\psi\cos\phi \\ \cos\psi\sin\phi & \rho\cos\psi\cos\phi & -\rho\sin\psi\sin\phi \\ \sin\psi & 0 & \rho\cos\psi \end{bmatrix} \tag{2.3.14}$$

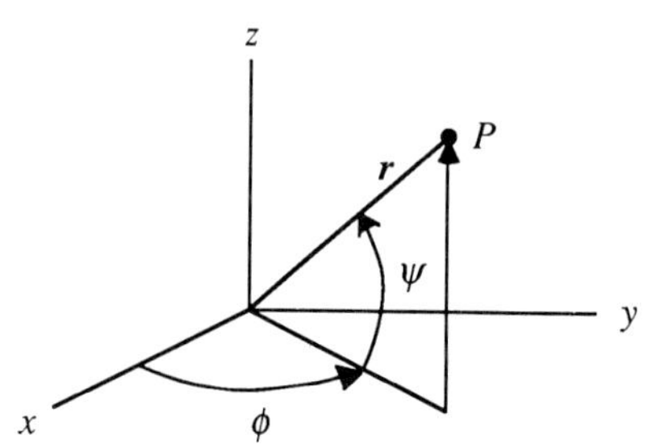

Figure 2.3.2 Spherical coordinates of point on unit sphere.

The determinant of this Jacobian is

$$|\mathbf{r_q}| = \rho^2 \cos^3\psi + \rho^2 \sin^2\psi \cos\psi = \rho^2 \cos\psi$$

which is singular at all points on the z axis (i.e., when $\psi = \pm \pi/2$). The singularity is most pronounced at the origin of the Cartesian reference frame where $\rho = 0$. Here, the second and third columns of the Jacobian of Eq. 2.3.14 are zero and the rank of $\mathbf{r_q}$ is only 1. Thus kinematic equations that are written in terms of spherical coordinates will encounter difficulties if the trajectory of a particle being analyzed intersects the z axis. This same difficulty was noted in Section 2.1.

2.4 PARTICLES CONSTRAINED TO MOVE ON CURVES AND SURFACES

In special cases, particles may represent machine components that are constrained to move along curves or surfaces that are fixed in space. Thus one- or two-dimensional motion of individual particles is permitted. This is the most elementary form of constrained motion. In contrast to the approach used in Section 1.2, the goal in kinematic analysis here is to describe the motion of particles in terms of generalized coordinates that define the curve or surface.

Generalized Coordinates for Particles on Curves

The position of a point P on a curve in space can be represented by the vector from the origin of a Cartesian reference frame to the point, expressed in terms of a scalar parameter. The general *parametric representation of a curve* is in terms of a scalar *generalized coordinate* q that determines Cartesian coordinates of a point in space, as shown in Fig. 2.4.1, in the form

$$\mathbf{r} = \mathbf{r}(q) = \begin{bmatrix} x(q) \\ y(q) \\ z(q) \end{bmatrix} \tag{2.4.1}$$

Figure 2.4.1 Parameterization of curve.

The velocity of point P is obtained by differentiating Eq. 2.4.1 to obtain the *velocity equation*

$$\dot{\mathbf{r}} = \mathbf{r}_q\dot{q} \equiv \boldsymbol{\tau}\dot{q} \tag{2.4.2}$$

where $\boldsymbol{\tau} = \mathbf{r}_q$ is *tangent* to the curve (see Fig. 2.4.1) and $\dot{q}$ is the *generalized velocity*. Note that this relation is linear in the generalized velocity, but the coefficient matrix $\mathbf{r}_q$ on the right depends on the generalized coordinate.

To see that $\boldsymbol{\tau} = \mathbf{r}_q$ is tangent to the curve defined by Eq. 2.4.1, note first that if ϵ is a small real constant,

$$\begin{aligned}\delta\mathbf{r} &= \mathbf{r}(q + \epsilon) - \mathbf{r}(q)\\ &\approx \mathbf{r}(q) + \mathbf{r}_q\epsilon - \mathbf{r}(q)\\ &= \mathbf{r}_q\epsilon\end{aligned}$$

Thus, as ϵ approaches zero, $\delta\mathbf{r}$ approaches the tangent to the curve and $\mathbf{r}_q$ is indeed tangent to the curve.

Differentiating Eq. 2.4.2 yields the *acceleration equation*

$$\begin{aligned}\ddot{\mathbf{r}} &= \mathbf{r}_q\ddot{q} + \mathbf{r}_{qq}\dot{q}^2\\ &= \boldsymbol{\tau}\ddot{q} + \mathbf{r}_{qq}\dot{q}^2\end{aligned} \tag{2.4.3}$$

where $\ddot{q}$ is the *generalized coordinate acceleration* and $\mathbf{r}_{qq}$ is the second derivative of $\mathbf{r}$ with respect to the scalar variable q. Note that the first term on the right of this equation is linear in the generalized acceleration, with the coefficient depending on q. The second term is quadratic in the generalized velocity, with a coefficient $\mathbf{r}_{qq}$ that depends on position. This is typical of the quadratic nature of acceleration, as a function of generalized velocity.

If $\mathbf{r}_q \neq \mathbf{0}$ for all q, then $x_q(q^0) \neq 0$, $y_q(q^0) \neq 0$, or $z_q(q^0) \neq 0$. By the implicit function theorem, q can be uniquely determined as a function of x, y, or z, at least in a neighborhood of q^0. Thus points on the curve defined by the mapping of Eq. 2.4.1 are uniquely defined by the generalized coordinate q. This uniqueness is lost only at points for which $\mathbf{r}_q = \mathbf{0}$ (i.e., points for which the Jacobian is column rank deficient; see Appendix A). Such points are singular points of the mapping of Eq. 2.4.1, or singular values of the generalized coordinate q. Note that in the singular case, a knowledge of $\dot{\mathbf{r}}$ in Eq. 2.4.2 fails to determine $\dot{q}$.

Apart from singular points, one generalized coordinate q uniquely determines the position vector $\mathbf{r}$ along the curve, hence the particle is said to have *one degree of freedom*.

Example 2.4.1: A spacecraft, modeled as a particle P, moves along a line in space that is defined by a unit vector $\mathbf{u}$ that passes through a point that is located by the Cartesian vector $\mathbf{r}_0$, as shown in Fig. 2.4.2. As in Example 1.2.1, q is the distance from the point defined by $\mathbf{r}_0$ to P, positive in the direction $\mathbf{u}$.

The position of the particle P in space is thus defined by the vector relation

$$\mathbf{r} = \mathbf{r}_0 + q\mathbf{u} \tag{2.4.4}$$

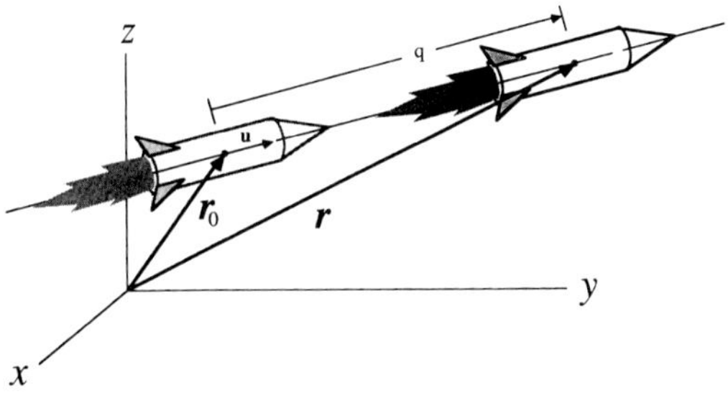

Figure 2.4.2 Particle on straight line.

Differentiating this equation twice with respect to time, with $\mathbf{r}_0$ and $\mathbf{u}$ constant, the velocity and acceleration equations are

$$\dot{\mathbf{r}} = \dot{q}\mathbf{u} \tag{2.4.5}$$

$$\ddot{\mathbf{r}} = \ddot{q}\mathbf{u} \tag{2.4.6}$$

Note that in this special case, the mappings from generalized velocity and acceleration to Cartesian velocity and acceleration are linear. This is true only because of lack of curvature of the path being followed by the particle.

Example 2.4.2: Consider the disk and balancing mass at point P of Example 2.2.1, with the disk forced to advance along the z axis, due to a lead screw with unit pitch, as shown in Fig. 2.4.3(a). Thus particle P moves on the helix shown in Fig. 2.4.3(b). The radius of the helix is 1 and the advance along the helix due to one revolution is also 1, which is a special case of Example 1.2.3. The position of a

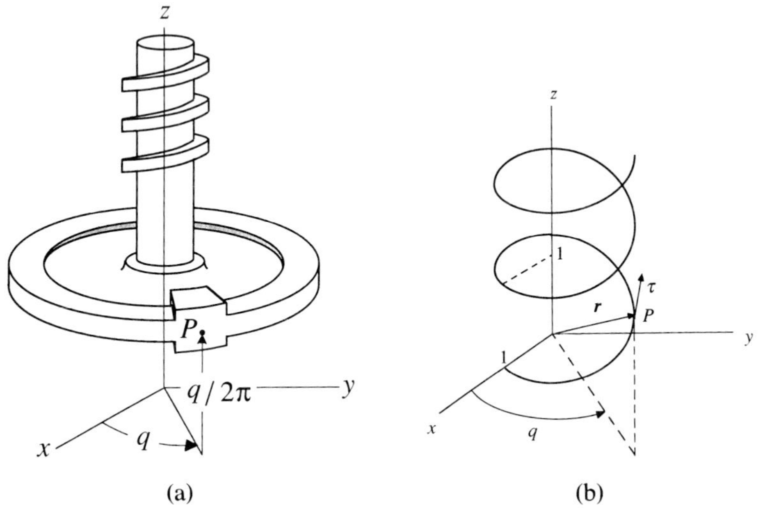

Figure 2.4.3 Particle on helix.

particle located at point P is thus

$$\mathbf{r} = \begin{bmatrix} \cos q \\ \sin q \\ \dfrac{q}{2\pi} \end{bmatrix} \tag{2.4.7}$$

The Jacobian of the mapping of Eq. 2.2.7 is

$$\mathbf{r}_q = \begin{bmatrix} -\sin q \\ \cos q \\ \dfrac{1}{2\pi} \end{bmatrix} \equiv \boldsymbol{\tau} \tag{2.4.8}$$

which is never zero, so there is no singular point for this mapping. The velocity of point P on the helix is obtained by differentiating Eq. 2.4.7, yielding

$$\dot{\mathbf{r}} = \mathbf{r}_q \dot{q} = \begin{bmatrix} -\sin q \\ \cos q \\ \dfrac{1}{2\pi} \end{bmatrix} \dot{q} \equiv \boldsymbol{\tau}\dot{q} \tag{2.4.9}$$

The acceleration of point P can be obtained by differentiating Eq. 2.4.9,

$$\begin{aligned} \ddot{\mathbf{r}} &= \mathbf{r}_q \ddot{q} + \dot{\mathbf{r}}_q \dot{q} \\ &= \boldsymbol{\tau}\ddot{q} + \begin{bmatrix} -\cos q \\ -\sin q \\ 0 \end{bmatrix} \dot{q}^2 \end{aligned} \tag{2.4.10}$$

Note that the coefficient of the quadratic term in the generalized velocity on the right of Eq. 2.4.10 is the projection onto the x-y plane of the negative of the vector $\mathbf{r}$. Thus the second term in Eq. 2.4.10 is the *centripetal acceleration* parallel to the x-y plane of a particle moving on the helix.

Although neither of the preceding examples involves singular points, singular mappings are possible if care is not taken in defining the parameterization of the curve.

Example 2.4.3: A particle P is to move along a parabola that lies in the x-y plane, as shown in Fig. 2.4.4. The equations of the parabola are $y = x^2$ and $z = 0$. Consider two different parameterizations of the parabola.

(1) *Parameterization by* $q = x$ leads to

$$\mathbf{r} = \begin{bmatrix} q \\ q^2 \\ 0 \end{bmatrix} \tag{2.4.11}$$

whose Jacobian is

$$\mathbf{r}_q = \begin{bmatrix} 1 \\ 2q \\ 0 \end{bmatrix} \tag{2.4.12}$$

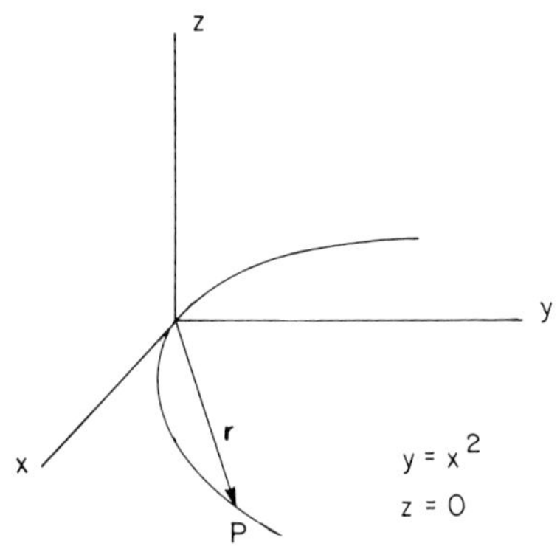

Figure 2.4.4 Particle on parabola.

which is never **0**. Hence this parameterization has no singular points. Velocity and acceleration are obtained by differentiating Eq. 2.2.11 with respect to time, to obtain

$$\begin{aligned}\dot{\mathbf{r}} &= \begin{bmatrix} 1 \\ 2q \\ 0 \end{bmatrix} \dot{q} \\ \ddot{\mathbf{r}} &= \begin{bmatrix} 1 \\ 2q \\ 0 \end{bmatrix} \ddot{q} + \begin{bmatrix} 0 \\ 2 \\ 0 \end{bmatrix} \dot{q}^2\end{aligned} \tag{2.4.13}$$

Note that the acceleration equation has the usual form of a term that is linear in acceleration, with coefficients depending on position, and a second term that is quadratic in the generalized coordinate velocity.

(2) *Parameterization by* $q = y$ leads to

$$\mathbf{r} = \begin{bmatrix} \pm\sqrt{q} \\ q \\ 0 \end{bmatrix} \tag{2.4.14}$$

Clearly, there is an ambiguity, or multiple branch of the parameterized path. Furthermore, the Jacobian of this mapping is

$$\mathbf{r}_q = \begin{bmatrix} \pm\ 1/(2\sqrt{q}) \\ 1 \\ 0 \end{bmatrix} \tag{2.4.15}$$

which is singular at $q = 0$. In fact, two forms of pathological behavior occur with this parameterization at $q = 0$. First, there is a loss of uniqueness as q is varied from zero to any small positive quantity. Second, if q is varied to become negative, loss of existence of **r** results. As usually occurs, at points where the implicit function theorem fails, in some directions of parameter variation existence is lost and in other directions uniqueness is lost.

This elementary example illustrates that selection of a proper parameterization is more a mathematical issue than a physical one. It is crucial that care be taken to choose generalized coordinates to parameterize motion of a system to avoid singular behavior (Probs. 2.4.1 and 2.4.4).

Generalized Coordinates for Particles on Surfaces

Basic considerations in defining kinematics of particles that move on surfaces are identical to those for particles moving on curves. The principal difference is that two independent parameters are required to define position on a surface; hence a particle moving on a smooth surface is said to have two *degrees of freedom*.

Let $\mathbf{q} = [q_1, q_2]^T$ be a two-dimensional generalized coordinate vector that is used to locate a point P on a surface in a Cartesian reference frame, with the mapping

$$\mathbf{r} = \mathbf{r}(\mathbf{q}) \tag{2.4.16}$$

The form of this relation is identical to Eq. 2.4.1 for a particle on a curve, with the exception that $\mathbf{q}$ is a 2-vector. Differentiating Eq. 2.4.16, using matrix calculus notation, leads to the *velocity equation*

$$\dot{\mathbf{r}} = \mathbf{r}_{\mathbf{q}}\dot{\mathbf{q}} \tag{2.4.17}$$

which is linear in the generalized velocity, but with a coefficient matrix that depends on generalized coordinates. Differentiating Eq. 2.4.17, making careful use of matrix calculus, leads to the *acceleration equation*

$$\ddot{\mathbf{r}} = \mathbf{r}_{\mathbf{q}}\ddot{\mathbf{q}} + (\mathbf{r}_{\mathbf{q}}\dot{\mathbf{q}})_{\mathbf{q}}\dot{\mathbf{q}} \tag{2.4.18}$$

which has a term that is linear in generalized acceleration, but quadratic in generalized velocity. The term $(\mathbf{r}_{\mathbf{q}}\dot{\mathbf{q}})_{\mathbf{q}}\dot{\mathbf{q}}$ in Eq. 2.4.18 may be expanded using Eq. 2.2.18 as

$$(\mathbf{r}_{\mathbf{q}}\dot{\mathbf{q}})_{\mathbf{q}}\dot{\mathbf{q}} = \left[\frac{\partial^2 r_i}{\partial q_1^2}\dot{q}_1^2 + 2\frac{\partial^2 r_i}{\partial q_1 \partial q_2}\dot{q}_1\dot{q}_2 + \frac{\partial^2 r_i}{\partial q_2^2}\dot{q}_2^2\right]_{3\times 1}$$

Note that Eqs. 2.4.16 to 2.4.18 are in fact valid for arbitrary dimensions of $\mathbf{r}$ and $\mathbf{q}$. The expansion of the quadratic term on the right of Eq. 2.4.18, of course, must be consistent with the dimension of $\mathbf{q}$ (Prob. 2.4.3).

Similar to the case of a particle that moves on a curve, vectors $\mathbf{r}_{q1}$ and $\mathbf{r}_{q2}$ are tangent to the surface defined by Eq. 2.4.16, as shown in Fig. 2.4.5. This is

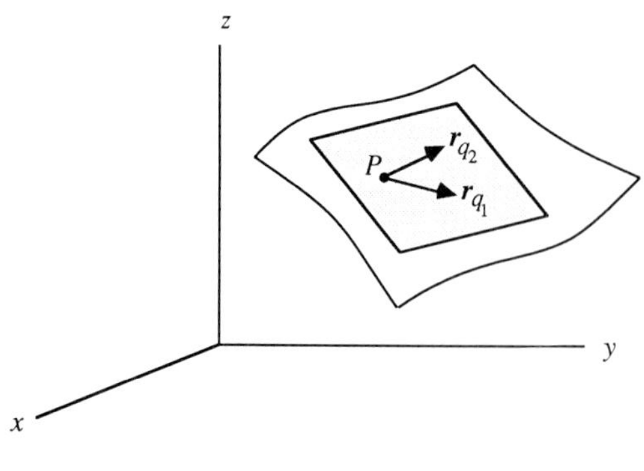

Figure 2.4.5 Tangent plane to surface.

true since $\mathbf{r}_{q_1}$ is the derivative of $\mathbf{r}(\mathbf{q})$ with respect to q_1, with q_2 held constant. If $q_2 = c$, then $\mathbf{r} = \mathbf{r}(q_1, c)$ defines a curve and $\mathbf{r}_{q_1}$ is tangent to that curve. A similar argument shows that $\mathbf{r}_{q_2}$ is tangent to the surface.

According to Eq. 2.4.17, if $\mathbf{r}_{q_1}$ and $\mathbf{r}_{q_2}$ are not parallel, any velocity vector $\dot{\mathbf{r}}$ in the plane tangent to the surface can be realized by unique values of $\dot{q}_1$ and $\dot{q}_2$; that is,

$$\dot{\mathbf{r}} = \mathbf{r}_{q_1}\dot{q}_1 + \mathbf{r}_{q_2}\dot{q}_2$$

However, if $\mathbf{r}_{q_1}$ and $\mathbf{r}_{q_2}$ are parallel; i.e., if $\mathbf{r}_{\mathbf{q}}$ fails to have full column rank, a velocity $\dot{\mathbf{r}}$ that is not parallel to $\mathbf{r}_{q_1}$ and $\mathbf{r}_{q_2}$ cannot be achieved by any values of $\dot{q}_1$ and $\dot{q}_2$. Thus, at a singular point, velocity in only one direction can be achieved.

Example 2.4.4: A plane may be defined by a pair of noncollinear vectors $\mathbf{g}$ and $\mathbf{h}$, as shown in Fig. 2.4.6. Any vector $\mathbf{r}_1$ that is a linear combination of $\mathbf{g}$ and $\mathbf{h}$, as shown in Fig. 2.4.6, lies in the plane defined by $\mathbf{g}$ and $\mathbf{h}$ at their point of intersection.

A typical point P in the plane may be located in a Cartesian reference frame by the vector

$$\begin{aligned} \mathbf{r} &= \mathbf{r}_0 + q_1\mathbf{g} + q_2\mathbf{h} \\ &= \mathbf{r}_0 + [\mathbf{g}, \mathbf{h}]\mathbf{q} \end{aligned} \tag{2.4.19}$$

Since vectors $\mathbf{g}$ and $\mathbf{h}$ are constant, the velocity and acceleration of point P are simply

$$\dot{\mathbf{r}} = [\mathbf{g}, \mathbf{h}]\dot{\mathbf{q}} \tag{2.4.20}$$

$$\ddot{\mathbf{r}} = [\mathbf{g}, \mathbf{h}]\ddot{\mathbf{q}} \tag{2.4.21}$$

As long as vectors $\mathbf{g}$ and $\mathbf{h}$ are noncollinear, the matrix $\mathbf{r}_{\mathbf{q}} = [\mathbf{g}, \mathbf{h}]$ has full column rank and Eq. 2.4.19 defines a nonsingular mapping everywhere (i.e., the mapping from $\mathbf{q}$ to the plane is one-to-one and the point P has two degrees of freedom). Note that if $\mathbf{g}$ and $\mathbf{h}$ were collinear, the mapping would be singular and only a one-degree-of-freedom system would result. This would amount to collapsing the plane to a single line and reducing the movement of a particle to a straight line, as in Example 2.4.1.

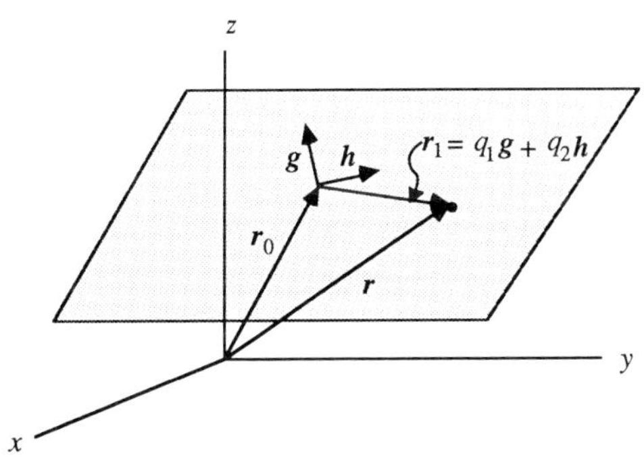

Figure 2.4.6 Particle on plane.

Example 2.4.5: Consider a particle P that moves on a sphere of unit radius and center at the origin of a Cartesian reference frame, as shown in Fig. 2.4.7. Using the generalized coordinates $\mathbf{q} = [\phi, \psi]^T$, the mapping of Eq. 2.1.6, with $\rho = 1$, that locates point P may be written as

$$\mathbf{r} = \begin{bmatrix} \cos q_1 \ \cos q_2 \\ \sin q_1 \ \cos q_2 \\ \sin q_2 \end{bmatrix} \tag{2.4.22}$$

The Jacobian of the mapping of Eq. 2.4.22 is

$$\mathbf{r_q} = \begin{bmatrix} -\sin q_1 \ \cos q_2 & -\cos q_1 \ \sin q_2 \\ \cos q_1 \ \cos q_2 & -\sin q_1 \ \sin q_2 \\ 0 & \cos q_2 \end{bmatrix} \tag{2.4.23}$$

When $q_2 = \psi = \pm \pi/2$, the first column of the Jacobian is zero, so the Jacobian is column-rank deficient and the points at the poles of the sphere on the z axis are singular. This is a special case of the singularity in spherical coordinates that was noted in Section 2.1. Thus singularities in the relationships between $\dot{\mathbf{r}}$ and $\dot{\mathbf{q}}$ and $\ddot{\mathbf{r}}$ and $\ddot{\mathbf{q}}$ also exist at these points.

To see geometrically what goes wrong when $q_2 = \pi/2$, consider the curve shown in Fig. 2.4.8, defined by $q_1 = \phi = 0$ on the sphere. As q_2 varies, from Eq. 2.4.23,

$$\mathbf{r}_{q1} = \begin{bmatrix} 0 \\ \cos q_2 \\ 0 \end{bmatrix}, \qquad \mathbf{r}_{q2} = \begin{bmatrix} -\sin q_2 \\ 0 \\ \cos q_2 \end{bmatrix}$$

Samples of these tangent vectors at points on the curve that approach the pole on the z axis are shown in Fig. 2.4.8. Since $\mathbf{r}_{q_1}$ approaches $\mathbf{0}$ at the pole, only a velocity in the direction $\mathbf{r}_{q_2}$ is achievable at the pole.

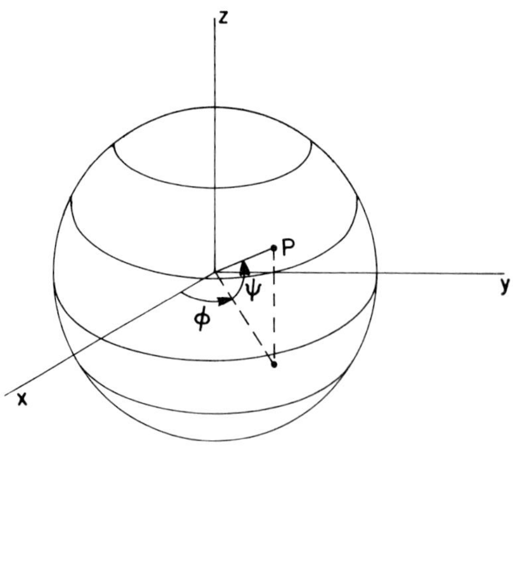

Figure 2.4.7 Particle on unit sphere.

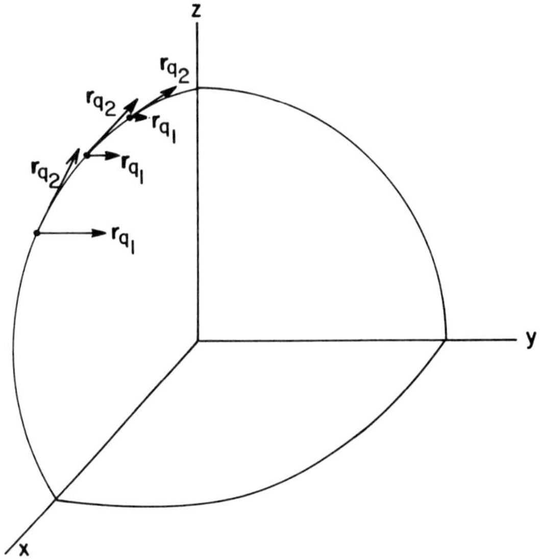

Figure 2.4.8 Tangent vectors to sphere near pole along curve $q_1 = 0$.

2.5 TRANSFORMATION OF COORDINATES

As shown in Section A.3 of Appendix A, a geometric vector is uniquely represented by an algebraic vector that contains the components of the geometric vector in a Cartesian reference frame. The components of a vector, however, must be defined in a specific Cartesian reference frame, denoted x-y-z in the preceding. Consider a second Cartesian x'-y'-z' frame, with the same origin as the x-y-z frame, as shown in Fig. 2.5.1. Unit x', y', and z' coordinate vectors are denoted $\vec{f}$, $\vec{g}$, and $\vec{h}$, respectively, and unit x, y, and z coordinate vectors are denoted $\vec{i}$, $\vec{j}$, and $\vec{k}$.

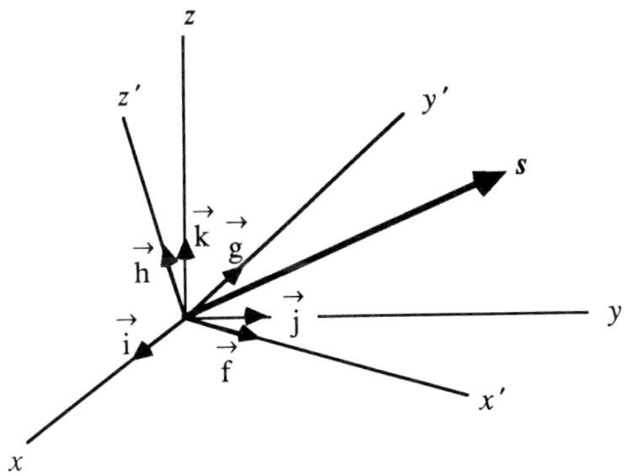

Figure 2.5.1 Two Cartesian reference frames.

A vector $\vec{s}$ in space can be represented in either of the frames as

$$\vec{s} = s_x\vec{i} + s_y\vec{j} + s_z\vec{k} \tag{2.5.1}$$

or

$$\vec{s} = s_{x'}\vec{f} + s_{y'}\vec{g} + s_{z'}\vec{h} \tag{2.5.2}$$

where the components are given as scalar products of $\vec{s}$ with the respective unit vectors (see Appendix A),

$$s_x = \vec{s}\cdot\vec{i}, \qquad s_y = \vec{s}\cdot\vec{j}, \qquad s_z = \vec{s}\cdot\vec{h} \tag{2.5.3}$$

and

$$s_{x'} = \vec{s}\cdot\vec{f}, \qquad s_{y'} = \vec{s}\cdot\vec{g}, \qquad s_{z'} = \vec{s}\cdot\vec{h} \tag{2.5.4}$$

The algebraic vectors that define $\vec{s}$ in the two frames are thus

$$\mathbf{s} = [s_x, s_y, s_z]^T \tag{2.5.5}$$

in the x-y-z frame (Prob 2.5.1) and

$$\mathbf{s}' = [s_{x'}, s_{y'}, s_{z'}]^T \tag{2.5.6}$$

in the x'-y'-z' frame (Prob. 2.5.2).

It is clear that there is a relation between $\mathbf{s}$ and $\mathbf{s}'$, since they are defined by the same geometric vector $\vec{s}$. To establish this relation, expand the $\vec{f}$, $\vec{g}$, and $\vec{h}$ unit vectors in terms of the $\vec{i}$, $\vec{j}$, and $\vec{k}$ unit vectors as

$$\begin{aligned}
\vec{f} &= a_{11}\vec{i} + a_{21}\vec{j} + a_{31}\vec{k} \\
\vec{g} &= a_{12}\vec{i} + a_{22}\vec{j} + a_{32}\vec{k} \\
\vec{h} &= a_{13}\vec{i} + a_{23}\vec{j} + a_{33}\vec{k}
\end{aligned} \tag{2.5.7}$$

where a_{ij} are the following *direction cosines*:

$$\begin{aligned}
a_{11} &= \vec{i}\cdot\vec{f} = \cos\theta(\vec{i}, \vec{f}) \\
a_{12} &= \vec{i}\cdot\vec{g} = \cos\theta(\vec{i}, \vec{g}) \\
a_{13} &= \vec{i}\cdot\vec{h} = \cos\theta(\vec{i}, \vec{h}) \\
a_{21} &= \vec{j}\cdot\vec{f} = \cos\theta(\vec{j}, \vec{f}) \\
a_{22} &= \vec{j}\cdot\vec{g} = \cos\theta(\vec{j}, \vec{g}) \\
a_{23} &= \vec{j}\cdot\vec{h} = \cos\theta(\vec{j}, \vec{h}) \\
a_{31} &= \vec{k}\cdot\vec{f} = \cos\theta(\vec{k}, \vec{f}) \\
a_{32} &= \vec{k}\cdot\vec{g} = \cos\theta(\vec{k}, \vec{g}) \\
a_{33} &= \vec{k}\cdot\vec{h} = \cos\theta(\vec{k}, \vec{h})
\end{aligned} \tag{2.5.8}$$

and cos $\theta(\vec{a}, \vec{b})$ is the angle between vectors $\vec{a}$ and $\vec{b}$ (see Appendix A). Substituting from Eq. 2.5.7 into Eq. 2.5.2 yields

$$\vec{s} = (a_{11}s_{x'} + a_{12}s_{y'} + a_{13}s_{z'})\vec{i} + (a_{21}s_{x'} + a_{22}s_{y'} + a_{23}s_{z'})\vec{j} + (a_{31}s_{x'} + a_{32}s_{y'} + a_{33}s_{z'})\vec{k}$$

Equating the right sides of this representation of $\vec{s}$ and Eq. 2.5.1,

$$\begin{aligned} s_x &= a_{11}s_{x'} + a_{12}s_{y'} + a_{13}s_{z'} \\ s_y &= a_{21}s_{x'} + a_{22}s_{y'} + a_{23}s_{z'} \\ s_z &= a_{31}s_{x'} + a_{32}s_{y'} + a_{33}s_{z'} \end{aligned} \tag{2.5.9}$$

or, in matrix form,

$$\mathbf{s} = \mathbf{A}\mathbf{s}' \tag{2.5.10}$$

where $\mathbf{A}$ is called the *direction cosine matrix*,

$$\mathbf{A} = \begin{bmatrix} a_{11} & a_{12} & a_{13} \\ a_{21} & a_{22} & a_{23} \\ a_{31} & a_{32} & a_{33} \end{bmatrix} \tag{2.5.11}$$

whose entries are defined in Eq. 2.5.8.

The direction cosine matrix $\mathbf{A}$ has a very special property. If the x-y-z component form of unit vectors $\vec{f}$, $\vec{g}$, and $\vec{h}$ are denoted by $\mathbf{f}$, $\mathbf{g}$, and $\mathbf{h}$ and the x-y-z components of vectors $\vec{i}$, $\vec{j}$, and $\vec{k}$, are denoted by $\mathbf{i}$, $\mathbf{j}$, and $\mathbf{k}$, then

$$\begin{aligned} \mathbf{i} &= [1, 0, 0]^T \\ \mathbf{j} &= [0, 1, 0]^T \\ \mathbf{k} &= [0, 0, 1]^T \end{aligned} \tag{2.5.12}$$

Equation 2.5.8 shows that

$$\begin{aligned} \mathbf{f} &= [a_{11}, a_{21}, a_{31}]^T \\ \mathbf{g} &= [a_{12}, a_{22}, a_{32}]^T \\ \mathbf{h} &= [a_{13}, a_{23}, a_{33}]^T \end{aligned}$$

Therefore, the matrix $\mathbf{A}$ of Eq. 2.5.11 can be written as

$$\mathbf{A} = [\mathbf{f}, \mathbf{g}, \mathbf{h}] \tag{2.5.13}$$

Since the unit vectors $\mathbf{f}$, $\mathbf{g}$, and $\mathbf{h}$ are orthogonal, expansion of the matrix product shows that (Prob. 2.5.3)

$$\mathbf{A}^T\mathbf{A} = \mathbf{I} \tag{2.5.14}$$

Thus the inverse of $\mathbf{A}$ is $\mathbf{A}^{-1} = \mathbf{A}^T$ and the direction cosine matrix $\mathbf{A}$ is an *or-*

thogonal matrix (see Appendix A). This special property permits an easy inversion of Eq. 2.5.10, to obtain

$$\mathbf{s}' = \mathbf{A}^T\mathbf{s} \tag{2.5.15}$$

Thus transforming between algebraic vectors that represent the same physical vector in the x-y-z and x'-y'-z' Cartesian reference frames is a trivial matter.

Example 2.5.1: The z' axis of the x'-y'-z' Cartesian frame is fixed in the x-y plane and the x' axis is rotated by an angle ϕ above the x-y plane, as shown in Fig. 2.5.2. The **f**, **g**, and **h** unit vectors are thus

$$\mathbf{f} = \left[\frac{\sqrt{2}}{2}\cos\phi, \frac{\sqrt{2}}{2}\cos\phi, \sin\phi\right]^T$$

$$\mathbf{g} = \left[-\frac{\sqrt{2}}{2}\sin\phi, -\frac{\sqrt{2}}{2}\sin\phi, \cos\phi\right]^T$$

$$\mathbf{h} = \left[\frac{\sqrt{2}}{2}, -\frac{\sqrt{2}}{2}, 0\right]^T$$

From Eq. 2.5.13, the transformation matrix from the x'-y'-z' frame to the x-y-z frame is

$$\mathbf{A} = \begin{bmatrix} \frac{\sqrt{2}}{2}\cos\phi & -\frac{\sqrt{2}}{2}\sin\phi & \frac{\sqrt{2}}{2} \\ \frac{\sqrt{2}}{2}\cos\phi & -\frac{\sqrt{2}}{2}\sin\phi & -\frac{\sqrt{2}}{2} \\ \sin\phi & \cos\phi & 0 \end{bmatrix}$$

A direct calculation verifies that **A** is orthogonal.

When the origins of the x-y-z and x'-y'-z' frames do not coincide, the foregoing analysis is applied between the x'-y'-z' and the translated (dashed) x-y-z frame with common origins, as shown in Fig. 2.5.3. If the algebraic vector $\mathbf{s}'^P$

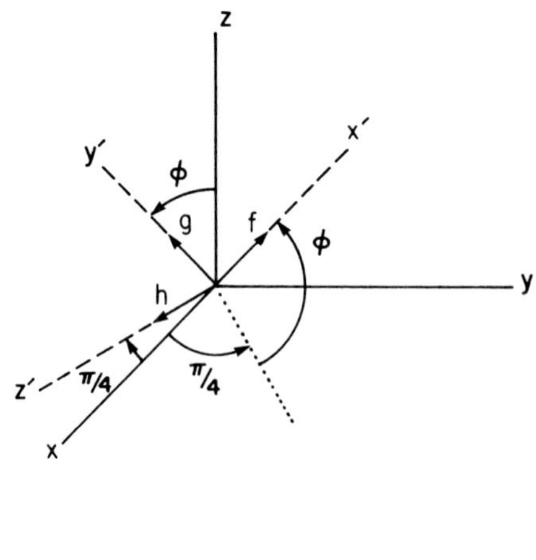

Figure 2.5.2 Orientation of reference frame.

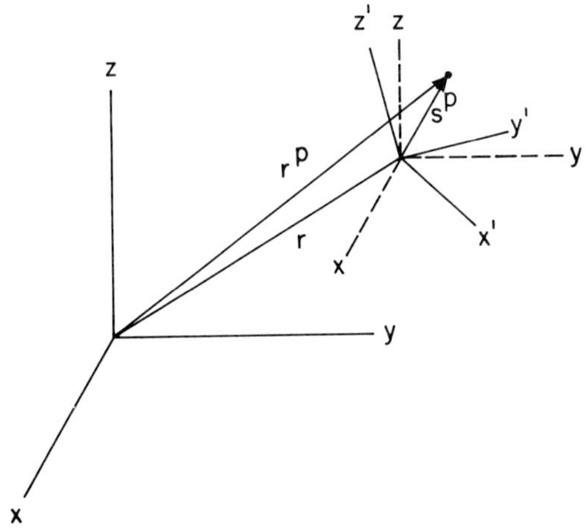

Figure 2.5.3 Translation and rotation of reference frame.

locates point P in the x'-y'-x' frame, then in the translated x-y-z frame this vector is $\mathbf{As}'^P$. Thus

$$\mathbf{r}^P = \mathbf{r} + \mathbf{As}'^P \tag{2.5.16}$$

where $\mathbf{r}$ is the vector from the origin of the x-y-z reference frame to the origin of the x'-y'-z' frame, as shown in Fig. 2.5.3.

The significance of Eq. 2.5.16 in the study of dynamics should not be underestimated. A body that moves in space (e.g., the aircraft shown in Fig. 2.5.4) is only precisely located and oriented in space when all points or particles that make up the body can be located. Consider the body as being rigid and attach an x'-y'-z' frame to the body (i.e., it must translate and rotate with the body). Points on the body (e.g., the tips P and Q of the wing and tail of the aircraft in Fig. 2.5.4) may be defined by vectors $\mathbf{s}'^P$ and $\mathbf{s}'^Q$ in the x'-y'-z' frame. Since P and Q are fixed in the rigid body, they do not change position relative to the x'-y'-z' frame (i.e., they are constant).

As the body moves and rotates in space, a point P that is fixed to the body is located in the stationary x-y-z reference frame by the vector $\mathbf{r}^P(t)$, which varies with time. Clearly, the vector $\mathbf{r}$ in Fig. 2.5.4 depends on time, since it locates the origin 0 of the x'-y'-z' frame as the body moves (i.e., $\mathbf{r} = \mathbf{r}(t)$). Finally, the orientation of the body, which is defined relative to the fixed x-y-z frame, changes

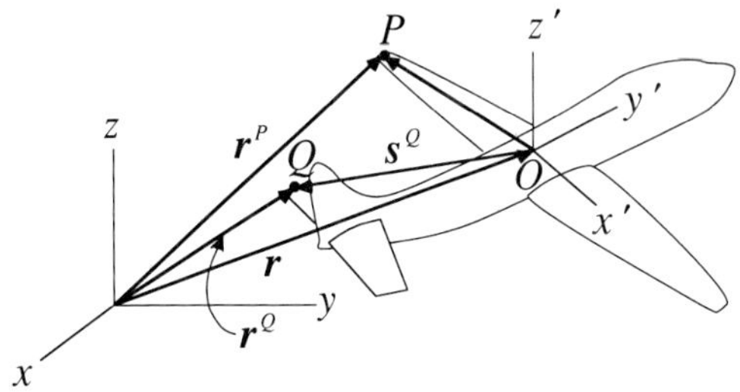

Figure 2.5.4 Locating and orienting aircraft in space.

in time and is defined by the direction cosine matrix $\mathbf{A}(t)$, which must be a function of time. Since vectors such as $\mathbf{s}'^P$ and $\mathbf{s}'^Q$ do not change with time, the same vector $\mathbf{r}(t)$ and matrix $\mathbf{A}(t)$ define the position and orientation of the body as it moves through space. They may thus serve as generalized coordinates for position and orientation of a body in space.

The nine direction cosines in matrix $\mathbf{A}$ define the orientation of the x'-y'-z' frame relative to the x-y-z reference frame, but they are not independent. Equation 2.5.14 provides just six independent [12] equations among the nine direction cosines, since in component form Eq. 2.5.14 is

$$\sum_{k=1}^{3} a_{ki}a_{kj} = \begin{cases} 1, & i = j \\ 0, & i < j, \end{cases} \qquad i, j = 1, 2, 3$$

and interchanging $i \neq j$ yields the same equation. Thus only three direction cosines are independent. While the nine direction cosines, subject to the constraints of Eq. 2.5.14, could be adopted as generalized coordinates that define orientation of the x'-x'-z' frame, this is neither practical nor convenient. Thus other orientation generalized coordinates are sought.

➧ **Example 2.5.2:** As a special case that serves as the foundation of planar kinematics and dynamics, consider the x-y-z and x'-y'-z' frames shown in Fig. 2.5.5, with z and z' axes coincident. The angle of rotation of the x' axis relative to the x axis, with counterclockwise as positive, is denoted as ϕ. From Eq. 2.5.8,

$$a_{11} = a_{22} = \cos\phi$$

$$a_{21} = \cos\left(\frac{\pi}{2} - \phi\right) = \sin\phi$$

$$a_{12} = \cos\left(\frac{\pi}{2} + \phi\right) = -\sin\phi$$

$$a_{33} = 1$$

$$a_{31} = a_{32} = a_{13} = a_{23} = 0$$

Thus the direction cosine matrix of Eq. 2.5.11 yields the transformation

$$\mathbf{s} = \begin{bmatrix} s_x \\ s_y \\ s_z \end{bmatrix} = \begin{bmatrix} \cos\phi & -\sin\phi & 0 \\ \sin\phi & \cos\phi & 0 \\ 0 & 0 & 1 \end{bmatrix} \begin{bmatrix} s_{x'} \\ s_{y'} \\ s_{z'} \end{bmatrix} = \mathbf{As}' \tag{2.5.17}$$

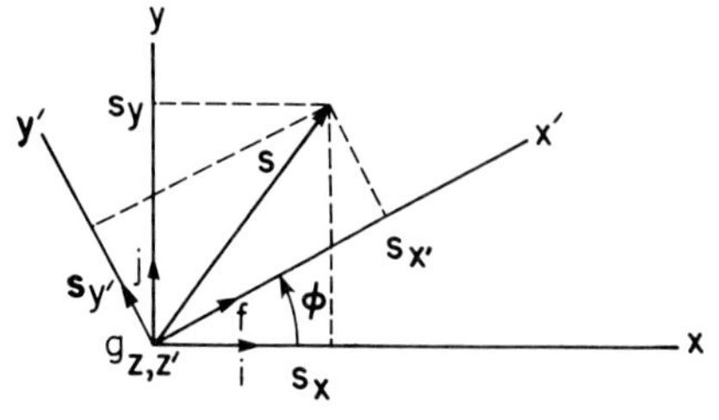

Figure 2.5.5 Reference frames with coincident z and z' axes.

This contains the special planar rotation relations

$$\begin{bmatrix} s_x \\ s_y \end{bmatrix} = \begin{bmatrix} \cos\phi & -\sin\phi \\ \sin\phi & \cos\phi \end{bmatrix} \begin{bmatrix} s_{x'} \\ s_{y'} \end{bmatrix} \tag{2.5.18}$$

$$\begin{bmatrix} s_{x'} \\ s_{y'} \end{bmatrix} = \begin{bmatrix} \cos\phi & \sin\phi \\ -\sin\phi & \cos\phi \end{bmatrix} \begin{bmatrix} s_x \\ s_y \end{bmatrix} \tag{2.5.19}$$

Comparing Eqs. 2.5.10 and 2.5.18, for rotation of coordinates in the plane (see Fig. 2.5.4), the planar orthogonal rotation transformation matrix is

$$\mathbf{A} = \begin{bmatrix} \cos\phi & -\sin\phi \\ \sin\phi & \cos\phi \end{bmatrix} \tag{2.5.20}$$

This result will be used with Eq. 2.5.16 in analysis of planar kinematics.

It is shown in Section A.3 of Appendix A that the vector $\mathbf{c}$ of components of the vector product $\vec{c} = \vec{a} \times \vec{b}$ of two vectors $\vec{a}$ and $\vec{b}$ is given in terms of the component vectors $\mathbf{a}$ and $\mathbf{b}$ of $\vec{a}$ and $\vec{b}$ as

$$\mathbf{c} = \tilde{\mathbf{a}}\mathbf{b}$$

where

$$\tilde{\mathbf{a}} \equiv \begin{bmatrix} 0 & -a_z & a_y \\ a_z & 0 & -a_x \\ -a_y & a_x & 0 \end{bmatrix} \tag{2.5.21}$$

is a 3×3 antisymmetric matrix that is formed from the components of $\mathbf{a}$:

$$\tilde{\mathbf{a}}^T = -\tilde{\mathbf{a}} \tag{2.5.22}$$

This agrees with the well-known fact that $\vec{a} \times \vec{b} = -\vec{b} \times \vec{a}$; that is,

$$\tilde{\mathbf{a}}\mathbf{b} = -\tilde{\mathbf{b}}\mathbf{a} \tag{2.5.23}$$

Another classical property of vector product is $\vec{a} \times \vec{a} = \vec{0}$, for any vector $\vec{a}$. The component vector counterpart of this identity is

$$\tilde{\mathbf{a}}\mathbf{a} = \mathbf{0} \tag{2.5.24}$$

Finally, useful identities that involve the vector product are repeated here, from Eqs. A.3.12 to A.3.16:

$$\tilde{\mathbf{a}}\tilde{\mathbf{b}} = \mathbf{b}\mathbf{a}^T - \mathbf{a}^T\mathbf{b}\mathbf{I} \tag{2.5.25}$$

$$\widetilde{(\tilde{\mathbf{a}}\mathbf{b})} = \mathbf{b}\mathbf{a}^T - \mathbf{a}\mathbf{b}^T \tag{2.5.26}$$

$$= \tilde{\mathbf{a}}\tilde{\mathbf{b}} - \tilde{\mathbf{b}}\tilde{\mathbf{a}} \tag{2.5.27}$$

$$\widetilde{(\mathbf{a} + \mathbf{b})} = \tilde{\mathbf{a}} + \tilde{\mathbf{b}} \tag{2.5.28}$$

Other useful relations may be obtained by noting that for any vectors $\mathbf{s}'$ and $\mathbf{v}'$ in the x'-y'-z' frame, the vector product of $\mathbf{s}$ and $\mathbf{v}$ in the x-y-z frame may be obtained by first forming the vector product of $\mathbf{s}'$ and $\mathbf{v}'$ and then transforming the result to the x-y-z frame:

$$\tilde{\mathbf{s}}\mathbf{v} = \mathbf{A}(\tilde{\mathbf{s}}'\mathbf{v}') \tag{2.5.29}$$

Using $\mathbf{v} = \mathbf{A}\mathbf{v}'$ yields

$$(\tilde{\mathbf{s}}\mathbf{A})\mathbf{v}' = (\mathbf{A}\tilde{\mathbf{s}}')\mathbf{v}'$$

Since this must hold for arbitrary $\mathbf{v}'$ (Prob. 2.5.4),

$$\tilde{\mathbf{s}}\mathbf{A} = \mathbf{A}\tilde{\mathbf{s}}' \tag{2.5.30}$$

Postmultiplying by $\mathbf{A}^T$ and using Eq. 2.5.14 yields the desired result:

$$(\widetilde{\mathbf{A}\mathbf{s}'}) = \tilde{\mathbf{s}} = \mathbf{A}\tilde{\mathbf{s}}'\mathbf{A}^T \tag{2.5.31}$$

Similarly,

$$(\widetilde{\mathbf{A}^T\mathbf{s}}) = \tilde{\mathbf{s}}' = \mathbf{A}^T\tilde{\mathbf{s}}\mathbf{A} \tag{2.5.32}$$

Consider a pair of Cartesian x_i'-y_i'-z_i' and x_j'-y_j'-z_j' frames shown in Fig. 2.5.6. An arbitrary vector $\mathbf{s}$ in the x-y-z frame has representations $\mathbf{s}_i'$ and $\mathbf{s}_j'$ in the x_i'-y_i'-z_i' and x_j'-y_j'-z_j' frames, respectively, so

$$\mathbf{s} = \mathbf{A}_i\mathbf{s}_i' = \mathbf{A}_j\mathbf{s}_j' \tag{2.5.33}$$

where $\mathbf{A}_i$ and $\mathbf{A}_j$ are the transformation matrices from the x_i'-y_i'-z_i' and x_j'-y_j'-z_j' frames to the x-y-z frame, respectively. Since $\mathbf{A}_i$ and $\mathbf{A}_j$ are orthogonal matrices, Eq. 2.5.33 yields

$$\mathbf{s}_i' = \mathbf{A}_i^T\mathbf{A}_j\mathbf{s}_j' \equiv \mathbf{A}_{ij}\mathbf{s}_j' \tag{2.5.34}$$

Since $\mathbf{s}_j'$ is an arbitrary vector,

$$\mathbf{A}_{ij} = \mathbf{A}_i^T\mathbf{A}_j \tag{2.5.35}$$

is the transformation matrix from the x_j'-y_j'-z_j' frame to the x_i'-y_i'-z_i' frame. A direct calculation shows that $\mathbf{A}_{ij}$ is an orthogonal matrix (Prob. 2.5.5).

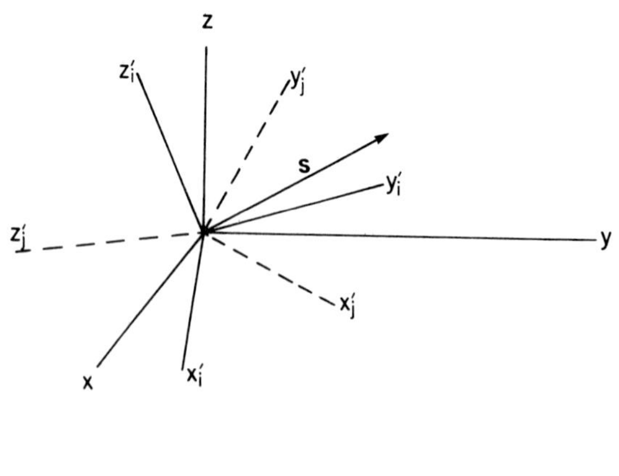

Figure 2.5.6 Reference frames with coincident origins.

➧ **Example 2.5.3:** In numerous applications, kinematic constraints dictate that certain axes of the x_i'-y_i'-z_i' and x_j'-y_j'-z_j' frames remain parallel (e.g., the vectors $\mathbf{h}_i$ and $\mathbf{h}_j$ shown in Fig. 2.5.7 are to be parallel). The angle θ of rotation, measured positive as counterclockwise from $\mathbf{f}_i$ to $\mathbf{f}_j$, is a key quantity that is to be calculated. From the definition of scalar product and the fact that the coordinate vectors shown in Fig. 2.5.7 are unit vectors,

$$\mathbf{f}_i^T\mathbf{f}_j = \cos\theta \tag{2.5.36}$$

Similarly, from the definition of vector product,

$$\tilde{\mathbf{f}}_i\mathbf{f}_j = \mathbf{h}_i \sin\theta$$

Taking the scalar product of both sides of this equation with $\mathbf{h}_i$ and using the fact that $\tilde{\mathbf{f}}_i\mathbf{h}_i = -\mathbf{g}_i$,

$$\sin\theta = \mathbf{h}_i^T\tilde{\mathbf{f}}_i\mathbf{f}_j = \mathbf{g}_i^T\mathbf{f}_j \tag{2.5.37}$$

Writing the unit vectors in terms of the respective reference frames in which they are fixed, and using the transformation matrices from these frames to the x-y-z frame, Eq. 2.5.36 becomes

$$\cos\theta = \mathbf{f}_i'^T\mathbf{A}_i^T\mathbf{A}_j\mathbf{f}_j' \tag{2.5.38}$$

Similarly, Eq. 2.5.37 becomes

$$\sin\theta = \mathbf{g}_i'^T\mathbf{A}_i^T\mathbf{A}_j\mathbf{f}_j' \tag{2.5.39}$$

If $\sin\theta$ and $\cos\theta$ are known, the value of θ, $0 \leq \theta < 2\pi$, may be uniquely determined by taking the arcsin of both sides of Eq. 2.5.39 and using the algebraic sign of $\cos\theta$ from Eq. 2.5.38 to uniquely evaluate θ. To simplify notation, denote

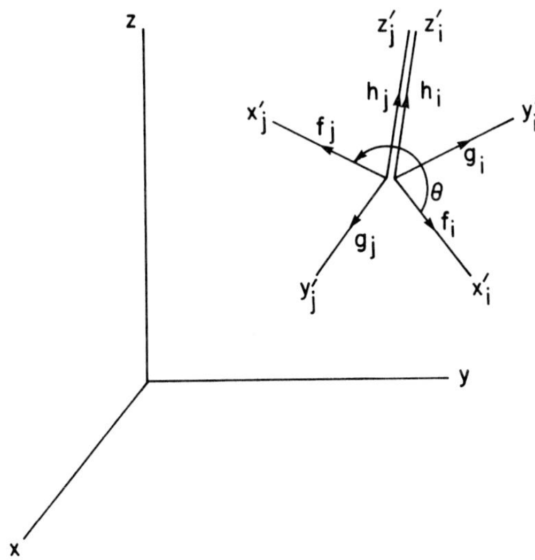

Figure 2.5.7 Triads with z' axes parallel.

$s = \sin\theta$ from Eq. 2.5.39 and $c = \cos\theta$ from Eq. 2.5.38. Then with

$$\frac{-\pi}{2} \leq \arcsin s \leq \frac{\pi}{2} \tag{2.5.40}$$

the angle θ is (Prob. 2.5.6)

$$\theta = \begin{cases} \arcsin s, & \text{if } s \geq 0 \text{ and } c \geq 0 \\ \pi - \arcsin s, & \text{if } s \geq 0 \text{ and } c < 0 \\ \pi - \arcsin s, & \text{if } s < 0 \text{ and } c < 0 \\ 2\pi + \arcsin s, & \text{if } s < 0 \text{ and } c \geq 0 \end{cases} \tag{2.5.41}$$

The angle calculated in Eq. 2.5.41 is in the range $0 \leq \theta < 2\pi$. If the number of revolutions that have occurred is important, logic must be supplied to determine the number n of revolutions that have occurred. The angle $2n\pi$ is then added to θ calculated in Eq. 2.5.41.

An important result that is used in defining relative orientation of reference frames is the following.

Euler's Theorem: If the origins of two Cartesian reference frames coincide, they may be brought into coincidence by a rotation about some axis that passes through the common origin.

This theorem is proved in Refs. 9 and 10 using orthogonality of the transformation matrix **A** between the reference frames and eigenvalue theory of matrices. A physical interpretation of Euler's theorem is given in Fig. 2.5.8, where the orthogonal projections of **i-f**, **j-g**, and **k-h** pairs onto the axis of rotation are equal. The axis about which rotation is to take place is called the *orientation axis*, defined by a unit vector **u**. The angle χ of rotation, measured in the plane perpendicular to **u**, is positive counterclockwise (i.e., using the right-hand sign con-

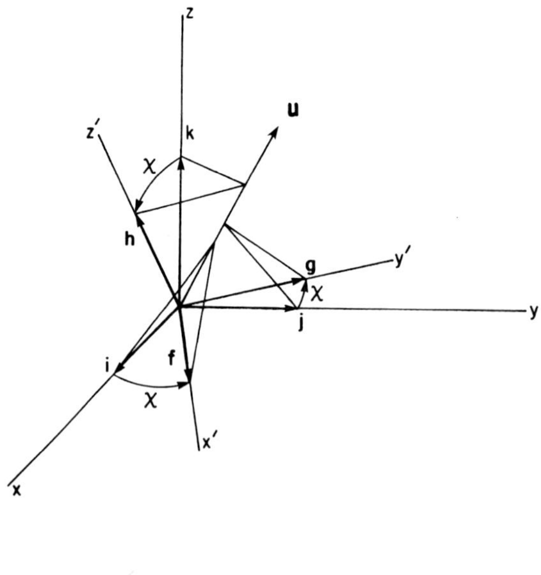

Figure 2.5.8 Euler rotation of reference frames.

vention; see Appendix A). Note that if the vector **u** had been taken in the opposite sense along the orientation axis, the angle χ shown in Fig. 2.5.8 would be replaced by $2\pi - \chi$.

Euler's theorem states that any orientation of the x'-y'-z' frame can be achieved by a rotation from the reference orientation about some axis. It is natural, therefore, to seek a representation of the coordinate transformation in terms of parameters of this rotation, namely the angle χ of rotation and the direction cosines of the orientation axis.

Figure 2.5.9 shows the orientation axis, the x'-y'-z' frame, the x-y-z reference frame, an arbitrary vector **s**, and the projections of **s** on each frame. The objective is to obtain relationships between the components **s** and **s**$'$ of the vector in the x-y-z and x'-y'-z' frames, respectively.

The direction cosine matrix **A** that provides the desired relation between the x'-y'-z' and x-y-z components of a vector **s** is obtained if unit vectors **f**, **g**, and **h** can be written in terms of χ and **u** (i.e., Eq. 2.5.13 yields $\mathbf{A} = [\mathbf{f}, \mathbf{g}, \mathbf{h}]$).

Consider first the relation between **h** and **k**. Since **h** is obtained in the x-y-z frame by a rotation of **k** about **u**, its tip lies in the plane that is perpendicular to **u** and contains the tip of **k**, as shown in Fig. 2.5.10. The vector product of **u** and **k**, $\tilde{\mathbf{u}}\mathbf{k}$, lies in a parallel plane and has length equal to the radius of the circle formed by intersecting the plane perpendicular to **u** and the cone formed by rotating **k** about the **u** axis. The center N of this circle is located by the vector $(\mathbf{u}^T\mathbf{k})\mathbf{u}$.

A vector **a** is constructed to pass through the tip of **h** and to be perpendicular to the vector $\mathbf{k} - (\mathbf{u}^T\mathbf{k})\mathbf{u}$ from point N to the tip of **k**, as shown in Fig. 2.5.10. The vector $\mathbf{k} - (\mathbf{u}^T\mathbf{k})\mathbf{u}$ lies in the plane formed by **u** and **k**, so it is perpendicular to $\tilde{\mathbf{u}}\mathbf{k}$.

Vectors **a** and $\tilde{\mathbf{u}}\mathbf{k}$ are parallel. Furthermore, the lengths of vectors $\mathbf{h} - (\mathbf{u}^T\mathbf{k})\mathbf{u}$ and $\tilde{\mathbf{u}}\mathbf{k}$ are equal, since they are radii of the same circle. Thus by simple trigonometry,

$$\mathbf{a} = \tilde{\mathbf{u}}\mathbf{k} \sin \chi$$

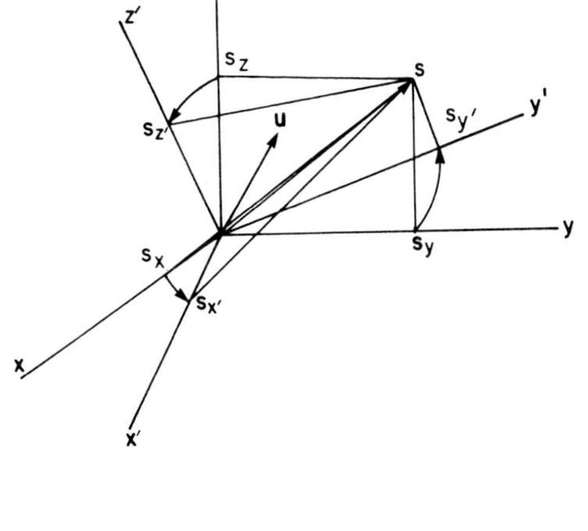

Figure 2.5.9 Vector **s** in x'-y'-z' and x-y-z frames.

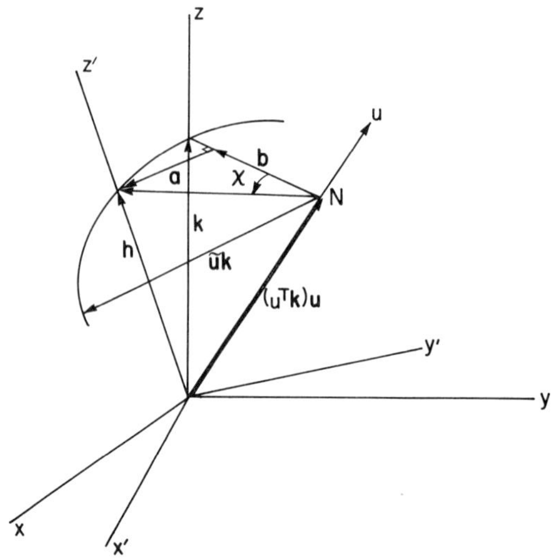

Figure 2.5.10 Relation between **k** and **h**.

Similarly, the vector **b** in Fig. 2.5.10 is

$$\mathbf{b} = [\mathbf{k} - (\mathbf{u}^T\mathbf{k})\mathbf{u}] \cos \chi$$

Thus

$$\begin{aligned}\mathbf{h} &= (\mathbf{u}^T\mathbf{k})\mathbf{u} + \mathbf{b} + \mathbf{a} \\ &= (\mathbf{u}^T\mathbf{k})\mathbf{u} + [\mathbf{k} - (\mathbf{u}^T\mathbf{k})\mathbf{u}] \cos \chi + \tilde{\mathbf{u}}\mathbf{k} \sin \chi\end{aligned}$$

A rearrangement of terms leads to

$$\mathbf{h} = \mathbf{k} \cos \chi + (\mathbf{u}^T\mathbf{k})\mathbf{u}(1 - \cos \chi) + (\tilde{\mathbf{u}}\mathbf{k}) \sin \chi \tag{2.5.42}$$

By means of the trigonometric identities

$$1 - \cos \chi = 2 \sin^2 \frac{\chi}{2}$$

$$\sin \chi = 2 \sin \frac{\chi}{2} \cos \frac{\chi}{2}$$

$$\cos \chi = 2 \cos^2 \frac{\chi}{2} - 1$$

Eq. 2.5.42 can be written in the form

$$\mathbf{h} = \mathbf{k}\left(2 \cos^2 \frac{\chi}{2} - 1\right) + 2\mathbf{u}(\mathbf{u}^T\mathbf{k}) \sin^2 \frac{\chi}{2} + 2\tilde{\mathbf{u}}\mathbf{k} \sin \frac{\chi}{2} \cos \frac{\chi}{2} \tag{2.5.43}$$

and manipulated to yield the desired result

$$\mathbf{h} = \left[\left(2 \cos^2 \frac{\chi}{2} - 1\right)\mathbf{I} + 2\left(\mathbf{u}\mathbf{u}^T \sin^2 \frac{\chi}{2} + \cos \frac{\chi}{2} \sin \frac{\chi}{2} \tilde{\mathbf{u}}\right)\right]\mathbf{k} \tag{2.5.44}$$

Since the geometric relations between **f**-**i** and **g**-**j** are identical to that between **h**-**k**, the relation of Eq. 2.5.44 is valid for these pairs (Prob. 2.5.7); that is,

$$\begin{aligned}\mathbf{f} &= \left[\left(2\cos^2\frac{\chi}{2} - 1\right)\mathbf{I} + 2\left(\mathbf{u}\mathbf{u}^T\sin^2\frac{\chi}{2} + \cos\frac{\chi}{2}\sin\frac{\chi}{2}\,\tilde{\mathbf{u}}\right)\right]\mathbf{i} \\ \mathbf{g} &= \left[\left(2\cos^2\frac{\chi}{2} - 1\right)\mathbf{I} + 2\left(\mathbf{u}\mathbf{u}^T\sin^2\frac{\chi}{2} + \cos\frac{\chi}{2}\sin\frac{\chi}{2}\,\tilde{\mathbf{u}}\right)\right]\mathbf{j}\end{aligned} \tag{2.5.45}$$

From Eq. 2.5.13, using Eqs. 2.5.44 and 2.5.45 and the fact that $[\mathbf{i}, \mathbf{j}, \mathbf{k}] = \mathbf{I}$,

$$\begin{aligned}\mathbf{A} &= [\mathbf{f}, \mathbf{g}, \mathbf{h}] \\ &= \left(2\cos^2\frac{\chi}{2} - 1\right)\mathbf{I} + 2\left(\mathbf{u}\mathbf{u}^T\sin^2\frac{\chi}{2} + \cos\frac{\chi}{2}\sin\frac{\chi}{2}\,\tilde{\mathbf{u}}\right)\end{aligned} \tag{2.5.46}$$

Equation 2.5.46 is a fundamental result that defines the effect of a general rotation of a reference frame relative to another. It will be used directly in Chapters 2 and 3. In Chapter 4 it is used to introduce generalized coordinates of rotation for rigid bodies, which form the foundation of kinematics and dynamics of multibody systems in Chapters 5 and 6.

Example 2.5.4: Let the x'-y'-z' frame of Fig. 2.5.8 have its origin in common with that of the x-y-z reference frame and rotate it about a fixed vector $\mathbf{u} = (1/\sqrt{3})\,[1, 1, 1]^T$, by angle ϕ. The direction cosine transformation matrix, from Eq. 2.5.46, is

$$\mathbf{A} = 2\begin{bmatrix} c^2 + \dfrac{s^2}{3} - \dfrac{1}{2} & \dfrac{s^2}{3} - \dfrac{sc}{\sqrt{3}} & \dfrac{s^2}{3} + \dfrac{sc}{\sqrt{3}} \\ \dfrac{s^2}{3} + \dfrac{sc}{\sqrt{3}} & c^2 + \dfrac{s^2}{3} - \dfrac{1}{2} & \dfrac{s^2}{3} - \dfrac{sc}{\sqrt{3}} \\ \dfrac{s^2}{3} - \dfrac{sc}{\sqrt{3}} & \dfrac{s^2}{3} + \dfrac{sc}{\sqrt{3}} & c^2 + \dfrac{s^2}{3} - \dfrac{1}{2} \end{bmatrix}$$

where $c \equiv \cos\phi/2$ and $s \equiv \sin\phi/2$. Geometrically, a fixed vector $\mathbf{s}'$ that emanates from the origin of the x'-y'-z' frame sweeps out a cone in the x-y-z frame, with axis of revolution $\mathbf{u}$ (Prob. 2.5.8).

2.6 SYSTEMS MODELED AS PARTICLES WITH CONSTRAINED RELATIVE MOTION

As introduced in Chapter 1, models of mechanical systems are often constructed by idealizing bodies with distributed mass as particles with mass concentrated at isolated points. If a total of nb such particles are considered in a Cartesian reference frame, particle P_i is located in space by the vector $\mathbf{r}_i$ from the origin of a Cartesian reference frame to the particle P_i, $i = 1, \ldots, nb$. A *composite vector of Cartesian coordinates* of the system of particles may then be written in column matrix form as

$$\mathbf{r} = [\mathbf{r}_1^T, \ldots, \mathbf{r}_{nb}^T]^T \tag{2.6.1}$$

which is a vector of dimension $3nb$.

Restrictions on positions of the particles that can be written in the form

$$\mathbf{\Phi}(\mathbf{r}, t) \equiv [\Phi_1(\mathbf{r}, t), \ldots, \Phi_m(\mathbf{r}, t)]^T = \mathbf{0} \tag{2.6.2}$$

are called *holonomic constraints*, or simply *kinematic constraints*. If time does not appear explicitly in the constraint equations, these holonomic constraints are called *stationary* (classically called *scleronomic*). Otherwise, they are called *time dependent* (classically called *rheonomic*). Constraints that cannot be written in the form of Eq. 2.6.2 are called *nonholonomic constraints*. Examples of nonholonomic constraints include inequalities and nonintegrable differential constraints.

Differentiating Eq. 2.6.2, using the chain rule of differentiation, yields

$$\mathbf{\Phi}_{\mathbf{r}}\dot{\mathbf{r}} = -\mathbf{\Phi}_t \tag{2.6.3}$$

which is a condition on velocities $\dot{\mathbf{r}}$ that are consistent with holonomic constraints, called the *constraint velocity equation*. Differentiating Eq. 2.6.3 yields the *constraint acceleration equation*

$$\mathbf{\Phi}_{\mathbf{r}}\ddot{\mathbf{r}} = -[(\mathbf{\Phi}_{\mathbf{r}}\dot{\mathbf{r}})_{\mathbf{r}}\dot{\mathbf{r}} + 2\mathbf{\Phi}_{\mathbf{r}t}\dot{\mathbf{r}} + \mathbf{\Phi}_{tt}] \equiv \boldsymbol{\gamma} \tag{2.6.4}$$

where

$$(\mathbf{\Phi}_{\mathbf{r}}\dot{\mathbf{r}})_{\mathbf{r}}\dot{\mathbf{r}} = \sum_{k=1}^{3nb} \mathbf{\Phi}_{\mathbf{r}r_k}\dot{\mathbf{r}}\dot{r}_k = \left[\sum_{j=1}^{3nb}\sum_{k=1}^{3nb} \mathbf{\Phi}_{i_{r_j r_k}}\dot{r}_j\dot{r}_k\right]_{m\times 1}$$

If the *constraint Jacobian* $\mathbf{\Phi}_{\mathbf{r}}$ of a constraint equation has full row rank for $\mathbf{r}^0$ and t^0 that satisfy Eq. 2.6.2, then the coordinate partitioning theorem of Section 2.3 implies that m of the Cartesian coordinates may be determined as functions of the remaining $3nb - m$ Cartesian coordinates. In this case, the system is said to have $3nb - m$ *degrees of freedom*, often abreviated *DOF*, in a neighborhood of $(\mathbf{r}^0, t^0)$. If at any point the constraint Jacobian fails to have full row rank, then at that point the system has an increased number of degrees of freedom. Computational methods for determining the rank of the Jacobian, hence the number of degrees of freedom and candidate sets of independent generalized coordinates, are presented in Appendix B.

Two distinct methods of parameterizing a system may be considered. First, the coordinate partitioning theorem may be used to identify an independent subset of components of the vector $\mathbf{r}$ that are treated as *independent Cartesian coordinates*. Second, a physically different set of generalized coordinates may be defined that locates each particle in the system and satisfies Eq. 2.6.2. That is, if a vector $\mathbf{q}$ of $3nb - m$ generalized coordinates can be found such that all possible positions of particles in the system can be uniquely represented as

$$\mathbf{r} = \mathbf{r}(\mathbf{q}) \tag{2.6.5}$$

and if

$$\mathbf{\Phi}(\mathbf{r}(\mathbf{q}), t) = \mathbf{0} \tag{2.6.6}$$

is satisfied for all $\mathbf{q}$, the generalized coordinates $\mathbf{q}$ are said to be *consistent with constraints*. In order that a unique $\mathbf{q}$ defines $\mathbf{r}$ in Eq. 2.6.5 (i.e., the mapping is one-to-one), the Jacobian $\mathbf{r_q}$ of the mapping of Eq. 2.6.5 must have full column rank. In this case the dimension of $\mathbf{q}$ is the number of degrees of freedom of the system and $\mathbf{q}$ forms a set of *independent generalized coordinates*. If Eq. 2.6.6 is not identically satisfied for all values of $\mathbf{q}$, then $\mathbf{q}$ forms a set of *dependent generalized coordinates*.

Example 2.6.1: An elementary example of a particle that is constrained to move in the Cartesian x-y-z reference frame is the *simple pendulum* shown in Fig. 2.6.1 that rotates about the y axis, with its mass concentrated at point P. Point P is constrained to move in the x-z plane, on a circle of radius 1, by the pivot arrangement shown in Fig. 2.6.1. These restrictions on motion of point P may be characterized by the stationary holonomic constraint equations

$$\mathbf{\Phi}(\mathbf{r}) \equiv \begin{bmatrix} x^2 + z^2 - 1 \\ y \end{bmatrix} = 0$$

The Jacobian of these constraint equations is

$$\mathbf{\Phi_r} = \begin{bmatrix} 2x & 0 & 2z \\ 0 & 1 & 0 \end{bmatrix}$$

Since x and z cannot simultaneously be 0, this Jacobian has full row rank. Therefore, the system has one degree of freedom. At positions for which $z \neq 0$, the coordinate partitioning theorem of Section 2.3 guarantees that z and y can be uniquely determined locally as a function of x. In neighboring positions, x may be selected

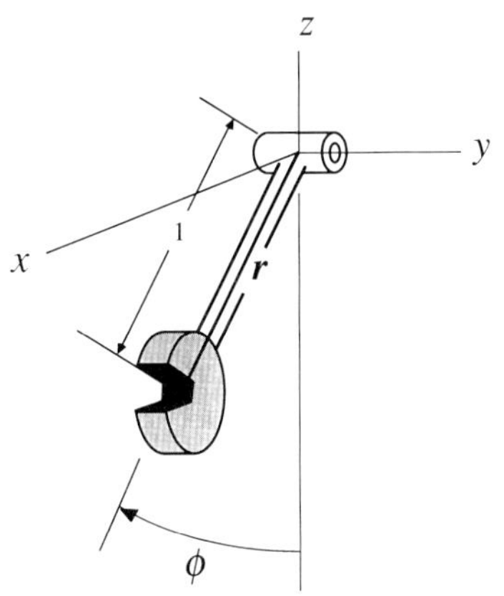

Figure 2.6.1 Simple pendulum.

as the independent generalized coordinate. This representation is not global however, since when the pendulum is horizontal, $z = 0$ and a different partitioning of coordinates is required. If only small amplitude motion of the pendulum about the vertical position is to be analyzed, then $z \neq 0$ and x may be selected as the independent generalized coordinate for analysis.

While no single Cartesian coordinate of point P can serve as an independent generalized coordinate for all positions of the simple pendulum, consider the angle ϕ shown in Fig. 2.6.1 as a generalized coordinate $q = \phi$. In terms of this coordinate,

$$\mathbf{r} = \begin{bmatrix} \sin q \\ 0 \\ -\cos q \end{bmatrix} \tag{2.6.7}$$

An easy calculation shows that the constraints are identically satisfied if the representation of Eq. 2.6.7 is substituted for $\mathbf{r}$, so q is consistent with constraints. Furthermore, the Jacobian of Eq. 2.6.7 is

$$\mathbf{r}_q = \begin{bmatrix} \cos q \\ 0 \\ \sin q \end{bmatrix}$$

Since $\cos q$ and $\sin q$ are not simultaneously zero, this Jacobian has full column rank and q is an independent generalized coordinate that is consistent with constraints. Thus, in this example, an independent generalized coordinate has been found that is valid for all positions and the system has one degree of freedom. Differentiating Eq. 2.6.7 with respect to time yields the velocity and acceleration of the particle, in terms of $\dot{q}$ and $\ddot{q}$, as

$$\begin{aligned} \dot{\mathbf{r}} &= \mathbf{r}_q \dot{q} \\ \ddot{\mathbf{r}} &= \mathbf{r}_q \ddot{q} + \begin{bmatrix} -\sin q \\ 0 \\ \cos q \end{bmatrix} \dot{q}^2 = \mathbf{r}_q \ddot{q} - \mathbf{r} \dot{q}^2 \end{aligned} \tag{2.6.8}$$

➧ **Example 2.6.2:** A two degree of freedom material handling device is shown in Fig. 2.6.2(a), with two booms of unit length. The first body pivots about the inertial z axis. An x'-y'-z' frame is fixed in the first bar at point m_1, where its mass is idealized as a particle, with its x' axis along the first bar and its z' axis parallel to the global z axis. Masses of bodies 1 and 2 are modeled as concentrated at the ends of the booms, as shown in Fig. 2.6.2(b).

In the x'-y'-z' frame, $\mathbf{s}_2' = [0, \sin \psi, -\cos \psi]^T$. From Eq. 2.5.17,

$$\mathbf{s}_2 = \begin{bmatrix} \cos \phi & -\sin \phi & 0 \\ \sin \phi & \cos \phi & 0 \\ 0 & 0 & 1 \end{bmatrix} \begin{bmatrix} 0 \\ \sin \psi \\ -\cos \psi \end{bmatrix} = \begin{bmatrix} -\sin \phi \ \sin \psi \\ \cos \phi \ \sin \psi \\ -\cos \psi \end{bmatrix}$$

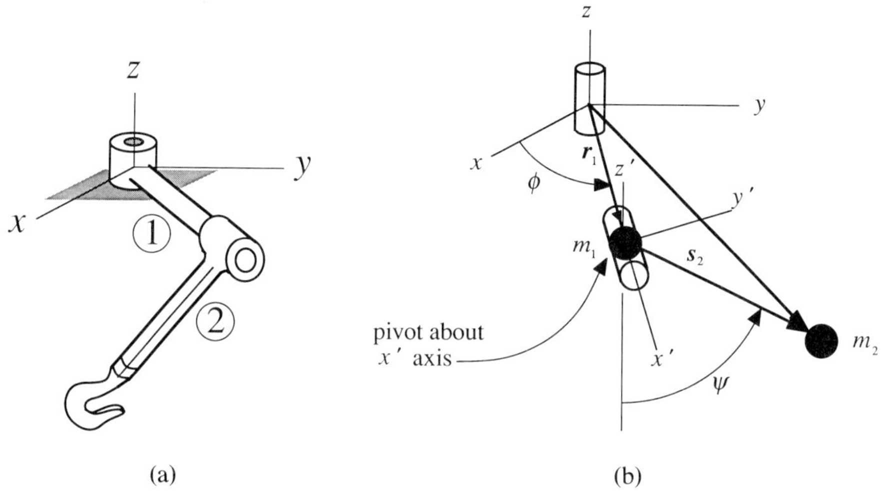

Figure 2.6.2 Two degree of freedom material handling device.

with $\mathbf{q} = [\phi, \psi]^T$,

$$\mathbf{r}_1 = \begin{bmatrix} \cos q_1 \\ \sin q_1 \\ 0 \end{bmatrix}$$

$$\mathbf{r}_2 = \mathbf{r}_1 + \mathbf{s}_2 = \begin{bmatrix} \cos q_1 - \sin q_1 \sin q_2 \\ \sin q_1 + \cos q_1 \sin q_2 \\ -\cos q_2 \end{bmatrix}$$

The composite Cartesian coordinate vector, written in terms of generalized coordinates, is

$$\mathbf{r} = \begin{bmatrix} \mathbf{r}_1 \\ \mathbf{r}_2 \end{bmatrix} = \begin{bmatrix} \cos q_1 \\ \sin q_1 \\ 0 \\ \cos q_1 - \sin q_1 \sin q_2 \\ \sin q_1 + \cos q_1 \sin q_2 \\ -\cos q_2 \end{bmatrix}$$

The Jacobian of this mapping is

$$\mathbf{r_q} = \begin{bmatrix} -\sin q_1 & 0 \\ \cos q_1 & 0 \\ 0 & 0 \\ -\sin q_1 - \cos q_1 \sin q_2 & -\sin q_1 \cos q_2 \\ \cos q_1 - \sin q_1 \sin q_2 & \cos q_1 \cos q_2 \\ 0 & \sin q_2 \end{bmatrix}$$

The first two equations of $\mathbf{r_q}\boldsymbol{\alpha} = \mathbf{0}$ are $-\alpha_1 \sin q_1 = 0 = \alpha_1 \cos q_1$. Since $\sin q_1$ and $\cos q_1$ are not simultaneously zero, $\alpha_1 = 0$. Similarly, the last three equations show that $\alpha_2 = 0$. Thus $\mathbf{r_q}$ has full column rank and $\mathbf{q}$ is an independent set of generalized coordinates.

The Cartesian velocity and acceleration, in terms of $\dot{\mathbf{q}}$ and $\ddot{\mathbf{q}}$, are obtained by differentiation as

$$\dot{\mathbf{r}} = \mathbf{r_q}\dot{\mathbf{q}}$$

$$\ddot{\mathbf{r}} = \mathbf{r_q}\ddot{\mathbf{q}} + \begin{bmatrix} -\cos q_1\dot{q}_1^2 \\ -\sin q_1\dot{q}_1^2 \\ 0 \\ -(\cos q_1 - \sin q_1 \sin q_2)\dot{q}_1^2 \\ -2\cos q_1 \cos q_2\dot{q}_1\dot{q}_2 + \sin q_1 \sin q_2\dot{q}_2^2 \\ -(\sin q_1 + \cos q_1 \sin q_2)\dot{q}_2^2 \\ -2\sin q_1 \cos q_2\dot{q}_1\dot{q}_2 - \cos q_1 \sin q_1\dot{q}_2^2 \\ \cos q_2\dot{q}_2^2 \end{bmatrix}$$

Example 2.6.3: A three degree of freedom material handling device is shown in Fig. 2.6.3(a). Body 1 is pivoted about the global z axis, as in Example 2.6.2. Body 2 slides along and rotates about the x_1' axis in body 1. The vector that locates the endpoint P in the x_2'-y_2'-z_2' frame is $\mathbf{s}_2'^P = [1, 1, 1]^T$. Masses of bodies 1 and 2 are modeled as concentrated at the center of body 1 and at point P on body 2, as shown in Fig. 2.6.3(b).

Generalized coordinates for this two particle system are $\mathbf{q} = [\phi, \alpha, \theta]^T$. To

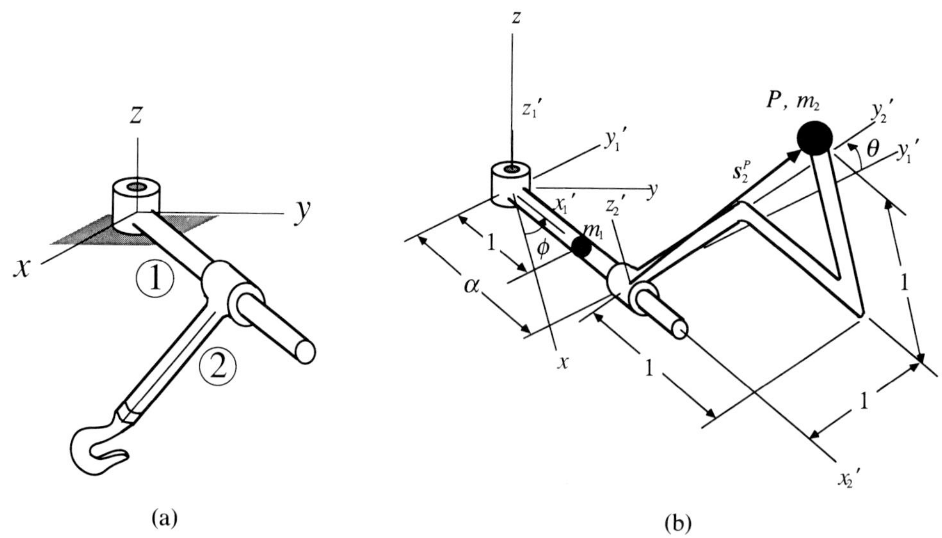

Figure 2.6.3 Three degree of freedom material handling device.

locate the particles in space, the transformation matrix from the x_1'-y_1'-z_1' frame to the x-y-z frame is

$$\mathbf{A}_1 = \begin{bmatrix} \cos q_1 & \sin q_1 & 0 \\ \sin q_1 & \cos q_1 & 0 \\ 0 & 0 & 1 \end{bmatrix}$$

and the transformation matrix from the x_2'-y_2'-z_2' frame to the x_1'-y_1'-z_1' frame is

$$\mathbf{A}_{12} = \begin{bmatrix} 1 & 0 & 0 \\ 0 & \cos q_3 & -\sin q_3 \\ 0 & \sin q_3 & \cos q_3 \end{bmatrix}$$

Vectors from the origin of the x-y-z frame to masses m_1 and m_2 are

$$\mathbf{r}_1 = \mathbf{A}_1 \begin{bmatrix} 1 \\ 0 \\ 0 \end{bmatrix} = \begin{bmatrix} \cos q_1 \\ \sin q_1 \\ 0 \end{bmatrix}$$

$$\begin{aligned} \mathbf{r}_2 &= q_2\mathbf{r}_1 + \mathbf{s}_2^P \\ &= q_2\mathbf{r}_1 + \mathbf{A}_1\mathbf{A}_{12}\begin{bmatrix} 1 \\ 1 \\ 1 \end{bmatrix} \\ &= \begin{bmatrix} (q_2 + 1)\cos q_1 - \sin q_1 \cos q_3 - \sin q_3) \\ (q_2 + 1)\sin q_1 + \cos q_1 (\cos q_3 - \sin q_3) \\ \sin q_3 - \cos q_3 \end{bmatrix} \end{aligned}$$

The composite Cartesian coordinate vector, in terms of $\mathbf{q}$, is

$$\mathbf{r} = \begin{bmatrix} \mathbf{r}_1 \\ \mathbf{r}_2 \end{bmatrix} = \begin{bmatrix} \cos q_1 \\ \sin q_1 \\ 0 \\ (q_2 + 1)\cos q_1 - \sin q_1 (\cos q_3 - \sin q_3) \\ (q_2 - 1)\sin q_1 + \cos q_1 (\cos q_3 - \sin q_3) \\ \sin q_3 + \cos q_3 \end{bmatrix}$$

The Jacobian of this mapping is

$$\mathbf{r_q} = \begin{bmatrix} -\sin q_1 & 0 & 0 \\ \cos q_1 & 0 & 0 \\ 0 & 0 & 0 \\ -(q_2 + 1)\sin q_1 - \cos q_1(\cos q_3 - \sin q_3) & \cos q_1 & \sin q_1(\sin q_3 + \cos q_3) \\ (q_2 + 1)\cos q_1 - \sin q_1(\cos q_3 - \sin q_3) & \sin q_1 & -\cos q_1(\sin q_3 + \cos q_3) \\ 0 & 0 & \cos q_3 - \sin q_3 \end{bmatrix}$$

The first two equations of $\mathbf{r_q}\boldsymbol{\alpha} = \mathbf{0}$ are $-\alpha_1 \sin q_1 = 0 = \alpha_1 \cos q_1$, so $\alpha_1 = 0$. Supressing α_1, the determinant of the coefficient matrix in the third and fourth equations for α_2 and α_3 is $-(\sin q_3 + \cos q_3)$. Thus unless this quantity is zero, $\alpha_2 = \alpha_3 = 0$. In the special case $\sin q_3 + \cos q_3 = 0$, $\cos q_3 - \sin q_3 \neq 0$, so the last equation of $\mathbf{r_q}\boldsymbol{\alpha} = \mathbf{0}$ implies that $\alpha_3 = 0$ and the third and fourth equations reduce to $\alpha_2 \cos q_1 = 0 = \alpha_2 \sin q_1$, so $\alpha_2 = 0$. Thus, for all $\mathbf{q}$, $\boldsymbol{\alpha} = \mathbf{0}$ and the Jacobian $\mathbf{r_q}$ has full column rank. Thus $\mathbf{q} = [\phi, \alpha, \theta]^T$ is a vector of independent generalized coordinates.

By direct differentiation, the Cartesian velocity and acceleration, in terms of $\dot{\mathbf{q}}$ and $\ddot{\mathbf{q}}$, are

$$\dot{\mathbf{r}} = \mathbf{r_q}\dot{\mathbf{q}}$$

$$\ddot{\mathbf{r}} = \mathbf{r_q}\ddot{\mathbf{q}} + \begin{bmatrix} -\cos q_1 \dot{q}_1^2 \\ -\sin q_1 \dot{q}_1^2 \\ 0 \\ -(q_2 + 1)\cos q_1 \dot{q}_1^2 + \sin q_1(\cos q_3 - \sin q_3)(\dot{q}_1^2 + \dot{q}_3^2) \\ -2\sin q_1 \dot{q}_2 \dot{q}_1 + 2\cos q_1(\sin q_3 + \cos q_3)\dot{q}_3\dot{q}_1 \\ -(q_2 + 1)\sin q_1 \dot{q}_1^2 - \cos q_1(\cos q_3 - \sin q_3)(\dot{q}_3^2 + \dot{q}_1^2) \\ +2\cos q_1 \dot{q}_2 \dot{q}_1 + 2\sin q_1(\cos q_3 + \sin q_3)\dot{q}_1\dot{q}_3 \\ -(\sin q_3 + \cos q_3)\dot{q}_3^2 \end{bmatrix}$$

Example 2.6.4: A pair of points P_1 and P_2 may be considered as two particles that make up a *dumbbell* with a bar of unit length connecting them, as shown in Fig. 2.6.4 and studied earlier in Example 1.3.1. The composite vector of Cartesian coordinates $\mathbf{r} = [\mathbf{r}_1^T, \mathbf{r}_2^T]^T$ must satisfy the condition that the distance between points remains unity; that is,

$$\begin{aligned} \Phi(\mathbf{r}) &= (\mathbf{r}_2 - \mathbf{r}_1)^T(\mathbf{r}_2 - \mathbf{r}_1) - 1 \\ &= (x_2 - x_1)^2 + (y_2 - y_1)^2 - (z_2 - z_1)^2 - 1 = 0 \end{aligned} \tag{2.6.9}$$

The Jacobian of this holonomic constraint equation with respect to Cartesian coordinates is

$$\begin{aligned} \Phi_{\mathbf{r}} &= [\Phi_{\mathbf{r}_1}, \Phi_{\mathbf{r}_2}] = 2[-(\mathbf{r}_2 - \mathbf{r}_1)^T, (\mathbf{r}_2 - \mathbf{r}_1)^T] \\ &= 2[-(x_2 - x_1), -(y_2 - y_1), -(z_2 - z_1), \\ &\quad (x_2 - x_1), (y_2 - y_1), (z_2 - z_1)] \end{aligned} \tag{2.6.10}$$

Since all entries in this row matrix are not simultaneously zero, it has full row rank. By the coordinate partitioning theorem, provided that $x_2 \neq x_1$, either x_2 or x_1 may be selected as dependent. Similar reasoning holds for y_2 and y_1 and for z_2 and z_1. However, since each of the three coordinate differences is zero in some position of

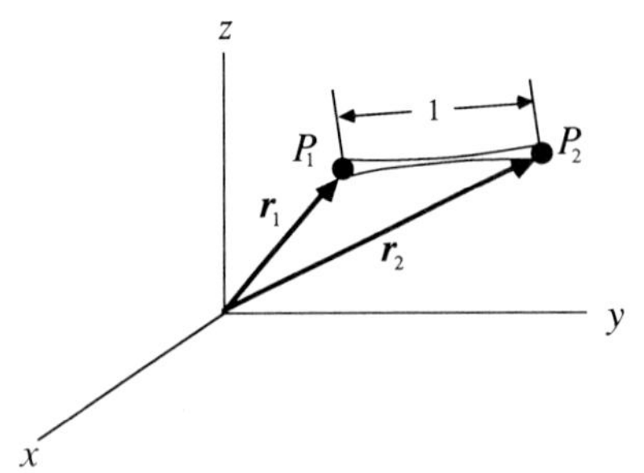

Figure 2.6.4 Dumbbell.

the dumbbell, no single Cartesian coordinate may be selected as a globally independent Cartesian coordinate.

As an alternative, consider using the pair of angles q_1 and q_2 shown in Fig. 2.6.5 to define the orientation of the vector from point P_1 to P_2. Defining the components of $\mathbf{r}_1$ as generalized coordinates q_3 through q_5, the mapping from generalized coordinates $\mathbf{q} \equiv [q_1, \ldots, q_5]^T$ to the composite vector of Cartesian coordinates is

$$\mathbf{r}(\mathbf{q}) \equiv \begin{bmatrix} \mathbf{r}_1(\mathbf{q}) \\ \mathbf{r}_2(\mathbf{q}) \end{bmatrix} = \begin{bmatrix} q_3 \\ q_4 \\ q_5 \\ q_3 + \cos q_2 \cos q_1 \\ q_4 + \cos q_2 \sin q_1 \\ q_5 + \sin q_2 \end{bmatrix} \tag{2.6.11}$$

A direct calculation verifies that this mapping identically satisfies the constraint equation of Eq. 2.6.9, so $\mathbf{q}$ is consistent with constraints. It may therefore be considered for analysis of the system.

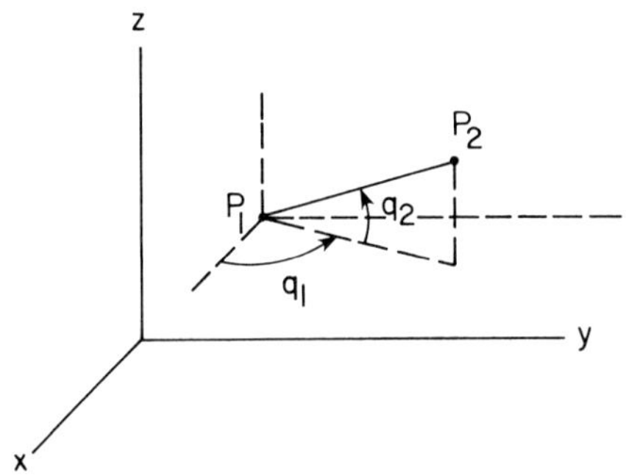

Figure 2.6.5 Orientation angles for dumbbell.

To determine whether the mapping of Eq. 2.6.11 is nonsingular, consider its Jacobian,

$$\mathbf{r_q} = \begin{bmatrix} 0 & 0 & 1 & 0 & 0 \\ 0 & 0 & 0 & 1 & 0 \\ 0 & 0 & 0 & 0 & 1 \\ -\cos q_2 \times \sin q_1 & -\sin q_2 \times \cos q_1 & 1 & 0 & 0 \\ \cos q_2 \times \cos q_1 & -\sin q_2 \times \sin q_1 & 0 & 1 & 0 \\ 0 & \cos q_2 & 0 & 0 & 1 \end{bmatrix} \tag{2.6.12}$$

Note that when $q_2 = \pm\pi/2$, the first column is zero. Hence in this vertical position of the dumbbell, the generalized coordinate mapping of Eq. 2.6.11 becomes singular and difficulties will arise in analysis. As long as this singular point is avoided, $\mathbf{q}$ forms an independent set of generalized coordinates that are consistent with constraints.

Although independent generalized coordinates are desirable, as in Example 2.6.4, it is not always practical or even possible to define independent generalized coordinates (i.e., such that all constraints on $\mathbf{r}$ in Eq. 2.6.6 are satisfied for all $\mathbf{q}$, where $\mathbf{r} = \mathbf{r}(\mathbf{q})$). Thus consider the case in which nh holonomic constraints of the form

$$\mathbf{\Phi}(\mathbf{q}, t) = \mathbf{\Phi}(\mathbf{r}(\mathbf{q}), t) = [\Phi_1(\mathbf{r}(\mathbf{q}), t), \ldots, \Phi_{nh}(\mathbf{r}(\mathbf{q}), t)]^T = \mathbf{0} \quad (2.6.13)$$

are not identically satisfied by all values of $\mathbf{q}$ (i.e., the generalized coordinates are dependent). Using the chain rule of differentiation, the Jacobian of Eq. 2.6.13 with respect to $\mathbf{q}$ is

$$\mathbf{\Phi_q} = \mathbf{\Phi_r r_q} \quad (2.6.14)$$

It must have full row rank if the constraints of Eq. 2.6.13 on $\mathbf{q}$ are independent. If $\mathbf{\Phi_q}$ has full row rank, the coordinate partitioning theorem implies that there are $k - nh$ independent generalized coordinates, where $\mathbf{q} = [q_1, \ldots, q_k]^T$. In this case, the system has $k - nh$ degrees of freedom.

Constraint velocity equations and *constraint acceleration equations* that must be satisfied by $\dot{\mathbf{q}}$ and $\ddot{\mathbf{q}}$, analogous to Eqs. 2.6.3 and 2.6.4 in $\dot{\mathbf{r}}$ and $\ddot{\mathbf{r}}$, are obtained by differentiating Eq. 2.6.14 with respect to time, using the chain rule of differentiation, to obtain

$$\begin{aligned} \mathbf{\Phi_q}\dot{\mathbf{q}} &= -\mathbf{\Phi}_t \\ \mathbf{\Phi_q}\ddot{\mathbf{q}} &= -[(\mathbf{\Phi_q}\dot{\mathbf{q}})_{\mathbf{q}}\dot{\mathbf{q}} + 2\mathbf{\Phi}_{\mathbf{q}t}\dot{\mathbf{q}} + \mathbf{\Phi}_{tt}] \equiv \boldsymbol{\gamma} \end{aligned} \quad (2.6.15)$$

where $\mathbf{\Phi_q}$ is defined in Eq. 2.6.14 and

$$\mathbf{\Phi}_{\mathbf{q}t} = \mathbf{\Phi}_{\mathbf{r}t}\mathbf{r_q} \quad (2.6.16)$$

$$\begin{aligned} (\mathbf{\Phi_q}\dot{\mathbf{q}})_{\mathbf{q}}\dot{\mathbf{q}} &= (\mathbf{\Phi_r r_q}\dot{\mathbf{q}})_{\mathbf{q}}\dot{\mathbf{q}} \quad (2.6.17) \\ &= \mathbf{\Phi_r}(\mathbf{r_q}\dot{\mathbf{q}})_{\mathbf{q}}\dot{\mathbf{q}} + (\mathbf{\Phi_r r_q}\dot{\mathbf{q}})_{\mathbf{r}}\mathbf{r_q}\dot{\mathbf{q}} \end{aligned}$$

where $\mathbf{r_q}$ depends on $\mathbf{q}$, but $\dot{\mathbf{q}}$ is treated as constant for partial differentiation in the first term on the right of Eq. 2.6.17. For the second term on the right of Eq. 2.6.17, $\mathbf{\Phi_r}$ depends on $\mathbf{r}$, but $\mathbf{r_q}$ and $\dot{\mathbf{q}}$ are treated as constants for partial differentiation.

To be more explicit regarding calculation of terms on the right of Eq. 2.6.17, note that the term $(\mathbf{r_q}\dot{\mathbf{q}})_{\mathbf{q}}\dot{\mathbf{q}}$ is given in Eq. 2.2.18. The partial derivative in the second term on the right of Eq. 2.6.17 can be expanded as

$$(\mathbf{\Phi_r\, r_q}\dot{\mathbf{q}})_{\mathbf{r}} = [\mathbf{\Phi}_{\mathbf{r}r_j}\mathbf{r_q}\dot{\mathbf{q}}]_{nh \times 3nb} \quad (2.6.18)$$

Example 2.6.5: The two particle model of a slider–crank mechanism, studied in Example 1.3.2 and shown in Fig. 2.6.6, permits rotation of the crank that connects point P_1 with the origin of the Cartesian reference frame about the z axis. Point P_2 is constrained to be a distance l from point P_1 and to slide along the x axis of the Cartesian reference frame.

Much as in the simple pendulum of Example 2.6.1, the angle $\phi = q_1$ of rotation of the first link may be selected as a generalized coordinate. Similarly, the x coordinate of point P_2 may be selected as a generalized coordinate $x_2 = q_2$. The composite

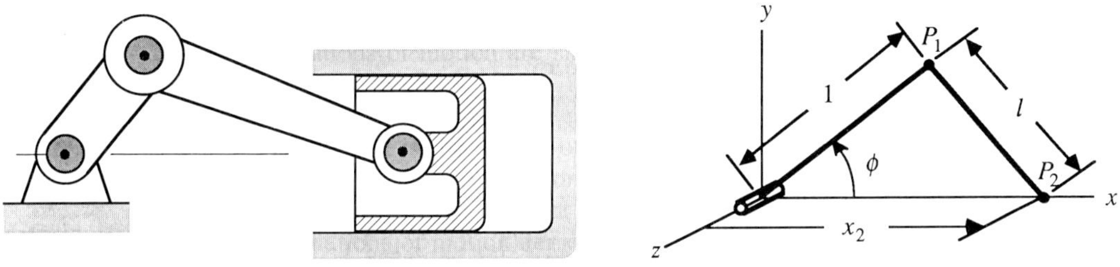

Figure 2.6.6 Two particle model of slider–crank mechanism.

Cartesian coordinate vector of the system may thus be written as

$$\mathbf{r}(\mathbf{q}) \equiv \begin{bmatrix} \mathbf{r}_1(\mathbf{q}) \\ \mathbf{r}_2(\mathbf{q}) \end{bmatrix} = [\cos q_1, \sin q_1, 0, q_2, 0, 0]^T$$

Since the distance between points P_1 and P_2 must be l, these generalized coordinates are not independent, but must satisfy

$$\begin{aligned} \Phi(\mathbf{r}(\mathbf{q})) &= (\mathbf{r}_2(\mathbf{q}) - \mathbf{r}_1(\mathbf{q}))^T(\mathbf{r}_2(\mathbf{q}) - \mathbf{r}_1(\mathbf{q})) - l^2 \\ &= (q_2 - \cos q_1)^2 + \sin^2 q_1 - l^2 = 0 \end{aligned}$$

The Jacobian of this constraint equation with respect to the generalized coordinate vector $\mathbf{q}$ is

$$\Phi_{\mathbf{q}} = 2[q_2 \sin q_1, (q_2 - \cos q_1)]$$

If $l > 1$, then, provided that the mechanism is assembled as shown in Fig. 2.6.6, point P_2 will always lie to the right of point P_1, so $q_2 - \cos q_1 > 0$ and $\mathbf{\Phi_q}$ has full row rank. In this case, $q_1 = \phi$ may be selected as the independent generalized coordinate and q_2 will be determined uniquely from the constraint equation. This is the usual situation with slider–crank mechanisms that are found in internal combustion engines and machine tools.

If $l < 1$, the second column of the constraint Jacobian will be zero when the line between points P_1 and P_2 becomes vertical. At this position, the constraint Jacobian with respect to q_2 becomes singular and q_2 may not be uniquely determined as a function of q_1. To see that this is not just a mathematical curiosity, Fig. 2.6.7 shows such a position. If an attempt is made to rotate the crank in a counterclockwise

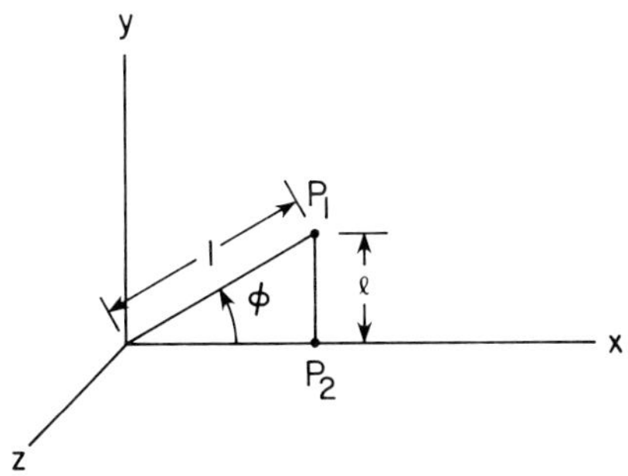

Figure 2.6.7 Lock-up position of slider–crank.

direction from the configuration shown in Fig. 2.6.7, as shown in Fig. 2.6.8(a), no solution of the constraint equation exists. This is because the mechanism has *locked-up* and cannot continue to turn, which is reflected in the mathematical loss of existence of a solution for q_2. Conversely, if the crank is rotated clockwise from the position shown in Fig. 2.6.7, as shown in Fig. 2.6.8(b), the coupler between points P_1 and P_2 could rotate either clockwise or counterclockwise. Hence uniqueness of solution for q_2 is lost. As discussed in Section 2.3, this is a typical consequence of the situation that occurs when hypotheses of the implicit function theorem are not satisfied. Note that this is not just a mathematical curiosity; it is associated with physical failure of the machine to perform a desired function.

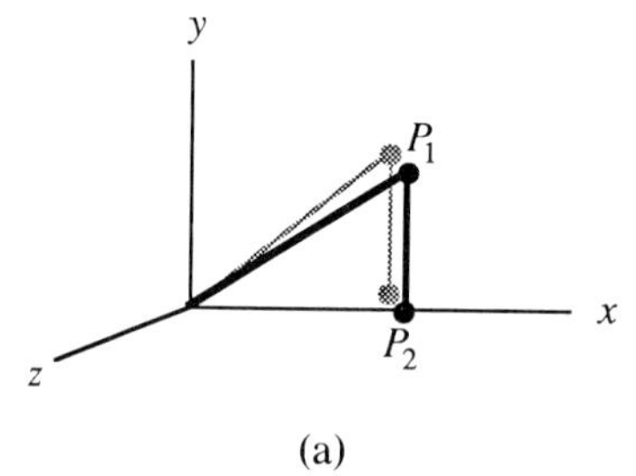

Figure 2.6.8 Behavior at lock-up of slider–crank.

It is interesting to note that if $l < 1$, then in the lock-up position of Fig. 2.6.7, the first column of the Jacobian is nonzero and the implicit function theorem guarantees that $q_1 = \phi$ may be selected as the dependent coordinate and determined uniquely as a function of $q_2 = x_2$, at least locally. This suggests that if control is awarded to q_2, then behavior of the system in controlling the position of P_2 is reasonable.

The complexity of the lock-up that occurs when $l < 1$ pales in light of the pathological behavior of the mechanism when $l = 1$. With $l = 1$, the slider–crank can reach the configuration shown in Fig. 2.6.9(a) in which point P_2 coincides with the origin of the reference frame (i.e., when $q_1 = \phi = \pi/2$). For $q_1 > \pi/2$, either of the configurations shown in Fig. 2.6.9(b) and (c) can occur. If the configuration of Fig. 2.6.9(b) occurs, particle P_2 moves to the left, as an apparent natural continuation

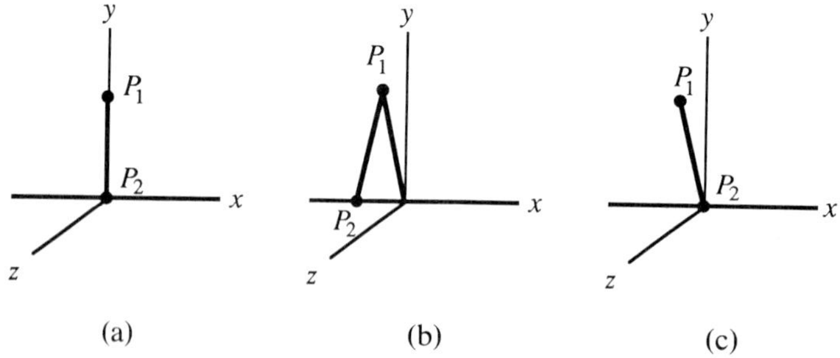

Figure 2.6.9 Behavior of slider–crank with $l = 1$.

of motion to the left if $\dot{q}_1 = \dot{\phi}_1 > 0$. Conversely, P_2 could remain at the origin of the x-y-z frame as q_1 continues to increase beyond the value $\pi/2$, in which case the two bars remain superposed and can only separate when $q_1 = \pi/2$ or $3\pi/2$. It is enlightening to answer the following question: Who is responsible for this mess, the engineer who designed the mechanism or the analyst who is studying it? In this case the problem is caused by the engineer. This is simply a bad design (Prob. 2.6.1).

The reader may have noted that in Examples 2.6.4 and 2.6.5, the generalized coordinates selected were not independent, as they had been in all previous examples. Although it may be theoretically possible to reduce the dimension of the generalized coordinate vector to the number of degrees of freedom of the system (i.e., to find independent generalized coordinates), it is often impractical or undesirable to do so. In such cases, analysis may be carried out using a set of *dependent generalized coordinates* and the associated holonomic constraint equations.

2.7 DIFFERENTIAL NONHOLONOMIC CONSTRAINTS

Occasionally, kinematic constraints are naturally specified in terms of particle velocities, for example, in the form of a scalar first order differential equation

$$\mathbf{b}^T(\mathbf{r}, t)\dot{\mathbf{r}} + c(\mathbf{r}, t) = 0 \tag{2.7.1}$$

or as the equivalent *differential constraint*

$$\mathbf{b}^T(\mathbf{r}, t)\, d\mathbf{r} + c(\mathbf{r}, t)\, dt = 0 \tag{2.7.2}$$

whose left side is called a *differential form* in the variables $\mathbf{r}$ and t.

Three natural questions arise in consideration of a constraint equation of the form of Eq. 2.7.2:

1. Is its differential form an *exact differential*; that is, is it the differential of some function $\Phi(\mathbf{r}, t)$ such that $d\Phi \equiv \Phi_{\mathbf{r}}\, d\mathbf{r} + \Phi_t\, dt = \mathbf{b}^T\, d\mathbf{r} + c\, dt$? If it is, then Eq. 2.7.2 is equivalent to the holonomic constraint $\Phi(\mathbf{r}, t) =$ constant.
2. Can an *integrating factor* $g(\mathbf{r}, t)$ be found so that $g(\mathbf{r}, t)$ times the differential form in Eq. 2.7.2 is an exact differential? If the answer is yes, then Eq. 2.7.2 is said to be an *integrable differential constraint* and is thus equivalent to a holonomic constraint.
3. Even if the differential is exact, or if the constraint is integrable, can a closed form solution $\Phi(\mathbf{r}, t) = 0$ be found that can be used as a holonomic constraint?

Classically, a constraint of the form of Eqs. 2.7.1 or 2.7.2 is called a *nonholonomic constraint* if it is not integrable. From a practical point of view, it makes little difference whether an integral exists or not, if its closed form cannot easily be found. Formulations that treat differential constraints of the form of Eq. 2.7.2 are presented in this book and are applicable whether or not the constraint is integrable. In this book, however, the historical definition of a nonholonomic constraint as a nonintegrable differential constraint will be preserved. A differential constraint for which an integral exists, but its closed form cannot easily be found, will be referred to as a differential constraint.

To simplify the statement of conditions for exactness of differential forms and integrability of differential constraints, let $\mathbf{u} = [u_1, \ldots, u_k]^T$ be a k-vector of variables and consider the differential constraint

$$\sum_{i=1}^{k} f_i(\mathbf{u})\, du_i = 0 \tag{2.7.3}$$

The following analytical condition from multivariable calculus determines whether a differential form with two or more variables is exact [1, 3, 4]:

Exact Differential Theorem: The differential form on the left side of Eq. 2.7.3, with continuously differentiable coefficients in two or more variables, is exact if and only if

$$\frac{\partial f_i}{\partial u_j} = \frac{\partial f_j}{\partial u_i}, \qquad i, j = 1, \ldots, n \tag{2.7.4}$$

for all values of $\mathbf{u}$.

Example 2.7.1: The differential form

$$u_2\, du_1 + du_2 - du_3$$

is not exact, since

$$\frac{\partial f_1}{\partial u_2} = 1 \neq 0 = \frac{\partial f_2}{\partial u_1}$$

There is no need to test the remaining conditions of Eq. 2.7.4, since if any fails, the differential form is not exact.

Example 2.7.2: The differential form

$$2u_2u_3\, du_1 + u_1u_3\, du_2 + u_1u_2\, du_3$$

is not exact, since

$$\frac{\partial f_1}{\partial u_2} = 2u_3 \neq u_3 = \frac{\partial f_2}{\partial u_1}$$

The statement of conditions for integrability of a differential constraint is more complex [1, 3, 4].

Integrable Differential Constraint Theorem: The differential constraint of Eq. 2.7.3 in three or more variables, with continuously differentiable coefficients $f_i(\mathbf{u})$, is integrable if and only if

$$f_\gamma\left(\frac{\partial f_\beta}{\partial u_\alpha} - \frac{\partial f_\alpha}{\partial u_\beta}\right) + f_\beta\left(\frac{\partial f_\alpha}{\partial u_\gamma} - \frac{\partial f_\gamma}{\partial u_\alpha}\right) + f_\alpha\left(\frac{\partial f_\gamma}{\partial u_\beta} - \frac{\partial f_\beta}{\partial u_\gamma}\right) = 0 \qquad (2.7.5)$$

for all α, β, and γ and for all values of $\mathbf{u}$. If the differential constraint of Eq. 2.7.3 satisfies Eq. 2.7.5, there exists an *integrating factor* $g(\mathbf{u}) \neq 0$ such that the differential form

$$\sum_{i=1}^{k} g(\mathbf{u}) f_i(\mathbf{u})\, du_i$$

is exact.

The reason for restricting consideration to three or more variables in the integrable differential constraint theorem is that the differential constraint of Eq. 2.7.2 in two variables r and t with $b(r, t)$ and $c(r, t)$ continuously differentiable is always integrable. To see this, consider the differential constraint of Eq. 2.7.2, with r a scalar variable. Provided that $b(r, t) \neq 0$, Eq. 2.7.2 can be written as

$$\frac{dr}{dt} = -\frac{c(r, t)}{b(r, t)}$$

Since the right side of this first order differential equation is continuously differentiable, there exists a solution (see Appendix C)

$$r = f(t)$$

The differential constraint can thus be reduced to a functional relation, whose differential yields the differential constraint.

For $k = 3$, there is only one condition to evaluate in Eq. 2.7.5. For $k > 3$, however, there are $k(k - 1)(k - 2)/6$ conditions, which becomes oppressive for k large. Even for $k = 4$, there are four conditions.

♦ **Example 2.7.3:** Applying the test of Eq. 2.7.5 to a differential constraint that is defined by the differential form of Example 2.7.1,

$$(-1)(0 - 1) + (0 - 0) + u_2(0 - 0) = 1 \neq 0$$

Thus the differential constraint is not integrable (i.e., it is nonholonomic).

♦ **Example 2.7.4:** Applying the test of Eq. 2.7.5 to the differential form of Example 2.7.2 yields

$$u_1u_2(u_3 - 2u_3) + u_1u_3(2u_2 - u_2) + 2u_2u_3(u_1 - u_1) = -u_1u_2u_3 + u_1u_2u_3 \equiv 0$$

Thus a differential constraint defined by the differential form of Example 2.7.2 is integrable. It remains, however, to find an integrating factor, in order to take practical advantage of this information.

To see that $g(\mathbf{u}) = u_1$ is an integrating factor, first multiply through the differential constraint defined by the differential form of Example 2.7.2 by u_1, to obtain

$$2u_1u_2u_3\,du_1 + u_1^2u_3\,du_2 + u_1^2u_2\,du_3 = 0$$

Direct application of the exact differential theorem confirms that the differential form on the left is exact. By inspection, this differential constraint is the differential of the holonomic constraint

$$u_1^2u_2u_3 = c$$

where the constant c is determined by the initial value u^0 of u (i.e., $c = (u_1^0)^2u_2^0u_3^0$).

Even if the integrable differential constraint theorem verifies that a differential constraint is integrable, in general it is very difficult to find an integrating factor and to find the closed-form expression for the associated holonomic constraint. Methods of integrating differential constraints are presented in Ref. 4.

➧ **Example 2.7.5:** A missile that flies near the ground has a guidance system that is powerful enough to keep the missile's velocity vector aimed at a target that moves along the x axis of a horizontal plane, as shown in Fig. 2.7.1(a). The missile is modeled as a particle P that is constrained to move in the x-y plane in such a way that its velocity is directed toward a point $[\xi(t), 0, 0]^T$ that moves along the x axis, where $\dot{\xi}(t) \neq 0$, as shown in Fig. 2.7.1(b).

Constraints on the vector $\mathbf{r}$ that locates point P are

$$z = 0$$

and

$$\dot{\mathbf{r}} = k\left(\mathbf{r} - \begin{bmatrix} \xi \\ 0 \\ 0 \end{bmatrix}\right)$$

where k is a negative constant. Multiplying the first of the velocity equations above by y and the second by $x - \xi(t)$ and subtracting yields

$$y\dot{x} - (x - \xi(t))\dot{y} = 0 \tag{2.7.6}$$

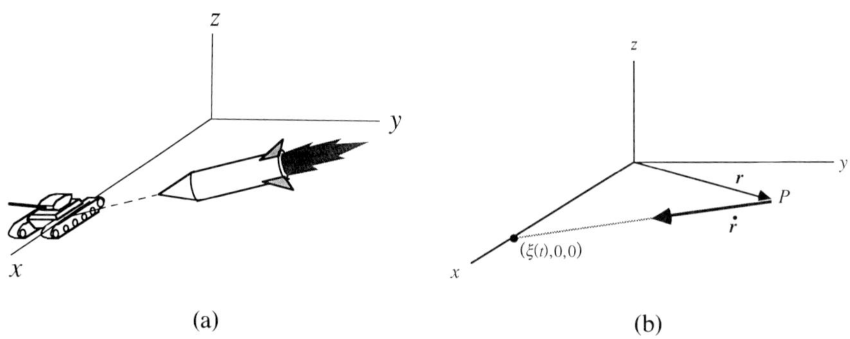

Figure 2.7.1 Point P pursuing $(\xi(t), 0, 0)$.

or the differential constraint

$$y\,dx + (\xi(t) - x)\,dy = 0 \tag{2.7.7}$$

Since $c = 0$ and $b_2 = \xi(t) - x$ in Eq. 2.7.2,

$$\frac{\partial b_2}{\partial t} = \dot{\xi} \neq 0 = \frac{\partial c}{\partial y}$$

Thus the differential form in the differential constraint of Eq. 2.7.7 is not exact.

Applying the test of Eq. 2.7.5 to the differential constraint of Eq. 2.7.7 yields

$$0 - (\xi(t) - x)(0 - 0) + y(0 + \dot{\xi}) = y\dot{\xi} \neq 0$$

Thus it is not integrable (i.e., it is nonholonomic).

If the differential form on the left side of Eq. 2.7.2 is exact or if the differential constraint is integrable, then all curves $\mathbf{r}(t)$ that satisfy Eq. 2.7.2 are on a surface of the form [4]

$$\psi(\mathbf{r}, t) = d$$

where d is a constant of integration. Thus if a differential constraint is integrable and if $\mathbf{r}(0) = \mathbf{r}^0$ is given, only curves on the surface

$$\psi(\mathbf{r}(t), t) = \psi(\mathbf{r}^0, 0)$$

are solutions. At time t_1, then, any point $\mathbf{r}^1$ for which $\psi(\mathbf{r}^1, t_1) \neq \psi(\mathbf{r}^0, 0)$ cannot be reached by moving along a curve that satisfies Eq. 2.7.2 and passes through point $\mathbf{r}^0$ at $t = 0$.

Conversely, it is shown in Ref. 4 (see Example 2.7.6) that if the differential constraint of Eq. 2.7.2 is not integrable (i.e., if it is a nonholonomic constraint), any point $\mathbf{r}^1$ can be reached at any time t_1 by passing along a curve that satisfies Eq. 2.7.2 and passes through $\mathbf{r}^0$ at $t = 0$, for any $\mathbf{r}^0$. Thus the collection of trajectories $\mathbf{r}(t)$ that satisfy a nonholonomic constraint fill the entire $\mathbf{r}$-space with trajectories that pass through a given point $\mathbf{r}^0$ at $t = 0$. From a purely kinematic point of view, therefore, a nonholonomic differential constraint admits a much larger family of solutions than an integrable differential constraint. The reason for this distinction is roughly the distinction between the character of solutions of ordinary and partial differential equations [4]. An integrable differential constraint can be reduced to a first order ordinary differential equation, whose general solution has a single constant of integration. In contrast, a nonholonomic differential constraint is a proper partial differential equation, whose general solution contains an arbitrary function, rather than an arbitrary constant (i.e., a much larger family of solutions).

Example 2.7.6: Consider the differential constraint

$$dx - y\,dt = 0$$

Applying the integrability test of Eq. 2.7.5,

$$-y(0 - 0) + 0 + (-1 - 0) = -1 \neq 0$$

so the differential constraint is not integrable. To see that any point $[x_1, y_1]^T$ can be reached at time t_1 from the point $[x_0, y_0]^T = 0$, consider the path

$$x(t) = f(t), \qquad y(t) = f'(t)$$

where f is any function that has three continuous derivatives. These expressions may be substituted into the differential constraint, to obtain

$$f'\,dt - f'\,dt \equiv 0$$

for any function f. Thus a function f is sought such that

$$f(0) = 0, \qquad f'(0) = 0$$

and

$$f(t_1) = x_1, \qquad f'(t_1) = y_1$$

A direct calculation of derivatives shows that

$$f(t) = (3x_1 - y_1t_1)\left(\frac{t}{t_1}\right)^2 - (2x_1 - y_1t_1)\left(\frac{t}{t_1}\right)^3 + At^2(t_1 - t)^2 + B\sin^2\frac{\pi t}{t_1}$$

satisfies these conditions for any constants A and B.

A general system of $l \leqslant 3nb$ differential constraints may be written as

$$\mathbf{B}(\mathbf{r}, t)\,d\mathbf{r} + \mathbf{c}(\mathbf{r}, t)\,dt = \mathbf{0} \tag{2.7.8}$$

where $\mathbf{B}$ is an $l \times 3nb$ matrix, with full row rank if the constraints are independent, and $\mathbf{c}$ is an l-vector. Let the same system of particles have an additional m holonomic constraints $\mathbf{\Phi}(\mathbf{r}, t) = \mathbf{0}$, with Jacobian $\mathbf{\Phi_r}$, where $m + l \geqslant 3nb$. If the *combined constraint Jacobian*

$$\mathbf{J} \equiv \begin{bmatrix} \mathbf{\Phi r}(\mathbf{r}, t) \\ \mathbf{B}(\mathbf{r}, t) \end{bmatrix} \tag{2.7.9}$$

has rank $l + m$, the system is said to have $3nb - (l + m)$ *degrees of freedom*.

In contrast to the holonomically constrained case, the coordinate partitioning theorem does not guarantee that $3nb - (l + m)$ independent generalized coordinates can be determined as algebraic functions of the remaining variables in a system with nonholonomic constraints. Since the constraints of Eq. 2.7.8 cannot be integrated to algebraic form, no associated algebraic equations exist to which the coordinate partitioning theorem may be applied. From a practical point of view, the same conclusion follows if differential constraints are integrable but no closed-form solution that satisfies initial conditions can be found.

➧ **Example 2.7.7:** Consider the wheel and axle assembly shown in Fig. 2.7.2(a), in which the wheels roll on the x-y plane without slip. Point P at the center of the axle is located at (x, y, R) in the x-y-z frame, where R is the radius of the

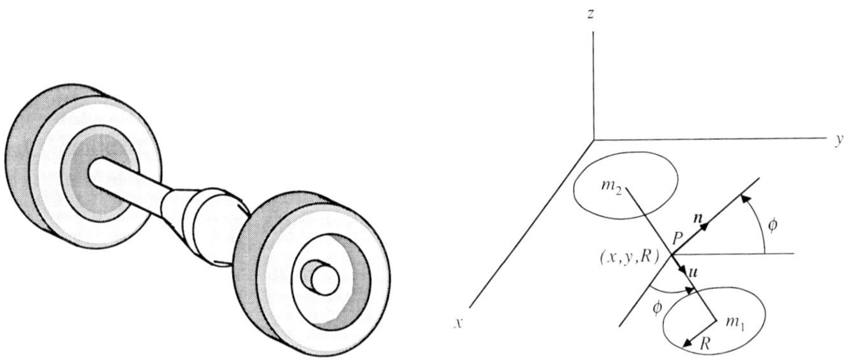

(a) (b)

Figure 2.7.2 Wheel–axle assembly.

wheels and the axle makes an angle ϕ with a line parallel to the x axis, as shown in Fig. 2.7.2(b).

The unit vector $\mathbf{n}$ that is perpendicular to the axle and parallel to the x-y plane is

$$\mathbf{n} = [-\sin\phi, \cos\phi, 0]^T$$

so a unit vector $\mathbf{u}$ along the axle, which is perpendicular to $\mathbf{n}$, is

$$\mathbf{u} = [\cos\phi, \sin\phi, 0]^T$$

Since the wheels do not slip, the component of the velocity $\dot{\mathbf{r}}$ of point P along the axle must be zero:

$$\mathbf{u}^T\dot{\mathbf{r}} = 0$$

or

$$\dot{x}\cos\phi + \dot{y}\sin\phi = 0 \tag{2.7.10}$$

Applying the test of the integrable differential constraint theorem, with (x, y, ϕ) corresponding to (α, β, γ) and $f_x = \cos\phi$, $f_y = \sin\phi$, and $f_\phi = 0$,

$$\sin\phi(-\sin\phi) + \cos\phi(-\cos\phi) = -1 \neq 0$$

Thus the constraint of Eq. 2.7.10 is not integrable (i.e., it is nonholonomic).

Example 2.7.8: Consider the particle in Example 2.7.5, with $\xi = -(t+1)^2$ and the additional holonomic constraint $y = t + 1$. The system of constraints is thus

$$\begin{aligned}\mathbf{\Phi}(\mathbf{r}, t) &\equiv \begin{bmatrix} y - t - 1 \\ z \end{bmatrix} = \mathbf{0} \\ \mathbf{b}^T\dot{\mathbf{r}} &\equiv y\dot{x} - (x + (t+1)^2)\dot{y} = 0\end{aligned} \tag{2.7.11}$$

The system Jacobian of Eq. 2.7.9 is

$$\mathbf{J} \equiv \begin{bmatrix} 0 & 1 & 0 \\ 0 & 0 & 1 \\ y & -(x + (t+1)^2) & 0 \end{bmatrix} \tag{2.7.12}$$

whose determinant is y. Since for $t \geq 0$, $y \neq 0$, this matrix has rank 3 and the system has zero degrees of freedom. Differentiating the first of Eqs. 2.7.11 with respect to time and combining it with the second yields the system of first order differential equations

$$\mathbf{J}(\mathbf{r}, t)\dot{\mathbf{r}} = \begin{bmatrix} 1 \\ 0 \\ 0 \end{bmatrix} \tag{2.7.13}$$

Since $|\mathbf{J}| \neq 0$,

$$\dot{\mathbf{r}} = [\mathbf{J}(\mathbf{r}, t)]^{-1} \begin{bmatrix} 1 \\ 0 \\ 0 \end{bmatrix} \tag{2.7.14}$$

With initial conditions that are consistent with the first of Eqs. 2.7.11:

$$\mathbf{r}(0) = \begin{bmatrix} x_0 \\ 1 \\ 0 \end{bmatrix} \tag{2.7.15}$$

Equations 2.7.13 and 2.7.15 have a unique solution (see Appendix C). Note that an initial position condition associated with the nonholonomic constraint must be specified to obtain a unique solution, since it is a differential relation; for example, if $x_0 = 1$, the unique solution is

$$\mathbf{r} = [(t+1)^2, t+1, 0]^T$$

Note that even though the constraint of Eq. 2.7.11 is nonholonomic (i.e., it is not integrable by itself to a functional equation in $\mathbf{r}$), when sufficient additional holonomic constraints are given and initial conditions are specified, the system is integrable. A similar result will be found when equations of motion are obtained, to be integrated with holonomic and nonholonomic equations of constraint.

The result of Example 2.7.8 is general. Regardless of whether a system of l independent differential constraints of the form of Eq. 2.7.8 is integrable or not, let $3nb - l$ additional holonomic constraints be specified, for which the combined constraint Jacobian of Eq. 2.7.9 has full rank. Then, with $3nb$ initial conditions $\mathbf{r}(t_0) = \mathbf{r}^0$ that are consistent with the holonomic constraints, a unique solution $\mathbf{r}(t)$ exists. The proof follows by applying the initial-value existence theorem of Appendix C to

$$\begin{aligned} \mathbf{J}(\mathbf{r}, t)\dot{\mathbf{r}} &= \begin{bmatrix} -\boldsymbol{\Phi}_t(\mathbf{r}, t) \\ -\mathbf{c}(\mathbf{r}, t) \end{bmatrix} \\ \mathbf{r}(t_0) &= \mathbf{r}^0 \end{aligned} \tag{2.7.16}$$

The same conclusion follows, as will be proved in Chapter 3, when equations of motion and initial conditions are added to a combination of differential and functional constraints, regardless of whether the differential constraints are nonholonomic or not.

The foregoing conclusion may appear to contradict the earlier observation that nonholonomic constraints admit larger families of trajectories $\mathbf{r}(t)$ than integrable differential constraints (e.g., see Example 2.7.6). However, holonomic constraints define surfaces (or *manifolds* [3]) in $(\mathbf{r}, t)$-space. Combining holonomic constraints with differential constraints has the effect of projecting solutions of the differential constraints onto the surface defined by the holonomic constraints. Thus, even though a nonholonomic differential constraint by itself admits a higher dimensional family of trajectories than an integrable differential constraint, the projection of each family onto the surface defined by holonomic constraints is of comparable dimension (i.e., the projection process eliminates the excess admissible solutions that are permitted by nonholonomic differential constraints).

2.8 VIRTUAL DISPLACEMENTS

The concept of virtual displacements plays a key role in variational methods for the study of dynamics. A *virtual displacement* $\delta\mathbf{r}$ of dimension n is defined as a small (or infinitesimal) displacement of particles in a system, with time held fixed. Virtual displacements may be thought of as small, hypothetical displacements that occur with time held fixed.

The concept of virtual displacements being *infinitesimal* can be made precise using the idea of *order of smallness*. A k-vector function $\mathbf{f}(\delta\mathbf{r})$ is said to be of order $\delta\mathbf{r}^m$, denoted $\mathbf{O}(\delta\mathbf{r}^m)$, if there is a finite constant $C > 0$ and a constant $\epsilon > 0$ such that

$$| f_i(\delta\mathbf{r}) | \leq C \sum_{j=1}^{n} | \delta r_j |^m, \qquad i = 1, \ldots, k \tag{2.8.1}$$

for all $\delta\mathbf{r}$ such that $| \delta r_j | \leq \epsilon$, $j = 1, \ldots, n$. Virtual displacements are in fact *partial differentials* of ordinary calculus, with the differential of time taken as zero. The operations of differential calculus and the concept of virtual displacement provide powerful tools for kinematic and dynamic analysis.

To relate virtual displacements and variations in generalized coordinates, consider a mapping from generalized coordinates $\mathbf{q}$ to Cartesian coordinates $\mathbf{r}$,

$$\mathbf{r} = \mathbf{r}(\mathbf{q}) \tag{2.8.2}$$

Using Taylor's theorem [3],

$$\mathbf{r}(\mathbf{q} + \delta\mathbf{q}) - \mathbf{r}(\mathbf{q}) = \mathbf{r}_\mathbf{q}(\mathbf{q})\, \delta\mathbf{q} + \mathbf{O}(\delta\mathbf{q}^2) \tag{2.8.3}$$

where the remainder term on the right contains terms that involve squares and higher powers of the *generalized coordinate variation* $\delta\mathbf{q}$. Calculations with virtual displacements ignore terms of order higher than 1; that is, in calculations with infinitesimals, quantities that are equal to within $\mathbf{O}(\delta\mathbf{q}^2)$ are declared to be *equal to first order*. More specifically, the differential of Eq. 2.8.2 with time held fixed yields

$$\delta\mathbf{r} = \mathbf{r_q}\,\delta\mathbf{q} \tag{2.8.4}$$

Thus, from Eq. 2.8.3,

$$\delta\mathbf{r} = \mathbf{r}(\mathbf{q} + \delta\mathbf{q}) - \mathbf{r}(\mathbf{q}) + \mathbf{O}(\delta\mathbf{q}^2) \tag{2.8.5}$$

➧ **Example 2.8.1:** The Cartesian coordinates of a particle P that represents the power tool, which moves on a cylinder of unit radius with its axis of revolution along the z axis, in Example 2.2.4, are given by

$$\mathbf{r} = \begin{bmatrix} \cos q_1 \\ \sin q_1 \\ q_2 \end{bmatrix} \tag{2.8.6}$$

where $\mathbf{q} = [\phi, z]^T$, as shown in Fig. 2.2.2. The relationship between virtual displacements $\delta\mathbf{r}$ of point P and variations $\delta\mathbf{q}$ in generalized coordinates is, from Eq. 2.8.4,

$$\delta\mathbf{r} = \begin{bmatrix} -\sin q_1 & 0 \\ \cos q_1 & 0 \\ 0 & 1 \end{bmatrix} \delta\mathbf{q} \tag{2.8.7}$$

Note that this relation between virtual displacement and generalized coordinate variations is identical in form to the relationship derived in Example 2.2.6 between Cartesian velocity $\dot{\mathbf{r}}$ and generalized coordinate time derivatives $\dot{\mathbf{q}}$. This is no accident, since the relationships between virtual displacements and velocities of Eqs. 2.8.4 and 2.2.16, respectively, are identical in form, having precisely the same coefficient matrix. The fact that both are linear relations in velocities will be shown later to have an important impact on the study of dynamics.

To be more general, consider a time dependent generalized coordinate mapping

$$\mathbf{r} = \mathbf{r}(\mathbf{q}, t) \tag{2.8.8}$$

Differentiating with respect to time yields

$$\dot{\mathbf{r}} = \mathbf{r_q}\dot{\mathbf{q}} + \mathbf{r}_t \tag{2.8.9}$$

which differs from Eq. 2.2.16 by the term associated with explicit dependence of the mapping on time.

➧ **Example 2.8.2:** Consider motion of a particle P on a cylinder whose centerline is the z axis but whose radius is equal to time (i.e., a cylinder that is expanding in radius with time). In terms of the same generalized coordinates used in Examples 2.2.6 and 2.8.1, with $R = t$,

$$\mathbf{r} = \begin{bmatrix} t \cos q_1 \\ t \sin q_1 \\ q_2 \end{bmatrix} \tag{2.8.10}$$

Since virtual displacements are defined with time held fixed, at any particular time t,

$$\delta\mathbf{r} = \begin{bmatrix} -t \sin q_1 & 0 \\ t \cos q_1 & 0 \\ 0 & 1 \end{bmatrix} \delta\mathbf{q} \tag{2.8.11}$$

To compare the behavior of virtual displacements and velocities in this example, the time derivative of Eq. 2.8.10 is calculated to obtain

$$\dot{\mathbf{r}} = \begin{bmatrix} -t \sin q_1 & 0 \\ t \cos q_1 & 0 \\ 0 & 1 \end{bmatrix} \dot{\mathbf{q}} + \begin{bmatrix} \cos q_1 \\ \sin q_1 \\ 0 \end{bmatrix} \tag{2.8.12}$$

Note that this relation is not identical to Eq. 2.8.11. It differs by terms that arise due to explicit dependence of the mapping of Eq. 2.8.10 on time.

It is helpful to interpret virtual displacements $\delta\mathbf{r}$ geometrically. Equation 2.8.4 may be viewed as defining kinematically admissible virtual displacements $\delta\mathbf{r}$ as a linear combination of the columns $\mathbf{r}_{q_j}$ of the coefficient matrix with variations δq_j in generalized coordinates as coefficients:

$$\delta\mathbf{r} = \mathbf{r}_{\mathbf{q}}\, \delta\mathbf{q} = \sum_{j=1}^{n} \mathbf{r}_{q_j}\, \delta q_j \tag{2.8.13}$$

If $\mathbf{r}_{\mathbf{q}}$ has full column rank, each kinematically admissible $\delta\mathbf{r}$ is tangent to the constraint surface with which generalized coordinates $\mathbf{q}$ are consistent (i.e., as defined by $\mathbf{\Phi}(\mathbf{q}, t) = \mathbf{0}$) and can be obtained from Eq. 2.8.13 for a unique $\delta\mathbf{q}$. At a singular point of the mapping of Eq. 2.8.2, however, the columns of $\mathbf{r}_{\mathbf{q}}$ are not linearly independent. Thus some kinematically admissible $\delta\mathbf{r}$ cannot be achieved, since the columns of $\mathbf{r}_{\mathbf{q}}$ do not span the tangent plane and other kinematically admissible $\delta\mathbf{r}$ can be obtained from an infinite number of $\delta\mathbf{q}$. The use of spherical coordinates to locate a point on a unit sphere in Example 2.4.5 encounters these difficulties at the poles on the z axis (Prob. 2.8.1).

Example 2.8.3: For the cylindrical coordinates of Example 2.8.1 on a cylinder of unit radius, vectors along the direction of the two columns of $\mathbf{r}_{\mathbf{q}}$ in Cartesian coordinate space are shown in Fig. 2.8.1. Note that $\mathbf{r}_{q_1}$ and $\mathbf{r}_{q_2}$ are noncollinear vectors that are tangent to the surface of the cylinder. Hence any vector $\delta\mathbf{r}$ in the tangent plane to the cylinder at point P can be obtained in terms of a unique $\delta\mathbf{q}$.

The foregoing physical interpretation of virtual displacements $\delta\mathbf{r}$ remains valid for the cylinder with time varying radius in Example 2.8.2. However, note that as time changes in that example, there is a component of motion of point P radially

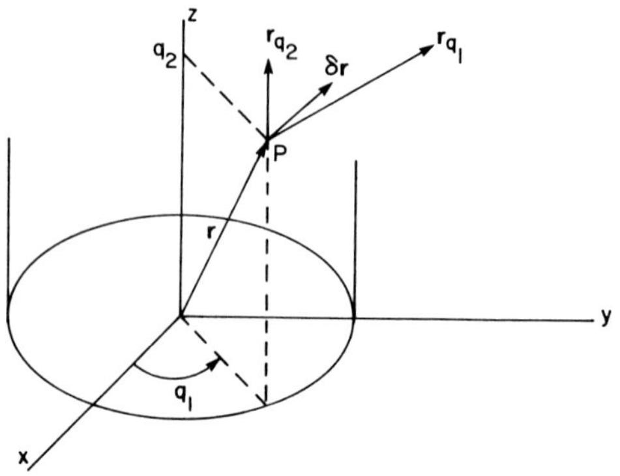

Figure 2.8.1 Virtual displacement of particle on cylinder.

outward from the z axis, which is not in the tangent plane to the cylinder at a fixed time t.

Let $d\mathbf{r}$ denote an *actual differential displacement* of particles, which from Eq. 2.8.8 satisfies

$$d\mathbf{r} = \mathbf{r_q}\, d\mathbf{q} + \mathbf{r}_t\, dt \tag{2.8.14}$$

For equal generalized coordinate *partial differentials* $\delta\mathbf{q}$ and *total differentials* $d\mathbf{q}$, the resulting virtual displacements $\delta\mathbf{r}$ of Eq. 2.8.13 and the actual differential displacements $d\mathbf{r}$ of Eq. 2.8.14 are different if the mapping of Eq. 2.8.8 is time dependent (i.e., if $\mathbf{r}_t \neq 0$). Thus for time dependent mappings, actual differential displacements $d\mathbf{r}$ that may occur over time that are not in directions of kinematically admissible virtual displacements $\delta\mathbf{r}$.

Consider next holonomic constraint equations of the form

$$\mathbf{\Phi}(\mathbf{r}, t) = \mathbf{0} \tag{2.8.15}$$

which are time dependent. A virtual displacement $\delta\mathbf{r}$ is said to be *consistent with constraints* if

$$\mathbf{\Phi_r}\, \delta\mathbf{r} = \mathbf{0} \tag{2.8.16}$$

(i.e., if it satisfies Eq. 2.8.15 to first order, with time held fixed). Equation 2.8.16 defines *kinematically admissible virtual displacements* as tangent to the constraint surface, with time held fixed. Using Taylor's theorem [3] and Eq. 2.8.16 to expand the constraint equation of Eq. 2.8.15 about $\mathbf{r}$, with time t held fixed, for any kinematically admissible $\delta\mathbf{r}$,

$$\mathbf{\Phi}(\mathbf{r} + \delta\mathbf{r}, t) = \tfrac{1}{2}\, (\mathbf{\Phi_r}\, \delta\mathbf{r})_{\mathbf{r}}\, \delta\mathbf{r}\big|_{(\mathbf{r}+\alpha\delta\mathbf{r},t)} + \mathbf{O}(\delta\mathbf{r}^3) \tag{2.8.17}$$

where $0 < \alpha < 1$. Thus, the constraint error due to a kinematically admissible virtual displacement is of order $\delta\mathbf{r}^2$.

➧ **Example 2.8.4:** The cylinder with unit radius of Fig. 2.8.1 can be characterized by the single holonomic constraint equation

$$\Phi(\mathbf{r}) = x^2 + y^2 - 1 = 0 \tag{2.8.18}$$

which is stationary. A kinematically admissible virtual displacement of Eq. 2.8.16 must thus satisfy

$$\Phi_{\mathbf{r}}\, \delta\mathbf{r} = [2x, 2y, 0]\, \delta\mathbf{r} = 0 \tag{2.8.19}$$

Noting that the vector $\Phi_{\mathbf{r}}^T$ is normal to the constraint surface, Eq. 2.8.19 simply implies that a kinematically admissible virtual displacement $\delta\mathbf{r}$ is tangent to the constraint surface, just as a kinematically admissible virtual displacement of Eq. 2.8.7 was tangent to the same constraint surface.

Examples 2.8.1 and 2.8.4 illustrate the consistency of generalized coordinate and Cartesian coordinate respresentations of constraint surfaces and the associated kinematically admissible virtual displacements. When independent generalized coordinates can be defined, consideration may be limited to variations in independent generalized coordinates, which lead automatically to kinematically admissible virtual displacements. In many cases, however, finding independent generalized coordinates is either impossible or impractical. In such cases, Cartesian coordinates, written in terms of generalized coordinates, must satisfy holonomic constraint equations, as in Eq. 2.6.13. Thus variations in generalized coordinates must satisfy linearized forms of the constraint equations, with time held fixed, in order to be kinematically admissible; that is, $\delta\mathbf{q}$ is a *kinematically admissible generalized coordinate variation* if

$$\mathbf{\Phi}_{\mathbf{q}}\, \delta\mathbf{q} = \mathbf{0} \tag{2.8.20}$$

where $\mathbf{\Phi}_{\mathbf{q}} = \mathbf{\Phi}_{\mathbf{r}}\mathbf{r}_{\mathbf{q}}$, as in Eq. 2.6.14.

For a time dependent holonomic constraint of Eq. 2.8.15, taking the total differential yields

$$\mathbf{\Phi}_{\mathbf{r}}\, d\mathbf{r} + \mathbf{\Phi}_t\, dt = \mathbf{0} \tag{2.8.21}$$

which must be satisfied by actual displacements $d\mathbf{r}$. Thus if $\mathbf{\Phi}_t \neq \mathbf{0}$, kinematically admissible virtual displacements are not actual displacements of the constrained system. This distinction must be clearly understood in order to make effective use of virtual displacements in dynamic analysis.

Similarly, *actual differentials* $d\mathbf{q}$ of generalized coordinates must satisfy

$$\mathbf{\Phi}_{\mathbf{q}}\, d\mathbf{q} + \mathbf{\Phi}_t\, dt = \mathbf{\Phi}_{\mathbf{r}}\mathbf{r}_{\mathbf{q}}\, d\mathbf{q} + \mathbf{\Phi}_t\, dt = \mathbf{0} \tag{2.8.22}$$

Thus, if $\mathbf{\Phi}_t \neq \mathbf{0}$, kinematically admissible generalized coordinate variations $\delta\mathbf{q}$ that satisfy Eq. 2.8.20 are not actual differentials $d\mathbf{q}$.

Consider next constraints of the form

$$\mathbf{B}(\mathbf{r}, t)\dot{\mathbf{r}} + \mathbf{c}(\mathbf{r}, t) = \mathbf{0} \tag{2.8.23}$$

where the coefficient matrix $\mathbf{B}$ has full row rank. This is the equivalent to the differential constraint

$$\mathbf{B}(\mathbf{r}, t)\, d\mathbf{r} + \mathbf{c}(\mathbf{r}, t)\, dt = \mathbf{0} \tag{2.8.24}$$

where $d\mathbf{r}$ is viewed as a differential, or actual displacement of the Cartesian vector $\mathbf{r}$, and dt is differential time.

Since virtual displacements are defined with time held fixed, it is natural to suppress the differential of time in Eq. 2.8.24 and define a virtual displacement $\delta\mathbf{r}$ to be consistent with the nonholonomic constraint of Eq. 2.8.23, equivalently to Eq. 2.8.24, if

$$\mathbf{B}(\mathbf{r}, t)\, \delta\mathbf{r} = \mathbf{0} \tag{2.8.25}$$

Just as in the case of time-dependent holonomic constraints, if $\mathbf{c}(\mathbf{r}, t) \neq \mathbf{0}$, then kinematically admissible virtual displacements $\delta\mathbf{r}$ of Eq. 2.8.25 are not actual displacements $d\mathbf{r}$ of Eq. 2.8.24.

When Cartesian coordinates are defined in terms of generalized coordinates and time by Eq. 2.8.8, the constraint of Eq. 2.8.23 is

$$\mathbf{B}(\mathbf{r}(\mathbf{q}, t), t)\mathbf{r}_{\mathbf{q}}\dot{\mathbf{q}} + \mathbf{B}(\mathbf{r}(\mathbf{q}, t), t)\mathbf{r}_t + \mathbf{c}(\mathbf{r}(\mathbf{q}, t), t) = \mathbf{0} \tag{2.8.26}$$

or, in differential form,

$$\mathbf{B}(\mathbf{r}(\mathbf{q}, t), t)\mathbf{r}_{\mathbf{q}}\, d\mathbf{q} + [\mathbf{B}(\mathbf{r}(\mathbf{q}, t), t)\mathbf{r}_t + \mathbf{c}(\mathbf{r}(\mathbf{q}, t), t)]\, dt = \mathbf{0} \tag{2.8.27}$$

A kinematically admissible generalized coordinate variation $\delta\mathbf{q}$ that is associated with Eq. 2.8.27 is defined with time held fixed; that is, to satisfy

$$\mathbf{B}(\mathbf{r}(\mathbf{q}, t), t)\mathbf{r}_{\mathbf{q}}\, \delta\mathbf{q} = \mathbf{0} \tag{2.8.28}$$

Thus a kinematically admissible generalized coordinate variation $\delta\mathbf{q}$ is not generally an actual generalized coordinate differential $d\mathbf{q}$ of Eq. 2.8.27.

Although it is often helpful to think of virtual displacements in physical terms, it should be remembered that they are defined mathematically as variations in Cartesian coordinates or generalized coordinates, with time held fixed. For particles whose kinematically admissible motion is defined by a generalized coordinate vector $\mathbf{q}$, the relationship between Cartesian virtual displacements $\delta\mathbf{r}$ and generalized coordinate variations $\delta\mathbf{q}$ satisfy Eq. 2.8.4, regardless of whether the mapping from generalized coordinates to Cartesian coordinates in Eq. 2.8.2 is time dependent or not. Similarly, for holonomic constraints of Eq. 2.8.15 and differential constraints of Eq. 2.8.24, kinematically admissible virtual displacements $\delta\mathbf{r}$ are defined as satisfying Eqs. 2.8.16 and 2.8.25, respectively. Only these mathematical properties of virtual displacements will be used in the development that follows. No "pseudophysical" interpretation of virtual displacements is needed for their use in dynamics.

When both holonomic and differential constraints of Eqs. 2.6.2 and 2.7.8 are imposed, the combined Jacobian of Eq. 2.7.9 may be used to express conditions on kinematically admissible virtual displacements $\delta\mathbf{r}$ as

$$\mathbf{J}\,\delta\mathbf{r} = \mathbf{0} \tag{2.8.29}$$

which encompasses both Eqs. 2.8.16 and 2.8.25. Similarly, Eqs. 2.8.20 and 2.8.28 define conditions on kinematically admissible generalized coordinate variations $\delta\mathbf{q}$ as

$$\mathbf{J}\mathbf{r}_{\mathbf{q}}\,\delta\mathbf{q} = \mathbf{0} \tag{2.8.30}$$

PROBLEMS

Section 2.1

2.1.1 Derive expressions for spherical coordinates $\mathbf{q} = [\rho, \phi, \psi]^T$ that map into the vector $\mathbf{r} = [x, y, z]^T$, where $x^2 + y^2 \neq 0$.

2.1.2 Show that for a given velocity $\dot{\mathbf{r}} = [\dot{x}, \dot{y}, \dot{z}]^T$, at $\mathbf{q} = [0, \phi, z]^T$, the velocity equation associated with Eq. 2.1.4 has a solution for $\dot{\mathbf{q}}$ only if $\tan\phi = y/x$. If this is the case, find $\dot{\mathbf{q}}$ and interpret the result geometrically. Show that the acceleration equation associated with Eq. 2.1.4 has a solution for $\ddot{\mathbf{q}}$ only if $\dot{\phi}$ satisfies

$$\ddot{x}\sin\phi - \ddot{y}\cos\phi = -2\dot{s}\dot{\phi}$$

2.1.3 Derive relations between velocities $\dot{\mathbf{r}}$ and $\dot{\mathbf{q}}$ and accelerations $\ddot{\mathbf{r}}$ and $\ddot{\mathbf{q}}$ for spherical coordinates, using the mapping of Eq. 2.1.6.

2.1.4 Derive conditions on $\mathbf{q} = [\rho, \phi, \pi/2]^T$ for which the velocity and acceleration relations obtained in Problem 2.1.3 have solutions for $\dot{\mathbf{q}}$ and $\ddot{\mathbf{q}}$, with $\dot{\mathbf{r}}$ and $\ddot{\mathbf{r}}$ given.

Section 2.2

2.2.1 Use the definition of matrix addition and properties of differentiation to verify that Eq. 2.2.2 is valid.

2.2.2 Verify that Eqs. 2.2.3 and 2.2.4 are valid.

2.2.3 Verify that Eqs. 2.2.8 to 2.2.10 are valid.

2.2.4 Verify that Eqs. 2.2.20 to 2.2.22 are valid.

2.2.5 If the mass of the power tool in Examples 2.2.4 and 2.2.6 is $m = 10$ kg and if the input histories are $q_1(t) = \pi/4(1 - \cos\pi t)$ and $q_2 = \frac{1}{2}(1 - \cos\pi t)$, $0 \leq t \leq 1$, evaluate the total force that must be applied to the tool. Convert this resultant force to a force along the z axis and a torque about the z axis.

Section 2.3

2.3.1 Consider a particle that moves along the intersection of the paraboloid of revolution $x^2 + y^2 - z = 0$ and the plane $x + y - 1 = 0$. Show that the Jacobian of these equations has full row rank everywhere on the curve.

2.3.2 Use the coordinate partitioning theorem to determine which of the variables in Problem 2.3.1 can be selected as independent over the entire curve of intersection.

Section 2.4

2.4.1 Consider the spiral curve shown in Fig. P2.4.1, with parameterization

$$\mathbf{r} = \begin{bmatrix} q^2 \cos q \\ q^2 \sin q \\ q^2 \end{bmatrix}$$

Show that $q = 0$ is a singular point.

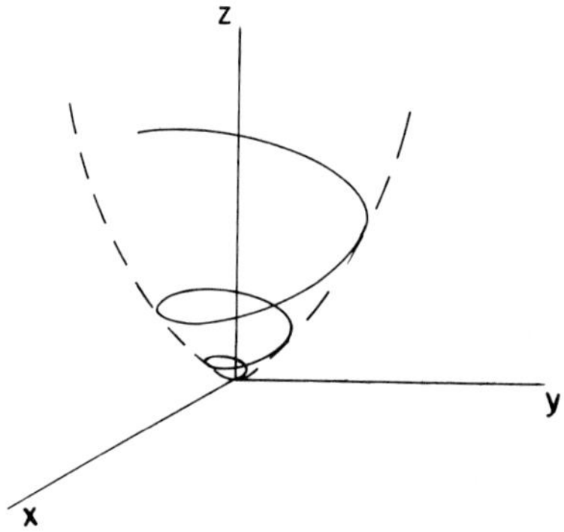

Figure P2.4.1

2.4.2 Show that the circle shown in Fig. P2.4.2 in the x-y plane cannot be parameterized by generalized coordinates $q = x$ or y to obtain a mapping that is everywhere nonsingular. Use $q = \phi$ as a generalized coordinate and show that the resulting mapping has no singular point.

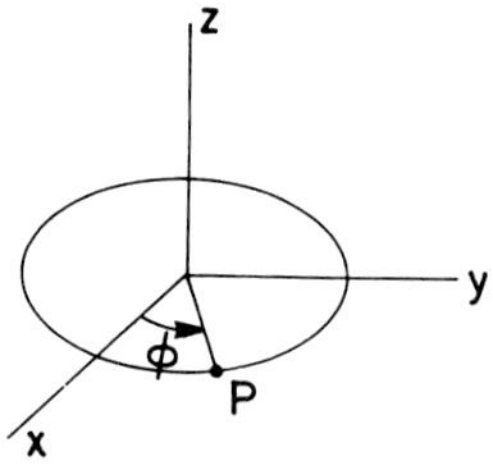

Figure P2.4.2

2.4.3 If $\mathbf{r}$ is n-dimensional and $\mathbf{q}$ is m-dimensional, carry out the expansion of the second term of Eq. 2.4.18 in terms of derivatives of $\mathbf{r}$ and q_i, $i = 1, \ldots, m$.

2.4.4 Show that the Jacobian $\mathbf{r_q}$ of Example 2.2.4 has full column rank, by showing that $\mathbf{r_q}\boldsymbol{\alpha} = \mathbf{0}$ implies that $\boldsymbol{\alpha} = \mathbf{0}$.

Section 2.5

2.5.1 Find the direction cosine representations of unit vectors in the same direction as the vectors $\mathbf{a} = [1, 1, 1]^T$ and $\mathbf{b} = [1, 1, 0]^T$.

2.5.2 If the vector **a** in Problem 2.5.1 is along the x' axis of a Cartesian x'-y'-z' reference frame and if **b** is in the first quadrant of its x'-y' plane, find the unit coordinate vectors **f**, **g**, and **h**. (*Hint*: Since **a** and **b** lie in the x'-y' plane, $\tilde{\mathbf{a}}\mathbf{b}$ must be along the z' axis. Form the transformation matrix from the x'-y'-z' frame to the x-y-z frame.)

2.5.3 Expand the product $\mathbf{A}^T\mathbf{A}$ using Eq. 2.5.13 to show that Eq. 2.5.14 is correct.

2.5.4 If $(\mathbf{B} - \mathbf{C})\mathbf{a} = \mathbf{0}$ for arbitrary **a**, show that $\mathbf{B} = \mathbf{C}$, where **B** and **C** are $m \times n$ matrices and **a** is an n-vector.

2.5.5 If $\mathbf{A}_i$ and $\mathbf{A}_j$ are orthogonal matrices, show that $\mathbf{A}_{ij}$ of Eq. 2.5.35 is orthogonal.

2.5.6 Verify that Eq. 2.5.41 uniquely defines the angle θ, $0 \leq \theta < 2\pi$, in Fig. 2.5.7.

2.5.7 Repeat the derivation of Eq. 2.5.44 for the **f-i** and **g-j** vectors, to show that Eq. 2.5.45 is correct.

2.5.8 Show that as ϕ is incremented by 120° in Example 2.5.4, beginning from $\phi = 0$, that the vector $\mathbf{s}' = [1, 0, 0]^T$ becomes $\mathbf{s}(0) = \mathbf{s}'$, $\mathbf{s}(120°) = [0, 1, 0]^T$, $\mathbf{s}(240°) = [0, 0, 1]^T$, and $\mathbf{s}(360°) = \mathbf{s}'$. Interpret these results physically, in terms of the cone that is swept out by **s** as ϕ varies.

Section 2.6

2.6.1 Show that if $l = 1$ in the slider–crank of Example 2.6.5, then when $q_1 = \pi/2$, the Jacobian is zero. Show that if $\dot{q}_1 = 1$, as q_1 passes through $\pi/2$, $\dot{q}_2$ remains continuous if P_2 continues to move to the left [Fig. 2.6.9(b)] and that $\dot{q}_2$ must be discontinuous if P_2 remains at the origin of the x-y-z frame [Fig. 2.6.9(c)]. If the mass of the particle P_2 is not zero, which case will occur?

2.6.2 Consider the nonlinear constraints

$$\mathbf{\Phi}(\mathbf{r}, t) = \begin{bmatrix} x^2 + y^2 - 1 \\ x + y - t \end{bmatrix} = \begin{bmatrix} 0 \\ 0 \end{bmatrix}$$

(a) Write the Jacobian matrix $\mathbf{\Phi}_\mathbf{r}$ of the constraints.

(b) Find all values of $\mathbf{r}(t)$ that satisfy the constraints and at which $\mathbf{\Phi}_\mathbf{r}$ fails to have full row rank.

(c) Write the velocity and acceleration equations associated with the constraints.

2.6.3 For the mechanism shown in Fig. P2.6.3, body 1, with m_1 at its tip, rotates about an axis in the x-y plane, where θ is fixed. The second body has mass m_2 at its tip and slides through a ball that rotates about point (0, 2, 0). The connection between bodies 1 and 2 at m_1 is a ball-and-socket joint. Body 1 is 1 unit long, and body 2 is 4 units long. Write vectors $\mathbf{r}_1$ and $\mathbf{r}_2$ that locate m_1 and m_2 in the x-y-z frame as functions of q. (*Hint*: Note that $\mathbf{a} = [0, 2, 0]^T - \mathbf{r}_1$ and **a** can never be **0**, so the vector from m_1 to m_2 is $4\mathbf{a}/|\mathbf{a}|$.) If $q = \sin t$, find $\dot{\mathbf{r}}_i$ and $\ddot{\mathbf{r}}_i$, $i = 1, 2$.

2.6.4 The point mass model of a spatial four-bar mechanism shown in Fig. P2.6.4 is formed by links of unit length, with concentrated masses that pivot about the x and y axes at points (0, 0, 0) and (0, 1, 0), respectively, and a massless link with spherical joints at m_1 and m_2 that has unit length.

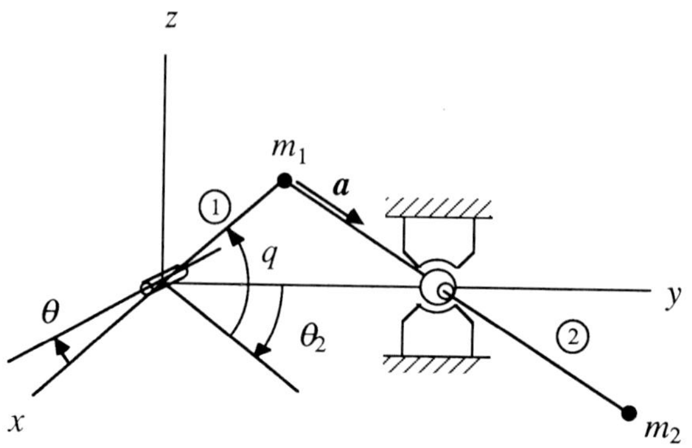

Figure P2.6.3

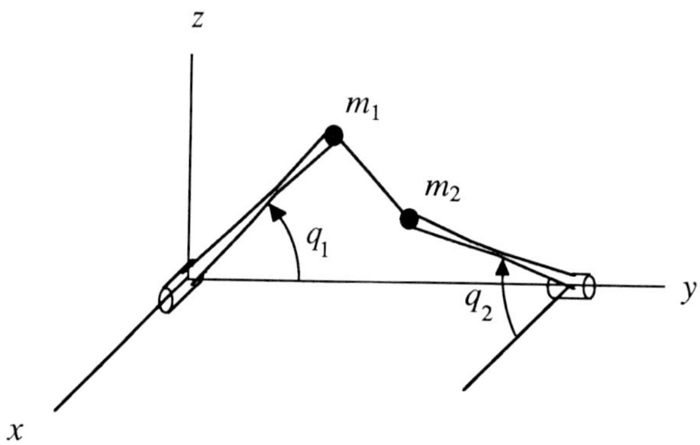

Figure P2.6.4

(a) Write $\mathbf{r}_1$ and $\mathbf{r}_2$, which locate masses m_1 and m_2, as functions of $\mathbf{q} = [q_1, q_2]^T$.
(b) Derive the equation of constraint $\Phi(\mathbf{q}) = 0$ that accounts for the fact that the bar between m_1 and m_2 has unit length.
(c) Derive the constraint velocity and acceleration equations in terms of $\dot{\mathbf{q}}$ and $\ddot{\mathbf{q}}$.

Section 2.7

2.7.1 Determine which of the following differential forms is exact:
(a) $2u_3\,du_1 + du_2 + 3\,du_3$
(b) $2u_1u_2\,du_1 + u_1^2\,du_2 + du_3 + 2u_4\,du_4$
(c) $2u_2u_3du_1 + u_1u_3\,du_2 + u_1u_2\,du_3$

2.7.2 Which of the differential forms of Problem 2.7.1 that are not exact are integrable?

2.7.3 Modeling the wheel and axle assembly of Example 2.7.7 as two particles in a plane, as shown in Fig. P2.7.3,

$$\mathbf{r} = \tfrac{1}{2}(\mathbf{r}_1 + \mathbf{r}_2)$$

$$\mathbf{u} = \mathbf{r}_2 - \mathbf{r}_1$$

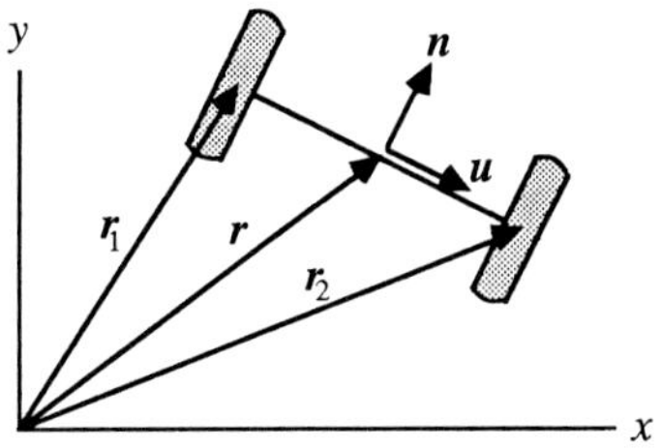

Figure P2.7.3

The differential constraint of Example 2.7.7 is

$$\tfrac{1}{2}(\mathbf{r}_2 - \mathbf{r}_1)^T(\dot{\mathbf{r}}_1 + \dot{\mathbf{r}}_2) = \mathbf{u}^T\dot{\mathbf{r}} = 0$$

or in component form,

$$(x_2 - x_1)\,dx_1 + (y_2 - y_1)\,dy_1 + (x_2 - x_1)\,dx_2 + (y_2 - y_1)\,dy_2 = 0$$

Determine whether this is a nonholonomic constraint in the variables x_1, y_1, x_2, and y_2.

Section 2.8

2.8.1 Show that at the singular point $\mathbf{q} = [\phi, \pi/2]^T$ of the generalized coordinate parameterization of Example 2.4.5, only the virtual displacement $\delta\mathbf{r} = [-\cos\phi, -\sin\phi, 0]^T\,\delta\phi$ can be achieved, for arbitrary $\delta\phi$. Show that each vector in the tangent plane to the sphere at this singular point can, however, be achieved for some value of ϕ, which is arbitrary at the singular point. Relate these singular properties of virtual displacements to singular properties of velocities studied in Prob. 2.1.4.

2.8.2 Write virtual displacements of m_1 and m_2 of Problem 2.6.3 in terms of δq.

2.8.3 Write virtual displacements of m_1 and m_2 of Problem 2.6.4 in terms of $\delta\mathbf{q}$. Write the equation that kinematically admissible $\delta\mathbf{q}$ must satisfy.

SUMMARY OF KEY FORMULAS

Algebraic Vectors

$$\tilde{\mathbf{a}} = \begin{bmatrix} 0 & -a_z & a_y \\ a_z & 0 & -a_x \\ -a_y & a_x & 0 \end{bmatrix} \qquad \text{(A.3.5)}$$

$$\tilde{\mathbf{a}}^T = -\tilde{\mathbf{a}}, \qquad \tilde{\mathbf{a}}\mathbf{b} = -\tilde{\mathbf{b}}\mathbf{a}, \qquad \tilde{\mathbf{a}}\mathbf{a} = \mathbf{0} \qquad \text{(A.3.7,9,10)}$$

$$\tilde{\mathbf{a}}\tilde{\mathbf{b}} = \mathbf{b}\mathbf{a}^T - \mathbf{a}^T\mathbf{b}\mathbf{I}, \qquad \widetilde{(\tilde{\mathbf{a}}\mathbf{b})} = \mathbf{b}\mathbf{a}^T - \mathbf{a}\mathbf{b}^T = \tilde{\mathbf{a}}\tilde{\mathbf{b}} - \tilde{\mathbf{b}}\tilde{\mathbf{a}} \qquad \text{(A.3.12,13,14)}$$

$$\tilde{\mathbf{a}}\tilde{\mathbf{b}} + \mathbf{a}\mathbf{b}^T = \tilde{\mathbf{b}}\tilde{\mathbf{a}} + \mathbf{b}\mathbf{a}^T, \qquad \widetilde{(\mathbf{a} + \mathbf{b})} = \tilde{\mathbf{a}} + \tilde{\mathbf{b}} \qquad \text{(A.3.15,16)}$$

Vector and Matrix Differentiation

$$\frac{d}{dt}(\alpha\mathbf{a}) = \dot{\alpha}\mathbf{a} + \alpha\dot{\mathbf{a}}, \qquad \frac{d}{dt}(\mathbf{a}^T\mathbf{b}) = \dot{\mathbf{a}}^T\mathbf{b} + \mathbf{a}^T\dot{\mathbf{b}} \tag{2.2.3,4}$$

$$\frac{dt}{dt}(\tilde{\mathbf{a}}\mathbf{b}) - \dot{\tilde{\mathbf{a}}}\mathbf{b} + \tilde{\mathbf{a}}\dot{\mathbf{b}} \tag{2.2.5}$$

$$\mathbf{a_q} \equiv \frac{\partial\mathbf{a}}{\partial\mathbf{q}} \equiv \left[\frac{\partial a_i}{\partial q_j}\right], \qquad (\mathbf{a}^T\mathbf{b})_\mathbf{q} = \mathbf{b}^T\mathbf{a_q} + \mathbf{a}^T\mathbf{b_q} \tag{2.2.14,15}$$

$$\mathbf{a_q} = \mathbf{a_b}\mathbf{b_q}, \qquad \dot{\mathbf{a}} = \mathbf{a_q}\dot{\mathbf{q}} + \mathbf{a}_t \tag{2.2.16}$$

Position Parameterization by Generalized Coordinates

$$\mathbf{r} = \mathbf{r}(\mathbf{q}), \qquad \dot{\mathbf{r}} = \mathbf{r_q}\dot{\mathbf{q}}, \qquad \ddot{\mathbf{r}} = \mathbf{r_q}\ddot{\mathbf{q}} + (\mathbf{r_q}\dot{\mathbf{q}})_\mathbf{q}\dot{\mathbf{q}} \tag{2.4.16,17,18}$$

Transformation of Coordinates

$$\mathbf{s} = \mathbf{A}\mathbf{s}', \qquad \mathbf{A} = [\mathbf{f}, \mathbf{g}, \mathbf{h}], \qquad \mathbf{A}^T\mathbf{A} = \mathbf{A}\mathbf{A}^T = \mathbf{I}, \qquad \mathbf{s}' = \mathbf{A}^T\mathbf{s} \tag{2.5.10,13,14,15}$$

$$\mathbf{r}^P = \mathbf{r} + \mathbf{s}^P = \mathbf{r}\ \mathbf{A}\mathbf{s}'^P \tag{2.5.16}$$

$$\tilde{\mathbf{s}} = \mathbf{A}\tilde{\mathbf{s}}'\mathbf{A}^T, \qquad \tilde{\mathbf{s}}' = \mathbf{A}^T\tilde{\mathbf{s}}\mathbf{A}, \qquad \mathbf{A}_{ij} = \mathbf{A}_i^T\mathbf{A}_j \tag{2.5.21,22,25}$$

Holonomic Constraints

$$\mathbf{\Phi}(\mathbf{r}, t) = \mathbf{0}, \qquad \mathbf{\Phi_r}\dot{\mathbf{r}} = -\mathbf{\Phi}_t, \tag{2.6.2,3}$$

$$\mathbf{\Phi_r}\ddot{\mathbf{r}} = -[(\mathbf{\Phi_r}\dot{\mathbf{r}})_\mathbf{r}\dot{\mathbf{r}} + 2\mathbf{\Phi}_{\mathbf{r}t}\dot{\mathbf{r}} + \mathbf{\Phi}_{tt}] \equiv \boldsymbol{\gamma} \tag{2.6.4}$$

$$\begin{aligned} &\mathbf{\Phi}(\mathbf{q}, t) = \mathbf{0}, \qquad \mathbf{\Phi_q}\dot{\mathbf{q}} = -\mathbf{\Phi}_t, \\ &\mathbf{\Phi_q}\ddot{\mathbf{q}} = -[(\mathbf{\Phi_q}\dot{\mathbf{q}})_\mathbf{q}\dot{\mathbf{q}} + 2\mathbf{\Phi}_{\mathbf{q}t}\dot{\mathbf{q}} + \mathbf{\Phi}_{tt}] \equiv \boldsymbol{\gamma} \end{aligned} \tag{2.6.13,15}$$

Differential Constraints

$$\mathbf{b}^T(\mathbf{r}, t)\dot{\mathbf{r}} + c(\mathbf{r}, t) = 0, \qquad \mathbf{b}^T(\mathbf{r}, t)\,d\mathbf{r} + c(\mathbf{r}, t)\,dt = 0 \tag{2.7.1,2}$$

$$\mathbf{B}(\mathbf{r}, t)\,d\mathbf{r} + \mathbf{c}(\mathbf{r}, t) = \mathbf{0} \tag{2.7.8}$$

Virtual Displacements

$$\delta\mathbf{r} = \mathbf{r_q}\,\delta\mathbf{q} \tag{2.8.4}$$

$$\mathbf{\Phi_r}\,\delta\mathbf{r} = \mathbf{0}, \qquad \mathbf{\Phi_q}\,\delta\mathbf{q} = \mathbf{0} \tag{2.8.16,20}$$

$$\mathbf{B}(\mathbf{r}, t)\,\delta\mathbf{r} = \mathbf{0}, \qquad \mathbf{B}(\mathbf{r}(\mathbf{q}, t), t)\mathbf{r_q}\,\delta\mathbf{q} = \mathbf{0} \tag{2.8.25,28}$$

3

Variational Methods for Dynamics of Systems Modeled as Particles

3.1 VARIATIONAL EQUATIONS OF MOTION OF A PARTICLE

Variational methods, or *virtual work methods* have been used extensively in the fields of dynamics and continuum mechanics. The basic idea on which variational methods are based is the following theorem.

Orthogonality Theorem: A vector $\mathbf{b} = [b_1, \ldots, b_n]^T$ in n-dimensional space R^n is the zero vector if and only if

$$\mathbf{s}^T\mathbf{b} = 0 \tag{3.1.1}$$

for all $\mathbf{s} = [s_1, \ldots, s_n]^T$ in R^n.

To understand the implications of this theorem, it is instructive to follow through its rather elementary proof. The "only if" part of the theorem follows directly, since if $\mathbf{b} = \mathbf{0}$, Eq. 3.1.1 holds for all $\mathbf{s}$ in R^n. The more important "if" part of the theorem may be proved by observing that if Eq. 3.1.1 holds for all $\mathbf{s}$ in R^n, it certainly holds for $\mathbf{s}_i = [0, \ldots, 1, \ldots, 0]^T$, where the unit value appears in position i. Evaluating Eq. 3.1.1 for $\mathbf{s}_i$ yields

$$\mathbf{s}_i^T\mathbf{b} = b_i = 0 \tag{3.1.2}$$

Since this equation holds for each $i = 1, \ldots, n$, $\mathbf{b} = \mathbf{0}$, and the theorem is proved.

As is clear from the proof of the orthogonality theorem, the conclusion of the theorem holds only if the vector $\mathbf{s}$ in Eq. 3.1.1 is arbitrary. It is possible for Eq. 3.1.1 to hold for a subset of vectors $\mathbf{s}$ in R^n and to have $\mathbf{b} \neq \mathbf{0}$. For example, if Eq. 3.1.1 held for the subset of all $\mathbf{s}$ in R^n with $s_1 = 0$, then b_1 could be different from zero and the conclusion of the theorem would fail.

Variational Equation of Motion in Cartesian Coordinates

To apply the orthogonality theorem to obtain the variational equation of motion for a particle, Newton's equations of motion for a single particle may be written in the form

$$m\ddot{\mathbf{r}} - \mathbf{F} = \mathbf{0} \tag{3.1.3}$$

where $\mathbf{F}$ is the resultant force that acts on the particle. Identifying the left side of Eq. 3.1.3 with the vector $\mathbf{b}$ in Eq. 3.1.1, the orthogonality theorem implies that Eq. 3.1.3 holds if and only if

$$\delta\mathbf{r}^T(m\ddot{\mathbf{r}} - \mathbf{F}) = 0 \tag{3.1.4}$$

for all *virtual displacements* $\delta\mathbf{r}$; that is, $\delta\mathbf{r}$ plays the role of $\mathbf{s}$ in the theorem. This *variational equation of motion* is equivalent to Newton's equations of motion for the particle of Eq. 3.1.3 if and only if $\delta\mathbf{r}$ is arbitrary.

Equation 3.1.4 may be written as

$$m\,\delta\mathbf{r}^T\ddot{\mathbf{r}} - \delta\mathbf{r}^T\mathbf{F} = 0 \tag{3.1.5}$$

If the virtual displacement $\delta\mathbf{r}$ is interpreted as a small physical displacement of the particle, the second term in Eq. 3.1.5 may be interpreted as the work done by the resultant force $\mathbf{F}$ that acts on the particle, as the particle moves through the virtual displacement $\delta\mathbf{r}$. This observation is the basis for defining the *virtual work* of force $\mathbf{F}$ as

$$\delta W = \delta\mathbf{r}^T\mathbf{F} \tag{3.1.6}$$

As shown in Sections 1.2 and 1.3, it is often advantageous to separate forces that act on a particle into *applied forces* $\mathbf{F}^A$ and *constraint forces* $\mathbf{F}^C$ (i.e., $\mathbf{F} = \mathbf{F}^A + \mathbf{F}^C$). Substituting this representation into Eq. 3.1.6 yields

$$\begin{aligned} \delta W &= \delta\mathbf{r}^T(\mathbf{F}^A + \mathbf{F}^C) = \delta\mathbf{r}^T\mathbf{F}^A + \delta\mathbf{r}^T\mathbf{F}^C \\ &\equiv \delta W^A + \delta W^C \end{aligned} \tag{3.1.7}$$

where δW^A and δW^C are the virtual work of the applied and constraint forces, respectively. Using Eqs. 3.1.6 and 3.1.7, Eq. 3.1.5 may be written in the form

$$m\,\delta\mathbf{r}^T\ddot{\mathbf{r}} - \delta W^A - \delta W^C = 0 \tag{3.1.8}$$

which is equivalent to Eq. 3.1.3, provided that virtual displacements $\delta\mathbf{r}$ are arbitrary.

If, as was the case in all examples treated in Chapter 1, the constraint forces on a particle do no work as the system undergoes a *kinematically admissible virtual displacement*, then

$$\delta\mathbf{r}^T\mathbf{F}^C = \delta W^C = 0 \tag{3.1.9}$$

for all kinematically admissible virtual displacements $\delta\mathbf{r}$. Constraints that act on a particle are called *workless constraints* if Eq. 3.1.9 holds for all kinematically admissible virtual displacements. The goal is now to use this fact to obtain a useful expression for the constraint force $\mathbf{F}^C$.

Let a kinematic constraint on a particle be written in the form

$$\mathbf{\Phi}(\mathbf{r}, t) = \mathbf{0} \tag{3.1.10}$$

According to Eq. 2.8.16, a virtual displacement $\delta\mathbf{r}$ is kinematically admissible if

$$\mathbf{\Phi}_{\mathbf{r}}\,\delta\mathbf{r} = \mathbf{0} \tag{3.1.11}$$

Thus Eq. 3.1.9 holds for all $\delta\mathbf{r}$ that satisfy Eq. 3.1.11.

The foregoing observation and the theorem that follows yield an explicit expression for $\mathbf{F}^C$ that can be used in the variational equation of motion.

Lagrange Multiplier Theorem: Let $\mathbf{b}$ be a vector in R^n and let $\mathbf{A}$ be an $m \times n$ matrix. If

$$\mathbf{s}^T\mathbf{b} = 0 \tag{3.1.12}$$

for all $\mathbf{s}$ in R^n such that

$$\mathbf{A}\mathbf{s} = \mathbf{0} \tag{3.1.13}$$

then there exists a vector $\boldsymbol{\lambda}$ in R^m, called a *Lagrange multiplier*, such that

$$\mathbf{s}^T\mathbf{b} + \mathbf{s}^T\mathbf{A}^T\boldsymbol{\lambda} = 0 \tag{3.1.14}$$

for all $\mathbf{s}$ in R^n. Due to the orthogonality theorem, this is equivalent to

$$\mathbf{b} + \mathbf{A}^T\boldsymbol{\lambda} = \mathbf{0} \tag{3.1.15}$$

If $\mathbf{A}$ has full row rank, then $\boldsymbol{\lambda}$ is unique.

Due to its importance in dynamics, this theorem is proved at the end of this section. For now it is used to obtain an explicit expression for workless constraint force $\mathbf{F}^C$. Identifying $\mathbf{s}$ with $\delta\mathbf{r}$, $\mathbf{b}$ with $\mathbf{F}^C$, and $\mathbf{A}$ with $\mathbf{\Phi}_{\mathbf{r}}$, the Lagrange multiplier theorem may be applied to Eq. 3.1.9, which must hold for all $\delta\mathbf{r}$ that satisfy Eq. 3.1.11. Thus there exists a Lagrange multiplier $\boldsymbol{\lambda}$ such that

$$\delta\mathbf{r}^T\mathbf{F}^C + \delta\mathbf{r}^T\mathbf{\Phi}_{\mathbf{r}}^T\boldsymbol{\lambda} = \mathbf{0} \tag{3.1.16}$$

for all $\delta\mathbf{r}$ in R^3, hence the orthogonality theorem implies that

$$\mathbf{F}^C + \mathbf{\Phi}_\mathbf{r}^T \boldsymbol{\lambda} = \mathbf{0}$$

or

$$\mathbf{F}^C = -\mathbf{\Phi}_\mathbf{r}^T \boldsymbol{\lambda} \tag{3.1.17}$$

Thus workless constraint forces are a linear combination of the transpose of rows of the constraint Jacobian, with the Lagrange multipliers as coefficients. The Lagrange multipliers thus uniquely characterize the constraint forces if $\mathbf{\Phi}_\mathbf{r}$ has full row rank (i.e., if the constraint equations of Eq. 3.1.10 are independent).

Example 3.1.1: Consider the vehicle that is constrained to move along the straight track in Example 1.2.1 and let mutually orthogonal unit vectors $\mathbf{v}$ and $\mathbf{w}$ be orthogonal to $\mathbf{u}$, as in Example 1.2.6. Then the constraint equations for the straight track are

$$\mathbf{\Phi}(\mathbf{r}) \equiv \begin{bmatrix} \mathbf{v}^T(\mathbf{r} - \mathbf{r}_0) \\ \mathbf{w}^T(\mathbf{r} - \mathbf{r}_0) \end{bmatrix} = \mathbf{0} \tag{3.1.18}$$

Since the Jacobian is

$$\mathbf{\Phi}_\mathbf{r} = \begin{bmatrix} \mathbf{v}^T \\ \mathbf{w}^T \end{bmatrix} \tag{3.1.19}$$

by Eq. 3.1.17, the constraint force is

$$\mathbf{F}^C = -[\mathbf{v}, \mathbf{w}]\boldsymbol{\lambda} = -\lambda_1 \mathbf{v} - \lambda_2 \mathbf{w} \tag{3.1.20}$$

which is indeed orthogonal to the track.

Example 3.1.2: Consider the particle that is constrained to move on a unit sphere in Example 2.4.5. The constraint equation of the sphere is

$$\Phi(\mathbf{r}) \equiv \mathbf{r}^T\mathbf{r} - 1 = 0 \tag{3.1.21}$$

and its Jacobian is

$$\Phi_\mathbf{r} = 2\mathbf{r}^T \tag{3.1.22}$$

Thus, by Eq. 3.1.17, the constraint force is just

$$\mathbf{F}^C = -2\lambda\mathbf{r} \tag{3.1.23}$$

which is indeed orthogonal to the surface of the sphere.

It is important to recall that a knowledge of both the Lagrange multipliers and the constraint Jacobian are required in Eq. 3.1.17 to determine constraint forces. It is sometimes stated in the literature on dynamics that Lagrange multipliers are constraint forces. As shown in general by Eq. 3.1.17 and specifically in Examples 3.1.1 and 3.1.2, this is not correct.

With the constraint force of Eq. 3.1.17, the resultant force on the particle is

$$\mathbf{F} = \mathbf{F}^A - \mathbf{\Phi}_\mathbf{r}^T \boldsymbol{\lambda} \tag{3.1.24}$$

Thus Eq. 3.1.4 is

$$\delta \mathbf{r}^T(m\ddot{\mathbf{r}} - \mathbf{F}^A + \boldsymbol{\Phi}_{\mathbf{r}}^T\boldsymbol{\lambda}) = 0 \tag{3.1.25}$$

which must hold for arbitrary $\delta\mathbf{r}$. For workless constraints, the *Lagrange multiplier form of the constrained equations of motion of a particle*, in Cartesian coordinates, is thus

$$m\ddot{\mathbf{r}} + \boldsymbol{\Phi}_{\mathbf{r}}^T\boldsymbol{\lambda} = \mathbf{F}^A \tag{3.1.26}$$

This equation, together with the holonomic constraint equation of Eq. 3.1.10, comprises a system of differential-algebraic equations of motion for the particle.

Example 3.1.3: With gravity acting in the negative z direction on the vehicle of Example 3.1.1 (i.e., $\mathbf{F}^A = -mg\mathbf{k}$), Eqs. 3.1.19 and 3.1.26 yield the constrained equation of motion

$$m\ddot{\mathbf{r}} + \mathbf{v}\lambda_1 + \mathbf{w}\lambda_2 = -mg\mathbf{k}$$

where the constraint equations of Eq. 3.1.18 must hold. Note that this is equivalent to Eq. 1.2.4 in Example 1.2.1, but with the explicit form of $\mathbf{F}^C$ known.

Example 3.1.4: Gravity in the negative z direction acts on the particle of Examples 2.4.5 and 3.1.2 that moves on a unit sphere (i.e., a spherical pendulum). Using Eqs. 3.1.22 and 3.1.26, its constrained equations of motion are

$$m\ddot{\mathbf{r}} + 2\mathbf{r}\lambda = -mg\mathbf{k}$$

where the constraint equation of Eq. 3.1.21 must hold.

Since the equations of motion derived here and the constraint of Eq. 3.1.21 comprise a system of differential-algebraic equations, the method presented in Appendix D is used to obtain numerical results. To compare results with the spring model of the spherical pendulum studied in Example 1.1.3, the same numerical data are used here.

Consider first the initial conditions $x(0) = \dot{x}(0) = 0$, $y(0) = 0$, and $\dot{y}(0) = 0.05$, 0.1, and 0.15 m/s. Note that no initial conditions on z and $\dot{z}$ are needed, since they are determined by the constraint equations. Numerical results confirm that $x(t) = 0$, as expected, and that deviation of $z(t)$ from $z = -1$ m is very small. Plots of $y(t)$ in Fig. 3.1.1(a) show the periodic nature of pendulum motion. These results for the rigid pendulum confirm the validity of the spring model in Example 1.1.3.

Consider next the initial conditions $x(0) = 0.1$ m, $\dot{x}(0) = 0.1$ m/s, $y(0) = 0.1$ m, and $\dot{y}(0) = 0$. Numerical results plotted in Fig. 3.1.1(b) and (c) show that periodic motion of the spherical pendulum occurs, in fact more precisely than is predicted by the spring model of Example 1.1.3.

Finally, with initial conditions $x(0) = \dot{x}(0) = 0$, $y(0) = 0$, and $\dot{y}(0) = 4.42896$ m/s, the pendulum should swing up to a horizontal position, come to rest, and drop back. Numerical results confirm that $x(t) = 0$ and predict the $y(t)$ and $z(t)$ response shown in Fig. 3.1.1(d). Note that the theoretical result expected occurs and that these results confirm the validity of the stiff spring spherical pendulum model of Example 1.1.3, specifically Fig. 1.1.6(e).

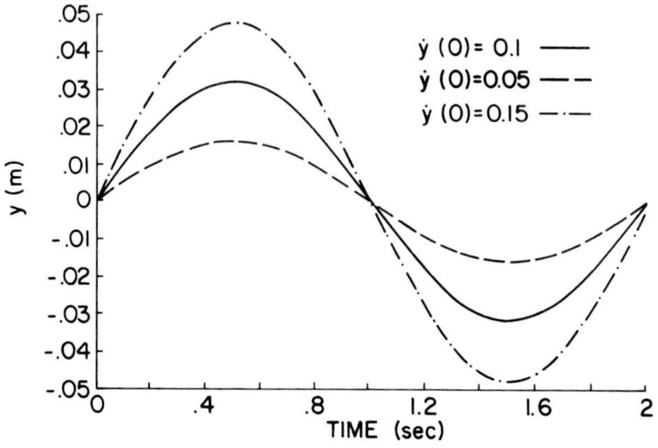

(a) y vs time for planar motion

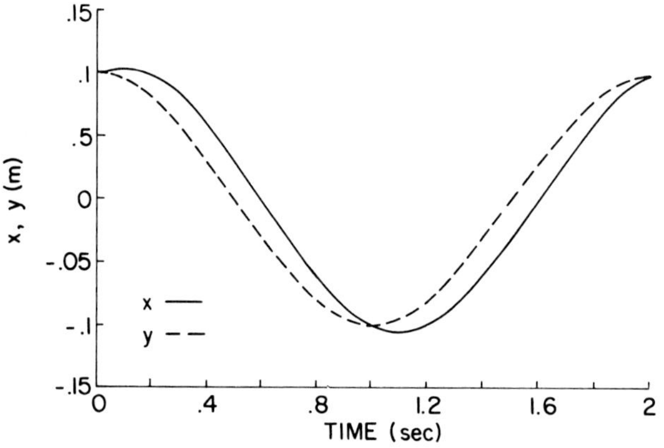

(b) x and y vs time for spatial motion

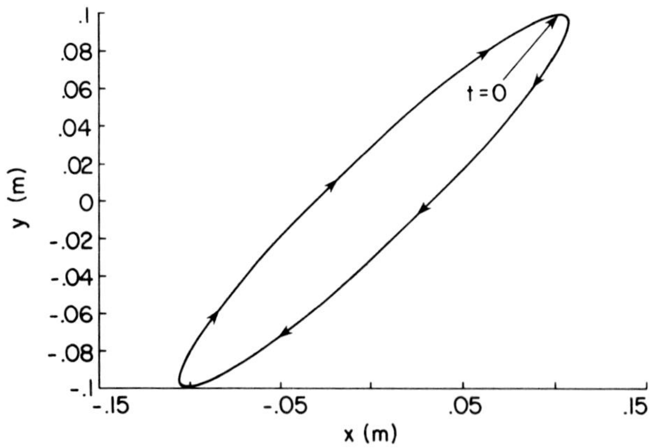

(c) y vs x for spatial motion

Figure 3.1.1 Numerical results for spherical pendulum.

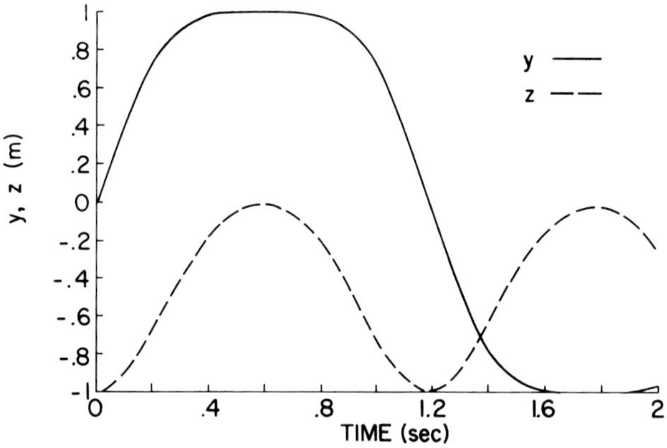

(d) y and z vs time for large amplitude planar motion **Figure 3.1.1** *(continued)*

For differential constraints of the form of Eq. 2.8.23 or Eq. 2.8.24,

$$\mathbf{B}(\mathbf{r}, t)\, d\mathbf{r} + \mathbf{c}(\mathbf{r}, t)\, dt = \mathbf{0}$$

virtual displacements must satisfy Eq. 2.8.25,

$$\mathbf{B}(\mathbf{r}, t)\, \delta\mathbf{r} = \mathbf{0}$$

which is true whether the differential constraint is nonholonomic or holonomic.

If the constraint force $\mathbf{F}^C$ associated with the differential constraint is workless,

$$\delta W^C = \delta\mathbf{r}^T\mathbf{F}^C = 0$$

for all $\delta\mathbf{r}$ that satisfy Eq. 2.8.25. Thus there exists a Lagrange multiplier $\boldsymbol{\mu}$ such that

$$\mathbf{F}^C = -\mathbf{B}^T\boldsymbol{\mu} \tag{3.1.27}$$

Following the same arguments used in deriving Eqs. 3.1.24 to 3.1.26, the *Lagrange multiplier form of the constrained equations of motion of a particle with differential constraints* is

$$m\ddot{\mathbf{r}} + \mathbf{B}^T\boldsymbol{\mu} = \mathbf{F}^A \tag{3.1.28}$$

It is tempting to apply this result to the nonholonomic missile pursuit problem of Example 2.7.5. The resulting equations are likely to be wrong, however, since it is not clear that the nonholonomic constraint that controls the velocity of the missile is workless.

Variational Equations of Motion in Generalized Coordinates

Consider motion of a particle whose kinematics are parameterized by a generalized coordinate vector $\mathbf{q}$, which may be a 1-, 2-, or 3-vector that defines the

Cartesian coordinate vector of the particle by the mapping

$$\mathbf{r} = \mathbf{r}(\mathbf{q}) \tag{3.1.29}$$

As in Section 2.8, a virtual displacement $\delta\mathbf{r}$ of the particle is related to a *generalized coordinate variation* $\delta\mathbf{q}$ by

$$\delta\mathbf{r} = \mathbf{r}_{\mathbf{q}}\,\delta\mathbf{q} \tag{3.1.30}$$

Differentiating Eq. 3.1.29 yields the velocity and acceleration relations

$$\begin{aligned} \dot{\mathbf{r}} &= \mathbf{r}_{\mathbf{q}}\dot{\mathbf{q}} \\ \ddot{\mathbf{r}} &= \mathbf{r}_{\mathbf{q}}\ddot{\mathbf{q}} + (\mathbf{r}_{\mathbf{q}}\dot{\mathbf{q}})_{\mathbf{q}}\dot{\mathbf{q}} \end{aligned} \tag{3.1.31}$$

Using Eq. 3.1.30, the virtual work of a force $\mathbf{F}$ in Eq. 3.1.6 may be written as

$$\delta W = \delta\mathbf{r}^T\mathbf{F} = \delta\mathbf{q}^T\mathbf{r}_{\mathbf{q}}^T\mathbf{F} \tag{3.1.32}$$

In analogy with the definition of virtual work as the scalar product of a force and the virtual displacement through which it is moved, the right side of Eq. 3.1.32 may be interpreted as the scalar product of a generalized coordinate variation $\delta\mathbf{q}$ and a *generalized force* $\mathbf{Q}$:

$$\delta W = \delta\mathbf{q}^T\mathbf{Q} \tag{3.1.33}$$

where the generalized force is defined as

$$\mathbf{Q} \equiv \mathbf{r}_{\mathbf{q}}^T\mathbf{F} \tag{3.1.34}$$

The procedure for calculating a generalized force in terms of a physical force is first to write the virtual work of the physical force in terms of a Cartesian virtual displacement, and second to use relationships between variations in generalized coordinates and virtual displacements to rewrite the virtual work in the form of Eq. 3.1.33. The coefficient of the generalized coordinate variation $\delta\mathbf{q}^T$ is then defined as the generalized force $\mathbf{Q}$. Since energy is a scalar quantity, the virtual work of the resultant of a set of forces is the sum of the virtual work of the individual forces. Thus the generalized force associated with each physical force that acts on a particle can be calculated and the total generalized force is obtained by adding individual generalized forces. Readers are encouraged to try calculating generalized forces (Prob. 3.1.1).

Distinguishing between applied force $\mathbf{F}^A$ and constraint force $\mathbf{F}^C$,

$$\begin{aligned} \mathbf{Q} &= \mathbf{Q}^A + \mathbf{Q}^C \\ &= \mathbf{r}_{\mathbf{q}}^T\mathbf{F}^A + \mathbf{r}_{\mathbf{q}}^T\mathbf{F}^C \end{aligned} \tag{3.1.35}$$

The virtual work is thus

$$\begin{aligned} \delta W &= \delta\mathbf{q}^T\mathbf{Q}^A + \delta\mathbf{q}^T\mathbf{Q}^C \\ &= \delta W^A + \delta W^C \end{aligned} \tag{3.1.36}$$

Substituting Eqs. 3.1.30 to 3.1.36 into the variational equation of motion of Eq. 3.1.5 yields

$$m\,\delta\mathbf{q}^T\mathbf{r}_\mathbf{q}^T(\mathbf{r}_\mathbf{q}\ddot{\mathbf{q}} + (\mathbf{r}_\mathbf{q}\dot{\mathbf{q}})_\mathbf{q}\dot{\mathbf{q}}) - \delta W = 0 \tag{3.1.37}$$

which must hold for all generalized coordinate variations $\delta\mathbf{q}$.

If generalized coordinates are consistent with constraints, then $\mathbf{q}$ is a vector of independent generalized coordinates and $\delta\mathbf{q}$ is arbitrary. If constraints that act on the particle are workless and if the generalized coordinates are independent,

$$\delta W^C = \delta\mathbf{q}^T\mathbf{Q}^C = 0 \tag{3.1.38}$$

for all $\delta\mathbf{q}$. By the orthogonality theorem, $\mathbf{Q}^C = \mathbf{0}$. Expanding δW, using Eq. 3.1.36 in this case, Eq. 3.1.37 reduces to

$$\delta\mathbf{q}^T(m\mathbf{r}_\mathbf{q}^T\mathbf{r}_\mathbf{q}\ddot{\mathbf{q}} + m\mathbf{r}_\mathbf{q}^T(\mathbf{r}_\mathbf{q}\dot{\mathbf{q}})_\mathbf{q}\dot{\mathbf{q}} - \mathbf{Q}^A) = 0 \tag{3.1.39}$$

which must hold for all generalized coordinate variations $\delta\mathbf{q}$. Thus if the generalized coordinates are independent and constraints are workless, the orthogonality theorem, applied to Eq. 3.1.39, yields the *differential equations of motion in independent generalized coordinates*,

$$m\mathbf{r}_\mathbf{q}^T\mathbf{r}_\mathbf{q}\ddot{\mathbf{q}} + m\mathbf{r}_\mathbf{q}^T(\mathbf{r}_\mathbf{q}\dot{\mathbf{q}})_\mathbf{q}\dot{\mathbf{q}} = \mathbf{Q}^A \tag{3.1.40}$$

As will be shown in general in Section 3.3, the matrix $\mathbf{r}_\mathbf{q}^T\mathbf{r}_\mathbf{q}$ is nonsingular if and only if $\mathbf{r}_\mathbf{q}$ has full column rank (i.e., if and only if the generalized coordinate mapping of Eq. 3.1.29 is nonsingular). Thus if care is taken to define independent generalized coordinates without singularities, the resulting equations of motion are well behaved.

The relatively simple form of Eq. 3.1.40 and the fact that it is a differential equation in independent generalized coordinates $\mathbf{q}$ are very significant. It is these factors that motivate the search for independent generalized coordinates in the study of dynamics of machines. Some words of caution are in order, however. First, there is no systematic analytical method of finding independent generalized coordinates (i.e., a measure of art and, more realistically, a great deal of experience is needed). Second, when frictional forces act, a knowledge of constraint forces is required to describe and introduce friction forces into the equations of motion. Lagrange multipliers and constraint equations, hence dependent generalized coordinates, are of substantial value in this case.

Example 3.1.5: The generalized coordinate q in Example 1.2.1 is independent and, from Eq. 1.2.1,

$$\mathbf{r}_q = \mathbf{u}$$

With $\mathbf{F}^A = -mg\mathbf{k}$, Eq. 3.1.34 yields

$$\mathbf{Q}^A = -mg\mathbf{u}^T\mathbf{k}$$

Since $(\mathbf{r}_q\dot{q})_q = (\mathbf{u}\dot{q})_q = \mathbf{0}$, Eq. 3.1.40 yields the differential equation of motion

$$m\ddot{q} = -mg\mathbf{u}^T\mathbf{k}$$

This is Eq. 1.2.7 of Example 1.2.2, which was derived in Chapter 1 using ad hoc methods.

➧ **Example 3.1.6:** For the vehicle on the helical parking ramp of Example 1.2.3, from Eq. 1.2.11, with $\phi = q$,

$$\mathbf{r}_q = \begin{bmatrix} -R\sin q \\ R\cos q \\ \dfrac{a}{2\pi} \end{bmatrix}$$

With $\mathbf{F}^A$ of Eq. 1.2.14,

$$\begin{aligned} \mathbf{Q}^A &= \left[-R\sin q,\ R\cos q,\ \frac{a}{2\pi}\right]\left\{-mg\mathbf{k} + \frac{T}{\sqrt{R^2 + a^2/(2\pi)^2}}\begin{bmatrix} -R\sin q \\ R\cos q \\ \dfrac{a}{2\pi} \end{bmatrix}\right\} \\ &= T\sqrt{R^2 + \frac{a^2}{(2\pi)^2}} - \frac{mga}{2\pi} \end{aligned}$$

Finally,

$$(\mathbf{r}_q\dot{q})_q\dot{q} = \begin{bmatrix} -R\cos q \\ -R\sin q \\ 0 \end{bmatrix}\dot{q}^2$$

so $\mathbf{r}_q^T(\mathbf{r}_q\dot{q})_q\dot{q} = 0$ and Eq. 3.1.40 yields

$$m\left[R^2 + \frac{a^2}{(2\pi)^2}\right]\ddot{q} = T\sqrt{R^2 + \frac{a^2}{(2\pi)^2}} - \frac{mga}{2\pi}$$

which is Eq. 1.2.16 of Example 1.2.3.

➧ **Example 3.1.7:** Consider the particle of Examples 2.4.5, 3.1.2, and 3.1.4 that moves on a unit sphere, under the action of gravity (i.e., $\mathbf{F}^A = -mg\mathbf{k}$). Using the independent generalized coordinates of Example 2.4.5, with the Jacobian of Eq. 2.4.23,

$$\mathbf{Q}^A = -mg\begin{bmatrix} 0 \\ \cos q_2 \end{bmatrix}$$

with careful manipulation (Prob. 3.1.3),

$$\mathbf{r}_\mathbf{q}^T\mathbf{r}_\mathbf{q} = \begin{bmatrix} \cos^2 q_2 & 0 \\ 0 & 1 \end{bmatrix}$$

$$\mathbf{r}_\mathbf{q}^T(\mathbf{r}_\mathbf{q}\dot{\mathbf{q}})_\mathbf{q}\dot{\mathbf{q}} = \sin q_2 \cos q_2 \begin{bmatrix} -2\dot{q}_1\dot{q}_2 \\ \dot{q}_1^2 \end{bmatrix}$$

Thus the differential equations of motion of Eq. 3.1.40 are

$$\begin{aligned} m\cos^2 q_2\ddot{q}_1 - 2m\sin q_2\cos q_2\dot{q}_1\dot{q}_2 &= 0 \\ m\ddot{q}_2 + m\sin q_2\cos q_2\dot{q}_1^2 &= -mg\cos q_2 \end{aligned} \tag{3.1.41}$$

Note that at $q_2 = \pm\pi/2$, the coefficient of $\ddot{q}_1$ in the first of Eqs. 3.1.41 is zero and the equations of motion are singular. This might have been expected, since it was shown in Example 2.4.5 that $\mathbf{r_q}$ is column rank deficient there, due to the essential singularity of spherical coordinates. This fact implies that $\mathbf{r_q}^T\mathbf{r_q}$ is singular when $q_2 = \pm\pi/2$, which is the source of the difficulty encountered here. The difficulty encountered here is due to selection of a singular set of generalized coordinates, not any flaw of nature. In this case, it may be preferable to work with the differential-algebraic equations of motion derived in Example 3.1.4 to avoid the singularity.

As in the case of Cartesian coordinates, if generalized coordinates $\mathbf{q}$ must satisfy constraint equations of the form

$$\mathbf{\Phi}(\mathbf{r}(\mathbf{q}), t) \equiv \mathbf{\Phi}(\mathbf{q}, t) = \mathbf{0} \tag{3.1.42}$$

they are dependent. Thus, generalized coordinate variations must satisfy Eq. 2.8.20; i.e.,

$$\mathbf{\Phi_q}\delta\mathbf{q} = \mathbf{0} \tag{3.1.43}$$

If constraint forces are workless, then from Eq. 3.1.17 they may be written as

$$\mathbf{F}^C = -\mathbf{\Phi}_\mathbf{r}^T\boldsymbol{\lambda}$$

In terms of generalized coordinates, using Eq. 3.1.30, the virtual work of the constraint reaction force is

$$\delta W^C = \delta\mathbf{r}^T\mathbf{F}^C = \delta\mathbf{q}^T\mathbf{r}_\mathbf{q}^T\mathbf{F}^C = -\delta\mathbf{q}^T\mathbf{r}_\mathbf{q}^T\mathbf{\Phi}_\mathbf{r}^T\boldsymbol{\lambda} = -\delta\mathbf{q}^T\mathbf{\Phi}_\mathbf{q}^T\boldsymbol{\lambda} = \delta\mathbf{q}^T\mathbf{Q}^C \tag{3.1.44}$$

for all $\delta\mathbf{q}$. Thus

$$\delta\mathbf{q}^T(\mathbf{Q}^C + \mathbf{\Phi}_\mathbf{q}^T\boldsymbol{\lambda}) = 0$$

for all $\delta\mathbf{q}$, so

$$\mathbf{Q}^C = -\mathbf{\Phi}_\mathbf{q}^T\boldsymbol{\lambda} \tag{3.1.45}$$

Note that the Lagrange multiplier $\boldsymbol{\lambda}$ gives both the generalized constraint force in Eq. 3.1.45 and the Cartesian constraint force in Eq. 3.1.17, with only different constraint Jacobians. This fact will be of great value in the study of the effects of Coulomb friction.

Substituting from Eqs. 3.1.44 and 3.1.45 into Eq. 3.1.37,

$$\delta\mathbf{q}^T[m\mathbf{r}_\mathbf{q}^T\mathbf{r}_\mathbf{q}\ddot{\mathbf{q}} + m\mathbf{r}_\mathbf{q}^T(\mathbf{r}_\mathbf{q}\dot{\mathbf{q}})_\mathbf{q}\dot{\mathbf{q}} + \mathbf{\Phi}_\mathbf{q}^T\boldsymbol{\lambda} - \mathbf{Q}^A] = 0$$

for all $\delta\mathbf{q}$. The orthogonality theorem yields the *Lagrange multiplier form of the constrained equations of motion* in generalized coordinates as

$$m\mathbf{r}_\mathbf{q}^T\mathbf{r}_\mathbf{q}\ddot{\mathbf{q}} + m\mathbf{r}_\mathbf{q}^T(\mathbf{r}_\mathbf{q}\dot{\mathbf{q}})_\mathbf{q}\dot{\mathbf{q}} + \mathbf{\Phi}_\mathbf{q}^T\boldsymbol{\lambda} = \mathbf{Q}^A \tag{3.1.46}$$

Equations 3.1.46 and 3.1.42 comprise a system of differential-algebraic equations of motion.

♦ **Example 3.1.8:** To illustrate the constrained generalized coordinate formulation with a single particle, consider again the vehicle on the helical ramp of Example 3.1.6. Using cylindrical coordinates of Section 2.1, with $s = R$, and $\mathbf{q} \equiv [q_1, q_2]^T = [\phi, z]^T$,

$$\mathbf{r} = \begin{bmatrix} R\cos q_1 \\ R\sin q_1 \\ q_2 \end{bmatrix} \tag{3.1.47}$$

which must satisfy the relation $q_2 = (a/2\pi)q_1$; that is, the constraint

$$\Phi(\mathbf{q}) \equiv q_2 - \frac{aq_1}{2\pi} = 0 \tag{3.1.48}$$

with

$$\mathbf{r_q} = \begin{bmatrix} -R\sin q_1 & 0 \\ R\cos q_1 & 0 \\ 0 & 1 \end{bmatrix}$$

The generalized applied force is due to gravity only, so

$$\mathbf{Q}^A = \mathbf{r_q}^T(-mg\mathbf{k}) = \begin{bmatrix} 0 \\ -mg \end{bmatrix}$$

and the essential terms in Eq. 3.1.46 are

$$\mathbf{r_q}^T\mathbf{r_q} = \begin{bmatrix} R^2 & 0 \\ 0 & 1 \end{bmatrix}$$

$$\mathbf{r_q}^T(\mathbf{r_q}\dot{\mathbf{q}})_\mathbf{q}\dot{\mathbf{q}} = \mathbf{0}$$

$$\Phi_\mathbf{q} = \begin{bmatrix} -\frac{a}{2\pi}, & 1 \end{bmatrix}$$

Thus Eq. 3.1.46 yields

$$\begin{aligned} mR^2\ddot{q}_1 - \frac{a}{2\pi}\lambda &= 0 \\ m\ddot{q}_2 + \lambda &= -mg \end{aligned} \tag{3.1.49}$$

Equations 3.1.49 and 3.1.48 comprise the differential-algebraic equations of motion for the vehicle.

In terms of generalized coordinates, the differential constraint of Eq. 2.8.24 may be written, using Eq. 2.8.14, as

$$\mathbf{Br_q}\, d\mathbf{q} + (\mathbf{c} + \mathbf{Br}_t)\, dt = \mathbf{0}$$

or

$$\mathbf{D}d\mathbf{q} + \mathbf{e}dt = \mathbf{0} \tag{3.1.50}$$

where

$$\begin{aligned} \mathbf{D} &= \mathbf{Br_q} \\ \mathbf{e} &= \mathbf{c} + \mathbf{Br}_t \end{aligned} \tag{3.1.51}$$

For a workless differential constraint, using Eqs. 3.1.27 and 3.1.51 yields

$$\begin{aligned}\delta W^C &= \delta \mathbf{r}^T \mathbf{F}^C \\ &= \delta \mathbf{q}^T \mathbf{r}_\mathbf{q}^T \mathbf{F}^C \\ &= -\delta \mathbf{q}^T \mathbf{r}_\mathbf{q}^T \mathbf{B}^T \boldsymbol{\mu} \\ &= -\delta \mathbf{q}^T \mathbf{D}^T \boldsymbol{\mu} \\ &\equiv \delta \mathbf{q}^T \mathbf{Q}^C\end{aligned}$$

so

$$\mathbf{Q}^C = -\mathbf{D}^T \boldsymbol{\mu}$$

Repeating the same steps taken in deriving Eq. 3.1.46, the *Lagrange multiplier form of the constrained equations of motion* in generalized coordinates, for differential constraints of Eq. 3.1.50, is

$$m\mathbf{r}_\mathbf{q}^T \mathbf{r}_\mathbf{q} \ddot{\mathbf{q}} + m\mathbf{r}_\mathbf{q}^T (\mathbf{r}_\mathbf{q} \dot{\mathbf{q}})_\mathbf{q} \dot{\mathbf{q}} + \mathbf{D}^T \boldsymbol{\mu} = \mathbf{Q}^A \tag{3.1.52}$$

Caution must be exercised in interpreting constraint forces that are predicted by Lagrange multipliers in the generalized coordinate formulation. Generalized coordinates are often defined so that motion that is admitted by Eq. 3.1.29 automatically satisfies some of the physical constraints that act on the particle. The remaining physical constraints that are defined by the constraint equations of Eq. 3.1.42 thus comprise only *partial constraints* on the particle. The associated *partial constraint force* $\mathbf{F}^C = -\boldsymbol{\Phi}_\mathbf{r}^T \boldsymbol{\lambda}$, therefore, provides only the component of the constraint force associated with the partial constraint of Eq. 3.1.42. Additional analysis in a Cartesian coordinate setting is required to recover the remaining components of constraint force. The example that follows illustrates this feature of the generalized coordinate formulation.

Example 3.1.9: Consider the simple pendulum of Fig. 3.1.2, with unit length and with gravity acting in the direction $\mathbf{u} = (1/\sqrt{3})[1, 1, 1]^T$. In Cartesian coordinates, the constraint equations are

$$\boldsymbol{\Phi}(\mathbf{r}) \begin{bmatrix} x^2 + y^2 - 1 \\ z \end{bmatrix} = \mathbf{0}$$

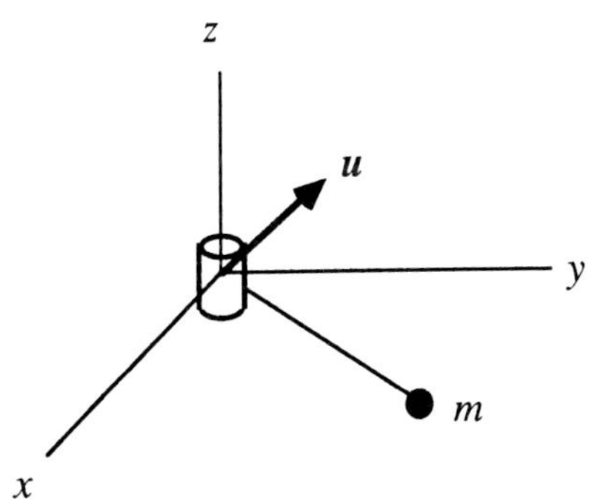

Figure 3.1.2 Simple pendulum.

Thus the full vector of constraint forces is

$$\mathbf{F}^C = -\begin{bmatrix} 2x & 0 \\ 2y & 0 \\ 0 & 1 \end{bmatrix}\begin{bmatrix} \lambda_1 \\ \lambda_2 \end{bmatrix}$$

and the constrained equations of motion in generalized coordinates are

$$m\ddot{\mathbf{r}} + \begin{bmatrix} 2x & 0 \\ 2y & 0 \\ 0 & 1 \end{bmatrix}\boldsymbol{\lambda} = \frac{mg}{\sqrt{3}}\begin{bmatrix} 1 \\ 1 \\ 1 \end{bmatrix}$$

Using generalized coordinates $q_1 = x$ and $q_2 = y$, with

$$\mathbf{r} = r(\mathbf{q}) \equiv \begin{bmatrix} q_1 \\ q_2 \\ 0 \end{bmatrix}$$

the second Cartesian constraint equation $z = 0$ is automatically satisfied. The partial constraint on $\mathbf{q}$ is thus

$$\begin{aligned} \hat{\Phi} &= x^2 + y^2 - 1 \\ &= q_1^2 + q_2^2 - 1 = 0 \end{aligned}$$

The associated partial constraint force is thus

$$\hat{\mathbf{F}}^C = -\hat{\Phi}_{\mathbf{r}}^T\hat{\lambda} = -\begin{bmatrix} 2x \\ 2y \\ 0 \end{bmatrix}\hat{\lambda}$$

With

$$\mathbf{r_q} = \begin{bmatrix} 1 & 0 \\ 0 & 1 \\ 0 & 0 \end{bmatrix}$$

$$\hat{\Phi}_{\mathbf{q}} = \hat{\Phi}_{\mathbf{r}}\mathbf{r_q} = [2x, 2y] = [2q_1, 2q_2]$$

$$\mathbf{Q}^A = \mathbf{r}_{\mathbf{q}}^T\mathbf{F}^A = \frac{mg}{\sqrt{3}}[1, 1]^T$$

the constrained equations of motion in generalized coordinates, from Eq. 3.1.46, are

$$m\ddot{\mathbf{q}} + 2\mathbf{q}\hat{\lambda} = \frac{mg}{\sqrt{3}}\begin{bmatrix} 1 \\ 1 \end{bmatrix}$$

In this elementary example, $\hat{\lambda} = \lambda_1$ is half of the tension in the bar of the pendulum and λ_2 is the axial constraint force in the hinge of the pendulum. The latter component of constraint force is suppressed in the generalized coordinate formulation, since the associated constraint is automatically satisfied. Interpretation of partial constraint forces may not be as obvious in more complex applications.

While illustrating the difficulty of suppressed components of constraint forces, this example illustrates an advantage of the generalized coordinate formulation. In the Cartesian formulation, there are three equations of motion and two constraint equations, for five variables. In the generalized coordinate formulation, in contrast, there are only two equations of motion and one constraint equation, for three variables.

The power of the variational approach presented here is that it is broadly applicable in a systematic way, whereas direct manipulation with Newton's equations of motion is dependent on ad hoc methods for success. Furthermore, the variational form of Eq. 3.1.39 is valid for constrained systems, provided that generalized coordinate variations $\delta\mathbf{q}$ are kinematically admissible. This form of the variational equations of motion is valuable in treating constrained systems for which the definition of an independent generalized coordinate parameterization is difficult or impossible.

Proof of the Lagrange Multiplier Theorem

If the matrix **A** of Eq. 3.1.13 does not have full row rank, one or more of its rows is a linear combination of the remaining rows. Presume for now that such rows have been eliminated so that the reduced matrix, denoted as **A** for convenience, has full row rank.

By the coordinate partitioning theorem, columns of **A** and elements of **s** can be interchanged so that Eq. 3.1.13 is of the form

$$[\mathbf{C}, \mathbf{D}]\begin{bmatrix}\mathbf{u}\\ \mathbf{v}\end{bmatrix} = \mathbf{0}$$

or

$$\mathbf{Cu} = -\mathbf{Dv} \tag{3.1.53}$$

where **C** is nonsingular. Thus

$$\mathbf{u} = -\mathbf{C}^{-1}\mathbf{Dv} \tag{3.1.54}$$

Reordering elements of **b** in the same way as **s**, Eq. 3.1.12 may be written in the form

$$[\mathbf{u}^T, \mathbf{v}^T]\begin{bmatrix}\mathbf{e}\\ \mathbf{f}\end{bmatrix} = 0$$

or

$$\mathbf{u}^T\mathbf{e} + \mathbf{v}^T\mathbf{f} = 0$$

Substituting for **u** from Eq. 3.1.54, this is

$$\mathbf{v}^T(-\mathbf{D}^T\mathbf{C}^{-1^T}\mathbf{e} + \mathbf{f}) = 0$$

which must hold for all **v**. The orthogonality theorem thus implies that

$$\mathbf{f} = \mathbf{D}^T\mathbf{C}^{-1^T}\mathbf{e} \tag{3.1.55}$$

Defining

$$\boldsymbol{\mu} = -\mathbf{C}^{-1^T}\mathbf{e} \tag{3.1.56}$$

Eq. 3.1.55 may be written in the form

$$\mathbf{f} = -\mathbf{D}^T\boldsymbol{\mu} \tag{3.1.57}$$

Multiplying both sides of Eq. 3.1.56 on the left by $\mathbf{C}^T$,

$$\mathbf{e} = -\mathbf{C}^T\boldsymbol{\mu} \tag{3.1.58}$$

Equations 3.1.58 and 3.1.57 may be written in matrix form as

$$\begin{bmatrix} \mathbf{e} \\ \mathbf{f} \end{bmatrix} = -\begin{bmatrix} \mathbf{C}^T \\ \mathbf{D}^T \end{bmatrix}\boldsymbol{\mu} = -[\mathbf{C}, \mathbf{D}]^T\boldsymbol{\mu} \tag{3.1.59}$$

Reordering the columns of $[\mathbf{C}, \mathbf{D}]$ and the rows of $\begin{bmatrix} \mathbf{e} \\ \mathbf{f} \end{bmatrix}$ and $\boldsymbol{\mu}$ to recover $\mathbf{A}$ and $\mathbf{b}$, $\boldsymbol{\mu}$ becomes $\boldsymbol{\lambda}$ and Eq. 3.1.59 yields the desired result,

$$\mathbf{b} + \mathbf{A}^T\boldsymbol{\lambda} = \mathbf{0} \tag{3.1.60}$$

To see that $\boldsymbol{\lambda}$ is unique if $\mathbf{A}$ has full row rank, assume that there is a second vector $\boldsymbol{\lambda}'$ such that Eq. 3.1.60 holds; that is,

$$\mathbf{b} + \mathbf{A}^T\boldsymbol{\lambda}' = \mathbf{0}$$

Subtracting these equations yields

$$\mathbf{A}^T(\boldsymbol{\lambda} - \boldsymbol{\lambda}') = \mathbf{0}$$

the left side of which is a linear combination of the columns of $\mathbf{A}^T$. Since the columns of $\mathbf{A}^T$ are linearly independent, $\boldsymbol{\lambda} - \boldsymbol{\lambda}' = \mathbf{0}$. Thus $\boldsymbol{\lambda}' = \boldsymbol{\lambda}$ and $\boldsymbol{\lambda}$ is in fact unique. By the orthogonality theorem, Eq. 3.1.60 is equivalent to Eq. 3.1.14, for all $\mathbf{s}$ in R^n. This completes the proof if $\mathbf{A}$ has full row rank.

If $\mathbf{A}$ fails to have full row rank, the proof above is valid, with $\mathbf{A}$ replaced by a matrix $\mathbf{A}'$ that is formed from $\mathbf{A}$ by deleting dependent rows $\mathbf{A}''$ of $\mathbf{A}$. In this case, Eq. 3.1.60 may be written as

$$\mathbf{b} = -\mathbf{A}'^T\boldsymbol{\lambda}^* \tag{3.1.61}$$

the right side of which is a unique linear combination of columns of $\mathbf{A}'^T$. Since the columns in $\mathbf{A}'^T$ and $\mathbf{A}''^T$ are linearly dependent, there are vectors $\boldsymbol{\mu}$ and $\boldsymbol{\nu}$, that are not zero such that

$$\mathbf{A}'^T\boldsymbol{\mu} + \mathbf{A}''^T\boldsymbol{\nu} = \mathbf{0} \tag{3.1.62}$$

Multiplying both sides on the left by $\mathbf{A}'$ and noting that $\mathbf{A}'\mathbf{A}'^T$ is nonsingular (Prob. 3.1.4),

$$\boldsymbol{\mu} = -(\mathbf{A}'\mathbf{A}'^T)^{-1}\mathbf{A}'\mathbf{A}''\boldsymbol{\nu}$$

and Eq. 3.1.62 becomes

$$-\mathbf{A}'^T[(\mathbf{A}'\mathbf{A}'^T)^{-1}\mathbf{A}'\mathbf{A}'']\boldsymbol{\nu} + \mathbf{A}''^T\boldsymbol{\nu} = \mathbf{0} \tag{3.1.63}$$

for any $\boldsymbol{\nu}$. Adding Eq. 3.1.63 to Eq. 3.1.61,

$$\begin{aligned}\mathbf{b} &= -\mathbf{A}'^T\{\boldsymbol{\lambda}^* + (\mathbf{A}'\mathbf{A}'^T)^{-1}\mathbf{A}'\mathbf{A}''\boldsymbol{\nu}\} + \mathbf{A}''^T\boldsymbol{\nu} \\ &= -\mathbf{A}^T\boldsymbol{\lambda}\end{aligned}$$

where, presuming $\mathbf{A}''$ are the bottom rows of $\mathbf{A}$,

$$\boldsymbol{\lambda} = \begin{bmatrix}\boldsymbol{\lambda}^* + (\mathbf{A}'\mathbf{A}'^T)^{-1}\mathbf{A}'\mathbf{A}''\boldsymbol{\nu} \\ -\boldsymbol{\nu}\end{bmatrix}$$

Thus $\boldsymbol{\lambda}$ contains the unique $\boldsymbol{\lambda}^*$ from Eq. 3.1.61 and the arbitrary vector $\boldsymbol{\nu}$. This shows that even if $\mathbf{A}$ fails to have full row rank, there exists a $\boldsymbol{\lambda}$ that satisfies Eqs. 3.1.14 and 3.1.15. However, it is not unique.

3.2 VARIATIONAL EQUATIONS OF MOTION FOR SYSTEMS OF PARTICLES

A system of nb particles in three-dimensional space is shown in Fig. 3.2.1. Particle masses are m_i, $i = 1, \ldots, nb$, and the particles are located by their Cartesian coordinate vectors $\mathbf{r}_i$ in an inertial x-y-z Cartesian reference frame. Two types of forces act on each particle. The applied resultant force $\mathbf{F}_i^A$ that acts on particle i

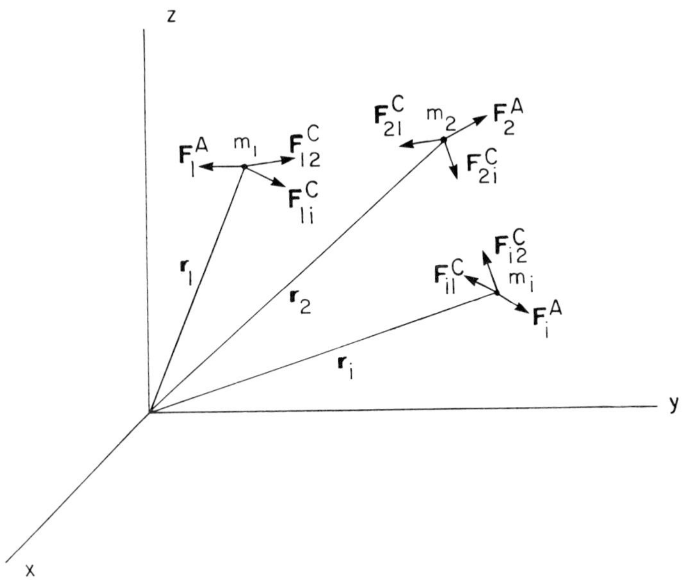

Figure 3.2.1 System of particles.

includes the effect of springs, dampers, and other force elements between particles that do work on the system. Constraint forces $\mathbf{F}_{ij}^C$ act on particle i, due to the influence of particle j, with $\mathbf{F}_{ii}^C \equiv \mathbf{0}$, and do no work on the system. Recall further that Newton's third law yields

$$\mathbf{F}_{ij}^C = -\mathbf{F}_{ji}^C \tag{3.2.1}$$

Holonomic constraints between pairs of particles in the system (e.g., particles i and j) that can be written in algebraic form are

$$\mathbf{\Phi}^{ij}(\mathbf{r}_i, \mathbf{r}_j, \mathbf{t}) = \mathbf{0} \tag{3.2.2}$$

Thus virtual displacements of particles i and j must satisfy

$$\mathbf{\Phi}_{\mathbf{r}_i}^{ij}\,\delta\mathbf{r}_i + \mathbf{\Phi}_{\mathbf{r}_j}^{ij}\,\delta\mathbf{r}_j = \mathbf{0} \tag{3.2.3}$$

Differential constraints, whether nonholonomic or holonomic, between particles i and j are written in the form

$$\mathbf{B}^{ij}\dot{\mathbf{r}}_i + \mathbf{B}^{ji}\dot{\mathbf{r}}_j + \mathbf{c}^{ij} = \mathbf{0} \tag{3.2.4}$$

Thus virtual displacements of particles i and j must satisfy

$$\mathbf{B}^{ij}\,\delta\mathbf{r}_i + \mathbf{B}^{ji}\,\delta\mathbf{r}_j = \mathbf{0} \tag{3.2.5}$$

By Newton's second law, the equations of motion of particle i are

$$m_i\ddot{\mathbf{r}}_i = \mathbf{F}_i^A + \sum_{j=1}^{nb} \mathbf{F}_{ij}^C, \quad i = 1, \ldots, nb \tag{3.2.6}$$

where the resultant force on the right side of the equation accounts for both the known applied forces and the unknown constraint forces that act on each particle. This constitutes a system of $3nb$ second order differential equations in $3nb$ Cartesian coordinates of the particles and the unknown constraint forces, which appear algebraically. Thus Eqs. 3.2.6, 3.2.2, and 3.2.4 comprise a system of *differential-algebraic equations* of motion for the Cartesian coordinates of the particles and the constraint forces.

Variational Equations of Motion in Cartesian Coordinates

As in Section 3.1, the variational equations of motion for the system of particles in Fig. 3.2.1 may be obtained by forming the scalar product of both sides of Eq. 3.2.6 with an arbitrary virtual displacement $\delta\mathbf{r}_i$ of particle i and summing over all bodies in the system, to obtain

$$\sum_{i=1}^{nb} \delta\mathbf{r}_i^T \left(m_i\ddot{\mathbf{r}}_i - \mathbf{F}_i^A - \sum_{j=1}^{nb} \mathbf{F}_{ij}^C \right) = 0 \tag{3.2.7}$$

This *variational equation of motion* holds for arbitrary virtual displacements $\delta\mathbf{r}_i$, so it is equivalent to the system of equations of Eq. 3.2.6.

Throughout the remainder of this book, forces between particles that do virtual work are included in $\mathbf{F}_i^A$ and are distinguished from constraint forces that do no virtual work for kinematically admissible virtual displacements (i.e., for $\delta\mathbf{r}_i$ and $\delta\mathbf{r}_j$ that satisfy Eqs. 3.2.3 and 3.2.5).

Consider a pair of particles m_i and m_j in Fig. 3.2.1. The total virtual work done by constraint forces $\mathbf{F}_{ij}^C$ and $\mathbf{F}_{ji}^C$ on particles i and j is

$$\delta W_{ij}^C = \delta\mathbf{r}_i^T\mathbf{F}_{ij}^C + \delta\mathbf{r}_j^T\mathbf{F}_{ji}^C = 0 \tag{3.2.8}$$

for all $\delta\mathbf{r}_i$ and $\delta\mathbf{r}_j$ that are consistent with constraints on the particles.

For the holonomic constraint of Eq. 3.2.2, Eq. 3.2.8 must hold for all $\delta\mathbf{r}_i$ and $\delta\mathbf{r}_j$ that satisfy Eq. 3.2.3. By the Lagrange multiplier theorem, there exists a multiplier $\boldsymbol{\lambda}^{ij}$ such that

$$\delta\mathbf{r}_i^T(\mathbf{F}_{ij}^C + \boldsymbol{\Phi}_{\mathbf{r}_i}^{ij^T}\boldsymbol{\lambda}^{ij}) + \delta\mathbf{r}_j^T(\mathbf{F}_{ji}^C + \boldsymbol{\Phi}_{\mathbf{r}_j}^{ij^T}\boldsymbol{\lambda}^{ij}) = 0$$

for all $\delta\mathbf{r}_i$ and $\delta\mathbf{r}_j$. Thus the orthogonality theorem implies that

$$\begin{aligned}\mathbf{F}_{ij}^C &= -\boldsymbol{\Phi}_{\mathbf{r}_i}^{ij^T}\boldsymbol{\lambda}^{ij}\\ \mathbf{F}_{ji}^C &= -\boldsymbol{\Phi}_{\mathbf{r}_j}^{ij^T}\boldsymbol{\lambda}^{ij}\end{aligned} \tag{3.2.9}$$

For the differential constraint of Eq. 3.2.4, Eq. 3.2.8 must hold for all $\delta\mathbf{r}_i$ and $\delta\mathbf{r}_j$ that satisfy Eq. 3.2.5. Thus there exists a Lagrange multiplier $\boldsymbol{\mu}^{ij}$ such that

$$\delta\mathbf{r}_i^T(\mathbf{F}_{ij}^C + \mathbf{B}^{ij^T}\boldsymbol{\mu}^{ij}) + \delta\mathbf{r}_j^T(\mathbf{F}_{ji}^C + \mathbf{B}^{ji^T}\boldsymbol{\mu}^{ij}) = 0$$

for all $\delta\mathbf{r}_i$ and $\delta\mathbf{r}_j$. This yields

$$\begin{aligned}\mathbf{F}_{ij}^C &= -\mathbf{B}^{ij^T}\boldsymbol{\mu}^{ij}\\ \mathbf{F}_{ji}^C &= -\mathbf{B}^{ji^T}\boldsymbol{\mu}^{ij}\end{aligned} \tag{3.2.10}$$

Using the conventions $\boldsymbol{\lambda}^{ij} \equiv \boldsymbol{\lambda}^{ji}$, $\boldsymbol{\mu}^{ij} \equiv \boldsymbol{\mu}^{ji}$, $\boldsymbol{\lambda}^{ij} = \mathbf{0}$ if there is no holonomic constraint of the form of Eq. 3.2.2 between particles i and j, and $\boldsymbol{\mu}^{ij} = \mathbf{0}$ if there is no differential constraint of the form of Eq. 3.2.4, Eq. 3.2.7 can be written in the form

$$\sum_{i=1}^{nb} \delta\mathbf{r}_i^T\left(m_i\ddot{\mathbf{r}}_i - \mathbf{F}_i^A + \sum_{j=1}^{nb} \boldsymbol{\Phi}_{\mathbf{r}_i}^{ij^T}\boldsymbol{\lambda}^{ij} + \sum_{j=1}^{nb} \mathbf{B}^{ij^T}\boldsymbol{\mu}^{ij}\right) = 0 \tag{3.2.11}$$

for all $\delta\mathbf{r}_i$, $i = 1, \ldots, nb$. The orthogonality theorem thus yields the *Lagrange multiplier form of the equations of motion for a constrained system of particles*,

$$m_i\ddot{\mathbf{r}}_i + \sum_{j=1}^{nb} \boldsymbol{\Phi}_{\mathbf{r}_i}^{ij^T}\boldsymbol{\lambda}^{ij} + \sum_{j=1}^{nb} \mathbf{B}^{ij^T}\boldsymbol{\mu}^{ij} = \mathbf{F}_i^A, \qquad i = 1, \ldots, nb \tag{3.2.12}$$

Equations 3.2.12, 3.2.2, and 3.2.4 comprise a system of differential-algebraic equations of motion for the system. It is important to note that this system has the same number of equations as unknowns. This is in contrast to the system of

Eqs. 3.2.6, 3.2.2, and 3.2.4, since there are more unknown components of constraint forces $\mathbf{F}_{ij}^C$ than components of Lagrange multipliers $\boldsymbol{\lambda}^{ij}$ or $\boldsymbol{\mu}^{ij}$. This important fact follows from Eqs. 3.2.9 and 3.2.10. If there are $k = 1$ or 2 constraints between particles, then there are just k components of $\boldsymbol{\lambda}^{ij}$ or $\boldsymbol{\mu}^{ij}$. In contrast, there are three unknown components of $\mathbf{F}_{ij}^C = \mathbf{F}_{ji}^C$. The reduction in dimension of the unknowns is due to the fact that the constraint forces defined by Eqs. 3.2.9 and 3.2.10 are consistent with the constraints (e.g., they are orthogonal to surfaces of contact in physical joints).

Example 3.2.1: The equations of motion of the dumbbell of Examples 1.3.1 and 2.6.4, modeled as two point masses connected by a massless bar of unit length, may now be written using Lagrange multipliers. Kinematics of the dumbbell was treated in Example 2.6.4. This model of the dumbbell is shown in Fig. 3.2.2, with externally applied forces.

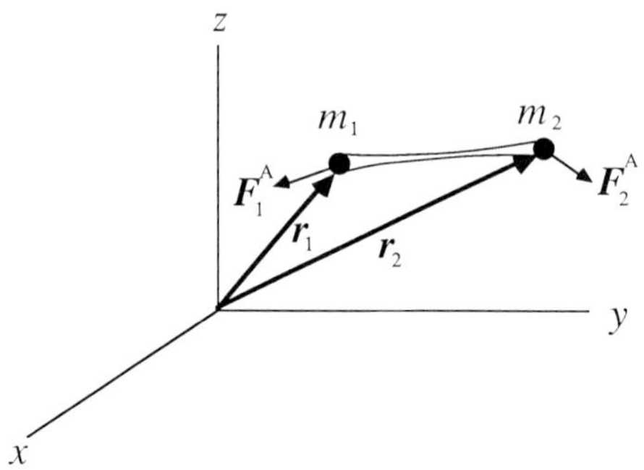

Figure 3.2.2 Dumbbell.

Using the constraint Jacobians of Eq. 2.6.10 in Eq. 3.2.12 yields

$$\begin{aligned} m_1\ddot{\mathbf{r}}_1 - 2(\mathbf{r}_2 - \mathbf{r}_1)\lambda &= \mathbf{F}_1^A \\ m_2\ddot{\mathbf{r}}_2 + 2(\mathbf{r}_2 - \mathbf{r}_1)\lambda &= \mathbf{F}_2^A \end{aligned} \tag{3.2.13}$$

Comparing results with Eq. 1.3.2 of Example 1.3.1, the tensile constraint force in the bar of the dumbbell is $\mathbf{F}^C = 2\lambda$.

Example 3.2.2: The wheel and axle assembly of Example 2.7.7 can be modeled as a two particle system in a plane, as shown in Fig. 3.2.3, where the axle

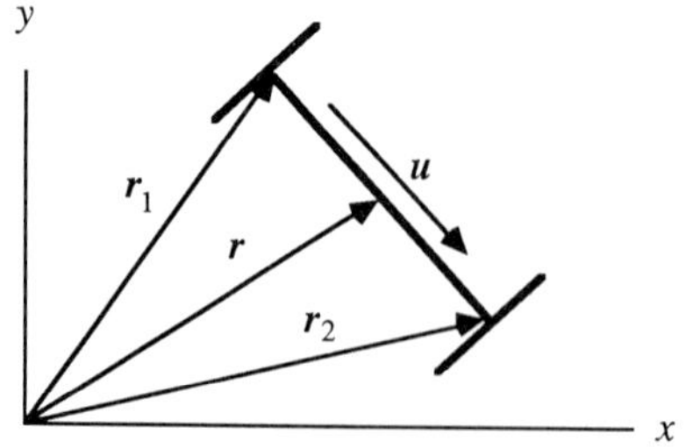

Figure 3.2.3 Wheel and axle set.

has unit length. The vectors **r** and **u** used in the formulation of Example 2.7.7 are

$$\mathbf{r} = \tfrac{1}{2}(\mathbf{r}_1 + \mathbf{r}_2)$$

$$\mathbf{u} = \mathbf{r}_2 - \mathbf{r}_1$$

The nonholonomic constraint $\mathbf{u}^T\dot{\mathbf{r}} = 0$ of Example 2.7.7, in the present formulation, is the nonholonomic differential constraint (see Prob. 2.7.3)

$$(\mathbf{r}_2 - \mathbf{r}_1)^T(\dot{\mathbf{r}}_1 + \dot{\mathbf{r}}_2) = 0 \tag{3.2.14}$$

where the factor $\frac{1}{2}$ has been canceled. In the notation of Eq. 3.2.4,

$$\mathbf{B}^{12} = \mathbf{B}^{21} = (\mathbf{r}_2 - \mathbf{r}_1)^T$$

$$c^{12} = 0$$

In addition to the differential constraint of Eq. 3.2.14, the distance constraint between particles is

$$\Phi \equiv (\mathbf{r}_2 - \mathbf{r}_1)^T(\mathbf{r}_2 - \mathbf{r}_1) - 1 = 0 \tag{3.2.15}$$

The Jacobian of this constraint, for $\mathbf{r} = [\mathbf{r}_1^T, \mathbf{r}_2^T]^T$, is

$$\Phi_{\mathbf{r}} = [\Phi_{\mathbf{r}_1}, \Phi_{\mathbf{r}_2}] = 2[-(\mathbf{r}_2 - \mathbf{r}_1)^T, (\mathbf{r}_2 - \mathbf{r}_1)^T]$$

Thus the equations of motion of Eq. 3.2.12 are

$$\begin{aligned} m\ddot{\mathbf{r}}_1 - 2(\mathbf{r}_2 - \mathbf{r}_1)\lambda + (\mathbf{r}_2 - \mathbf{r}_1)\mu &= 0 \\ m\ddot{\mathbf{r}}_2 + 2(\mathbf{r}_2 - \mathbf{r}_1)\lambda + (\mathbf{r}_2 - \mathbf{r}_1)\mu &= 0 \end{aligned} \tag{3.2.16}$$

Equations 3.2.14, 3.2.15, and 3.2.16 comprise a system of six differential-algebraic equations for the six variables $\mathbf{r}_1$, $\mathbf{r}_2$, λ, and μ. Note that it involves both holonomic and nonholonomic constraints.

♦ **Example 3.2.3:** The elementary model of the slider–crank mechanism of Example 1.3.2 (see Fig. 1.3.2) may be treated directly using the constrained equations of motion. Denoting the base of the mechanism (ground) as particle 3, which does not move and hence requires no equations of motion, the constraints of Eq. 1.3.4 may be written in the form

$$\begin{aligned} \Phi^{13} &= \mathbf{r}_1^T\mathbf{r}_1 - \ell_1^2 = 0 \\ \Phi^{23} &= \mathbf{r}_2^T\mathbf{j} = 0 \\ \Phi^{12} &= (\mathbf{r}_2 - \mathbf{r}_1)^T(\mathbf{r}_2 - \mathbf{r}_1) - 1 = 0 \end{aligned} \tag{3.2.17}$$

With $\mathbf{F}_1^A = \mathbf{0}$ and $\mathbf{F}_2^A = F_2^A\mathbf{i}$, the equations of motion of Eq. 3.2.12 are

$$\begin{aligned} m_1\ddot{\mathbf{r}}_1 + 2\mathbf{r}_1\lambda^{13} - 2(\mathbf{r}_2 - \mathbf{r}_1)\lambda^{12} &= \mathbf{0} \\ m_2\ddot{\mathbf{r}}_2 + 2(\mathbf{r}_2 - \mathbf{r}_1)\lambda^{12} + \mathbf{j}\lambda^{23} &= F_2^A\mathbf{i} \end{aligned} \tag{3.2.18}$$

Comparing Eqs. 1.3.8 and 3.2.18, it is clear that $F^{C_1}/\ell_1 = 2\lambda^{13}$, $F^{C_2} = 2\lambda^{12}$, and $F^{C_3} = \lambda^{23}$.

♦ **Example 3.2.4:** Consider the blocks shown in Fig. 3.2.4, modeled as particles, with coefficients of friction μ_1 and μ_2 between sliding surfaces. The blocks move only in the x-y plane and gravity acts in the negative y direction.

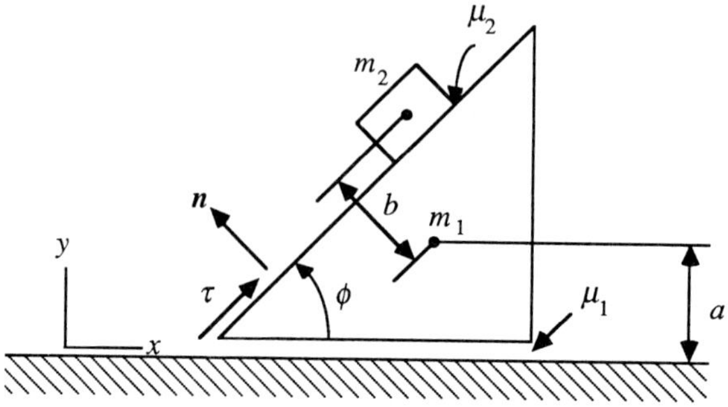

Figure 3.2.4 Sliding blocks.

Retaining x and y coordinates of each particle as generalized coordinates,

$$\mathbf{r}_1 = \begin{bmatrix} x_1 \\ y_1 \end{bmatrix}, \quad \mathbf{r}_2 = \begin{bmatrix} x_2 \\ y_2 \end{bmatrix}$$

The constraint on motion of particle m_1 relative to ground, denoted as particle 3, is simply

$$\Phi^{13} = y_1 - a = 0$$

Using the unit vector $\mathbf{n} = [-\sin\phi, \cos\phi]^T$, the condition that the directed distance between particles perpendicular to the surface along which the second block slides is

$$\begin{aligned} \Phi^{12} &= \mathbf{n}^T(\mathbf{r}_2 - \mathbf{r}_1) - b \\ &= (x_1 - x_2)\sin\phi - (y_1 - y_2)\cos\phi = 0 \end{aligned}$$

The constraint force between m_1 and ground is, by Eq. 3.2.9,

$$\mathbf{F}_{13}^C \equiv \mathbf{F}_1^C = -\Phi_{\mathbf{r}_1}^{13^T}\lambda^{13} = -[0, 1]^T\lambda^{13}$$

Similarly, for the constraint between m_1 and m_2,

$$\begin{aligned} \mathbf{F}_{12}^C &= -\Phi_{\mathbf{r}_1}^{12^T}\lambda^{12} \\ &= -[\sin\phi, -\cos\phi]^T\lambda^{12} \\ &= \mathbf{n}\lambda^2 \end{aligned}$$

which is normal to the surfaces along which sliding occurs. Thus the magnitudes of the constraint forces, which are normal to the sliding surfaces, are

$$\begin{aligned} F_1^C &= |\lambda^{13}| \\ F_2^C &= |\lambda^{12}| \end{aligned}$$

Using the unit vector $\boldsymbol{\tau} = [\cos\phi, \sin\phi]^T$ along the sliding surfaces, the coulomb friction forces, while sliding of both surfaces occurs, are

$$\begin{aligned} \mathbf{F}_1^f &= -\mu_1 |\lambda^{13}| \operatorname{sgn}(\dot{x}_1)\mathbf{i} + \mu_2 |\lambda^{12}| \operatorname{sgn}[(\dot{\mathbf{r}}_2 - \dot{\mathbf{r}}_1)^T\boldsymbol{\tau}]\boldsymbol{\tau} \\ \mathbf{F}_2^f &= -\mu_2 |\lambda^{12}| \operatorname{sgn}[(\dot{\mathbf{r}}_2 - \dot{\mathbf{r}}_1)^T\boldsymbol{\tau}]\boldsymbol{\tau} \end{aligned}$$

From Eq. 3.2.12, with $\mathbf{F}_1^f$ and $\mathbf{F}_2^f$ as applied forces, the equations of motion are thus

$$m_1\ddot{x}_1 + \mu_1 \mid \lambda^{13} \mid \text{sgn}(\dot{x}_1) - \mu_2 \mid \lambda^{12} \mid \text{sgn}([(\dot{\mathbf{r}}_2 - \dot{\mathbf{r}}_1)^T\boldsymbol{\tau}]) \cos\phi + \lambda^{12}\sin\phi = 0$$

$$m_1\ddot{y}_1 - \mu_2 \mid \lambda^{12} \mid \text{sgn}([(\dot{\mathbf{r}}_2 - \dot{\mathbf{r}}_1)^T\boldsymbol{\tau}]) \sin\phi + m_1 g + \lambda^{13} - \lambda^{12}\cos\phi = 0$$

$$m_2\ddot{x}_2 + \mu_2 \mid \lambda^{12} \mid \text{sgn}([(\dot{\mathbf{r}}_2 - \dot{\mathbf{r}}_1)^T\boldsymbol{\tau}]) \cos\phi - \lambda^{12}\sin\phi = 0$$

$$m_2\ddot{y}_2 + \mu_2 \mid \lambda^{12} \mid \text{sgn}([(\dot{\mathbf{r}}_2 - \dot{\mathbf{r}}_1)^T\boldsymbol{\tau}]) \sin\phi + m_2 g + \lambda^{12}\cos\phi = 0$$

Variational Equations of Motion in Generalized Coordinates

Let the kinematics of a constrained system of particles be parameterized by a generalized coordinate vector $\mathbf{q}$, whose components may be independent or dependent, but which determine the Cartesian coordinates of each particle in the system as

$$\mathbf{r}_i = \mathbf{r}_i(\mathbf{q}), \qquad i = 1, \ldots, nb \tag{3.2.19}$$

Taking the differential of Eq. 3.2.19 yields the relationship between generalized coordinate variations $\delta\mathbf{q}$ and virtual displacements $\delta\mathbf{r}_i$ as

$$\delta\mathbf{r}_i = \mathbf{r}_{i_\mathbf{q}}\,\delta\mathbf{q} \tag{3.2.20}$$

Taking two derivatives of Eq. 3.2.19 with respect to time yields the acceleration relation

$$\ddot{\mathbf{r}}_i = \mathbf{r}_{i_\mathbf{q}}\ddot{\mathbf{q}} + (\mathbf{r}_{i_\mathbf{q}}\dot{\mathbf{q}})_\mathbf{q}\dot{\mathbf{q}} \tag{3.2.21}$$

The virtual work of applied forces on the system may be written, using Eq. 3.2.20, as

$$\begin{aligned}\delta W^A &= \sum_{i=1}^{nb} \delta\mathbf{r}_i^T\mathbf{F}_i^A \\ &= \delta\mathbf{q}^T \sum_{i=1}^{nb} \mathbf{r}_{i_\mathbf{q}}^T\mathbf{F}_i^A \\ &\equiv \delta\mathbf{q}^T\mathbf{Q}^A\end{aligned}$$

which defines the *generalized applied force*

$$\mathbf{Q}^A = \sum_{i=1}^{nb} \mathbf{r}_{i_\mathbf{q}}^T\mathbf{F}_i^A \tag{3.2.22}$$

Summing Eq. 3.1.39 over each particle in the system and using Eq. 3.2.22 yields

$$\delta\mathbf{q}^T \left\{ \sum_{i=1}^{nb} m_i\, \mathbf{r}_{i_\mathbf{q}}^T\mathbf{r}_{i_\mathbf{q}}\ddot{\mathbf{q}} - \mathbf{Q}^A + \sum_{i=1}^{nb} m_i\mathbf{r}_{i_\mathbf{q}}^T(\mathbf{r}_{i_\mathbf{q}}\dot{\mathbf{q}})_\mathbf{q}\dot{\mathbf{q}} \right\} = 0 \tag{3.2.23}$$

which must hold for all *kinematically admissible generalized coordinate variations* $\delta\mathbf{q}$.

If the generalized coordinates are independent, then $\delta\mathbf{q}$ is arbitrary and the coefficient of $\delta\mathbf{q}^T$ in Eq. 3.2.23 must be zero, yielding the *differential equations of motion in independent generalized coordinates* as

$$\sum_{i=1}^{nb} m_i \mathbf{r}_{i_\mathbf{q}}^T \mathbf{r}_{i_\mathbf{q}} \ddot{\mathbf{q}} + \sum_{i=1}^{nb} m_i \mathbf{r}_{i_\mathbf{q}}^T (\mathbf{r}_{i_\mathbf{q}} \dot{\mathbf{q}})_\mathbf{q} \dot{\mathbf{q}} = \mathbf{Q}^A \tag{3.2.24}$$

Example 3.2.5: The double pendulum shown in Fig. 3.2.5 is modeled as two point masses, m_1 and m_2, that are constrained by massless links of unit length. The link to which m_1 is attached is constrained to rotate about the z axis, so m_1 moves on the circular arc of unit radius in the x-y plane. Mass m_2 is attached to a rigid link that is pivoted to the first link at m_1, about an axis parallel to the z axis. Therefore, m_2 also moves in the x-y plane. Gravitational force acts in the negative y direction.

In terms of the angles q_1 and q_2 that the links make with the negative y axis, Cartesian coordinate vectors that locate m_1 and m_2 are

$$\mathbf{r}_1 = \begin{bmatrix} \sin q_1 \\ -\cos q_1 \\ 0 \end{bmatrix} \tag{3.2.25}$$

$$\mathbf{r}_2 = \begin{bmatrix} \sin q_1 + \sin q_2 \\ -\cos q_1 - \cos q_2 \\ 0 \end{bmatrix} \tag{3.2.26}$$

By direct calculation,

$$\mathbf{r}_{1_\mathbf{q}} = \begin{bmatrix} \cos q_1 & 0 \\ \sin q_1 & 0 \\ 0 & 0 \end{bmatrix}$$

$$\mathbf{r}_{2_\mathbf{q}} = \begin{bmatrix} \cos q_1 & \cos q_2 \\ \sin q_1 & \sin q_2 \\ 0 & 0 \end{bmatrix}$$

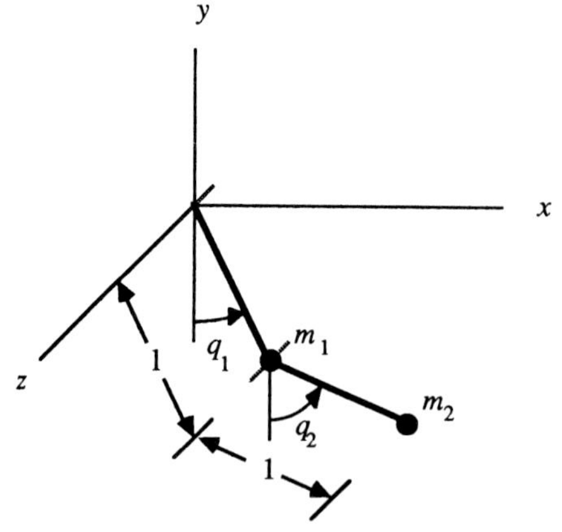

Figure 3.2.5 Double pendulum.

Using Eq. 3.2.22, the generalized force that corresponds to the gravitational forces on the particles is

$$\mathbf{Q}^A = -m_1 g \mathbf{r}_{1_q}^T \mathbf{j} - m_2 g \mathbf{r}_{2_q}^T \mathbf{j} = \begin{bmatrix} -(m_1 + m_2) g \sin q_1 \\ -m_2 g \sin q_2 \end{bmatrix}$$

The quadratic velocity terms that are needed for Eq. 3.2.24 are

$$\mathbf{r}_{1_q}^T (\mathbf{r}_{1_q} \dot{\mathbf{q}})_q \dot{\mathbf{q}} = \mathbf{0}$$

$$\mathbf{r}_{2_q}^T (\mathbf{r}_{2_q} \dot{\mathbf{q}})_q \dot{\mathbf{q}} = \begin{bmatrix} (\sin q_1 \cos q_2 - \cos q_1 \sin q_2) \dot{q}_2^2 \\ (\cos q_1 \sin q_2 - \sin q_1 \cos q_2) \dot{q}_1^2 \end{bmatrix} = \begin{bmatrix} -\sin(q_2 - q_1) \dot{q}_2^2 \\ \sin(q_2 - q_1) \dot{q}_1^2 \end{bmatrix}$$

Since the components of $\mathbf{q}$ are independent, the differential equations of motion are obtained from Eq. 3.2.24 as

$$\begin{aligned} (m_1 + m_2)\ddot{q}_1 + m_2 \cos(q_2 - q_1)\ddot{q}_2 &= -(m_1 + m_2) g \sin q_1 + m_2 \sin(q_2 - q_1)\dot{q}_2^2 \\ m_2 \ddot{q}_2 + m_2 \cos(q_2 - q_1)\ddot{q}_1 &= -m_2 g \sin q_2 - m_2 \sin(q_2 - q_1)\dot{q}_1^2 \end{aligned} \tag{3.2.27}$$

To assure that these equations can be solved for $\ddot{q}_1$ and $\ddot{q}_2$, the determinant of the coefficient matrix for the accelerations in Eq. 3.2.27 may be evaluated as

$$\begin{vmatrix} m_1 + m_2 & m_2 \cos(q_2 - q_1) \\ m_2 \cos(q_2 - q_1) & m_2 \end{vmatrix} = m_1 m_2 + m_2^2[1 - \cos^2(q_2 - q_1)] \geqslant m_1 m_2 > 0$$

Thus the coefficient matrix is nonsingular and Eq. 3.2.27 may be solved for generalized coordinate accelerations.

To investigate the behavior of the solution of these nonlinear differential equations, let $m_1 = m_2 = 1$ kg and begin the simulation with initial conditions $q_1(0) = q_2(0) = 0.5$ rad and $\dot{q}_1(0) = \dot{q}_2(0) = 0$. This means that the bars of the pendulum are initially parallel, inclined about 28.6° from the negative z axis, and at rest. The initial-value problem is solved numerically, using the method of Appendix D. Plots of $q_i(t)$, $\dot{q}_i(t)$, and $\ddot{q}_i(t)$ are shown in Fig. 3.2.6(a), (b), and (c) respectively. Note that the bars deviate from their initially parallel configuration, due to nonlinear coupling effects in the equations of motion. Further, the motion is not simply periodic.

As a second numerical experiment, m_2 is changed to 10 kg. All other data and initial conditions are the same as in the first case above. Plots of results given in Fig. 3.2.7 indicate that motion of the double pendulum is dominated by the large mass m_2. The motion predicted is nearly periodic, with the smaller mass m_1 contributing primarily a high-frequency oscillatory purturbation.

♦ **Example 3.2.6:** The equations of motion of the particle model of the two degree of freedom material handling device of Example 2.6.2 can now be written. With gravity acting in the negative z direction as the only applied force,

$$\mathbf{Q}^A = -m_1 g \mathbf{r}_{1_q}^T \mathbf{k} - m_2 g \mathbf{r}_{2_q}^T \mathbf{k} = [0, -m_2 g \sin q_2]^T$$

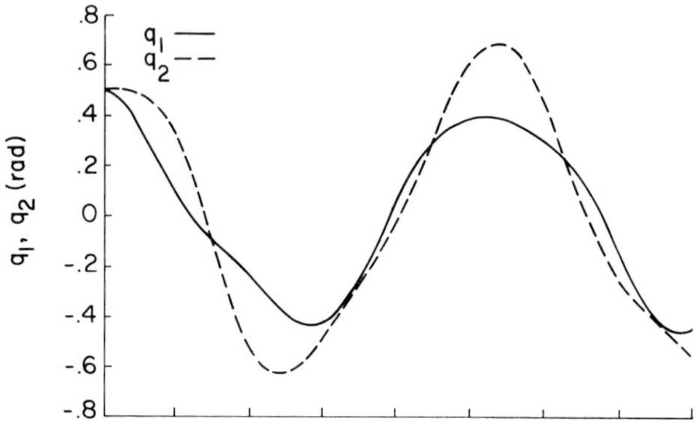

(a) q_1 and q_2 vs time

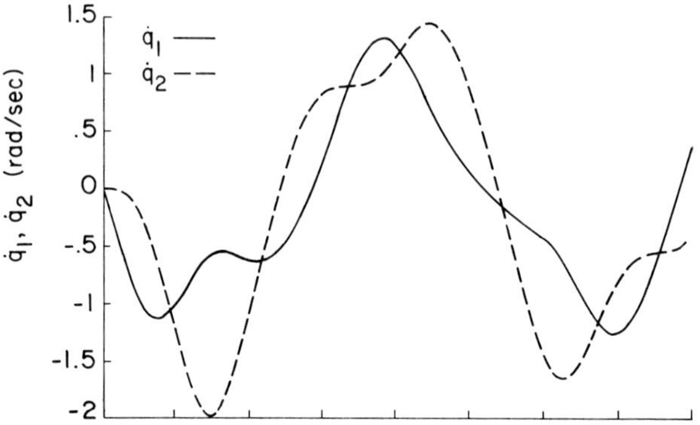

(b) $\dot{q}_1$ and $\dot{q}_2$ vs time

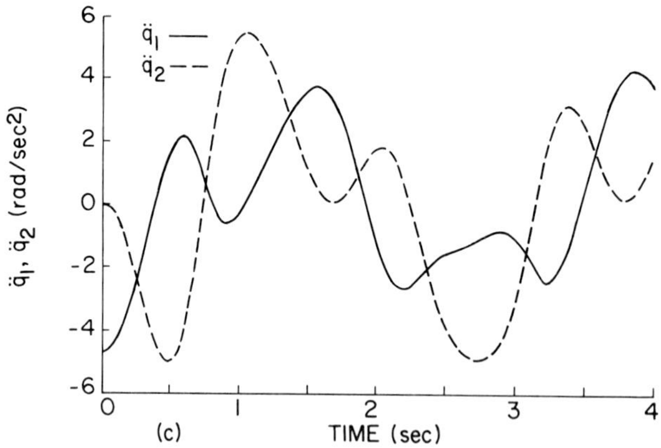

(c) $\ddot{q}_1$ and $\ddot{q}_2$ vs time

Figure 3.2.6 Motion of double pendulum versus time: equal masses.

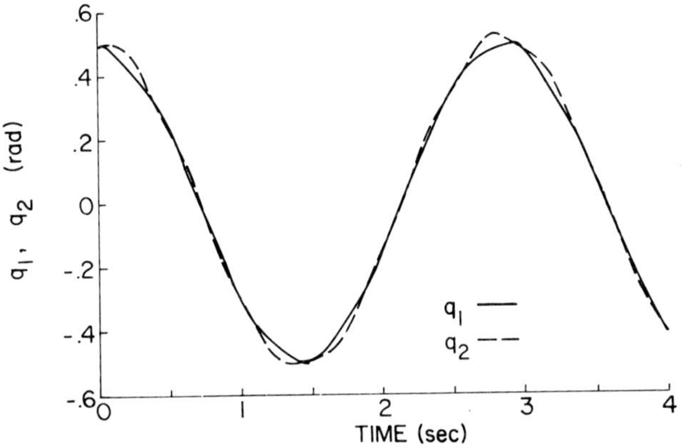

Figure 3.2.7 q_1 and q_2 versus time: double pendulum with unequal masses.

where expressions from Example 2.6.2 are used. Since the components of $\mathbf{q}$ are independent, the differential equations of motion of Eq. 3.2.24 may be applied. Using expressions derived in Example 2.6.2 and expanding terms in Eq. 3.2.24 (Prob. 3.2.3), the equations of motion of the system are

$$\begin{bmatrix} m_1 + m_2(1 + \sin^2 q_2) & m_2 \cos q_2 \\ m_2 \cos q_2 & m_2 \end{bmatrix} \begin{bmatrix} \ddot{q}_1 \\ \ddot{q}_2 \end{bmatrix}$$

$$= \begin{bmatrix} 0 \\ -m_2 g \sin q_2 \end{bmatrix} + m_2 \begin{bmatrix} \sin q_2(\dot{q}_2^2 - 2\cos q_2 \dot{q}_1 \dot{q}_2) \\ \cos q_2 \sin q_2 \dot{q}_1^2 \end{bmatrix} \qquad (3.2.28)$$

The determinant of the acceleration coefficient matrix is

$$\begin{vmatrix} m_1 + m_2(1 + \sin^2 q_2) & m_2 \cos q_2 \\ m_2 \cos q_2 & m_2 \end{vmatrix} = m_1 m_2 + m_2^2(1 + \sin^2 q_2 - \cos^2 q_2) \geqslant m_1 m_2 > 0$$

so the matrix is nonsingular and the equations of motion may be solved for $\ddot{q}$.

For numerical experimentation, let $m_1 = m_2 = 1$ kg. With the initial conditions $q_1(0) = \pi/4$ rad, $q_2(0) = \pi/2$ rad, and $\dot{q}_1(0) = \dot{q}_2(0) = 0$, the mechanism is at rest in approximately the configuration shown in Fig. 2.6.2(b). The motion predicted by solving the initial-value problem numerically, using the method presented in Appendix D, is shown in Fig. 3.2.8.

If m_2 is increased to 10 kg and all other data are retained, the motion shown in Fig. 3.2.9 is predicted. Note that while the same trends occur, the larger mass at the extremity leads to substantially higher velocities and accelerations.

➧ **Example 3.2.7:** As a final example with independent generalized coordinates, consider again the particle model of the three degree of freedom material handling device of Example 2.6.3. With gravity acting in the negative z direction as the only applied force,

$$\mathbf{Q}^A = -m_1 g \mathbf{r}_{1_\mathbf{q}}^T \mathbf{k} - m_2 g \mathbf{r}_{2_\mathbf{q}}^T \mathbf{k}$$

$$= [0, 0, -m_3 g(\cos q_3 - \sin q_3)]^T$$

where expressions derived in Example 2.6.3 are used. Since the generalized coor-

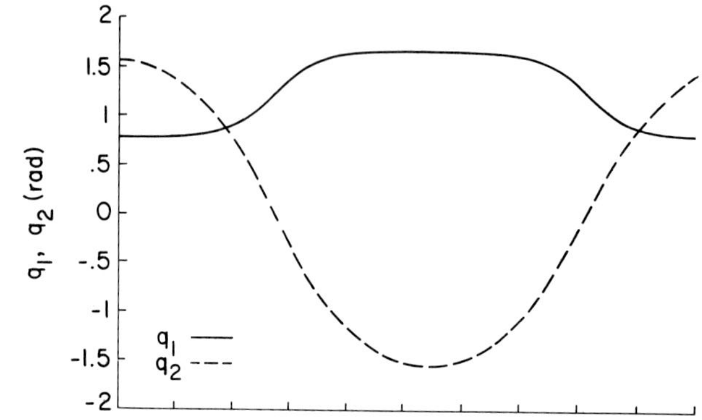

(a) q_1 and q_2 vs time

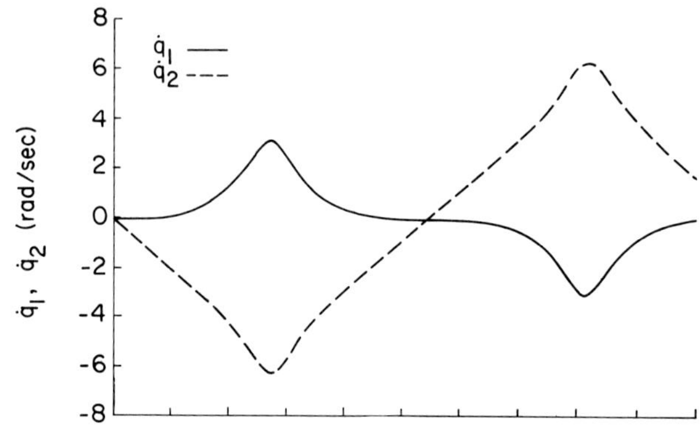

(b) $\dot{q}_1$ and $\dot{q}_2$ vs time

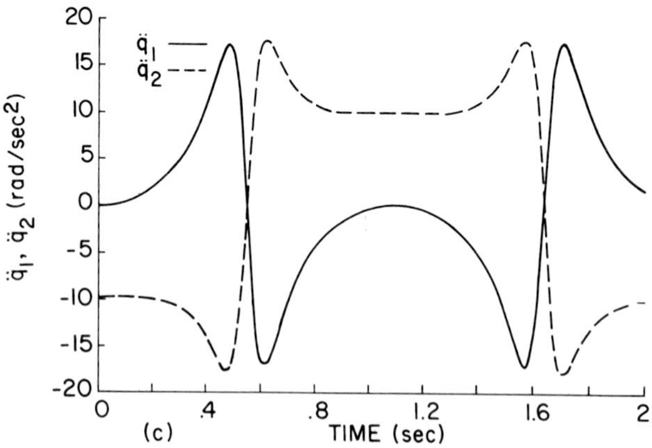

(c) $\ddot{q}_1$ and $\ddot{q}_2$ vs time

Figure 3.2.8 Numerical results for two-body material handler, $m = 1$ kg.

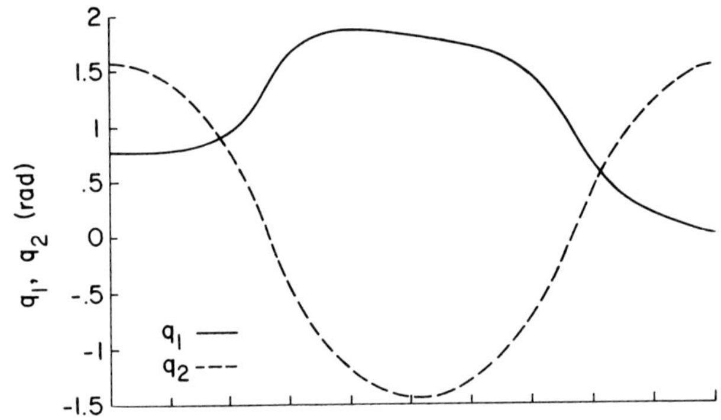

(a) q_1 and q_2 vs time

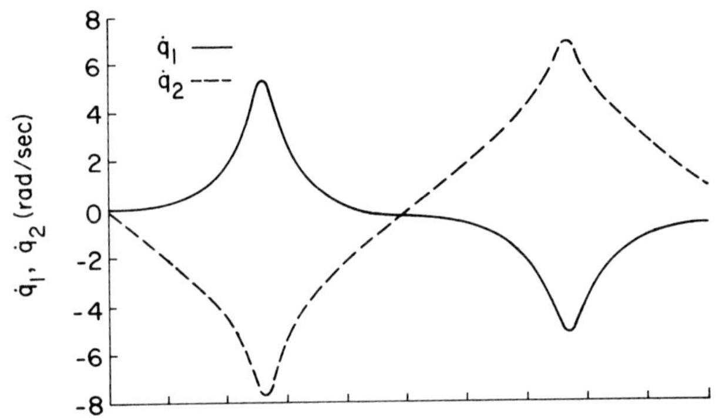

(b) $\dot{q}_1$ and $\dot{q}_2$ vs time

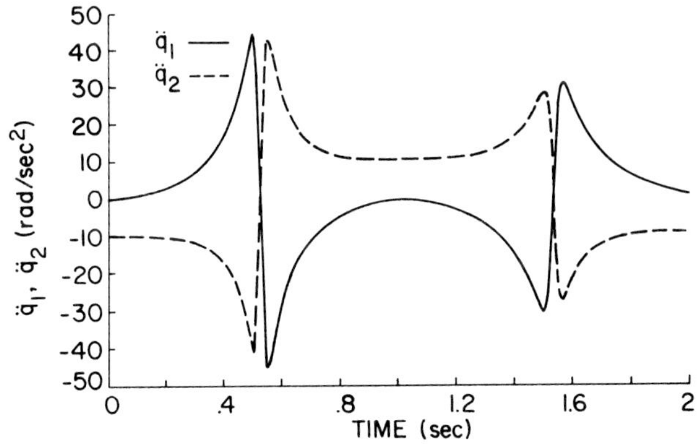

(c) $\ddot{q}_1$ and $\ddot{q}_2$ vs time

Figure 3.2.9 Numerical results for two-body material handler, $m_2 = 10$ kg.

dinates are independent, differential equations of motion may be obtained by applying Eq. 3.2.24. Using expressions from Example 2.6.3 to expand terms in Eq. 3.2.24, the equations of motion are obtained as (Prob. 3.2.4)

$$\begin{bmatrix} m_1 + m_2[(q_2+1)^2 + (\cos q_3 - \sin q_3)^2] & -m_2(\cos q_3 - \sin q_3) & -m_2(q_2+1)(\sin q_3 + \cos q_3) \\ -m_2(\cos q_3 - \sin q_3) & m_2 & 0 \\ -m_2(q_2+1)(\sin q_3 + \cos q_3) & 0 & 2m_2 \end{bmatrix} \begin{bmatrix} \ddot{q}_1 \\ \ddot{q}_2 \\ \ddot{q}_3 \end{bmatrix}$$

$$= \begin{bmatrix} 0 \\ 0 \\ -m_2 g(\cos q_3 - \sin q_3) \end{bmatrix}$$

$$+ m_2 \begin{bmatrix} (q_2+1)(\cos q_3 - \sin q_3)\dot{q}_3^2 - 2(q_2+1)\dot{q}_2\dot{q}_1 + 2(\cos^2 q_3 - \sin^2 q_3)\dot{q}_3\dot{q}_1 \\ (q_2+1)\dot{q}_1^2 - 2(\sin q_3 + \cos q_3)\dot{q}_3\dot{q}_1 \\ 2(\sin q_3 + \cos q_3)\dot{q}_2\dot{q}_1 - (\cos^2 q_3 - \sin^2 q_3)\dot{q}_1^2 \end{bmatrix} \tag{3.2.29}$$

The determinant of the acceleration coefficient matrix is

$$\begin{aligned} &2m_1 m_2^2 + m_2^3(q_2+1)^2[2 - (\sin q_3 + \cos q_3)^2] \\ &= 2m_1 m_2^2 + m_2^3(q_3+1)^2\,(1 - 2\sin q_3 \cos q_3) \\ &\geqslant 2m_1 m_2^2 > 0 \end{aligned}$$

so the matrix is nonsingular and the equations of motion can be solved for $\ddot{\mathbf{q}}$.

As a numerical experiment, let $m_1 = m_2 = m_3 = 1$ kg. Initial conditions are selected as $q_1(0) = \pi/4$ rad, $q_2(0) = 1$ m, $q_3(0) = 0$, and $\dot{q}_1(0) = \dot{q}_2(0) = \dot{q}_3(0) = 0$. The mechanism is thus at rest, in approximately the configuration shown in Fig. 2.6.3(b), at $t = 0$. The motion predicted by numerically solving the initial-value problem is shown in Fig. 3.2.10.

Note that even when independent generalized coordinates lead to differential equations of motion (e.g., in Examples 3.3.2, 3.3.4, and 3.3.5), highly nonlinear and often very complicated equations of motion result. There is virtually no hope of finding closed-form solutions, so numerical integration methods must be used (see Appendix D).

If generalized coordinates $\mathbf{q}$ that define the mappings of Eq. 3.2.19 are not independent, but must satisfy holonomic constraints of the form

$$\mathbf{\Phi}^{ij}(\mathbf{r}_i(\mathbf{q}), \mathbf{r}_j(\mathbf{q}), t) \equiv \mathbf{\Phi}^{ij}(\mathbf{q}, t) = \mathbf{0} \tag{3.2.30}$$

kinematically admissible generalized coordinate variations must satisfy

$$(\mathbf{\Phi}^{ij}_{\mathbf{r}_i}\mathbf{r}_{i\mathbf{q}} + \mathbf{\Phi}^{ij}_{\mathbf{r}_j}\mathbf{r}_{j\mathbf{q}})\,\delta\mathbf{q} \equiv \mathbf{\Phi}^{ij}_{\mathbf{q}}\,\delta\mathbf{q} = \mathbf{0} \tag{3.2.31}$$

If constraint forces that act on particles i and j due to the constraints of Eq.

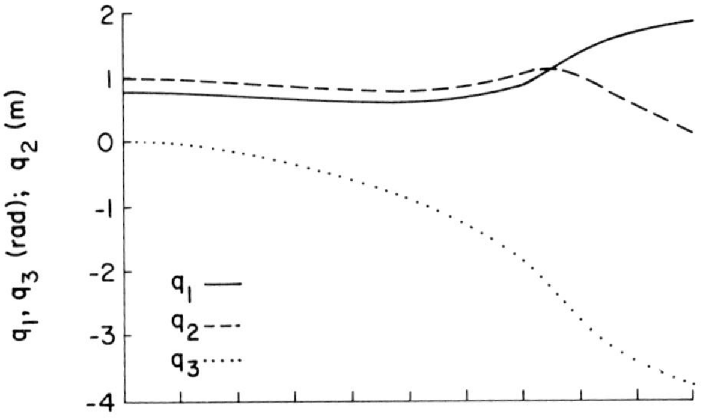

(a) q_1, q_2, and q_3 vs time

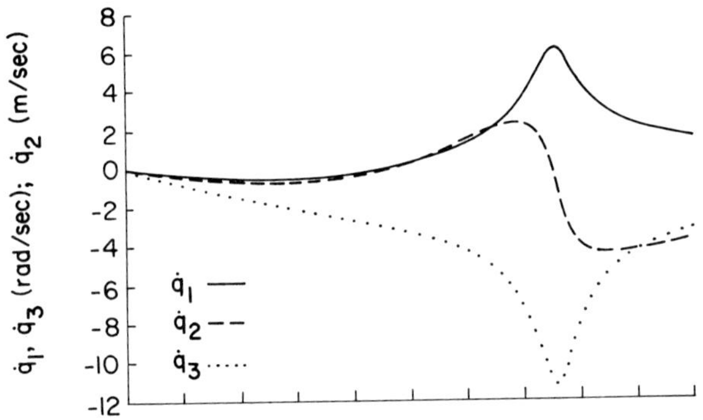

(b) $\dot{q}_1$, $\dot{q}_2$, and $\dot{q}_3$ vs time

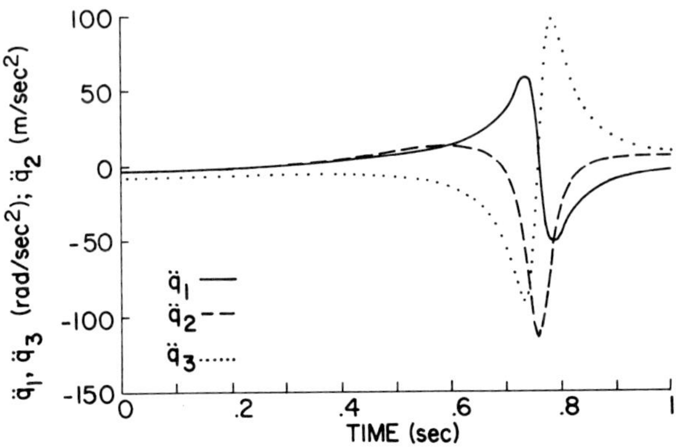

(c) $\ddot{q}_1$, $\ddot{q}_2$, and $\ddot{q}_3$ vs time

Figure 3.2.10 Numerical results for three-body material handler.

3.2.30 are workless, then

$$\begin{aligned}\delta W_{ij}^C &= \delta\mathbf{r}_i^T\mathbf{F}_{ij}^C + \delta\mathbf{r}_j^T\mathbf{F}_{ji}^C \\ &= -\delta\mathbf{r}_i^T\boldsymbol{\Phi}_{\mathbf{r}_i}^{ij^T}\boldsymbol{\lambda}^{ij} - \delta\mathbf{r}_j\boldsymbol{\Phi}_{\mathbf{r}_j}^{ij^T}\boldsymbol{\lambda}^{ij} \\ &= -\delta\mathbf{q}^T(\mathbf{r}_{i\mathbf{q}}^T\boldsymbol{\Phi}_{\mathbf{r}_i}^{ij} + \mathbf{r}_{j\mathbf{q}}^T\boldsymbol{\Phi}_{\mathbf{r}_j}^{ij^T})\boldsymbol{\lambda}^{ij} \\ &= -\delta\mathbf{q}^T\boldsymbol{\Phi}_{\mathbf{q}}^{ij^T}\boldsymbol{\lambda}^{ij} \\ &\equiv \delta\mathbf{q}^T\mathbf{Q}_{ij}^C\end{aligned} \tag{3.2.32}$$

where Eqs. 3.2.9, 3.2.20, and 3.2.31 have been used. Thus the generalized constraint force for the holonomic constraint of Eq. 3.2.30 is

$$\mathbf{Q}_{ij}^C \equiv -\boldsymbol{\Phi}_{\mathbf{q}}^{ij^T}\boldsymbol{\lambda}^{ij} \tag{3.2.33}$$

and $\boldsymbol{\lambda}^{ij}$ is a vector of Lagrange multipliers.

Similarly, if there are differential constraints of the form

$$\begin{aligned}\mathbf{B}^{ij}\dot{\mathbf{r}}_i + \mathbf{B}^{ji}\dot{\mathbf{r}}_j + \mathbf{c}^{ij} &= (\mathbf{B}^{ij}\mathbf{r}_{i_\mathbf{q}} + \mathbf{B}^{ji}\mathbf{r}_{j_\mathbf{q}})\dot{\mathbf{q}} + \mathbf{c}^{ij} \\ &\equiv \mathbf{D}^{ij}\dot{\mathbf{q}} + \mathbf{c}^{ij} = \mathbf{0}\end{aligned} \tag{3.2.34}$$

then kinematically admissible $\delta\mathbf{q}$ for the constraint of Eq. 3.2.34 must satisfy

$$(\mathbf{B}^{ij}\mathbf{r}_{i_\mathbf{q}} + \mathbf{B}^{ji}\mathbf{r}_{j_\mathbf{q}})\,\delta\mathbf{q} = \mathbf{0}$$

or simply

$$\mathbf{D}^{ij}\,\delta\mathbf{q} = \mathbf{0} \tag{3.2.35}$$

If constraint forces that act on particles i and j due to the constraints of Eq. 3.2.34 are workless, then

$$\begin{aligned}\delta W_{ij}^C &= \delta\mathbf{r}_i^T\mathbf{F}_{ij}^C + \delta\mathbf{r}_j^T\mathbf{F}_{ji}^C \\ &= -\delta\mathbf{r}_i^T\mathbf{B}^{ij^T}\boldsymbol{\mu}^{ij} - \delta\mathbf{r}_j^T\mathbf{B}^{ji^T}\boldsymbol{\mu}^{ij} \\ &= -\delta\mathbf{q}^T(\mathbf{r}_{i\mathbf{q}}^T\mathbf{B}^{ij^T} + \mathbf{r}_{j\mathbf{q}}^T\mathbf{B}^{ij^T})\boldsymbol{\mu}^{ij} \\ &= -\delta\mathbf{q}^T\mathbf{D}^{ij^T}\boldsymbol{\mu}^{ij} \\ &\equiv \delta\mathbf{q}^T\mathbf{Q}_{ij}^C\end{aligned} \tag{3.2.36}$$

where Eqs. 3.2.10, 3.2.20, and 3.2.34 have been used. Thus the generalized constraint force for the differential constraint of Eq. 3.2.34 is

$$\mathbf{Q}_{ij}^C = -\mathbf{D}^{ij^T}\boldsymbol{\mu}^{ij} \tag{3.2.37}$$

and $\boldsymbol{\mu}^{ij}$ is a vector of Lagrange multipliers.

Summing Eq. 3.2.37 over all particles in the system and accounting for both applied and constraint forces yields

$$\delta\mathbf{q}^T\left\{\sum_{i=1}^{nb} m_i\mathbf{r}_{i_\mathbf{q}}^T\mathbf{r}_{i_\mathbf{q}}\ddot{\mathbf{q}} + \sum_{i=1}^{nb} m_i\mathbf{r}_{i_\mathbf{q}}^T(\mathbf{r}_{i_\mathbf{q}}\dot{\mathbf{q}})_\mathbf{q}\dot{\mathbf{q}} - \mathbf{Q}^A - \sum_{i,j}\mathbf{Q}_{ij}^C\right\} = 0 \qquad (3.2.38)$$

for all $\delta\mathbf{q}$. Thus, by the orthogonality theorem,

$$\sum_{i=1}^{nb} m_i\mathbf{r}_{i_\mathbf{q}}^T\mathbf{r}_{i_\mathbf{q}}\ddot{\mathbf{q}} + \sum_{i=1}^{nb} m_i\mathbf{r}_{i_\mathbf{q}}^T(\mathbf{r}_{i_\mathbf{q}}\dot{\mathbf{q}})_\mathbf{q}\ddot{\mathbf{q}} - \sum_{i,j}\mathbf{Q}_{ij}^C = \mathbf{Q}^A \qquad (3.2.39)$$

For the case of workless constraints of the form of Eqs. 3.2.30 and 3.2.34, Eq. 3.2.39 is

$$\sum_{i=1}^{nb} m_i\mathbf{r}_{i_\mathbf{q}}^T\mathbf{r}_{i_\mathbf{q}}\ddot{\mathbf{q}} + \sum_{i=1}^{nb} m_i\mathbf{r}_{i_\mathbf{q}}^T(\mathbf{r}_{i_\mathbf{q}}\dot{\mathbf{q}})_\mathbf{q}\dot{\mathbf{q}} + \sum_{i,j}\mathbf{\Phi}_\mathbf{q}^{ij}\boldsymbol{\lambda}^{ij} + \sum_{i,j}\mathbf{D}^{ij^T}\boldsymbol{\mu}^{ij} = \mathbf{Q}^A \qquad (3.2.40)$$

where the summations $\sum_{i,j}$ are over i and j for which particles are constrained by Eq. 3.2.30 or 3.2.34. Equation 3.2.40 is the Lagrange multiplier form of the equations of motion for a constrained system of particles, in terms of generalized coordinates. Equations 3.2.40, 3.2.30, and 3.2.34 comprise a system of differential-algebraic equations for $\mathbf{q}$, $\boldsymbol{\lambda}^{ij}$, and $\boldsymbol{\mu}^{ij}$.

➧ **Example 3.2.8:** Consider again the two particle model of the wheel and axle set of Example 3.2.2. Defining an x'-y' frame that is fixed to the axle, as shown in Fig. 3.2.11, $\mathbf{r} = [x, y]^T$ and ϕ are used as generalized coordinates, $q_1 = x$, $q_2 = y$, and $q_3 = \phi$. The masses at the wheels are located by

$$\begin{aligned}\mathbf{r}_1 &= \begin{bmatrix} q_1 \\ q_2 \end{bmatrix} - 0.5\begin{bmatrix} \cos q_3 \\ \sin q_3 \end{bmatrix} \\ \mathbf{r}_2 &= \begin{bmatrix} q_1 \\ q_2 \end{bmatrix} + 0.5\begin{bmatrix} \cos q_3 \\ \sin q_3 \end{bmatrix}\end{aligned} \qquad (3.2.41)$$

The nonholonomic constraint is

$$\begin{aligned}(\mathbf{r}_2 - \mathbf{r}_1)^T\dot{\mathbf{r}} &= [\cos q_3, \sin q_3]\begin{bmatrix} \dot{q}_1 \\ \dot{q}_2 \end{bmatrix} \\ &= \dot{q}_1\cos q_3 + \dot{q}_2\sin q_3 = 0\end{aligned} \qquad (3.2.42)$$

Thus the matrix $\mathbf{D}^{12}$ of Eq. 3.2.34 is

$$\mathbf{D}^{12} = [\cos q_3, \sin q_3, 0]$$

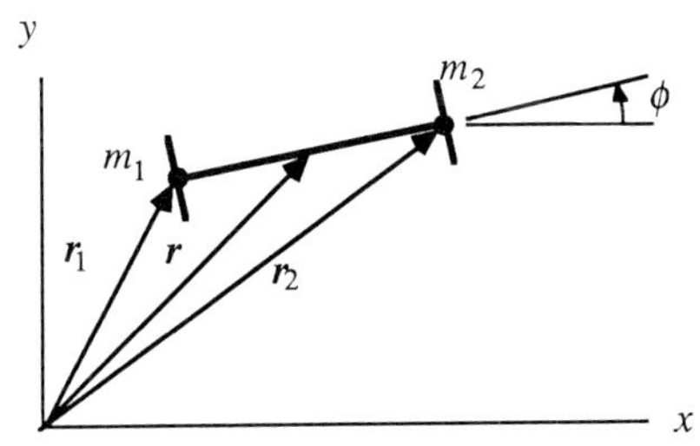

Figure 3.2.11 Wheel and axle generalized coordinates.

For use in the equations of motion,

$$\mathbf{r}_{1_\mathbf{q}} = \begin{bmatrix} 1 & 0 & 0.5 \sin q_3 \\ 0 & 1 & -0.5 \cos q_3 \end{bmatrix}$$

$$\mathbf{r}_{2_\mathbf{q}} = \begin{bmatrix} 1 & 0 & -0.5 \sin q_3 \\ 0 & 1 & 0.5 \cos q_3 \end{bmatrix}$$

$$(\mathbf{r}_{1_\mathbf{q}}\dot{\mathbf{q}})_\mathbf{q}\dot{\mathbf{q}} = 0.5\dot{q}_3^2 \begin{bmatrix} \cos q_3 \\ \sin q_3 \end{bmatrix}$$

$$(\mathbf{r}_{2_\mathbf{q}}\dot{\mathbf{q}})_\mathbf{q}\dot{\mathbf{q}} = -0.5\dot{q}_3^2 \begin{bmatrix} \cos q_3 \\ \sin q_3 \end{bmatrix}$$

$$\begin{bmatrix} m_1 + m_2 & 0 & 0.5(m_1 - m_2) \sin q_3 \\ 0 & m_1 + m_2 & -0.5(m_1 - m_2) \cos q_3 \\ 0.5(m_1 - m_2) \sin q_3 & -0.5(m_1 - m_2) \cos q_3 & 0.25(m_1 + m_2) \end{bmatrix} \begin{bmatrix} \ddot{q}_1 \\ \ddot{q}_2 \\ \ddot{q}_3 \end{bmatrix}$$
$$+ 0.5\dot{q}_3^2(m_1 - m_2) \begin{bmatrix} \cos q_3 \\ \sin q_3 \\ 0 \end{bmatrix} + \begin{bmatrix} \cos q_3 \\ \sin q_3 \\ 0 \end{bmatrix} \mu = 0 \tag{3.2.43}$$

Equations 3.2.43 and 3.2.42 comprise a system of four differential-algebraic equations for the four variables q_1, q_2, q_3, and μ. While the form of the equations of motion in generalized coordinates is somewhat more complicated than the equations derived in Example 3.2.2 in Cartesian coordinates, there is a significant reduction in the number of equations and variables. This is typical of use of generalized coordinates.

➧ **Example 3.2.9:** A generalized coordinate formulation for the slider–crank model of Examples 1.3.2 and 3.2.3, with bars of unit length, may be formed by using the double pendulum model of Example 3.2.5. As shown in Fig. 3.2.12, the constraint required is that the y coordinate of m_2 be zero; that is,

$$\Phi(y_2(\mathbf{q})) = -\cos q_1 - \cos q_2 = 0 \tag{3.2.44}$$

so

$$\Phi_\mathbf{q} = [\sin q_1, \sin q_2]$$

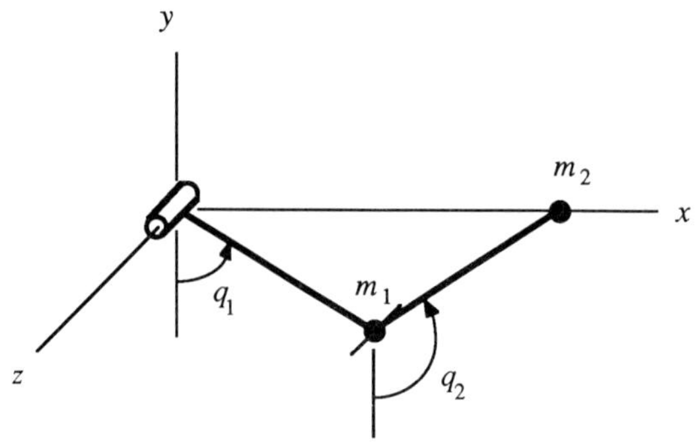

Figure 3.2.12 Double pendulum model of slider–crank.

The equations of motion are obtained by simply adding $\Phi_{\mathbf{q}}^T\lambda$ to Eqs. 3.2.27. In matrix form, this is

$$\begin{bmatrix} m_1 + m_2 & m_2\cos(q_2 - q_1) \\ m_2\cos(q_2 - q_1) & m_2 \end{bmatrix}\begin{bmatrix} \ddot{q}_1 \\ \ddot{q}_2 \end{bmatrix} + m_2\sin(q_2 - q_1)\begin{bmatrix} -\dot{q}_2^2 \\ \dot{q}_1^2 \end{bmatrix}$$

$$+ \begin{bmatrix} \sin q_1 \\ \sin q_2 \end{bmatrix}\lambda = \begin{bmatrix} -(m_1 + m_2)g\sin q_1 \\ -m_2 g\sin q_2 \end{bmatrix} \tag{3.2.45}$$

Equations 3.2.45 and 3.2.44 comprise a system of three differential-algebraic equations for the three variables q_1, q_2, and λ. Contrasting this system with the system of seven equations in seven variables in the Cartesian formulation of the same model in Example 3.2.3, the advantage of the generalized coordinate formulation becomes clear.

To illustrate the use of numerical methods of Appendix D for solving differential-algebraic equations, consider the following data for this example: $m_1 = m_2 = 1$ kg, $q_1(0) = 3\pi/4$ rad, and $\dot{q}_1(0) = 0$, so the mechanism starts from rest with the crank 45° above the x-z plane. The associated initial conditions on q_2 and $\dot{q}_2$ are determined by the holonomic constraint of Eq. 3.2.44 and its associated velocity equation. Plots of the generalized coordinates, their velocities, and their accelerations are shown in Fig. 3.2.13(a), (b), and (c). At about $t = 1.18$ s, $q_1 = 0$ and $q_2 = \pi$ rad. As seen from Fig. 3.2.12, m_2 is at the origin and the bars are coincident. Since this is a singular configuration, $\Phi_{\mathbf{q}} = \mathbf{0}$, so the Jacobian is row rank deficient and the DAE solver stops.

As a second case, let $m_2 = 10$ kg and all other data remain the same. As shown in Fig. 3.2.14, the same characteristics of motion occur, including the singular configuration that stops the integration. Note, however, that movement is slower, due to the larger slider mass.

It was shown in Example 1.3.3 that the distribution of mass in a body may substantially influence constraint forces that act in the system. Single particle models of bodies that move as components of machines are therefore inadequate for many applications. While Example 1.3.3 shows that additional particles can be included in a model to account for the effects of distributed mass, the multitude of generalized constraints and constraint equations that would be required in a Cartesian coordinate formulation would make computation with such models impractical. The use of generalized coordinates, however, is quite practical in such cases. This is actually the basis for *rigid body* modeling in Chapters 4 to 6. Rather than attempting to develop this concept in general, a slider–crank example is used to illustrate the basic concepts.

➧ **Example 3.2.10:** Consider the slider–crank model of the moving parts in an engine or compressor, as shown in Fig. 3.2.15. A large number of particles are distributed around the flywheel and along the axis of symmetry of the connecting rod, to represent the distribution of mass in these components. There are n_1 particles that are equally spaced around the flywheel and n_2 particles that are equally spaced along the connecting rod, denoted m_{ji}, $i = 1, \ldots, n_j$, $j = 1, 2$. The mass m_{2n_2} is the mass of the piston. The radius of the mass point circle on the flywheel is 5 units and the length of the connecting rod is 1 unit.

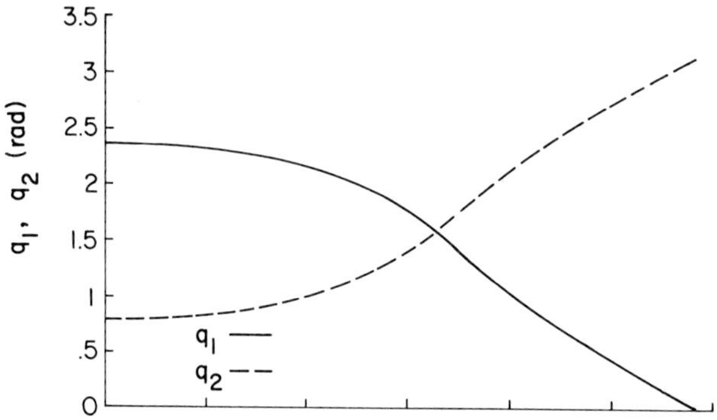

(a) q_1 and q_2 vs time

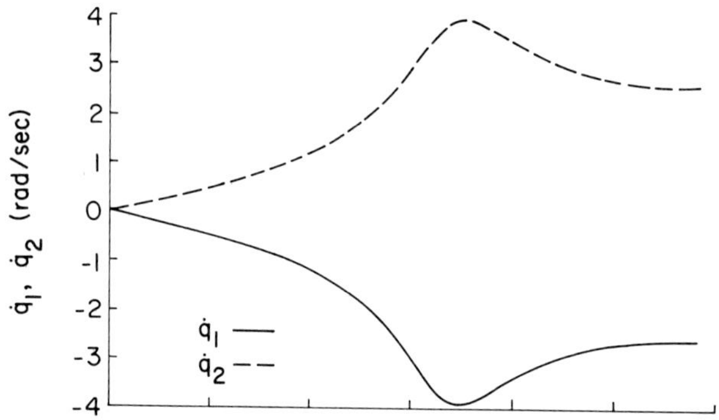

(b) $\dot{q}_1$ and $\dot{q}_2$ vs time

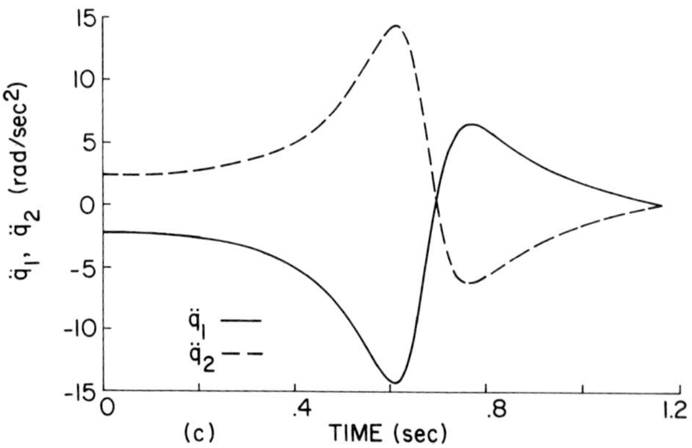

(c) $\ddot{q}_1$ and $\ddot{q}_2$ vs time

Figure 3.2.13 Dynamic response of slider crank; equal masses.

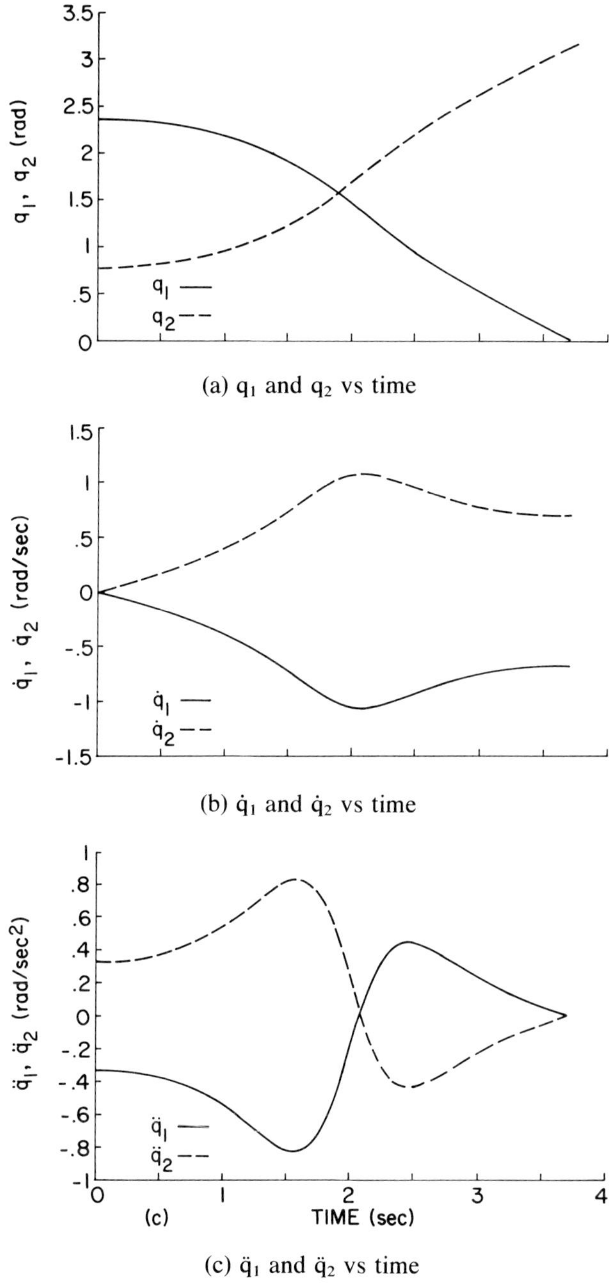

(a) q_1 and q_2 vs time

(b) $\dot{q}_1$ and $\dot{q}_2$ vs time

(c) $\ddot{q}_1$ and $\ddot{q}_2$ vs time

Figure 3.2.14 Dynamic response of slider–crank; $m_1 = 1$ kg, $m_2 = 10$ kg.

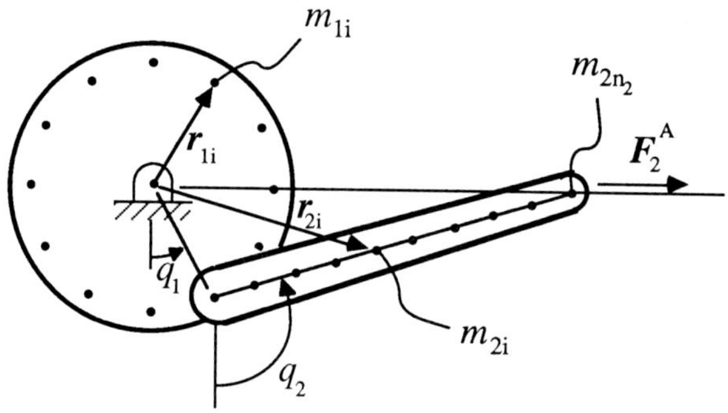

Figure 3.2.15 Many particle model of slider–crank.

Vectors that locate particles in the model in the x-y plane are

$$\mathbf{r}_{1i} = s\begin{bmatrix} \sin\left(q_1 + \dfrac{2\pi i}{n_1}\right) \\ -\cos\left(q_1 + \dfrac{2\pi i}{n_1}\right) \end{bmatrix}, \qquad i = 1, \ldots, n_1 \tag{3.2.46}$$

$$\mathbf{r}_{2i} = \begin{bmatrix} s\sin q_1 + \dfrac{i}{n_2}\sin q_2 \\ -s\cos q_1 - \dfrac{i}{n_2}\cos q_2 \end{bmatrix}, \qquad i = 1, \ldots, n_2 \tag{3.2.47}$$

The single constraint is

$$\Phi = -s\cos q_1 - \cos q_2 = 0 \tag{3.2.48}$$

whose Jacobian is

$$\Phi_{\mathbf{q}} = [s\sin q_1, \sin q_2] \tag{3.2.49}$$

For use in the equations of motion,

$$\begin{aligned} \mathbf{Q}^A &= \mathbf{r}_{2n_2\mathbf{q}}^T F_2^A \mathbf{i} \\ &= F_2^A \begin{bmatrix} s\cos q_1 \\ \cos q_2 \end{bmatrix} \end{aligned}$$

$$\mathbf{r}_{1i_\mathbf{q}} = s\begin{bmatrix} \cos\left(q_1 + \dfrac{2\pi i}{n_1}\right) & 0 \\ \sin\left(q_1 + \dfrac{2\pi}{n_1}\right) & 0 \end{bmatrix}$$

$$\mathbf{r}_{2i_\mathbf{q}} = \begin{bmatrix} s\cos q_1 & \dfrac{i}{n_2}\cos q_2 \\ s\sin q_1 & \dfrac{i}{n_2}\sin q_2 \end{bmatrix}$$

$$\mathbf{r}_{1i_\mathbf{q}}^T(\mathbf{r}_{1i_\mathbf{q}}\dot{\mathbf{q}})_\mathbf{q}\dot{\mathbf{q}} = \mathbf{0}$$

$$\mathbf{r}_{2i_\mathbf{q}}^T(\mathbf{r}_{2i_\mathbf{q}}\dot{\mathbf{q}})_\mathbf{q}\dot{\mathbf{q}} = \frac{si}{n_2}\sin(q_2 - q_1)\begin{bmatrix} -\dot{q}_2^2 \\ \dot{q}_1^2 \end{bmatrix}$$

The mass points that model the flywheel and connecting rod are constrained so that they cannot move relative to each other. Thus forces of constraint between the particles can do no work in any kinematically admissible virtual displacement. Thus Eq. 3.2.40 is applicable. Using the results above, it is expanded to obtain (Prob. 3.2.5)

$$\left(\sum_{i=1}^{n1} m_{1i}\right)\begin{bmatrix} s^2 & 0 \\ 0 & 0 \end{bmatrix}\ddot{\mathbf{q}} + \sum_{i=1}^{n2} m_{2i}\begin{bmatrix} s^2 & \frac{si}{n_2}\cos(q_2 - q_1) \\ \frac{si}{n_2}\cos(q_2 - q_1) & \left(\frac{i}{n_2}\right)^2 \end{bmatrix}\ddot{\mathbf{q}}$$

$$+ \sum_{i=1}^{n2} m_{2i}\frac{si}{n_2}\sin(q_2 - q_1)\begin{bmatrix} -\dot{q}_2^2 \\ \dot{q}_1^2 \end{bmatrix} + \begin{bmatrix} s\sin q_1 \\ \sin q_2 \end{bmatrix}\lambda = F_2^A\begin{bmatrix} s\cos q_1 \\ \cos q_2 \end{bmatrix} \tag{3.2.50}$$

Defining

$$\begin{aligned} J_{10} &= s^2\sum_{i=1}^{n1} m_{1i} \\ J_{20} &= s^2\sum_{i=1}^{n2} m_{2i} \\ J_{21} &= \sum_{i=1}^{n2}\frac{si}{n_2}m_{2i} \\ J_{22} &= \sum_{i=1}^{n2}\left(\frac{i}{n_2}\right)^2 m_{2i} \end{aligned} \tag{3.2.51}$$

Eq. 3.2.50 reduces to

$$\begin{bmatrix} J_{10} + J_{20} & J_{21}\cos(q_2 - q_1) \\ J_{21}\cos(q_2 - q_1) & J_{22} \end{bmatrix}\begin{bmatrix} \ddot{q}_1 \\ \ddot{q}_2 \end{bmatrix} + J_{21}\sin(q_2 - q_1)\begin{bmatrix} -\dot{q}_2^2 \\ \dot{q}_1^2 \end{bmatrix}$$

$$+ \begin{bmatrix} s\sin q_1 \\ \sin q_2 \end{bmatrix}\lambda = F_2^A\begin{bmatrix} s\cos q_1 \\ \cos q_2 \end{bmatrix} \tag{3.2.52}$$

Note that apart from the applied force vectors on the right sides of Eqs. 3.2.45 and 3.2.51, the forms of the equations are identical. The inertia constants of Eq. 3.2.51 that define the influence of the distribution of mass are properties of the bodies, since they may be computed once and stored for use as constants in Eq. 3.2.52. Even if n_1 and n_2 are large, evaluation of terms in Eq. 3.2.51 is cheap. Furthermore, it needs to be done only once, even though Eq. 3.2.52 must be evaluated at many instants during integration. Finally, note that if n_1 and n_2 approach ∞, m_{1i} and m_{2i} may be viewed as differential masses and the four terms in Eq. 3.2.51 become integrals. This is very similar to the process used in Chapter 5 to define inertial properties of rigid bodies.

To complete the development of variational equations of motion of systems of particles, the results of Example 3.2.10 are generalized to obtain the equations of motion of a *rigid body* that moves in the x-y plane. More specifically, consider the body shown in Fig. 3.2.16. The mass density, or mass per unit of area, is denoted $\gamma(x', y')$, where x' and y' are coordinates in an x'-y' frame that is fixed

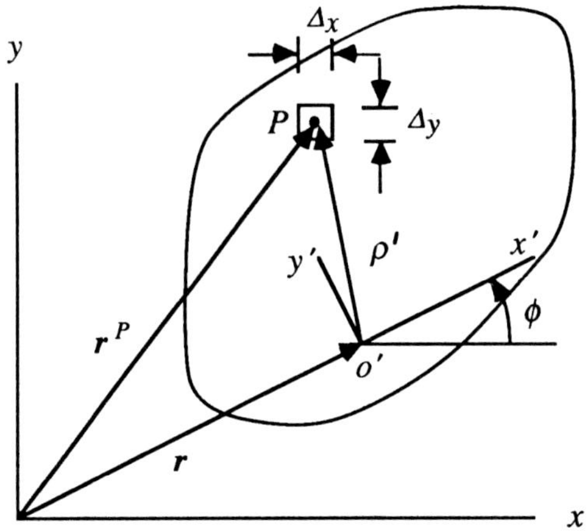

Figure 3.2.16 Planar rigid body.

to the body. Let the origin O' of the x'-y' frame be selected so that

$$\int_A \gamma(x', y')\boldsymbol{\rho}'\, dx'\, dy' = \mathbf{0} \tag{3.2.53}$$

where $\boldsymbol{\rho}' = [x', y']^T$ (i.e., O' is the *center of mass* of the body).

Let the body be thought of as made up of differential masses $\gamma(x', y')\, dx'\, dy'$ at points P shown in Fig. 3.2.16, located at $\boldsymbol{\rho}' = [x', y']^T$ in the x'-y' frame. Assume for now (see Chapter 5 for more details) that the forces of constraint between particles in the body do no work. Also denote external force per unit of mass on the body at point (x', y') as $\mathbf{f}(x', y')$, represented in the x-y frame.

In terms of $\mathbf{r} = [x, y]^T$ and ϕ, any point P, as shown in Fig. 3.2.16, can be located in an inertial x-y frame by the vector (Prob. 3.2.6)

$$\begin{aligned}\mathbf{r}^P &= \mathbf{r} + \begin{bmatrix} \cos\phi & -\sin\phi \\ \sin\phi & \cos\phi \end{bmatrix} \boldsymbol{\rho}' \\ &\equiv \mathbf{r} + \mathbf{A}\boldsymbol{\rho}' \end{aligned} \tag{3.2.54}$$

The velocity and acceleration of the point are thus

$$\begin{aligned}\dot{\mathbf{r}}^P &= \dot{\mathbf{r}} + \dot{\phi} \begin{bmatrix} -\sin\phi & -\cos\phi \\ \cos\phi & -\sin\phi \end{bmatrix} \boldsymbol{\rho}' \\ &= \dot{\mathbf{r}} + \dot{\phi}\mathbf{B}\boldsymbol{\rho}' \end{aligned} \tag{3.2.55}$$

$$\begin{aligned}\ddot{\mathbf{r}}^P &= \ddot{\mathbf{r}} + \ddot{\phi} \begin{bmatrix} -\sin\phi & -\cos\phi \\ \cos\phi & -\sin\phi \end{bmatrix} \boldsymbol{\rho}' + \dot{\phi}^2 \begin{bmatrix} -\cos\phi & \sin\phi \\ -\sin\phi & -\cos\phi \end{bmatrix} \boldsymbol{\rho}' \\ &= \ddot{\mathbf{r}} + \ddot{\phi}\mathbf{B}\boldsymbol{\rho}' - \dot{\phi}^2\mathbf{A}\boldsymbol{\rho}' \end{aligned} \tag{3.2.56}$$

since $\boldsymbol{\rho}'$ is constant.

Consider the body as modeled by a large number of small rectangular particles located at points P_i, $i = 1, \ldots, n$. The mass of the particle at P_i is

$$m_i = \gamma(x_i', y_i')\,\Delta x_i'\,\Delta y_i' \tag{3.2.57}$$

The generalized force of Eq. 3.2.22 on the particle at P_i is (Prob. 3.2.7)

$$Q_i^A = \begin{bmatrix} f(x_i', y_i') \\ n'(x_i' y_i') \end{bmatrix} \gamma(x_i', y_i')\,\Delta x_i'\,\Delta y_i' \tag{3.2.58}$$

where (Prob. 3.2.8)

$$n'(x_i', y_i') = (-x_i' \sin\phi - y' \cos\phi) f_x + (x' \cos\phi - y' \sin\phi) f_y \tag{3.2.59}$$

is the *moment* of force density $\mathbf{f} = [f_x, f_y]^T$ about the origin O' of the x'-y' frame.

Using the fact that

$$\mathbf{r}_{i\mathbf{q}} = [\mathbf{I}, \mathbf{B}\boldsymbol{\rho}']$$

where $\mathbf{q} = [\mathbf{r}^T, \phi]^T$ are independent generalized coordinates, Eq. 3.2.24 yields the equations of motion of the system of particles as (Prob. 3.2.9)

$$\begin{bmatrix} \left(\sum_{i=1}^{n} \gamma(x_i', y_i')\,\Delta x_i'\,\Delta y_i'\right)\mathbf{I} & \mathbf{B}\left(\sum_{i=1}^{n} \gamma(x_i', y_i')\boldsymbol{\rho}_i'\,\Delta x_i'\,\Delta y_i'\right) \\ \left(\sum_{i=1}^{n} \gamma(x_i', y_i')\boldsymbol{\rho}_i'\,\Delta x_i',\,\Delta y_i'\right)^T \mathbf{B}^T & \sum_{i=1}^{n} \gamma(x_i', y_i')\boldsymbol{\rho}_i'^T\boldsymbol{\rho}_i'\,\Delta x_i'\,\Delta y_i' \end{bmatrix} \begin{bmatrix} \ddot{\mathbf{r}} \\ \ddot{\phi} \end{bmatrix}$$

$$- \dot{\phi}^2 \begin{bmatrix} \mathbf{A}\left(\sum_{i=1}^{n} \gamma(x_i', y_i')\boldsymbol{\rho}_i'\,\Delta x_i'\,\Delta y_i'\right) \\ 0 \end{bmatrix} = \begin{bmatrix} \sum_{i=1}^{n} \mathbf{f}(x_i', y_i')\gamma(x_i', y_i')\,\Delta x_i'\,\Delta y_i' \\ \sum_{i=1}^{n} n'(x_i', y_i')\gamma(x_i', y_i')\,\Delta x_i'\,\Delta y_i' \end{bmatrix}$$

Taking the limit as n approaches ∞, the sums become integrals and

$$\sum_{i=1}^{\infty} \gamma(x_i', y_i')\,\Delta x_i'\,\Delta y_i' = \int_A \gamma(x', y')\,dx'\,dy' \equiv m \tag{3.2.61}$$

$$\sum_{i=1}^{\infty} \gamma(x_i', y_i')\boldsymbol{\rho}_i'^T\boldsymbol{\rho}_i'\,\Delta x_i'\,\Delta y_i' = \int_A \gamma(x', y')\boldsymbol{\rho}'^T\boldsymbol{\rho}'\,dx'\,dy' \equiv J_{O'} \tag{3.2.62}$$

$$\sum_{i=1}^{\infty} \gamma(x_i', y_i')\boldsymbol{\rho}_i\,\Delta x_i'\,\Delta y_i' = \int_A \gamma(x', y')\boldsymbol{\rho}'\,dx'\,dy' = \mathbf{0}$$

$$\sum_{i=1}^{\infty} \mathbf{f}(x_i', y_i')\gamma(x_i', y_i')\,\Delta x_i'\,\Delta y_i' = \int_A \mathbf{f}(x', y')\gamma(x', y')\,dx'\,dy' \equiv \mathbf{F}^A \tag{3.2.63}$$

$$\sum_{i=1}^{\infty} n'(x_i', y_i')\gamma(x_i', y_i')\,\Delta x_i'\,\Delta y_i' = \int_A n'(x', y')\gamma(x', y')\,dx'\,dy' \equiv N'^A \tag{3.2.64}$$

where m is the mass of the body, J'_0 the *centroidal polar moment of inertia* of the body, $\mathbf{F}^A$ the resultant force applied to the body, and N'^A the moment of the applied forces about the axis through O' that is perpendicular to the plane. Using these results in Eq. 3.2.60 yields the familiar *planar rigid body equations of motion*,

$$m\ddot{\mathbf{r}} = \mathbf{F}^A \tag{3.2.65}$$

$$J_{0'}\ddot{\phi} = N'^A \tag{3.2.66}$$

While these equations of motion are a special case of the rigid body equations of motion derived in Chapter 5, it is helpful to see that concepts of dynamics of systems of particles do indeed yield practical and important results.

3.3 INITIAL CONDITIONS FOR CONSTRAINED SYSTEMS OF PARTICLES

Using the notation of Sections 2.6 and 2.7, for a system of nb particles, the *composite vector of Cartesian coordinates* is the $3nb \times 1$ vector

$$\mathbf{r} = [\mathbf{r}_1^T, \ldots, \mathbf{r}_{nb}^T]^T \tag{3.3.1}$$

If the kinematics of the system is defined by a vector $\mathbf{q}$ of n generalized coordinates,

$$\mathbf{r} = \mathbf{r}(\mathbf{q}) \tag{3.3.2}$$

Consider first $nh \leq 3nb$ holonomic constraints of the form

$$\mathbf{\Phi}(\mathbf{r}, t) = \mathbf{0} \tag{3.3.3}$$

whose Jacobian $\mathbf{\Phi}_r$ is an $nh \times 3nb$ matrix that has full row rank (i.e., the system has $3nb - nh \geq 0$ *degrees of freedom*). The constraint velocity equation of Eq. 2.6.3 is

$$\mathbf{\phi}_\mathbf{r}\dot{\mathbf{r}} = -\mathbf{\phi}_t \tag{3.3.4}$$

Both Eqs. 3.3.3 and 3.3.4 must hold for all time $t \geq t_0$, where t_0 is the *initial time*.

If the system of particles is unconstrained, then Newton's equations of motion and the *unconstrained initial conditions*

$$\mathbf{r}(t_0) = \mathbf{r}^0 \tag{3.3.5}$$

$$\dot{\mathbf{r}}(t_0) = \mathbf{r}^0 \tag{3.3.6}$$

determine the motion of the particles (i.e., $\mathbf{r} = \mathbf{r}(t)$ for $t \geq t_0$; see Appendix C).

If the constraints of Eqs. 3.3.3 and 3.3.4 must hold at $t = t_0$, then $\mathbf{r}^0$ and $\dot{\mathbf{r}}^0$ cannot be selected freely (i.e., they must satisfy Eqs. 3.3.3 and 3.3.4, respectively). The number of *initial position conditions* that can be freely selected is

$3nb - nh$, specified in the general form

$$\mathbf{\Phi}^I(\mathbf{r}(t_0)) = \mathbf{0} \tag{3.3.7}$$

The complete set of initial conditions is Eq. 3.3.3, evaluated at t_0, and Eq. 3.3.7; that is,

$$\begin{bmatrix} \mathbf{\Phi}(\mathbf{r}(t_0), t_0) \\ \mathbf{\Phi}^I(\mathbf{r}(t_0)) \end{bmatrix} = \mathbf{0} \tag{3.3.8}$$

As shown by the implicit function theorem, Eq. 3.3.8 uniquely determines $\mathbf{r}(t_0)$ locally if it has a solution and if its Jacobian is nonsingular; that is, if

$$\left| \begin{bmatrix} \mathbf{\Phi}_{\mathbf{r}}(\mathbf{r}(t_0), t_0) \\ \mathbf{\Phi}^I_r(\mathbf{r}(t_0)) \end{bmatrix} \right| \neq 0 \tag{3.3.9}$$

Similarly, $3nb - nh$ *initial velocity conditions* of the form

$$\mathbf{C}\dot{\mathbf{r}}(t_0) - \mathbf{d} = \mathbf{0} \tag{3.3.10}$$

are specified, such that Eq. 3.3.4, evaluated at t_0, and Eq. 3.3.10 uniquely determine $\dot{\mathbf{r}}(t_0)$. A necessary and sufficient condition is that their combined coefficient matrix is nonsingular; that is,

$$\left| \begin{bmatrix} \mathbf{\Phi}_r(\mathbf{r}(t_0), t_0) \\ \mathbf{C} \end{bmatrix} \right| \neq 0 \tag{3.3.11}$$

Example 3.3.1: For the two particle dumbbell model of Examples 1.3.1, 2.6.4, and 3.2.1, there was just one holonomic constraint,

$$\Phi(\mathbf{r}) \equiv (\mathbf{r}_2 - \mathbf{r}_1)^T(\mathbf{r}_2 - \mathbf{r}_1) - 1 = 0 \tag{3.3.12}$$

and the constraint velocity equation is

$$2(\mathbf{r}_2 - \mathbf{r}_1)^T(\dot{\mathbf{r}}_2 - \dot{\mathbf{r}}_1) = 0 \tag{3.3.13}$$

The constraint Jacobian is

$$\Phi_{\mathbf{r}} = 2[-(\mathbf{r}_2 - \mathbf{r}_1)^T, (\mathbf{r}_2 - \mathbf{r}_1)^T] \tag{3.3.14}$$

The number of initial position and velocity conditions to be specified is $3 \times 2 - 1 = 5$. For example, consider initial position conditions

$$\mathbf{\Phi}^I(\mathbf{r}(t_0)) \equiv \begin{bmatrix} \mathbf{r}_1(t_0) \\ x_2(t_0) - \dfrac{\sqrt{2}}{2} \\ y_2(t_0) \end{bmatrix} = \mathbf{0} \tag{3.3.15}$$

Substituting the values from Eq. 3.3.15 into Eq. 3.3.12 yields

$$0.5 + z_2^2(t_0) - 1 = 0$$

or $z_2(t_0) = \pm\sqrt{2}/2$. To see that selecting $z_2(t_0) = \sqrt{2}/2$ yields a proper set of initial conditions, the test of Eq. 3.3.9 can be verified (Prob. 3.3.1). As initial velocity conditions, select

$$\mathbf{C}\dot{\mathbf{r}}(t_0) \equiv \begin{bmatrix} \dot{\mathbf{r}}_1(t_0) \\ \dot{x}_2(t_0) - 1 \\ \dot{y}_2(t_0) - 1 \end{bmatrix} = \mathbf{0} \tag{3.3.16}$$

Substituting these conditions and the initial position information calculated above into Eq. 3.3.13 yields

$$\sqrt{2}(1 + \dot{z}_2(t_0)) = 0$$

so $\dot{z}_2(t_0) = -1$ completes the definition of the initial conditions. To see that this is a proper set of initial conditions, the test of Eq. 3.3.11 can be verified (Prob. 3.3.2).

Example 3.3.2: To see that selection of initial conditions for constrained systems is delicate and fraught with potential difficulties, consider again the dumbbell of Example 3.3.1. If the initial position conditions were selected as

$$\mathbf{\Phi}^I(\mathbf{r}(t_0)) \equiv \begin{bmatrix} \mathbf{r}_1(t_0) \\ x_2(t_0) - 1 \\ y_2(t_0) - 1 \end{bmatrix} = \mathbf{0}$$

then troubles exist. To see this, these relations can be substituted into Eq. 3.3.12, at $t = t_0$,

$$1 + 1 + z_2^2(t_0) - 1 = 0$$

or

$$z_2^2(t_0) = -1$$

which has no solution. Thus initial position conditions for constrained systems must be selected carefully. Even if the initial position conditions of Example 3.3.1 were selected, if initial velocity conditions were selected as

$$\mathbf{C}\dot{\mathbf{r}}(t_0) - \mathbf{d} \equiv \begin{bmatrix} \dot{\mathbf{r}}_1(t_0) \\ \dot{x}_2(t_0) - 1 \\ \dot{z}_2(t_0) \end{bmatrix} = \mathbf{0}$$

trouble arises. To see this, substitute these values and the initial positions of Example 3.3.1 into Eq. 3.3.13, at t_0, to obtain

$$\sqrt{2}(1 + 0) = \sqrt{2} \neq 0$$

so there is no value of $\dot{y}_2(t_0)$ that satisfies Eq. 3.3.13. A check of the condition of Eq. 3.3.11 shows that it is not satisfied (Prob. 3.3.3), which explains the source of the problem.

Consider next the case in which, in addition to the nh holonomic constraints of Eq. 3.3.3, there are ℓ differential constraints of the form of Eq. 2.7.8, or equivalently,

$$\mathbf{B}(\mathbf{r}, t)\dot{\mathbf{r}} + \mathbf{c}(\mathbf{r}, t) = \mathbf{0} \tag{3.3.17}$$

Since Eq. 3.3.17 imposes no restrictions on $\mathbf{r}(t_0)$, $3nb - nh$ initial position conditions of the form of Eq. 3.3.7 may be specified, as in the case of holonomically constrained systems.

Initial velocities $\dot{\mathbf{r}}(t_0)$ must, however, satisfy both Eqs. 3.3.4 and 3.3.17. Using the combined constraint Jacobian of Eq. 2.7.9, $\dot{\mathbf{r}}(t_0)$ must satisfy

$$\mathbf{J}(\mathbf{r}(t_0), t_0)\,\dot{\mathbf{r}}(t_0) \equiv \begin{bmatrix} \mathbf{\Phi_r}(\mathbf{r}(t_0), t_0) \\ \mathbf{B}(\mathbf{r}(t_0), t_0) \end{bmatrix} \dot{\mathbf{r}}(t_0) = \begin{bmatrix} -\mathbf{\Phi}_t(\mathbf{r}(t_0), t_0) \\ -\mathbf{c}(\mathbf{r}(t_0), t_0) \end{bmatrix} \tag{3.3.18}$$

where $\mathbf{J}$ is assumed to have full row rank equal to $nh + \ell$. Thus only $3nb - nh - \ell$ initial velocity conditions, of the form of Eq. 3.3.10, can be specified. For Eqs. 3.3.10 and 3.3.18 to determine $\dot{\mathbf{r}}(t_0)$ uniquely, it is necessary and sufficient that

$$\left\| \begin{bmatrix} \mathbf{J}(\mathbf{r}(t_0), t_0) \\ \mathbf{C} \end{bmatrix} \right\| \neq 0 \tag{3.3.19}$$

Example 3.3.3: For the wheel and axle set of Examples 2.7.7 and 3.2.2, in the plane, $\mathbf{r} = [\mathbf{r}_1^T, \mathbf{r}_2^T]^T$ is a 4-vector and the single holonomic constraint of Eq. 3.2.15 must be satisfied. Thus three initial position conditions must be satisfied; for example,

$$\mathbf{\Phi}^I(\mathbf{r}(t_0)) \equiv \begin{bmatrix} \mathbf{r}_1(t_0) \\ x_2(t_0) \end{bmatrix} = \mathbf{0} \tag{3.3.20}$$

With this information, Eq. 3.2.15 at t_0 yields

$$y_2^2(t_0) - 1 = 0$$

so $y_2(t_0) = 1$ is an acceptable initial condition. To see that it is locally unique, the test of Eq. 3.3.9 may be verified (Prob. 3.3.4). The combined differential constraint of Eq. 3.2.14 and constraint velocity condition associated with Eq. 3.2.15 is

$$\mathbf{J}(\mathbf{r})\dot{\mathbf{r}} \equiv \begin{bmatrix} -2(\mathbf{r}_2 - \mathbf{r}_1)^T & 2(\mathbf{r}_2 - \mathbf{r}_1)^T \\ (\mathbf{r}_2 - \mathbf{r}_1)^T & (\mathbf{r}_2 - \mathbf{r}_1)^T \end{bmatrix} \begin{bmatrix} \dot{\mathbf{r}}_1 \\ \dot{\mathbf{r}}_2 \end{bmatrix} = \mathbf{0} \tag{3.3.21}$$

Thus, only two initial velocity conditions can be specified; e.g.,

$$\mathbf{C}\dot{\mathbf{r}}(t_0) + \mathbf{c} \equiv \dot{\mathbf{r}}_1(t_0) - [1, 1]^T = \mathbf{0} \tag{3.3.22}$$

To see if this is a proper set of initial conditions, the test of Eq. 3.3.19 may be made; that is,

$$\left\| \begin{bmatrix} 0 & -2 & 0 & 2 \\ 0 & 1 & 0 & 1 \\ 1 & 0 & 0 & 0 \\ 0 & 1 & 0 & 0 \end{bmatrix} \right\| = 0$$

There is no solution, since initial velocities that are inconsistent with constraints have been defined. To see why the difficulty arose, Fig. 3.3.1 shows the wheel and axle set in the prescribed initial condition. The initial velocity of m_1 specified by Eq. 3.3.22, as shown in Fig. 3.3.1, has a component along the axle, which is inconsistent

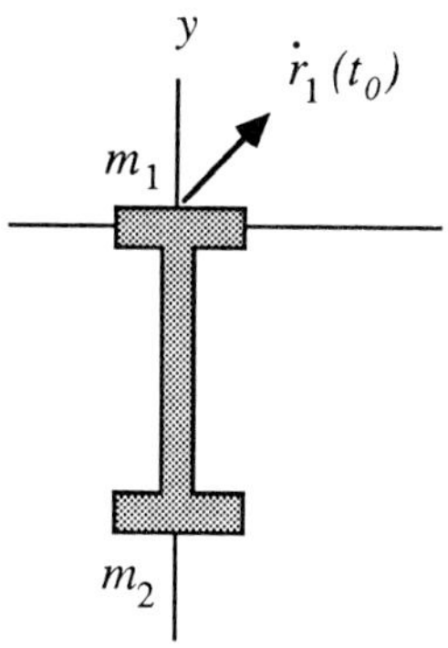

Figure 3.3.1 Initial position of wheel and axle set.

with the nonholonomic constraint of Eq. 3.2.14. Thus both mathematical and physical criteria agree and help in pinpointing the problem.

On physical grounds, it is clear that the initial velocities of both wheels (masses) must be in the x direction. Thus the following initial conditions should be acceptable:

$$\mathbf{C}\dot{\mathbf{r}}(t_0) - \mathbf{d} \equiv \begin{bmatrix} \dot{x}_1(t_0) - 1 \\ \dot{x}_2(t_0) + 1 \end{bmatrix} = \mathbf{0} \tag{3.3.23}$$

To make sure that these initial conditions are mathematically acceptable, the test of Eq. 3.3.19 can be applied; that is,

$$\left\| \begin{bmatrix} 0 & -2 & 0 & 2 \\ 0 & 1 & 0 & 1 \\ 1 & 0 & 0 & 0 \\ 0 & 0 & 1 & 0 \end{bmatrix} \right\| = 4 \neq 0$$

Happily, in Examples 3.3.1 to 3.3.3 both physical and mathematical criteria agree. Note, however, that physical reasoning gave the insights needed to specify acceptable sets of initial conditions. This should not be surprising, since it is in fact the physical system that is being modeled. The value of mathematical checks should, however, not be dismissed. It is easy to make mistakes in formulating models of constrained mechanical systems. A mathematical test can often alert the engineer to the presence of difficulty and cause reconsideration of the model, as in Example 3.3.3. Complementary physical and mathematical analysis is thus essential in modeling the dynamics of constrained systems.

In terms of an n vector of generalized coordinates $\mathbf{q}$ that define particle positions in Eq. 3.3.2, but must satisfy nh independent holonomic constraints of the form

$$\boldsymbol{\Phi}(\mathbf{q}, t) = \mathbf{0} \tag{3.3.24}$$

the velocity equation is

$$\boldsymbol{\Phi}_{\mathbf{q}}\dot{\mathbf{q}} = -\boldsymbol{\Phi}_t \tag{3.3.25}$$

An additional $n - nh$ position and velocity initial conditions must be specified. *Initial position conditions* of the form

$$\mathbf{\Phi}^I(\mathbf{q}(t_0)) = \mathbf{0} \tag{3.3.26}$$

are specified such that Eq. 3.3.24 at t_0 and Eq. 3.3.26 uniquely determine $\mathbf{q}(t_0)$ locally; that is, there is a solution $\mathbf{q}(t_0)$ such that

$$\left| \begin{bmatrix} \mathbf{\Phi}_\mathbf{q}(\mathbf{q}(t_0), t_0) \\ \mathbf{\Phi}^I_\mathbf{q}(\mathbf{q}(t_0)) \end{bmatrix} \right| \neq 0 \tag{3.3.27}$$

Finally, $n - nh$ *initial velocity conditions* of the form

$$\mathbf{C}\dot{\mathbf{q}}(t_0) - \mathbf{d} = \mathbf{0} \tag{3.3.28}$$

must be specified such that Eq. 3.3.25 at t_0 and Eq. 3.3.28 uniquely determine $\dot{\mathbf{q}}(t_0)$; that is, it is necessary and sufficient that

$$\left| \begin{bmatrix} \mathbf{\Phi}_\mathbf{q}(\mathbf{q}(t_0), t_0) \\ \mathbf{C} \end{bmatrix} \right| \neq 0 \tag{3.3.29}$$

Example 3.3.4: For the elementary slider–crank model of Example 3.2.9, the generalized coordinates $\mathbf{q} = [q_1, q_2]^T$ were required to satisfy the holonomic constraint

$$\Phi(\mathbf{q}) \equiv -\cos q_1 - \cos q_2 = 0 \tag{3.3.30}$$

hence also the constraint velocity condition

$$\Phi_\mathbf{q}\dot{\mathbf{q}} = [\sin q_1, \sin q_2]\dot{\mathbf{q}} = 0$$

Thus only one initial position condition can be specified; for example,

$$\Phi^I(\mathbf{q}(t_0)) \equiv x_2(t_0) - \sqrt{2} = \sin q_1(t_0) + \sin q_2(t_0) - \sqrt{2} = 0 \tag{3.3.31}$$

From Fig. 3.2.7, a solution of Eqs. 3.3.30 and 3.3.31 is observed as $q_1 = \pi/4$ and $q_2 = 3\pi/4$. An analytical check of the condition of Eq. 3.3.27 is

$$\left| \begin{bmatrix} 0.707 & 0.707 \\ 0.707 & -0.707 \end{bmatrix} \right| = -1 \neq 0$$

so the initial position condition is acceptable. As an initial velocity condition, select

$$\mathbf{C}\dot{\mathbf{q}}(t_0) - \mathbf{d} \equiv \dot{x}_2(t_0) - 1 = [\cos q_1(t_0), \cos q_2(t_0)]\dot{\mathbf{q}}(t_0) - 1 = 0 \tag{3.3.32}$$

Checking the condition of Eq. 3.3.29 yields

$$\left| \begin{bmatrix} 0.707 & 0.707 \\ 0.707 & -0.707 \end{bmatrix} \right| = -1 \neq 0$$

so all is well and the unique initial velocities are $\dot{q}_1(t_0) = 0.707$ and $\dot{q}_2(t_0) = -0.707$.

The final case to consider is nh holonomic constraints of Eqs. 3.3.24 and 3.3.25 and ℓ differential constraints of the form

$$\mathbf{D}(\mathbf{q}, t)\dot{\mathbf{q}} + \mathbf{c}(\mathbf{q}, t) = \mathbf{0} \tag{3.3.33}$$

where $n \geqslant nh + \ell$.

Initial position constraints are as in Eq. 3.3.26. Initial velocity conditions, however, must account for Eq. 3.3.33. Define the combined constraint Jacobian at t_0 as

$$\mathbf{J}(\mathbf{q}(t_0), t_0) = \begin{bmatrix} \mathbf{\Phi_q}(\mathbf{q}(t_0), t_0) \\ \mathbf{D}(\mathbf{q}(t_0)) \end{bmatrix} \tag{3.3.34}$$

A set of $n - nh - \ell$ initial velocity conditions of the form of Eq. 3.3.28 must be specified, such that

$$\left\| \begin{bmatrix} \mathbf{J}(\mathbf{q}(t_0), t_0) \\ \mathbf{C} \end{bmatrix} \right\| \neq 0 \tag{3.3.35}$$

so that Eqs. 3.3.25, 3.3.33, and 3.3.28 uniquely determine $\dot{\mathbf{q}}(t_0)$.

➧ **Example 3.3.5:** For the generalized coordinate formulation of the wheel and axle set in Example 3.2.8, there are no holonomic constraints. Thus the generalized coordinates $\mathbf{r} = [q_1, q_2]^T$ and $\phi = q_3$ can be selected freely (e.g., $q_1 = 0$, $q_2 = -0.5$, and $q_3 = 3\pi/2$). This yields the same initial position of the system as shown in Fig. 3.3.1. Initial velocity must satisfy the nonholonomic constraint of Eq. 3.2.42, evaluated at t_0:

$$\mathbf{D}(\mathbf{q}(t_0))\dot{\mathbf{q}}(t_0) = -\dot{q}_2(t_0) = 0$$

The same initial velocities specified by Eq. 3.3.23 can be obtained in this formulation with the initial velocity conditions

$$\mathbf{C}\dot{\mathbf{q}}(t_0) - \mathbf{d} \equiv \begin{bmatrix} \dot{q}_1(t_0) \\ \dot{q}_3(t_0) + 2 \end{bmatrix} = \mathbf{0}$$

Checking the condition of Eq. 3.3.35 yields

$$\left\| \begin{bmatrix} 0 & -1 & 0 \\ 1 & 0 & 0 \\ 0 & 0 & 1 \end{bmatrix} \right\| = 1 \neq 0$$

so again all is well and the initial velocities are uniquely determined as $\dot{q}_1(t_0) = 0$, $\dot{q}_2(t_0) = 0$, and $\dot{q}_3(t_0) = -2$.

3.4 MATRIX FORMULATION OF CONSTRAINED EQUATIONS OF MOTION

The Lagrange multiplier form of constrained equations of motion of Section 3.3, for both Cartesian and generalized coordinate formulations, may be written as matrix equations that involve accelerations and Lagrange multipliers. The fundamental form of these equations and physical properties of kinetic energy guarantee that, under reasonable physical conditions, a unique solution of the constrained equations of motion exists.

Cartesian Coordinate Formulation

Consider the Cartesian coordinate formulation with $\mathbf{r} = [\mathbf{r}_1^T, \ldots, \mathbf{r}_{nb}^T]^T$. Holonomic constraint equations are of the form

$$\mathbf{\Phi}(\mathbf{r}, t) = \mathbf{0} \tag{3.4.1}$$

Let the mass matrix of the system of particles be defined as

$$\mathbf{M} = \operatorname{diag}(m_1\mathbf{I}, \ldots, m_{nb}\mathbf{I}) \tag{3.4.2}$$

where $m_i \geq 0$, $i = 1, \ldots, nb$. Finally, let the vector of applied forces be

$$\mathbf{F}^A = [\mathbf{F}_1^{A^T}, \ldots, \mathbf{F}_{nb}^{A^T}]^T \tag{3.4.3}$$

The Lagrange multiplier form of the constrained equations of motion is given by Eq. 3.2.12, in matrix form, as

$$\mathbf{M}\ddot{\mathbf{r}} + \mathbf{\Phi}_{\mathbf{r}}^T\boldsymbol{\lambda} = \mathbf{F}^A \tag{3.4.4}$$

Since these equations involve more variables than equations, it is clear that additional equations must be found if a solution for accelerations and Lagrange multipliers is to be achieved.

Differentiating the holonomic constraint equations of Eq. 3.4.1 with respect to time yields the familiar *constraint velocity equation*

$$\mathbf{\Phi}_{\mathbf{r}}\dot{\mathbf{r}} = -\mathbf{\Phi}_t \tag{3.4.5}$$

Differentiating this result with respect to time yields the *constraint acceleration equation*

$$\mathbf{\Phi}_{\mathbf{r}}\ddot{\mathbf{r}} = -(\mathbf{\Phi}_{\mathbf{r}}\dot{\mathbf{r}})_{\mathbf{r}}\dot{\mathbf{r}} - 2\mathbf{\Phi}_{\mathbf{r}t}\dot{\mathbf{r}} - \mathbf{\Phi}_{tt} \equiv \boldsymbol{\gamma} \tag{3.4.6}$$

Since the applied force vector $\mathbf{F}^A$ may depend on velocities $\dot{\mathbf{r}}$, it is natural to view Eqs. 3.4.4 and 3.4.6 as equations in which the unknowns are the acceleration $\ddot{\mathbf{r}}$ and Lagrange multiplier $\boldsymbol{\lambda}$, where the coefficient matrices and right-side vectors depend on $\mathbf{r}$ and $\dot{\mathbf{r}}$. Writing Eqs. 3.4.4 and 3.4.6 in matrix form yields

$$\begin{bmatrix} \mathbf{M} & \mathbf{\Phi}_{\mathbf{r}}^T \\ \mathbf{\Phi}_{\mathbf{r}} & \mathbf{0} \end{bmatrix} \begin{bmatrix} \ddot{\mathbf{r}} \\ \boldsymbol{\lambda} \end{bmatrix} = \begin{bmatrix} \mathbf{F}^A \\ \boldsymbol{\gamma} \end{bmatrix} \tag{3.4.7}$$

Example 3.4.1: The matrix form of the constrained Cartesian equations of motion of the simple pendulum of Example 3.1.9 is, using Eq. 3.4.6,

$$\begin{bmatrix} m & 0 & 0 & 2x & 0 \\ 0 & m & 0 & 2y & 0 \\ 0 & 0 & m & 0 & 1 \\ 2x & 2y & 0 & 0 & 0 \\ 0 & 0 & 1 & 0 & 0 \end{bmatrix} \begin{bmatrix} \ddot{x} \\ \ddot{y} \\ \ddot{z} \\ \lambda_1 \\ \lambda_2 \end{bmatrix} = \begin{bmatrix} \dfrac{mg}{\sqrt{3}} \\ \dfrac{mg}{\sqrt{3}} \\ \dfrac{mg}{\sqrt{3}} \\ -2\dot{x}^2 - 2\dot{y}^2 \\ 0 \end{bmatrix}$$

The determinant of the coefficient matrix is $4m(x^2 + y^2) = 4m \neq 0$. Thus there is a unique solution for $\ddot{\mathbf{r}}$ and $\boldsymbol{\lambda}$.

For differential constraints of the form of Eq. 3.2.4,

$$\mathbf{B}(\mathbf{r}, t)\dot{\mathbf{r}} + \mathbf{c}(\mathbf{r}, t) = \mathbf{0} \tag{3.4.8}$$

an acceleration equation is obtained by differentiating once with respect to time; that is,

$$\mathbf{B}\ddot{\mathbf{r}} = -[(\mathbf{B}\dot{\mathbf{r}})_{\mathbf{r}}\dot{\mathbf{r}} + \mathbf{B}_t\dot{\mathbf{r}} + \mathbf{c}_{\mathbf{r}}\dot{\mathbf{r}} + \mathbf{c}_t] \equiv \boldsymbol{\gamma}^n \tag{3.4.9}$$

Using the combined constraint Jacobian $\mathbf{J}$ of Eq. 2.7.9, the system acceleration equation associated with the combined constraints of Eqs. 3.4.6 and 3.4.9 is

$$\mathbf{J}\ddot{\mathbf{r}} \equiv \begin{bmatrix} \boldsymbol{\Phi}_{\mathbf{r}} \\ \mathbf{B} \end{bmatrix} \ddot{\mathbf{r}} = \begin{bmatrix} \boldsymbol{\gamma} \\ \boldsymbol{\gamma}^n \end{bmatrix} \equiv \boldsymbol{\gamma}^c \tag{3.4.10}$$

Using the notation $\boldsymbol{\lambda} = [\boldsymbol{\lambda}^{12^T}, \ldots, \boldsymbol{\mu}^{12^T}, \ldots]^T$ and the combined constraint Jacobian, the constrained equations of motion of Eq. 3.2.12 can be written in matrix form as

$$\mathbf{M}\ddot{\mathbf{r}} + \mathbf{J}^T\boldsymbol{\lambda} = \mathbf{F}^A$$

Combining this with Eq. 3.4.10, the matrix form of the constrained equations of motion with differential constraints is

$$\begin{bmatrix} \mathbf{M} & \mathbf{J}^T \\ \mathbf{J} & \mathbf{0} \end{bmatrix} \begin{bmatrix} \ddot{\mathbf{r}} \\ \boldsymbol{\lambda} \end{bmatrix} = \begin{bmatrix} \mathbf{F}^A \\ \boldsymbol{\gamma}^c \end{bmatrix} \tag{3.4.11}$$

A natural question now arises: Are the coefficient matrices on the left sides of Eqs. 3.4.7 and 3.4.11 nonsingular? Since Eq. 3.4.7 is a special case of Eq. 3.4.11, the question will be answered for the latter, as follows.

Constrained Dynamic Existence Theorem (Cartesian Coordinates): If

$$\mathbf{a}^T\mathbf{M}\mathbf{a} > 0$$

for all *kinematically admissible virtual velocities* $\mathbf{a} \neq 0$; that is, for all $\mathbf{a}$ such that

$$\mathbf{J}\mathbf{a} = \mathbf{0}$$

and if the combined constraint Jacobian $\mathbf{J}$ has full row rank, the coefficient matrix on the left side of Eq. 3.4.11 (hence also Eq. 3.4.7) is nonsingular and the constrained equations of motion have a unique solution for $\ddot{\mathbf{r}}$ and $\boldsymbol{\lambda}$.

In order to see that both the positive definiteness property of $\mathbf{M}$ for kinematically admissible virtual velocities and the Jacobian rank condition of this theorem are essential, it is helpful to carry out a proof of the theorem. Recall that a square matrix $\mathbf{A}$ is nonsingular if and only if $\mathbf{A}\mathbf{x} = \mathbf{0}$ implies that $\mathbf{x} = \mathbf{0}$ [2]. To apply this result in proving the theorem, form the equation

$$\begin{bmatrix} \mathbf{M} & \mathbf{J}^T \\ \mathbf{J} & \mathbf{0} \end{bmatrix} \begin{bmatrix} \mathbf{c} \\ \mathbf{b} \end{bmatrix} = \mathbf{0} \tag{3.4.12}$$

where $\mathbf{c}$ has the dimension of $\ddot{\mathbf{r}}$ and $\mathbf{b}$ has the dimension of $\boldsymbol{\lambda}$. Expanding this matrix equation into a pair of equations yields

$$\begin{aligned} \mathbf{Mc} + \mathbf{J}^T\mathbf{b} &= \mathbf{0} \\ \mathbf{Jc} &= \mathbf{0} \end{aligned} \tag{3.4.13}$$

Multiplying both sides of the first of Eqs. 3.4.13 on the left by $\mathbf{c}^T$ yields

$$\mathbf{c}^T\mathbf{Mc} + \mathbf{c}^T\mathbf{J}^T\mathbf{b} = 0$$

From the second of Eqs. 3.4.13, it is observed that $\mathbf{c}^T\mathbf{J}^T = \mathbf{0}$, yielding

$$\mathbf{c}^T\mathbf{Mc} = 0$$

Since $\mathbf{c}$ satisfies the second of Eqs. 3.4.13, it is a kinematically admissible virtual velocity. Positive definiteness of $\mathbf{M}$ for kinematically admissible virtual velocities thus implies that $\mathbf{c} = \mathbf{0}$. Substituting this result into the first of Eqs. 3.4.13 yields

$$\mathbf{J}^T\mathbf{b} = \mathbf{0} \tag{3.4.14}$$

Since the Jacobian $\mathbf{J}$ has full row rank, its transpose has full column rank and Eq. 3.4.14 implies that a linear combination of the columns of the transpose of the Jacobian, with b_i as coefficients, is zero. This implies that $\mathbf{b} = 0$ and completes the proof of the theorem.

Certainly, the hypotheses of the theorem hold if the masses m_i, $i = 1, \ldots, nb$, are all positive, in which case $\mathbf{M}$ is positive definite, and if the constraints are independent. Even if some of the particles in a system have zero mass but are constrained in such a way that kinetic energy is positive for all nonzero kinematically admissible virtual velocities, then the theorem holds and Eqs. 3.4.7 and 3.4.11 have a unique solution for acceleration $\ddot{\mathbf{r}}$ and Lagrange multipliers $\boldsymbol{\lambda}$, once position and velocity of the system are determined.

It is important to note that the coefficient matrices on the left sides of Eqs. 3.4.7 and 3.4.11 depend on $\mathbf{r}$ and that terms on the right may depend on both $\mathbf{r}$ and $\dot{\mathbf{r}}$. Numerical methods for constructing solutions of dynamic equations, based on the solution of Eqs. 3.4.7 and 3.4.11 and the constraint equations of Eqs. 3.4.1 and 3.4.8, are discussed in Appendix D.

Generalized Coordinate Formulation

In case the Cartesian coordinate vector that locates all particles in a system is parameterized by a generalized coordinate vector $\mathbf{q}$, in the form

$$\mathbf{r} = \mathbf{r}(\mathbf{q}) \tag{3.4.15}$$

then the constraint equations of Eq. 3.4.1 may be written in terms of generalized coordinates as

$$\mathbf{\Phi}(\mathbf{q}, t) \equiv \mathbf{\Phi}(\mathbf{r}(\mathbf{q}), t) = \mathbf{0} \tag{3.4.16}$$

Defining the *reduced mass matrix*

$$\hat{\mathbf{M}} \equiv \mathbf{r}_{\mathbf{q}}^T \mathbf{M} \mathbf{r}_{\mathbf{q}} \tag{3.4.17}$$

the equations of motion of Eq. 3.2.40, without differential constraints, may be written in the form (Prob. 3.4.1)

$$\hat{\mathbf{M}}\ddot{\mathbf{q}} + \mathbf{\Phi}_{\mathbf{q}}^T \boldsymbol{\lambda} = \mathbf{Q}^A - \mathbf{r}_{\mathbf{q}}^T \mathbf{M}(\mathbf{r}_{\mathbf{q}}\dot{\mathbf{q}})_{\mathbf{q}}\dot{\mathbf{q}} \equiv \mathbf{R} \tag{3.4.18}$$

Since this equation involves more unknowns than equations, additional information is required in order that a solution may be obtained.

Differentiating Eq. 3.4.16 with respect to time yields the familiar constraint velocity equation

$$\mathbf{\Phi}_{\mathbf{q}}\dot{\mathbf{q}} = -\mathbf{\Phi}_t \tag{3.4.19}$$

Differentiating again yields the *constraint acceleration equation*

$$\mathbf{\Phi}_{\mathbf{q}}\ddot{\mathbf{q}} = -(\mathbf{\Phi}_{\mathbf{q}}\dot{\mathbf{q}})_{\mathbf{q}}\dot{\mathbf{q}} - 2\mathbf{\Phi}_{\mathbf{q}t}\dot{\mathbf{q}} - \mathbf{\Phi}_{tt} \equiv \boldsymbol{\gamma} \tag{3.4.20}$$

Combining Eqs. 3.4.18 and 3.4.20 in matrix form yields

$$\begin{bmatrix} \hat{\mathbf{M}} & \mathbf{\Phi}_{\mathbf{q}}^T \\ \mathbf{\Phi}_{\mathbf{q}} & \mathbf{0} \end{bmatrix} \begin{bmatrix} \ddot{\mathbf{q}} \\ \boldsymbol{\lambda} \end{bmatrix} = \begin{bmatrix} \mathbf{R} \\ \boldsymbol{\gamma} \end{bmatrix} \tag{3.4.21}$$

which is of the same form as the equations of motion in Cartesian coordinates in Eq. 3.4.7.

Example 3.4.2: The matrix form of the constrained equations of motion of the slider–crank model of Example 3.2.9, using Eqs. 3.2.44, 3.2.45, and 3.4.20, is

$$\begin{bmatrix} (m_1 + m_2) & m_2\cos(q_2 - q_1) & \sin q_1 \\ m_2\cos(q_2 - q_1) & m_2 & \sin q_2 \\ \sin q_1 & \sin q_2 & 0 \end{bmatrix} \begin{bmatrix} \ddot{q}_1 \\ \ddot{q}_2 \\ \lambda \end{bmatrix}$$

$$= \begin{bmatrix} -(m_1 + m_2)g\sin q_1 + m_2\dot{q}_2^2\sin(q_2 - q_1) \\ -m_2 g\sin q_2 - m_2\dot{q}_1^2\sin(q_2 - q_1) \\ -\dot{q}_1^2\cos q_1 - \dot{q}_2^2\cos q_2 \end{bmatrix}$$

The determinant of the coefficient matrix is

$$-[m_1\sin^2 q_2 + m_2\sin^2(q_2 - q_1)] \leq 0$$

In order for the determinant to be zero, $\sin q_2 = 0 = \sin(q_2 - q_1)$; that is, (1) $q_2 = 0$ and $q_1 = 0$ or π, or (2) $q_2 = \pi$ and $q_1 = \pi$ or 0. Note that in all these cases, $\Phi_{\mathbf{q}} = [\sin q_1, \sin q_2]$ is $\mathbf{0}$; that is, it fails to have full row rank and the hypotheses of the constrained dynamic existence theorem fail to be satisfied. This is not too surprising, since it was shown in Example 2.6.5 that the slider–crank mechanism with bars of equal length has extreme singular behavior at these same configurations. If the model of Example 3.2.9 were modified so that the length of the first bar is less than 1, the coefficient matrix in the matrix equations of motion is always nonsingular (Prob. 3.4.2).

As in the case of Cartesian coordinates, if differential constraints are present in the generalized coordinate formulation, Eq. 3.4.21 is valid if the system Jacobian $\mathbf{J}$ is used in place of $\mathbf{\Phi_q}$ and the appropriate modification is made to the right side of the acceleration equation.

Example 3.4.3: The acceleration equation for the nonholonomic constraint of Eq. 3.2.42, for the wheel and axle of Example 3.2.8, is

$$\ddot{q}_1 \cos q_3 + \ddot{q}_2 \sin q_3 = (\dot{q}_1 \sin q_3 - \dot{q}_2 \cos q_3)\dot{q}_3$$

Combining Eq. 3.2.43 with $m_1 = m_2 = m$ and this result, the matrix equations of motion are

$$\begin{bmatrix} 2m & 0 & 0 & \cos q_3 \\ 0 & 2m & 0 & \sin q_3 \\ 0 & 0 & \dfrac{m}{2} & 0 \\ \cos q_3 & \sin q_3 & 0 & 0 \end{bmatrix} \begin{bmatrix} \ddot{q}_1 \\ \ddot{q}_2 \\ \ddot{q}_3 \\ \boldsymbol{\mu} \end{bmatrix} = \begin{bmatrix} 0 \\ 0 \\ 0 \\ (\dot{q}_1 \sin q_3 - \dot{q}_2 \cos q_3)\dot{q}_3 \end{bmatrix}$$

The determinant of the coefficient matrix is just $-m^2 \neq 0$, so the matrix is always nonsingular. This conclusion agrees with the constrained dynamic existence theorem, since $\mathbf{M} = \text{diag}(2m, 2m, m/2)$ is positive definite and $\mathbf{J} = [\cos q_3, \sin q_3]$ has full row rank, since $\sin q_3$ and $\cos q_3$ cannot both be zero.

Constrained Dynamic Existence Theorem (Generalized Coordinates): The constrained dynamic existence theorem for Cartesian coordinates is also valid for the generalized coordinate formulation, provided that the parameterization of Eq. 3.4.15 in terms of generalized coordinates $\mathbf{q}$ is nonsingular (see Section 2.3) and the Jacobian $\mathbf{\Phi_q}$ or $\mathbf{J}$ has full row rank.

To see that this is true, form the equation

$$\begin{bmatrix} \hat{\mathbf{M}} & \mathbf{\Phi_q}^T \\ \mathbf{\Phi_q} & \mathbf{0} \end{bmatrix} \begin{bmatrix} \hat{\mathbf{c}} \\ \hat{\mathbf{b}} \end{bmatrix} = \mathbf{0} \tag{3.4.22}$$

If it can be shown that $\hat{\mathbf{c}}$ and $\hat{\mathbf{b}}$ are zero, the coefficient matrix in Eq. 3.4.22, hence in Eq. 3.4.21, is nonsingular.

The first of Eqs. 3.4.22 is

$$\hat{\mathbf{M}}\hat{\mathbf{c}} + \mathbf{\Phi_q}^T\hat{\mathbf{b}} = \mathbf{0} \tag{3.4.23}$$

Multiplying both sides of this equation on the left by $\hat{\mathbf{c}}^T$ yields

$$\hat{\mathbf{c}}^T\hat{\mathbf{M}}\hat{\mathbf{c}} + \hat{\mathbf{c}}^T\mathbf{\Phi_q}^T\hat{\mathbf{b}} = 0$$

The second term on the left of this equation is zero, due to the second of Eqs. 3.4.22. Therefore,

$$\hat{\mathbf{c}}^T \hat{\mathbf{M}} \hat{\mathbf{c}} = 0$$

Using the definition of $\hat{\mathbf{M}}$ of Eq. 3.4.17, this is

$$(\hat{\mathbf{c}}^T \mathbf{r}_\mathbf{q}^T)\mathbf{M}(\mathbf{r}_\mathbf{q}\hat{\mathbf{c}}) = 0 \tag{3.4.24}$$

Using the chain rule of differentiation in the second of Eqs. 3.4.22 yields

$$\mathbf{\Phi}_\mathbf{q}\hat{\mathbf{c}} = \mathbf{\Phi}_\mathbf{r}(\mathbf{r}_\mathbf{q}\hat{\mathbf{c}}) = \mathbf{0}$$

Thus the vector $\mathbf{r}_\mathbf{q}\hat{\mathbf{c}}$ is a kinematically admissible Cartesian virtual velocity, so by the hypothesis of the constrained dynamic existence theorem in Cartesian coordinates,

$$\mathbf{r}_\mathbf{q}\hat{\mathbf{c}} = \mathbf{0}$$

The matrix $\mathbf{r}_\mathbf{q}$ has full column rank, since the parameterization is nonsingular, so $\hat{\mathbf{c}} = \mathbf{0}$. Therefore, Eq. 3.4.23 reduces to

$$\mathbf{\Phi}_\mathbf{q}^T\hat{\mathbf{b}} = \mathbf{0}$$

Since $\mathbf{\Phi}_\mathbf{q}$ has full row rank, its transpose has full column rank and $\hat{\mathbf{b}} = \mathbf{0}$. Thus the extension of the constrained dynamic existence theorem is proved for the generalized coordinate formulation.

3.5 CONDITIONS OF EQUILIBRIUM

Attention has thus far been focused on obtaining equations for dynamic response of systems modeled as particles, under the action of applied forces. A very special form of dynamic response is a value of the generalized coordinate vector that is consistent with constraints and for which accelerations of all particles in the system, relative to the same stationary reference frame, are zero. Such a state of a system is called a state of *equilibrium*. Note that this is equivalent to the condition that all particles in the system have the same constant velocity, relative to the same stationary reference frame.

To be more specific, Newton's second law requires that the resultant of all forces that act on a particle be zero if its acceleration is to be zero; that is,

$$\mathbf{F}_i = \mathbf{F}_i^A + \mathbf{F}_i^C = \mathbf{0}, \qquad i = 1, \ldots, nb \tag{3.5.1}$$

Presuming that applied and constraint forces may be written in terms of generalized coordinates of the system, Eq. 3.5.1 can be solved to determine states of equilibrium.

Example 3.5.1: For the spherical pendulum of unit length shown in Fig. 3.5.1, the constraint force $\mathbf{F}^C$ that acts on the particle is in the $\mathbf{r}$ direction, where

$$\mathbf{r}(\mathbf{q}) = \begin{bmatrix} \cos q_2 \sin q_1 \\ \sin q_2 \sin q_1 \\ -\cos q_1 \end{bmatrix}$$

Thus

$$\mathbf{F}^C = -k \begin{bmatrix} \cos q_2 \sin q_1 \\ \sin q_2 \sin q_1 \\ -\cos q_1 \end{bmatrix}$$

where k is a constant if the pendulum is in equilibrium. The only applied force is due to gravity, which acts in the negative z direction (i.e., $\mathbf{F}^A = mg[0, 0, -1]^T$). Thus, for equilibrium, Eq. 3.5.1 requires that

$$-k \cos q_2 \sin q_1 = 0$$

$$-k \sin q_2 \sin q_1 = 0$$

$$k \cos q_1 - mg = 0$$

This implies that $\sin q_1 = 0$ and $k = mg/\cos q_1$. Two distinct equilibrium positions are thus $q_1 = 0$ and π. The first solution is the expected position in which the particle hangs vertically downward. In the second equilibrium position, the particle is on the positive z axis (i.e., the pendulum stands vertically upward). Although this seems unnatural, it is in fact a state of equilibrium.

Multiplying both sides of Eq. 3.5.1 on the left by $\delta\mathbf{r}_i^T$ and summing over all particles in the system yields

$$\delta W = \delta W^A + \delta W^C = \sum_{i=1}^{nb} (\delta\mathbf{r}_i^T\mathbf{F}_i^A + \delta\mathbf{r}_i^T\mathbf{F}_i^C) = 0$$

Provided that constraints that act on the particles are workless, this reduces to

$$\sum_{i=1}^{nb} \delta\mathbf{r}_i^T\mathbf{F}_i^A = 0 \tag{3.5.2}$$

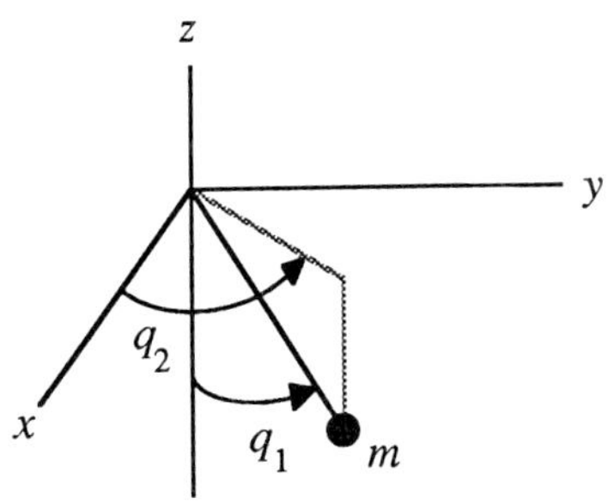

Figure 3.5.1 Spherical pendulum.

for all virtual displacements $\delta\mathbf{r}$ that are consistent with constraints $\mathbf{\Phi}(\mathbf{r}) = \mathbf{0}$ that act on the particles. This form of the conditions of equilibrium is called the *principle of virtual work.*

♦ **Example 3.5.2:** For the constrained particle in Example 3.5.1, the constraint force $\mathbf{F}^C$ in the bar of the pendulum is indeed workless. Furthermore, using $\delta\mathbf{r} = \mathbf{r_q}\,\delta\mathbf{q}$, Eq. 3.5.2 becomes

$$\delta\mathbf{r}^T\mathbf{F}^A = -mg \sin q_1\, \delta q_1 = 0$$

for arbitrary δq_1. Thus the condition of equilibrium is immediately obtained as $\sin q_1 = 0$, which is the same result as obtained in Example 3.5.1. Note, however, the ease and simplicity of the result obtained with the principle of virtual work.

Recall that virtual displacements $\delta\mathbf{r}_i$ are consistent with constraints $\mathbf{\Phi}(\mathbf{r}) = \mathbf{0}$ if they satisfy

$$\sum_{i=1}^{nb} \mathbf{\Phi}_{\mathbf{r}_i}\,\delta\mathbf{r}_i = \mathbf{0}$$

Since Eq. 3.5.2 must hold for all such $\delta\mathbf{r}_i$, the Lagrange multiplier theorem implies that there exists a Lagrange multiplier vector $\boldsymbol{\lambda}$ such that

$$\sum_{i=1}^{nb} \delta\mathbf{r}_i^T[\mathbf{F}_i^A - \mathbf{\Phi}_{\mathbf{r}_i}^T\boldsymbol{\lambda}] = 0$$

for all $\delta\mathbf{r}_i$, where the minus sign is selected to maintain consistency of the physical significance of $\boldsymbol{\lambda}$ with Eq. 3.2.9. This yields the *Lagrange multiplier form of the equations of equilibrium,*

$$\mathbf{F}_i^A - \mathbf{\Phi}_{\mathbf{r}_i}^T\boldsymbol{\lambda} = \mathbf{0}, \qquad i = 1, \ldots, nb \tag{3.5.3}$$

Equations 3.5.3 and the constraint equation $\mathbf{\Phi}(\mathbf{r}) = \mathbf{0}$ may be solved to determine equilibrium configurations $\mathbf{r}^e$.

Writing the virtual work of applied forces on a particle, or systems of particles, in terms of generalized coordinates $\mathbf{q}$, Eq. 3.5.2 may be summed over all particles in the system and rewritten as

$$\delta\mathbf{q}^T\mathbf{Q}^A = 0 \tag{3.5.4}$$

which must hold for all generalized coordinate variations $\delta\mathbf{q}$ that are consistent with workless kinematic constraints. Equation 3.5.4 is called the *virtual work condition of equilibrium.*

If the generalized coordinates are independent, $\delta\mathbf{q}$ in Eq. 3.5.4 is arbitrary and the *equilibrium equations in independent generalized coordinates* are

$$\mathbf{Q}^A(\mathbf{q}) = \mathbf{0} \tag{3.5.5}$$

Recalling the definition of generalized force in Eq. 3.2.22, this is just

$$\sum_{i=1}^{nb} \mathbf{r}_{i_\mathbf{q}}(\mathbf{q})^T \mathbf{F}_i^A(\mathbf{q}) = \mathbf{0} \tag{3.5.6}$$

which is valid only if the generalized coordinates are independent.

➧ **Example 3.5.3:** For the double pendulum of Example 3.2.5, $\mathbf{q} = [q_1, q_2]^T$ are independent generalized coordinates, so the equilibrium equation of Eq. 3.5.6 is valid. With gravity acting in the negative z direction and using calculations from Example 3.2.5, Eq. 3.5.6 yields

$$\mathbf{Q}^A(\mathbf{q}) = -\begin{bmatrix} (m_1 + m_2)g \sin q_1 \\ m_2 g \sin q_2 \end{bmatrix} = \mathbf{0}$$

Thus $q_1 = n\pi$, $n = 0, 1, \ldots$, and $q_2 = m\pi$, $m = 0, 1, \ldots$, are equilibrium configurations. The solution $q_1 = q_2 = 0$ is the expected configuration in which both bars hang vertically downward (i.e., along the negative z axis). Solutions such as $\mathbf{q} = [0, \pi]^T$ and $\mathbf{q} = [\pi, \pi]^T$ imply that one or both bars stands vertically upward, which appears to be unnatural.

Consider next the case in which generalized coordinates are not independent; that is, $\mathbf{q}$ must satisfy holonomic constraint equations

$$\mathbf{\Phi}(\mathbf{q}) = \mathbf{0} \tag{3.5.7}$$

In order to make use of the virtual work form of the equilibrium equations for a constrained system, recall that generalized coordinate variations $\delta\mathbf{q}$ must satisfy the linearized form of Eq. 3.5.7; that is,

$$\mathbf{\Phi}_\mathbf{q}\, \delta\mathbf{q} = \mathbf{0} \tag{3.5.8}$$

Since Eq. 3.5.4 must hold for all $\delta\mathbf{q}$ that satisfy Eq. 3.5.8, the Lagrange multiplier theorem guarantees existence of a Lagrange multiplier $\boldsymbol{\lambda}$ such that

$$\delta\mathbf{q}^T(\mathbf{Q}^A - \mathbf{\Phi}_\mathbf{q}^T\boldsymbol{\lambda}) = 0 \tag{3.5.9}$$

which must hold for arbitrary $\delta\mathbf{q}$. The minus sign is introduced to maintain consistency with Eq. 3.2.33. Thus the coefficient of $\delta\mathbf{q}^T$ in Eq. 3.5.9 must be zero, yielding the *Lagrange multiplier form of constrained equilibrium equations*,

$$\mathbf{Q}^A - \mathbf{\Phi}_\mathbf{q}^T\boldsymbol{\lambda} = \mathbf{0} \tag{3.5.10}$$

Note that Eqs. 3.5.7 and 3.5.10 together comprise a set of algebraic equations. The number of equations is equal to the sum of the number of generalized coordinates nc and the number of Lagrange multipliers nh. Thus there is some hope that Eqs. 3.5.7 and 3.5.10 will yield equilibrium configurations.

➧ **Example 3.5.4:** The elementary slider–crank mechanism of Example 3.2.9 is described by generalized coordinates $\mathbf{q} = [q_1, q_2]^T$ that must satisfy the constraint equation of Eq. 3.2.44. Using the generalized force expression from Example 3.2.5

and the constraint Jacobian from Example 3.2.9, the equilibrium equation for the slider–crank is, from Eq. 3.5.10,

$$-\begin{bmatrix}(m_1 + m_2)g \sin q_1 \\ m_2 g \sin q_2\end{bmatrix} - \begin{bmatrix}\sin q_1 \\ \sin q_2\end{bmatrix} \lambda = 0$$

With the constraint of Eq. 3.2.44, this is

$$[(m_1 + m_2)g + \lambda] \sin q_1 = 0$$

$$[m_2 g + \lambda] \sin q_2 = 0$$

$$\cos q_1 + \cos q_2 = 0$$

The first two equations cannot be satisfied if both $\sin q_1$ and $\sin q_2$ are nonzero. Thus one must be zero, say $\sin q_i = 0$, so $\cos q_i = \pm 1$. In order to satisfy the third equation, $\cos q_j = \mp 1$, $j \neq i$ is required. Thus $\sin q_j = 0$ and all equilibrium configurations satisfy $\sin q_1 = \sin q_2 = 0$. The general solution is $\mathbf{q} = [n\pi, (n+1)\pi]^T$, $n = 0, 1, \ldots$.

For the equilibrium configuration $\mathbf{q} = [0, \pi]^T$, m_2 is at the origin and m_1 is on the negative z axis (see Fig. 3.2.7); that is, the first bar hangs vertically downward, which seems reasonable. In the alternate equilibrium configuration $\mathbf{q} = [\pi, 2\pi]^T$, or equivalently, $\mathbf{q} = [\pi, 0]^T$, m_2 is at the origin and m_1 is on the positive z axis (i.e., the first bar stands vertically upward). This appears to be unnatural.

Since Eq. 3.5.5 and Eqs. 3.5.7 and 3.5.10 are generally nonlinear in the generalized coordinate vector $\mathbf{q}$, it is seldom that closed-form solutions for equilibrium configurations can be found for realistic applications. Iterative solutions (e.g., the Newton–Raphson iterative method presented in Appendix B) must be used to solve for equilibrium configurations. In addition to creating the need for iterative solution techniques, the nonlinear character of equilibrium equations may lead to difficulties associated with multiple equilibrium configurations, or perhaps no equilibrium configuration.

Example 3.5.5: For the spatial pendulum of Examples 3.5.1 and 3.5.2, the generalized coordinates $\mathbf{q} = [q_1, q_2]^T$ are independent. Thus, using the expression for $\mathbf{r}(\mathbf{q})$ derived in Example 3.5.1 and Eq. 3.5.5,

$$\mathbf{Q}^A = \mathbf{r}_{\mathbf{q}}^T(-mg\mathbf{k}) = \begin{bmatrix}-mg \sin q_1 \\ 0\end{bmatrix} = \mathbf{0}$$

so $\sin q_1 = 0$. This indicates that q_2 is arbitrary and q_1 is either 0 or π. In the equilibrium configuration $\mathbf{q} = [0, q_2]^T$ (see Fig. 3.5.1), the bar hangs vertically downward and q_2 is arbitrary, which is reasonable. If $q_1 = \pi$, the bar stands vertically upward, which is a physically unnatural equilibrium configuration. In either case, q_2 is arbitrary, which yields a large family of equilibrium configurations.

Example 3.5.6: From Example 1.1.3, Eq. 3.5.1 for equilibrium of the flexible pendulum with $k > mg$ and $\ell_0 = 1$ is

$$k\left(\frac{\ell - 1}{\ell}\right)\mathbf{r} + mg\mathbf{k} = \mathbf{0}$$

Thus $\mathbf{r} = \alpha\mathbf{k}$, for some constant α. Substituting $\ell = \sqrt{\mathbf{r}^T\mathbf{r}} = \sqrt{\alpha^2}$ in this equation and multiplying both sides by $\ell\mathbf{k}^T$ yields

$$k(\sqrt{\alpha^2} - 1)\alpha + \sqrt{\alpha^2}\, mg = 0$$

A positive solution for α (i.e., $\alpha_1 = \sqrt{\alpha^2}$) is

$$\alpha_1 = 1 - \frac{mg}{k}$$

whereas a negative solution (i.e., $\alpha_2 = -\sqrt{\alpha^2}$) is

$$\alpha_2 = -\left(1 + \frac{mg}{k}\right)$$

The positive solution α_1, or $\mathbf{r}_1 = \alpha_1\mathbf{k}$, indicates that the pendulum stands upward along the z axis and that the spring is in compression (shortened), which appears unnatural. Conversely, for the negative solution $\mathbf{r}_2 = \alpha_2\mathbf{k}$, the pendulum hangs along the negative z axis and the spring is stretched, which is expected on physical grounds.

If forces that act on a particle or a system of particles are conservative, the virtual work may be written as the negative of a variation in *potential energy*; that is,

$$\delta W = -\delta V = 0 \tag{3.5.11}$$

For an unconstrained system, the equilibrium condition $\delta W = 0$, hence $\delta V = V_{\mathbf{q}}\,\delta\mathbf{q} = 0$, for arbitrary $\delta\mathbf{q}$ requires that

$$\frac{\partial V}{\partial \mathbf{q}} \equiv V_{\mathbf{q}} = \mathbf{0} \tag{3.5.12}$$

Thus for an unconstrained conservative system, Eq. 3.5.12 yields the equations of equilibrium.

A form of pathological behavior that has not been encountered in dynamics plagues considerations in equilibrium analysis. Some equilibrium configurations are stable and some are unstable. A system is in a configuration of *stable equilibrium* if a small perturbation from the equilibrium configuration $\mathbf{q}^e$ may lead to only a small displacement of the system. It appears intuitively clear, and it is developed analytically in the literature [7, 9, 12], that if the potential energy at an equilibrium configuration is at a *strict relative minimum*, then to move the system away from the equilibrium configuration requires that a positive amount of work be done by an external agent. Conversely, if the potential energy at the equilibrium configuration is not a strict relative minimum, there must be a direction of movement $\delta\mathbf{q}$ such that V remains the same or decreases, leading to continued movement or an increase in kinetic energy, hence accelerating movement away from the equilibrium configuration. Thus, for an equilibrium configuration to be stable, the associated potential energy of the system must be at a strict relative minimum.

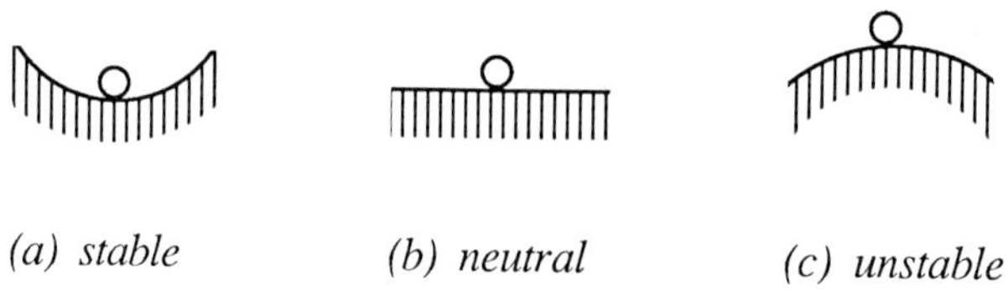

Figure 3.5.2 Equilibrium of balls on curves.

A visually clear illustration of the *minimum potential energy* criterion for stable equilibrium is shown in Fig. 3.5.2. A ball rolls on curves in a plane, with gravity acting vertically downward. Clearly, the ball shown in Fig. 3.5.2(a) is at a position of *stable equilibrium* and the ball shown in Fig. 3.5.2(c) is at a position of *unstable equilibrium*. The ball shown in Fig. 3.5.2(b), however, can move a finite distance with no work done by a lateral applied force. It is thus said to be in a configuration of *neutral equilibrium*.

To take advantage of the foregoing condition for stable equilibrium of a conservative system, consider the Taylor expansion of potential energy about an equilibrium configuration $\mathbf{q}^e$,

$$V(\mathbf{q}^e + \delta\mathbf{q}) = V(\mathbf{q}^e) + V_{\mathbf{q}}(\mathbf{q}^e)\,\delta\mathbf{q} + \tfrac{1}{2}\delta\mathbf{q}^T V_{\mathbf{qq}}(\mathbf{q}^e)\,\delta\mathbf{q} + O(\delta\mathbf{q}^3) \tag{3.5.13}$$

where the matrix of second derivatives of potential energy is

$$V_{\mathbf{qq}} \equiv (V_{\mathbf{q}}^T)_{\mathbf{q}} = \left[\frac{\partial^2 V}{\partial q_i\,\partial q_j}\right] \tag{3.5.14}$$

According to Eq 3.5.12, at an equilibrium configuration the second term on the right of Eq. 3.5.13 must be zero. This leads to the condition that for any sufficiently small generalized coordinate variation $\delta\mathbf{q} \neq \mathbf{0}$,

$$V(\mathbf{q}^e + \delta\mathbf{q}) - V(\mathbf{q}^e) = \tfrac{1}{2}\delta\mathbf{q}^T V_{\mathbf{qq}}\,\delta\mathbf{q} + O(\delta\mathbf{q}^3) > 0 \tag{3.5.15}$$

This condition is satisfied if the matrix of second derivatives of potential energy with respect to generalized coordinates is a *positive definite matrix*. To see that the term $O(\delta\mathbf{q}^3)$ in Eq. 3.5.15 causes no contradiction, note that if $\delta\bar{\mathbf{q}}^T V_{\mathbf{qq}}\,\delta\bar{\mathbf{q}} > 0$ for a vector $\delta\bar{\mathbf{q}}$, then for α small enough and $\delta\mathbf{q} = \alpha\,\delta\bar{\mathbf{q}}$, the quadratic term in Eq. 3.5.15 is larger in magnitude than $O(\delta\mathbf{q}^3)$.

There are numerous tests for positive definiteness of a matrix [2]. One of the simplest conceptual tests for positive definiteness of an $n \times n$ symmetric matrix $\mathbf{A}$ is written in terms of its *principal submatrices*:

$$\mathbf{A}_i = \begin{bmatrix} a_{11} & \cdots & a_{1i} \\ \vdots & & \\ a_{i1} & & a_{ii} \end{bmatrix}, \qquad i = 1, \ldots, n \tag{3.5.16}$$

where $\mathbf{A}_n = \mathbf{A}$. The ith principal submatrix of matrix $\mathbf{A}$ in Eq. 3.5.16 is obtained by retaining only the first i rows and i columns of $\mathbf{A}$. There are thus n principal submatrices of an $n \times n$ matrix. A simple, but computationally inefficient, test for positive definiteness is that a symmetric $n \times n$ matrix $\mathbf{A}$ is positive definite if and only if the determinants of all its principal submatrices are positive [2]; that is,

$$|\mathbf{A}_i| > 0, \qquad i = 1, \ldots, n \tag{3.5.17}$$

♦ **Example 3.5.7:** Consider the spherical pendulum of Examples 3.5.1, 3.5.2, and 3.5.3, with $q_2 = 0$; that is, the pendulum pivots about the y axis. From Example 3.5.2, $\delta W = -mg \sin q_1 \, \delta q_1$, so $W = mg \cos q_1$ and $V = -mg \cos q_1$. Since there is only one generalized coordinate, $V_{qq} = mg \cos q_1$ is a 1×1 matrix. At the solution $q_1 = 0$, its determinant is simply $|V_{qq}| = mg \cos 0 = mg > 0$. Thus this equilibrium configuration is stable, which agrees with the physical observation made in Example 3.5.1. At the solution $q = \pi$, $|V_{qq}| = mg \cos \pi = -mg < 0$, so this equilibrium configuration is unstable.

For the spherical pendulum of Examples 3.5.1, 3.5.2, and 3.5.5, with q_2 free to vary, $\delta W = -mg \sin q_1 \, \delta q_1$, so $W = mg \cos q_1$ and $V = -mg \cos q_1$. Since $\mathbf{q} = [q_1, q_2]^T$ in this case,

$$V_{\mathbf{qq}} = \begin{bmatrix} mg \cos q_1 & 0 \\ 0 & 0 \end{bmatrix}$$

Since $|V_{\mathbf{qq}}| = 0$, the matrix $V_{\mathbf{qq}}$ fails to be positive definite. In fact, q_2 was arbitrary in the equilibrium configurations found in Example 3.5.5. This shows that a large rotation about the z axis can be achieved by exerting a torque that does an arbitrarily small amount of work. Thus neither of the equilibrium configurations found in Example 3.5.5 is stable.

As indicated by these examples, equilibrium configurations found for conservative systems must be checked for stability. Unstable equilibrium configurations do not normally survive in nature and they may have disastrous consequences if allowed to occur in engineering systems.

In engineering applications, the complexity of forming and solving nonlinear equilibrium equations can be untenable. With the advent of powerful numerical minimization methods, the principle of minimum potential energy for conservative systems can be applied directly. One such numerical method is presented and illustrated in Ref. 12. It is not developed here, since the focus of this book is on fundamentals rather than on numerical methods. One major advantage of using a numerical method that seeks configurations of minimum potential energy is that it automatically rejects unstable configurations of equilibrium.

3.6 LAGRANGE'S EQUATIONS OF MOTION

As a final topic in dynamics of systems of particles, an alternative formulation of equations of motion that is due to Lagrange is presented, called *Lagrange's equations of motion*. Even though the form of Lagrange's equations of motion is elegantly simple, they are inefficient for actual use in constructing equations of motion. The value of Lagrange's equations stems from insights they yield into some problems and first integrals they yield in others.

The *kinetic energy* of a system of particles, as defined in Section 1.4, is

$$T = \sum_i \tfrac{1}{2} m_i \dot{\mathbf{r}}_i^T \dot{\mathbf{r}}_i \tag{3.6.1}$$

In terms of generalized coordinates for which $\mathbf{r}_i = \mathbf{r}_i(\mathbf{q})$, substituting for $\dot{\mathbf{r}}_i$ from Eq. 2.2.16 yields

$$T = \sum_i \tfrac{1}{2} m_i \dot{\mathbf{q}}^T \mathbf{r}_{i\mathbf{q}}^T \mathbf{r}_{i\mathbf{q}} \dot{\mathbf{q}} \tag{3.6.2}$$

In preparation for developing Lagrange's equations of motion, some analytical manipulations of derivatives of kinetic energy are needed. The partial derivative of kinetic energy with respect to $\dot{q}_j$ is, using Eq. 3.6.2,

$$\begin{aligned} \frac{\partial T}{\partial \dot{q}_j} &= \sum_i m_i \dot{\mathbf{q}}^T \mathbf{r}_{i\mathbf{q}}^T \mathbf{r}_{iq_j} \\ &= \sum_i m_i \mathbf{r}_{iq_j}^T \mathbf{r}_{i\mathbf{q}} \dot{\mathbf{q}} \end{aligned} \tag{3.6.3}$$

Taking the time derivative of both sides and expanding yields

$$\frac{d}{dt}\left(\frac{\partial T}{\partial \dot{q}_j}\right) = \sum_i m_i [\mathbf{r}_{iq_j}^T \mathbf{r}_{i\mathbf{q}} \ddot{\mathbf{q}} + \mathbf{r}_{iq_j}^T (\mathbf{r}_{i\mathbf{q}} \dot{\mathbf{q}})_{\mathbf{q}} \dot{\mathbf{q}} + \dot{\mathbf{q}}^T \mathbf{r}_{i\mathbf{q}}^T \mathbf{r}_{iq_j\mathbf{q}} \dot{\mathbf{q}}] \tag{3.6.4}$$

Next, note that

$$\frac{\partial T}{\partial q_j} = \sum_i m_i \dot{\mathbf{q}}^T \mathbf{r}_{i\mathbf{q}}^T \mathbf{r}_{i\mathbf{q}q_j} \dot{\mathbf{q}} \tag{3.6.5}$$

Presuming that $\mathbf{r}_i$ has two continuous derivatives with respect to generalized coordinates, the order of taking derivatives is immaterial and

$$\mathbf{r}_{i\mathbf{q}q_j} = \mathbf{r}_{iq_j\mathbf{q}} \tag{3.6.6}$$

This relation shows that the third term on the right of Eq. 3.6.4 is identical to the right side of Eq. 3.6.5. Therefore,

$$\frac{d}{dt}\left(\frac{\partial T}{\partial \dot{q}_j}\right) - \frac{\partial T}{\partial q_j} = \sum_i m_i [\mathbf{r}_{iq_j}^T \mathbf{r}_{i\mathbf{q}} \ddot{\mathbf{q}} + \mathbf{r}_{iq_j}^T (\mathbf{r}_{i\mathbf{q}} \dot{\mathbf{q}})_{\mathbf{q}} \dot{\mathbf{q}}] \tag{3.6.7}$$

Note that j may be regarded as the row index of a matrix on the right side of Eq. 3.6.7, yielding the matrix relation

$$\left[\frac{d}{dt}\left(\frac{\partial T}{\partial \dot{\mathbf{q}}}\right) - \frac{\partial T}{\partial \mathbf{q}}\right]^T = \sum_i m_i [\mathbf{r}_{i\mathbf{q}}^T \mathbf{r}_{i\mathbf{q}} \ddot{\mathbf{q}} + \mathbf{r}_{i\mathbf{q}}^T (\mathbf{r}_{i\mathbf{q}} \dot{\mathbf{q}})_{\mathbf{q}} \dot{\mathbf{q}}] \tag{3.6.8}$$

Multiplying both sides of Eq. 3.6.8 by $\delta\mathbf{q}^T$ and noting that the right side is just the expression involved in the variational equation of motion of Eq. 3.2.23 yields

$$\delta\mathbf{q}^T \left\{ \left[\frac{d}{dt}\left(\frac{\partial T}{\partial \dot{\mathbf{q}}}\right) - \frac{\partial T}{\partial \mathbf{q}}\right]^T - \mathbf{Q}^A \right\} = 0 \tag{3.6.9}$$

which must hold for all generalized coordinate variations $\delta\mathbf{q}$ that are consistent with constraints that act on the system. Equation 3.6.9 is called the *variational form of Lagrange's equations of motion*.

If generalized coordinates $\mathbf{q}$ are independent (i.e., if $\delta\mathbf{q}$ in Eq. 3.6.9 is arbitrary), its coefficient must be zero, yielding the *vector form of Lagrange's equations of motion* in independent generalized coordinates,

$$\left[\frac{d}{dt}(T_{\dot{\mathbf{q}}}) - T_{\mathbf{q}}\right]^T - \mathbf{Q}^A = \mathbf{0} \tag{3.6.10}$$

Expanding this vector form of the equation in terms of its components yields the more conventional form of *Lagrange's equations of motion* in independent generalized coordinates,

$$\frac{d}{dt}\left(\frac{\partial T}{\partial \dot{q}_j}\right) - \frac{\partial T}{\partial q_j} - Q_j^A = 0, \qquad j = 1, \ldots, n \tag{3.6.11}$$

As might be expected from the derivation of Lagrange's equations, when the left side of Eq. 3.6.11 is evaluated, the third term on the right of Eq. 3.6.4 is calculated, even if it is not recognized, and cancels terms in Eq. 3.6.5. At best, this is a wasteful manipulation. More likely, in realistic applications the equality of terms that should cancel one another may not be recognized and unnecessarily complicated expressions are retained in the resulting equations of motion. This suggests that even though Lagrange's equations of motion are elegant in form and easy to remember, they are not practical for actual derivation of general equations of motion. Equation 3.2.24 is a much more practical and effective form of the equations of motion for applications.

➧ **Example 3.6.1:** For the double pendulum of Example 3.2.5, from Eqs. 3.2.25 and 3.2.26,

$$\begin{aligned} T &= \tfrac{1}{2}\{m_1\dot{q}_1^2 + m_2[\dot{q}_1^2 + 2(\cos q_1 \cos q_2 + \sin q_1 \sin q_2)\dot{q}_1\dot{q}_2 + \dot{q}_2^2]\} \\ &= \tfrac{1}{2}\{m_1\dot{q}_1^2 + m_2[\dot{q}_1^2 + 2\cos(q_2 - q_1)\dot{q}_1\dot{q}_2 + \dot{q}_2^2]\} \end{aligned}$$

and

$$\mathbf{Q}^A = \begin{bmatrix} -(m_1 + m_2)g \sin q_1 \\ -m_2 g \sin q_2 \end{bmatrix}$$

The first of Eqs. 3.6.11 (i.e., with $j = 1$) is

$$\frac{d}{dt}\{(m_1 + m_2)\dot{q}_1 + m_2\cos(q_2 - q_1)\dot{q}_2\} - m_2\sin(q_2 - q_1)\dot{q}_1\dot{q}_2 + (m_1 + m_2)g\sin q_1 = 0$$

Expanding the first term in detail yields

$$(m_1 + m_2)\ddot{q}_1 + m_2\cos(q_2 - q_2)\ddot{q}_2 + m_2\sin(q_2 - q_1)\dot{q}_1\dot{q}_2 - m_2\sin(q_2 - q_1)\dot{q}_2^2 - m_2\sin(q_2 - q_1)\dot{q}_1\dot{q}_2 + (m_1 + m_2)g\sin q_1 = 0$$

As anticipated, terms that arose from $\partial T/\partial q_1$ cancel corresponding terms from $(d/dt)(\partial T/\partial \dot{q}_1)$, yielding

$$(m_1 + m_2)\ddot{q}_1 + m_2 \cos(q_2 - q_1)\ddot{q}_2 - m_2 \sin(q_2 - q_1)\dot{q}_2^2 + (m_1 + m_2)g \sin q_1 = 0$$

which is identical to the first of Eqs. 3.2.27.

If applied forces that act on a system of particles are conservative, there exists a *potential energy* V such that

$$\mathbf{Q}^A = -V_{\mathbf{q}}^T \tag{3.6.12}$$

where V depends only on $\mathbf{q}$ and not on $\dot{\mathbf{q}}$ or t. In terms of kinetic and potential energy, the *Lagrangian* is defined as

$$L \equiv T - V \tag{3.6.13}$$

Observe from Eq. 3.6.12 that the equation

$$\delta\mathbf{q}^T \left[\frac{d}{dt}\left(\frac{\partial L}{\partial \dot{\mathbf{q}}}\right) - \frac{\partial L}{\partial \mathbf{q}}\right]^T = 0 \tag{3.6.14}$$

expands to obtain Eq. 3.6.9, for all generalized coordinate variations $\delta\mathbf{q}$ that are consistent with constraints. In particular, if generalized coordinates $\mathbf{q}$ are independent, then $\delta\mathbf{q}$ is arbitrary and its coefficient in Eqs. 3.6.14 must be zero, yielding the even more elegant form of *Lagrange's equations of motion*,

$$\frac{d}{dt}\left(\frac{\partial L}{\partial \dot{q}_j}\right) - \frac{\partial L}{\partial q_j} = 0, \qquad j = 1, \ldots, nc \tag{3.6.15}$$

for independent generalized coordinates.

➧ **Example 3.6.2:** For the two degree of freedom material handling device of Examples 2.6.2 and 3.2.6, let $m_1 = 0$ and $m_2 = m$. Using expressions for $\dot{\mathbf{r}}_2$ derived in Example 2.6.2, the kinetic energy is (Prob. 3.6.2)

$$T = \frac{m}{2}[(1 + \sin^2 q_2)\dot{q}_1^2 + 2 \cos q_2 \dot{q}_1 \dot{q}_2 + \dot{q}_1^2]$$

The potential energy of the gravitational force, which acts in the negative z direction, is

$$V = mgz = -mg \cos q_2$$

Thus

$$\begin{aligned} L &= T - V \\ &= \frac{m}{2}[(1 + \sin^2 q_2)\dot{q}_1^2 + 2 \cos q_2 \dot{q}_1 \dot{q}_2 + \dot{q}_2^2] + mg \cos q_2 \end{aligned}$$

Since q_1 and q_2 are independent, Eq. 3.6.15 can be applied, with

$$\frac{d}{dt}\left(\frac{\partial L}{\partial \dot{q}_1}\right) = \frac{d}{dt} m[(1 + \sin^2 q_2)\dot{q}_1 + \cos q_2 \dot{q}_2]$$

$$= m[(1 + \sin^2 q_2)\ddot{q}_1 + \cos q_2 \ddot{q}_2 + 2 \cos q_2 \sin q_2 \dot{q}_1 \dot{q}_2 - \sin q_2 \dot{q}_2^2]$$

$$\frac{\partial L}{\partial q_1} = 0$$

$$\frac{d}{dt}\left(\frac{\partial L}{\partial \dot{q}_2}\right) = \frac{d}{dt} m[\cos q_2 \dot{q}_1 + \dot{q}_2]$$

$$= m[\cos q_2 \ddot{q}_1 + \ddot{q}_2 - \sin q_2 \dot{q}_1 \dot{q}_2]$$

$$\frac{\partial L}{\partial q_2} = m[\cos q_2 \sin q_2 \dot{q}_1^2 - \sin q_2 \dot{q}_1 \dot{q}_2 - g \sin q_2]$$

Substituting these expressions into Eq. 3.6.15 yields the same equations of motion that are obtained directly from Eq. 3.2.28 in Example 3.2.6.

Additional information is available here, however, as a result of using Lagrange's equations of motion. Since $\partial L/\partial q_1 = 0$, Eq. 3.6.15 with $j = 1$ is just

$$\frac{d}{dt}\left(\frac{\partial L}{\partial \dot{q}_1}\right) = 0$$

Thus

$$\frac{\partial L}{\partial \dot{q}_1} = C$$

where C is an arbitrary constant of integration. From the calculations above, this is

$$(1 + \sin^2 q_2)\dot{q}_1 + \cos q_2 \dot{q}_2 = C$$

which is called a *first integral* of the equations of motion.

The special result obtained in Example 3.6.2 illustrates one of the main values of Lagrange's equations of motion. In any conservative system with independent generalized coordinates, if

$$\frac{\partial L}{\partial q_j} = 0 \tag{3.6.16}$$

for some j, Eq. 3.6.15 yields the *first integral*,

$$\frac{\partial L}{\partial \dot{q}_j} = C_j \tag{3.6.17}$$

and the associated generalized coordinate q_j is called an *ignorable coordinate*. Thus, even though Lagrange's equations are not efficient for derivation of equations of motion, they sometimes yield first integrals that provide analytical information about the solution.

Since $L = L(\mathbf{q}, \dot{\mathbf{q}}, t)$, the constant C_j in Eq. 3.6.17 is determined by the initial conditions $\mathbf{q}(t_0) = \mathbf{q}^0$ and $\dot{\mathbf{q}}(t_0) = \dot{\mathbf{q}}^0$; that is,

$$C_j = \frac{\partial L}{\partial \dot{q}_j}(\mathbf{q}_0, \dot{\mathbf{q}}_0, t_0) \tag{3.6.18}$$

Then Eq. 3.6.17 provides an explicit algebraic relation between $\mathbf{q}$ and $\dot{\mathbf{q}}$, or a first order differential equation in $\mathbf{q}$.

For a constrained system of particles, the generalized coordinates may be required to satisfy holonomic kinematic constraints of the form

$$\mathbf{\Phi}(\mathbf{q}, t) = \mathbf{0} \tag{3.6.19}$$

Thus generalized coordinate variations must satisfy

$$\mathbf{\Phi}_{\mathbf{q}}\, \delta\mathbf{q} = \mathbf{0} \tag{3.6.20}$$

Since Eq. 3.6.14 must hold for all $\delta\mathbf{q}$ that satisfy Eq. 3.6.20, the Lagrange multiplier theorem guarantees existence of a Lagrange multiplier vector $\boldsymbol{\lambda}$ such that

$$\left[\frac{d}{dt}\left(\frac{\partial L}{\partial \dot{\mathbf{q}}}\right) - \frac{\partial L}{\partial \mathbf{q}}\right]^T + \mathbf{\Phi}_{\mathbf{q}}^T \boldsymbol{\lambda} = \mathbf{0} \tag{3.6.21}$$

While equivalent to Eq. 3.2.40 for conservative systems, this result suffers the same analytical deficiency noted above for Lagrange's equations with independent generalized coordinates. That is, $\partial L/\partial \mathbf{q}$ must cancel expressions in the first term on the left of Eq. 3.6.21. Furthermore, even if $\partial L/\partial \dot{q}_j = 0$ for some j, no first integral is obtained unless $\mathbf{\Phi}_{q_j}^T \boldsymbol{\lambda} = \mathbf{0}$, which is very unlikely. Thus the Lagrange multiplier form of Lagrange's equations in Eq. 3.6.21 may not have significant practical value.

3.7 RECIPES FOR WRITING EQUATIONS OF MOTION AND EQUILIBRIUM

The purpose of this section is to summarize procedures and equations that can be used to write equations of motion for dynamics and equilibrium of systems of particles. To facilitate use of the formulations that have been developed in this chapter in applications, equation numbers cited in this section are those from the sections where the equations are derived. While the variational formulation of equations of motion and equilibrium is the key theoretical tool used in deriving equations of motion and in interpreting the physical significance of Lagrange multipliers, the results obtained may be applied directly, without resorting to virtual displacement and variational analysis in individual applications.

This section is organized to present methods that are suitable for Cartesian coordinate and generalized coordinate formulations, both for independent and dependent coordinates. Finally, the use of Lagrange's equations is summarized for only independent generalized coordinates.

Independent Cartesian Coordinates

If a system of nb particles are kinematically unconstrained (i.e., all Cartesian coordinates are independent), *Newton's equations of motion* for each particle may be written as

$$m\ddot{\mathbf{r}}_i - \mathbf{F}_i = \mathbf{0}, \qquad i = 1, \ldots, nb \tag{1.1.1}$$

where the forces $\mathbf{F}_i = \mathbf{F}_i(\mathbf{r}_1, \ldots, \mathbf{r}_{nb}, \dot{\mathbf{r}}_1, \ldots, \dot{\mathbf{r}}_{nb}, t)$ may depend on the positions and velocities of all particles in the system. *Initial conditions* of motion are $\mathbf{r}_i(t_0) = \mathbf{r}_i^0$, $\dot{\mathbf{r}}_i(t_0) = \dot{\mathbf{r}}_i^0$, $i = 1, \ldots, nb$.

The *equilibrium equations* for an unconstrained system of particles are simply

$$\mathbf{F}_i(\mathbf{r}_1, \ldots, \mathbf{r}_{nb}) = 0, \qquad i = 1, \ldots, nb \tag{3.5.1}$$

where velocities are zero and time does not appear.

Dependent Cartesian Coordinates

For a system of nb particles, with the composite vector of Cartesian coordinates $\mathbf{r} = [\mathbf{r}_1^T, \ldots, \mathbf{r}_{nb}^T]^T$, that are subject to kinematic constraints of the form

$$\mathbf{\Phi}(\mathbf{r}, t) = \mathbf{0} \tag{3.4.1}$$

if constraint reaction forces do no work during motion of the system, the *Lagrange multiplier form of constrained equations of motion* is

$$\mathbf{M}\ddot{\mathbf{r}} + \mathbf{\Phi}_{\mathbf{r}}^T\boldsymbol{\lambda} = \mathbf{F}^A \tag{3.4.4}$$

where $\mathbf{M} = \text{diag}(m_1\mathbf{I}, \ldots, m_{nb}\mathbf{I})$ is the mass matrix, $\mathbf{F}^A = [\mathbf{F}_1^{A^T}, \ldots, \mathbf{F}_{nb}^{A^T}]^T$ is the vector of applied forces that may be functions of Cartesian coordinates, velocities, and time, and $\boldsymbol{\lambda}$ is the vector of *Lagrange multipliers*, whose dimension is equal to the dimension of $\mathbf{\Phi}$ in Eq. 3.4.1.

Combining Eq. 3.4.4 and the acceleration equation of Eq. 3.4.6 yields the *matrix form of constrained equations of motion* in Cartesian coordinates,

$$\begin{bmatrix} \mathbf{M} & \mathbf{\Phi}_{\mathbf{r}}^T \\ \mathbf{\Phi}_{\mathbf{r}} & \mathbf{0} \end{bmatrix} \begin{bmatrix} \ddot{\mathbf{r}} \\ \boldsymbol{\lambda} \end{bmatrix} = \begin{bmatrix} \mathbf{F}^A \\ \boldsymbol{\gamma} \end{bmatrix} \tag{3.4.7}$$

The equations of motion that include differential constraints of Eq. 3.2.4 are written by replacing $\mathbf{\Phi}_{\mathbf{q}}$ with the combined constraint Jacobian of Eq. 3.4.10 and modifying the right side of the equations.

Initial position conditions at t_0 are

$$\mathbf{\Phi}(\mathbf{r}(t_0), t_0) = \mathbf{0} \tag{3.3.3}$$

$$\mathbf{\Phi}^I(\mathbf{r}(t_0)) = \mathbf{0} \tag{3.3.7}$$

where it is required that

$$\left| \begin{bmatrix} \mathbf{\Phi_r}(\mathbf{r}(t_0), t_0) \\ \mathbf{\Phi^I_r}(\mathbf{r}(t_0)) \end{bmatrix} \right| \neq 0 \tag{3.3.9}$$

Initial velocity conditions at t_0 are

$$\mathbf{\Phi_r}\dot{\mathbf{r}}(t_0) = -\mathbf{\Phi}_t(t_0) \tag{3.3.4}$$

$$\mathbf{C}\dot{\mathbf{r}}(t_0) = \mathbf{d} \tag{3.3.10}$$

where it is required that

$$\left| \begin{bmatrix} \mathbf{\Phi_r}(\mathbf{r}(t_0), t_0) \\ \mathbf{C} \end{bmatrix} \right| \neq 0 \tag{3.3.11}$$

Equilibrium conditions for a constrained system are

$$\mathbf{F}_i^A - \mathbf{\Phi}_{\mathbf{r}_i}^T\boldsymbol{\lambda} = \mathbf{0}, \qquad i = 1, \ldots, nb \tag{3.5.3}$$

$$\mathbf{\Phi}(\mathbf{r}) = \mathbf{0} \tag{3.4.1}$$

where $\boldsymbol{\lambda}$ is a Lagrange multiplier vector.

The *constraint reaction force* on body i due to the holonomic constraint of Eq. 3.2.2 with body j, for both dynamics and equilibrium, is

$$\mathbf{F}_{ij}^C = -\mathbf{\Phi}_{\mathbf{r}_i}^{ij^T}\boldsymbol{\lambda}^{ij} \tag{3.2.9}$$

For differential constraints of Eq. 3.2.4, the constraint reaction forces are

$$\mathbf{F}_{ij}^C = -\mathbf{B}^{ij^T}\boldsymbol{\mu}^{ij} \tag{3.2.10}$$

Independent Generalized Coordinates

Generalized coordinates $\mathbf{q}$ may be defined and Cartesian coordinates of all nb particles in a system derived in the form

$$\mathbf{r}_i = \mathbf{r}_i(\mathbf{q}), \qquad i = 1, \ldots, nb \tag{3.2.19}$$

The transformation Jacobians $\mathbf{r}_{i_\mathbf{q}}$ are calculated and *generalized applied forces* are evaluated from

$$\mathbf{Q}^A = \sum_{i=1}^{nb} \mathbf{r}_{i_\mathbf{q}}^T \mathbf{F}_i^A \tag{3.2.22}$$

The *equations of motion in independent generalized coordinates* are

$$\sum_{i=1}^{nb} m_i \mathbf{r}_{i_\mathbf{q}}^T \mathbf{r}_{i_\mathbf{q}} \ddot{\mathbf{q}} - \mathbf{Q}^A + \sum_{i=1}^{nb} m_i \mathbf{r}_{i_\mathbf{q}}^T (\mathbf{r}_{i_\mathbf{q}}\dot{\mathbf{q}})_\mathbf{q}\dot{\mathbf{q}} = \mathbf{0} \tag{3.2.24}$$

Initial conditions for these equations of motion are $\mathbf{q}(t_0) = \mathbf{q}^0$ and $\dot{\mathbf{q}}(t_0) = \dot{\mathbf{q}}^0$

Equilibrium equations for a system of particles, in terms of independent generalized coordinates $\mathbf{q}$, are

$$\mathbf{Q}^A(\mathbf{q}) = \mathbf{0} \tag{3.5.5}$$

Dependent Generalized Coordinates

If generalized coordinates $\mathbf{q}$ are subject to constraints of the form

$$\mathbf{\Phi}(\mathbf{q}, t) = \mathbf{0} \tag{3.4.16}$$

then the *Lagrange multiplier form of constrained equations of motion* is

$$\mathbf{r}_\mathbf{q}^T\mathbf{M}\mathbf{r}_\mathbf{q}\ddot{\mathbf{q}} + \mathbf{r}_\mathbf{q}^T\mathbf{M}(\mathbf{r}_\mathbf{q}\dot{\mathbf{q}})_\mathbf{q}\dot{\mathbf{q}} - \mathbf{Q}^A + \mathbf{\Phi}_\mathbf{q}^T\boldsymbol{\lambda} = \mathbf{0} \tag{3.4.18}$$

where $\boldsymbol{\lambda}$ are Lagrange multipliers, $\mathbf{M} = \text{diag}(m_1\mathbf{I}, \ldots, m_{nb}\mathbf{I})$, and $\mathbf{r}(\mathbf{q}) = [\mathbf{r}_1^T(\mathbf{q}), \ldots, \mathbf{r}_{nb}^T(\mathbf{q})]^T$.

Combining Eq. 3.4.18 and the acceleration equation of Eq. 3.4.20 yields the *matrix form of constrained equations of motion* in generalized coordinates,

$$\begin{bmatrix} \hat{\mathbf{M}} & \mathbf{\Phi}_\mathbf{q}^T \\ \mathbf{\Phi}_\mathbf{q} & \mathbf{0} \end{bmatrix} \begin{bmatrix} \ddot{\mathbf{q}} \\ \boldsymbol{\lambda} \end{bmatrix} = \begin{bmatrix} \mathbf{R} \\ \boldsymbol{\gamma} \end{bmatrix} \tag{3.4.21}$$

where

$$\hat{\mathbf{M}} = \mathbf{r}_\mathbf{q}^T\mathbf{M}\mathbf{r}_\mathbf{q} \tag{3.4.17}$$

$$\mathbf{R} = \mathbf{Q}^A - \mathbf{r}_\mathbf{q}\mathbf{M}(\mathbf{r}_\mathbf{q}\dot{\mathbf{q}})_\mathbf{q}\dot{\mathbf{q}} \tag{3.4.18}$$

Initial position conditions at t_0 are

$$\mathbf{\Phi}(\mathbf{q}(t_0), t_0) = \mathbf{0} \tag{3.3.24}$$

$$\mathbf{\Phi}^I(\mathbf{q}(t_0)) = \mathbf{0} \tag{3.3.26}$$

where it is required that

$$\left|\begin{bmatrix} \mathbf{\Phi}_\mathbf{q}(\mathbf{q}(t_0), t_0) \\ \mathbf{\Phi}_\mathbf{q}^I(\mathbf{q}(t_0)) \end{bmatrix}\right| \neq 0 \tag{3.3.27}$$

Initial velocity conditions at t_0 are

$$\mathbf{\Phi}_\mathbf{q}\dot{\mathbf{q}}(t_0) = -\mathbf{\Phi}_t(t_0) \tag{3.3.25}$$

$$\mathbf{C}\dot{\mathbf{q}}(t_0) = \mathbf{d} \tag{3.3.28}$$

where it is required that

$$\left|\begin{bmatrix} \mathbf{\Phi}_\mathbf{q}(\mathbf{q}(t_0), t_0) \\ \mathbf{C} \end{bmatrix}\right| \neq 0 \tag{3.3.29}$$

The *Lagrange multiplier form of constrained equilibrium equations* is

$$\mathbf{Q}^A - \mathbf{\Phi}_\mathbf{q}^T\boldsymbol{\lambda} = \mathbf{0} \tag{3.5.10}$$

which must be solved with the constraint equation of Eq. 3.5.7 for $\mathbf{q}$ and the Langrange multiplier $\boldsymbol{\lambda}$.

Lagrange's Equations of Motion

For a system of nb particles, in terms of the generalized coordinate formulation, the kinetic energy is

$$T = \sum_{i=1}^{nb} \tfrac{1}{2} m_i \dot{\mathbf{q}}^T \mathbf{r}_{i_q}^T \mathbf{r}_{i_q} \dot{\mathbf{q}} \tag{3.6.2}$$

For a conservative system with potential energy $V(\mathbf{q})$, the *Lagrangian* is defined as

$$L \equiv T - V \tag{3.6.13}$$

For independent generalized coordinates, *Lagrange's equations of motion* are

$$\frac{d}{dt}\left(\frac{\partial L}{\partial \dot{q}_j}\right) - \frac{\partial L}{\partial q_j} = 0, \qquad j = 1, \ldots, nc \tag{3.6.15}$$

If

$$\frac{\partial L}{\partial q_j} = 0 \tag{3.6.16}$$

for some j, then q_j is called an *ignorable coordinate* and Lagrange's equations yield the *first integral*,

$$\frac{\partial L}{\partial \dot{q}_j} = C_j \tag{3.6.17}$$

where the constant of integration C_j is determined by the initial conditions

$$C_j = \frac{\partial L}{\partial \dot{q}_j}(\mathbf{q}^0, \dot{\mathbf{q}}^0, t_0) \tag{3.6.18}$$

PROBLEMS

Section 3.1

3.1.1 For the force $\mathbf{F} = [t, x, y]^T$ that acts on point P, find expressions for generalized forces $\mathbf{Q}$ for the following sets of generalized coordinates:
- **(a)** Cylindrical coordinates in Example 2.2.4
- **(b)** Spherical coordinates in Example 2.3.6
- **(c)** Particle on a straight line in Example 2.4.1
- **(d)** Particle on a helix in Example 2.4.2

3.1.2 Consider a particle that is constrained to move on the paraboloid of revolutions shown in Fig. P3.1.2.

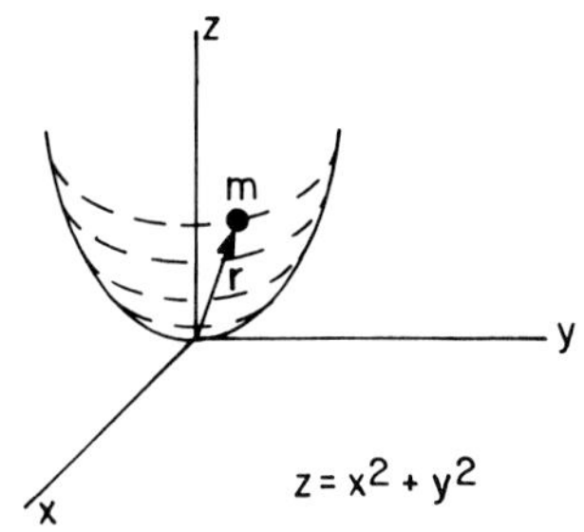

Figure P3.1.2

(a) Let $x = q_1$ and $y = q_2$ define a parameterization of the surface. Derive the mapping $\mathbf{r} = \mathbf{r}(\mathbf{q})$ and determine whether it has any singular points.

(b) Derive differential equations of motion in terms of $\mathbf{q}$ for a particle of mass m that moves on the surface of the paraboloid of revolution without friction, under the action of gravity, starting at rest from the point $\mathbf{r}(0) = [1, 1, 2]^T$.

3.1.3 Verify calculations that lead to Eq. 3.1.41 in Example 3.1.7.

3.1.4 If $\mathbf{A}'$ of Eq. 3.1.61 is a matrix with full row rank, show that the product $\mathbf{A}'\mathbf{A}'^T$ is nonsingular. (*Hint:* Show that $\mathbf{A}'\mathbf{A}'^T$ is positive definite, hence nonsingular.)

3.1.5 Write kinematic constraints of Eq. 3.1.10 for the vehicle in Example 1.2.3 and find constraint forces on the vehicle in terms of Lagrange multipliers, using Eq. 3.1.17. Compare this result to that obtained in Example 1.2.4 and find the Lagrange multipliers.

3.1.6 Write the constrained equation of motion of Eq. 3.1.26 for the vehicle of Problem 3.1.5 with gravity acting in the negative z direction.

3.1.7 Let $q = x$ and write the Lagrange multiplier form of the constrained equations of motion of Eq. 3.1.46 for the pendulum in Problem 1.2.1.

3.1.8 Let $q = y$ and write the Lagrange multiplier form of the constrained equation of motion of Eq. 3.1.46 for the particle in Problem 1.2.2.

Section 3.2

3.2.1 Consider the pendulum formed by a massless bar that rotates about the y axis and has mass m_1 at its tip (Fig. P3.2.1), with a second mass m_2 attached as shown by

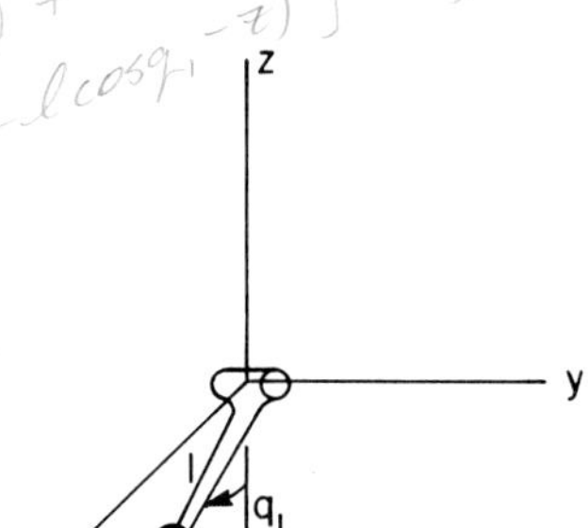

Figure P3.2.1

a spring with force given by $f = k(l - 1)$. Let the generalized coordinates for the system be q_1 and the x, y, and z coordinates of mass m_2. Derive differential equations of motion for the system.

3.2.2 Two particles are constrained to move as shown in Fig. P3.2.2, where there is no friction as m_2 slides along the bar that supports m_1. Force in the spring is $f = k(d - \frac{1}{2})$, tension positive. Write equations of motion of this system.

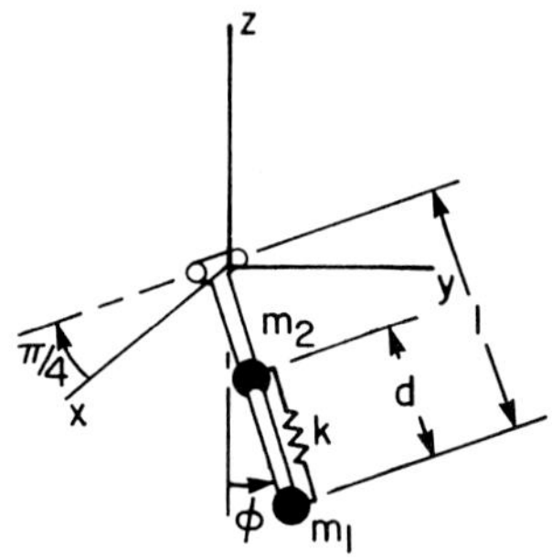

Figure P3.2.2

3.2.3 Carry out the expansion of Eq. 3.2.24, to obtain the equations of motion in Example 3.2.6.

3.2.4 Carry out the expansion of Eq. 3.2.24, to obtain the equations of motion in Example 3.2.7.

3.2.5 Carry out the expansion of Eq. 3.2.40, using relations derived in Example 3.2.10, to obtain the equations of motion of Eq. 3.2.50.

3.2.6 Use elementary trigonometry and Fig. 3.2.9 to derive Eq. 3.2.54.

3.2.7 Verify the expression of Eq. 3.2.58 for generalized force on P_i.

3.2.8 Verify the expression of Eq. 3.2.59 for the moment of force density **f** about the origin of the x'-y' frame.

3.2.9 Verify the equations of motion of Eq. 3.2.60.

3.2.10 Repeat Example 3.2.3 for the refined particle model of the slider–crank mechanism in Example 1.3.3.

3.2.11 Consider the mechanism shown in Fig. P3.2.11, where m_1 and m_2 are constrained to move on the smooth curve $y = x^2$ and the distance between m_1 and m_2 is constrained by a bar of length l. Write the equations of motion of Eq. 3.2.12 for the mechanism, with gravity acting in the negative y direction.

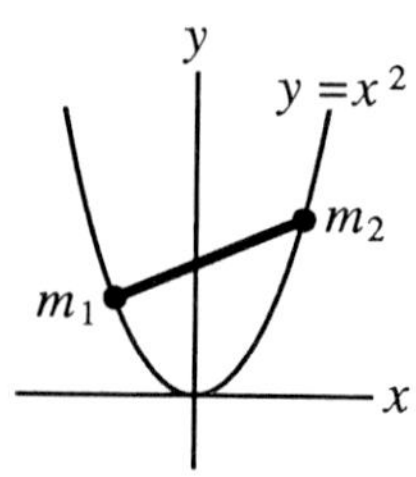

Figure P3.2.11

3.2.12 Write the equation of motion of Eq. 3.2.12 for the mechanism shown in Fig. P3.2.12, with gravity acting in the negative y direction. Discuss any possible singular configurations. (*Hint:* Check to see if there is any singular square sub-Jacobian of the constraint equations.)

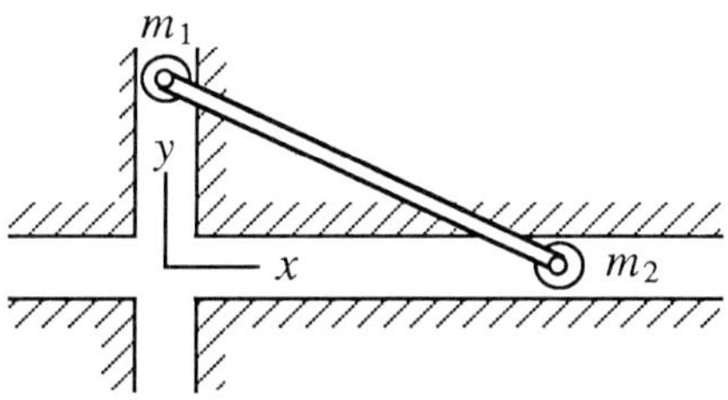

Figure P3.2.12

3.2.13 Write the equation of motion of Eq. 3.2.40 for the mechanism in Problem 1.3.3, using $q_1 = \theta_1$ and $q_2 = \theta_2$.

3.2.14 Write the equation of motion of Eq. 3.2.40 for the mechanism in Problem 3.2.12, with $q_1 = x_2$ and $q_2 = y_1$.

3.2.15 Repeat Problem 3.2.11 by replacing the two-particle mechanism by a rigid body whose mass and centroidal polar moment of inertia are m and $ml^2/12$, respectively.

Section 3.3

3.3.1 Verify that $z(t_0) = \sqrt{2}/2$ and Eq. 3.3.15 yield values of $\mathbf{r}_1(t_0)$ and $\mathbf{r}_2(t_0)$ that satisfy Eq. 3.3.12 and that the test of Eq. 3.3.9 is satisfied.

3.3.2 Verify that the initial velocity conditions defined in Eq. 3.3.16, with Eq. 3.3.13, satisfy the test of Eq. 3.3.11.

3.3.3 Verify that the test of Eq. 3.3.11 is not satisfied for the initial velocity conditions of Example 3.3.2.

3.3.4 Verify that the initial position conditions of Eq. 3.3.20 and the holonomic constraint of Eq. 3.2.15 satisfy the test of Eq. 3.3.9.

3.3.5 Write the initial position and velocity conditions of Eqs. 3.3.7 and 3.3.10 for the vehicle in Examples 1.2.3 and 3.1.6. Check if the initial position and velocity tests of Eqs. 3.3.9 and 3.3.11 are satisfied.

3.3.6 Consider the mechanism in Problem 3.2.12, with the bar initially at rest in a horizontal position [i.e., $y_1(0) = 0$]. Write the initial position and velocity conditions and check to see if the tests of Eqs. 3.3.9 and 3.3.11 are satisfied.

Section 3.4

3.4.1 Use the definitions of $\mathbf{M}$ and $\hat{\mathbf{M}}$ in Eqs. 3.4.2 and 3.4.17 to show that Eq. 3.2.40 reduces to Eq. 3.4.18.

3.4.2 Show that if the length of the crank (the first bar) is less than 1, the coefficient matrix on the left of the equations of motion in Example 3.4.2 is always nonsingular.

3.4.3 Write the matrix form of the constrained equations of motion for the mechanism in Problem 3.2.12, using Cartesian coordinates. Check to see if the matrix on the left of the equation satisfies the constrained dynamic existence theorem.

3.4.4 Write the matrix form of the constrained equations of motion with generalized coordinates for the mechanism in Problem 3.2.15. Check to see if the matrix on the left of the equation satisfies the constrained dynamic existence theorem for generalized coordinates.

Section 3.5

3.5.1 Determine which equilibrium configurations found in the following examples are stable:
(a) Example 3.5.4
(b) Example 3.5.6

3.5.2 Analyze equilibrium of the three degree of freedom system of Example 3.3.5. Determine whether each of the equilibrium configurations found is stable.

3.5.3 Repeat Problem 3.5.2, with gravity acting in the positive y direction.

3.5.4 Analyze the equilibrium of the system in Problem 3.2.2. Determine whether each of the equilibrium configurations found is stable.

3.5.5 Find equilibrium configurations and determine if they are stable, for:
(a) Problem 3.2.11
(b) Problem 3.2.12

Section 3.6

3.6.1 Derive Lagrange's equations of motion for the planar double pendulum of Example 3.2.5. Show that q_1 is ignorable and obtain a first integral, for initial conditions $q_1(0) = q_2(0) = \dot{q}_1(0) = \dot{q}_2(0) = 0$.

3.6.2 Use expressions for $\dot{\mathbf{r}}_2$ derived in Example 2.6.2 to derive the expression for kinetic energy given in Example 3.6.2.

3.6.3 A mass m (modeled as a particle) slides on a massless shaft that rotates about the z axis, as shown in Fig. P3.6.3. A spring with spring constant k is attached between the shaft and the mass. Gravity acts in the negative z direction. Derive Lagrange's equations of motion for the system, using the generalized coordinates shown. Show that q_1 is ignorable and obtain a first integral, for initial conditions $q_1(0) = 0$, $q_2(0) = 1$, and $\dot{q}_1(0) = \dot{q}_2(0) = 0$.

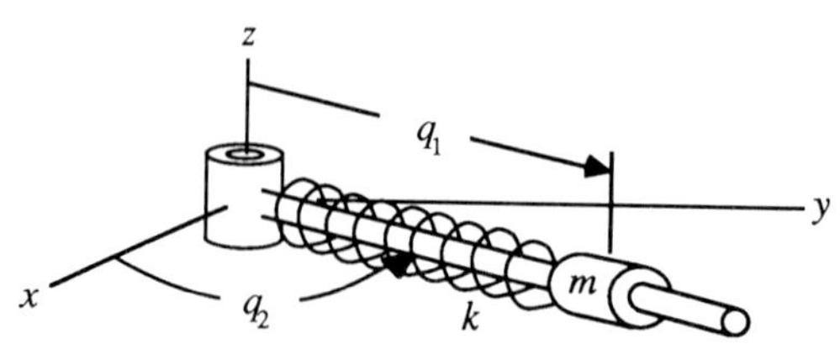

Figure P3.6.3

SUMMARY OF KEY FORMULAS

Variational Equations of Motion of a Particle

$$m\,\delta\mathbf{r}^T\ddot{\mathbf{r}} = \delta\mathbf{r}^T\mathbf{F} = \delta\mathbf{r}^T(\mathbf{F}^A + \mathbf{F}^C) = \delta W^A + \delta W^C = \delta W \qquad (3.1.5,\ 7)$$

$$\mathbf{Q} = \mathbf{Q}^A + \mathbf{Q}^C = \mathbf{r}_\mathbf{q}^T\mathbf{F}^A + \mathbf{r}_\mathbf{q}^T\mathbf{F}^C, \qquad \delta W = \delta\mathbf{q}^T\mathbf{Q}^A + \delta\mathbf{q}^T\mathbf{Q}^C \qquad (3.1.35,\ 36)$$

$$m\,\delta\mathbf{q}^T\mathbf{r}_\mathbf{q}^T(\mathbf{r}_\mathbf{q}\ddot{\mathbf{q}} + (\mathbf{r}_\mathbf{q}\dot{\mathbf{q}})_\mathbf{q}\dot{\mathbf{q}}) - \delta W = 0 \qquad (3.1.37)$$

Lagrange Multiplier Form of Equations of Motion of a Particle

$$\mathbf{F}^C = -\mathbf{\Phi}_\mathbf{r}^T\boldsymbol{\lambda}, \qquad m\ddot{\mathbf{r}} + \mathbf{\Phi}_\mathbf{r}^T\boldsymbol{\lambda} = \mathbf{F}^A \qquad (3.1.17,\ 26)$$

$$\mathbf{Q}^C = -\mathbf{\Phi}_\mathbf{q}^T\boldsymbol{\lambda}, \qquad m\mathbf{r}_\mathbf{q}^T\mathbf{r}_\mathbf{q}\ddot{\mathbf{q}} + m\mathbf{r}_\mathbf{q}^T(\mathbf{r}_\mathbf{q}\dot{\mathbf{q}})_\mathbf{q}\dot{\mathbf{q}} + \mathbf{\Phi}_\mathbf{q}^T\boldsymbol{\lambda} = \mathbf{Q}^A \qquad (3.1.45,\ 46)$$

Initial Conditions

$$\mathbf{\Phi}^I(\mathbf{r}(t_0)) = \mathbf{0}, \qquad \mathbf{C}\dot{\mathbf{r}}(t_0) - \mathbf{d} = \mathbf{0} \qquad (3.3.7,\ 10)$$

$$\mathbf{\Phi}^I(\mathbf{q}(t_0)) = \mathbf{0}, \qquad \mathbf{C}\dot{\mathbf{q}}(t_0) - \mathbf{d} = \mathbf{0} \qquad (3.3.26,\ 28)$$

Matrix Form of Equations of Motion

$$\begin{bmatrix} \mathbf{M} & \mathbf{\Phi}_\mathbf{r}^T \\ \mathbf{\Phi}_\mathbf{r} & \mathbf{0} \end{bmatrix} \begin{bmatrix} \ddot{\mathbf{r}} \\ \boldsymbol{\lambda} \end{bmatrix} = \begin{bmatrix} \mathbf{F}^A \\ \boldsymbol{\gamma} \end{bmatrix} \qquad (3.4.7)$$

$$\begin{bmatrix} \hat{\mathbf{M}} & \mathbf{\Phi}_\mathbf{q}^T \\ \mathbf{\Phi}_\mathbf{q} & \mathbf{0} \end{bmatrix} \begin{bmatrix} \ddot{\mathbf{q}} \\ \boldsymbol{\lambda} \end{bmatrix} = \begin{bmatrix} \mathbf{R} \\ \boldsymbol{\gamma} \end{bmatrix}, \qquad \begin{aligned} \hat{\mathbf{M}} &= \mathbf{r}_\mathbf{q}^T\mathbf{M}\mathbf{r}_\mathbf{q} \\ \mathbf{R} &= \mathbf{Q}^A - \mathbf{r}_\mathbf{q}^T\mathbf{M}(\mathbf{r}_\mathbf{q}\dot{\mathbf{q}})_\mathbf{q}\dot{\mathbf{q}} \end{aligned} \qquad (3.4.21,\ 17,\ 18)$$

Equations of Equilibrium

$$\mathbf{F}_i^A - \mathbf{\Phi}_{\mathbf{r}_i}^T\boldsymbol{\lambda} = \mathbf{0}, \qquad i = 1, \ldots, nb \qquad (3.5.3)$$

$$\mathbf{Q}^A - \mathbf{\Phi}_\mathbf{q}^T\boldsymbol{\lambda} = \mathbf{0} \qquad (3.5.10)$$

Lagrange's Equations of Motion

$$\frac{d}{dt}\left(\frac{\partial T}{\partial \dot{q}_j}\right) - \frac{\partial T}{\partial q_j} - Q_j^A = 0, \qquad j = 1, \ldots, nc \qquad (3.6.11)$$

$$\frac{d}{dt}\left(\frac{\partial L}{\partial \dot{q}_j}\right) - \left(\frac{\partial L}{\partial q_j}\right) = 0, \qquad j = 1, \ldots, nc, \quad L \equiv T - V \qquad (3.6.15,\ 13)$$

4 Kinematics of a Rigid Body

A *rigid body* is defined as being made up of a continuum of particles that are constrained not to move relative to one another. While actual bodies are never perfectly rigid, deformation effects often have negligible influence on the motion of a machine that is made up of multiple bodies. For this reason, most attention in the study of dynamics of machines is devoted to modeling individual bodies in a system as rigid. In this chapter, kinematic relations needed for rigid-body dynamics in Chapters 5 and 6 are developed.

Substantial technical differences arise in kinematic analysis of rigid bodies, as compared to particles. The concept of orientation of a rigid body, which has no counterpart in the context of particles arises naturally since bodies may both translate and rotate in space. Most of the attention in this chapter is on methods of defining orientation and angular velocity of rigid bodies, in preparation for introducing concepts of dynamics of rigid bodies in Chapter 5.

4.1 REFERENCE FRAMES FOR LOCATION AND ORIENTATION OF A RIGID BODY IN SPACE

Consider, for purposes of introduction, a rectangular rigid body shown in Fig. 4.1.1. The geometry (or shape) of the body may be defined in an x'-y'-z' reference

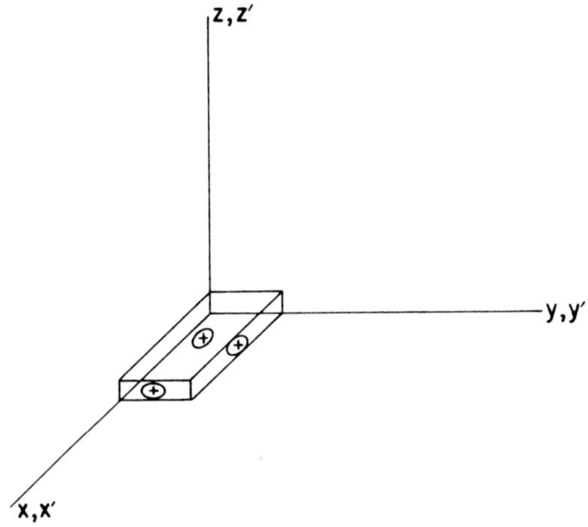

Figure 4.1.1 Rectangular body in reference position.

frame that is fixed to the body, called a *body-fixed reference frame*. The x'-y'-z' frame is coincident with the x-y-z global reference frame in the position shown in Fig. 4.1.1. This body-fixed reference frame might be thought of as a *drafting board reference frame* in engineering practice. That is, the engineer defines points, lines, and surfaces on a machine component in terms of front, side, and top views, relative to the x'-y'-z' frame; to create a three-dimensional definition of the component. Thus all data that define the geometry of the body are specified in the x'-y'-z' frame.

Consider next the translated and rotated position of the body shown in Fig. 4.1.2. Translation of the body is easily defined by the vector **r** from the origin of the x-y-z reference frame to the origin of the x'-y'-z' frame. Specificaton of orientation is, however, more complex. Nevertheless, once the x'-y'-z' frame is oriented relative to the x-y-z reference frame, all points of interest on the body may be located in the x-y-z reference frame.

To see that orientation of a body in space raises delicate technical questions, consider two rotation sequences for a body, from its orientation in Fig. 4.1.1. Counterclockwise rotations of the body by angles of $\pi/2$ are to be made about its

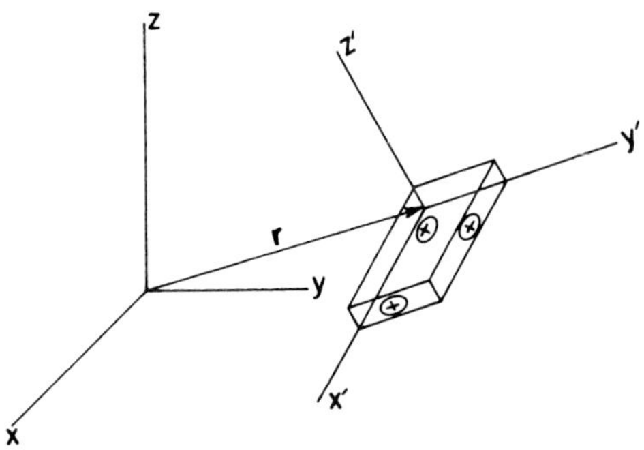

Figure 4.1.2 Body translated and rotated.

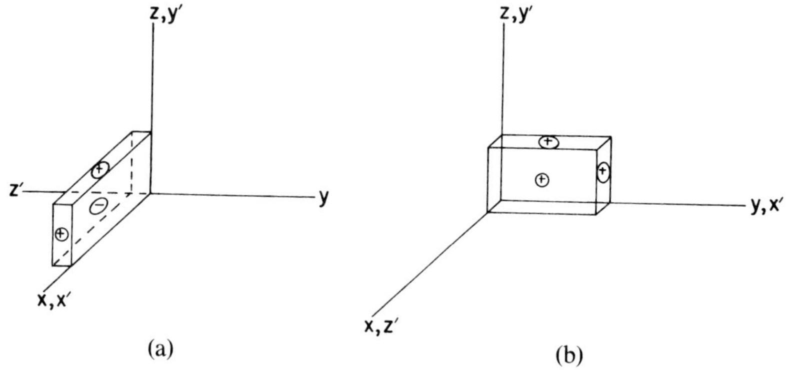

Figure 4.1.3 Rotations of body by $\pi/2$ about x' and then y' axes.

body-fixed x' and y' axes. Consider first the rotation sequence shown in Fig. 4.1.3. The first rotation is about the body-fixed x' axis, which initially coincides with the reference frame x axis. The second rotation is about the body-fixed y' axis, when it coincides with the reference frame z axis, as shown in Fig. 4.1.3(a). The resulting orientation is shown in Fig. 4.1.3(b).

Consider next the same counterclockwise rotations about body-fixed axes, but in reverse order. First, the body is rotated $\pi/2$ about its body-fixed y' axis, which initially coincides with the reference frame y axis, as shown in Fig. 4.1.4(a). The body is then rotated $\pi/2$ about its body-fixed x' axis, when it coincides with the negative reference frame z axis, as shown in Fig. 4.1.4(a). The resulting orientation is shown in Fig. 4.1.4(b). Since the orientations shown in Figs. 4.1.3(b) and 4.1.4(b) are distinctly different, it is clear that the order of rotation is important in defining orientation of a body in space. Much as in the case of matrix multiplication, in which the order of terms in the product of two matrices is critically

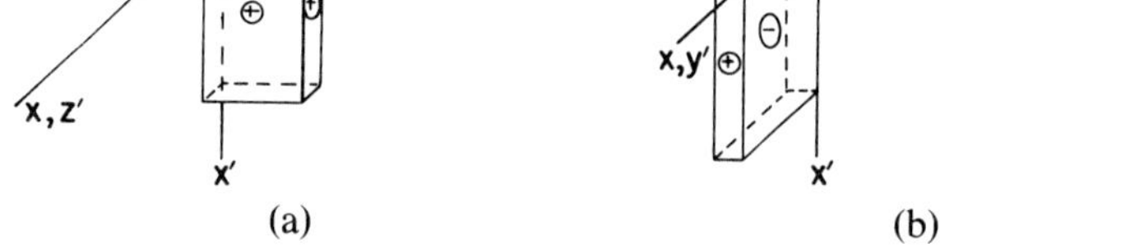

Figure 4.1.4 Rotations of body by $\pi/2$ about y' and then x' axes.

important, the order of rotations is similarly important. This shows clearly that large amplitude rotation is not a vector quantity, since the order of rotations is not commutative (i.e., orders of rotation cannot be inverted).

As shown schematically in Fig. 4.1.5, a general body can be located and oriented in space by defining the position vector **r** of the origin of its body-fixed x'-y'-z' frame, relative to the x-y-z reference frame and defining the orientation of the x'-y'-z' frame. As shown in Section 2.5, orientation of the x'-y'-z' frame is defined relative to the x-y-z reference frame by an orthogonal transformation matrix **A**. To see that the vector **r** and matrix **A** define the position of all points on the body in space, consider a typical point P on the body shown in Fig. 4.1.5. Denoting by $\mathbf{s}'^P$ the constant vector in the x'-y'-z' frame that locates point P, Eq. 2.5.16 yields

$$\begin{aligned} \mathbf{r}^P &= \mathbf{r} + \mathbf{s}^P \\ &= \mathbf{r} + \mathbf{A}\mathbf{s}'^P \end{aligned} \tag{4.1.1}$$

The vector $\mathbf{s}'^P$ may be thought of as being defined in a *drafting board coordinate system*, which is a natural way for the engineer to locate points on the body.

Properties of the orthogonal rotation matrix **A** developed in Section 2.5 will be used here to develop relations that define kinematics of a rigid body in space. Results derived in Section 2.5 are employed here to define orientation generalized coordinates for rigid bodies in space. Prior to introducing orientation generalized coordinates, however, it is of value to determine the number of rotational degrees of freedom.

The nine direction cosines in the orthogonal matrix **A** define the orientation of the x'-y'-z' frame, relative to the x-y-z reference frame, but they are not independent. Equation 2.5.14 provides only six possibly independent equations

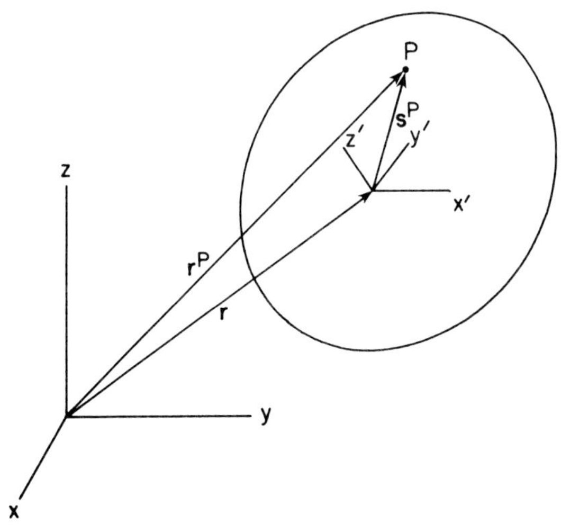

Figure 4.1.5 Position and orientation of body in space.

among the nine direction cosines, since in component form Eq. 2.5.14 is

$$\sum_{k=1}^{3} a_{ki}a_{kj} = \begin{cases} 1, & i = j \\ 0, & i \neq j \end{cases}, \qquad i, j = 1, 2, 3 \tag{4.1.2}$$

and interchanging $i \neq j$ yields the same equation, because $\mathbf{A}^T\mathbf{A} = \mathbf{I}$ is symmetric. To show that there are exactly three rotational degrees of freedom of a rigid body, it must be shown that Eqs. 4.1.2 impose six independent constraints on the nine components of matrix $\mathbf{A}$.

First recall, from Eq. 2.5.13, that $\mathbf{A} = [\mathbf{f}, \mathbf{g}, \mathbf{h}]$, where $\mathbf{f}$, $\mathbf{g}$, and $\mathbf{h}$ are unit vectors along the x', y', and z' axes, respectively. Denote the vector of nine direction cosines in $\mathbf{A}$ as *orientation generalized coordinates*:

$$\mathbf{q} = \begin{bmatrix} \mathbf{f} \\ \mathbf{g} \\ \mathbf{h} \end{bmatrix} \tag{4.1.3}$$

The six constraints imposed on these nine variables by Eq. 4.1.2, with $j \geq i$, may be written as

$$\mathbf{\Phi}(\mathbf{q}) \equiv \begin{bmatrix} \frac{1}{2}\mathbf{f}^T\mathbf{f} - \frac{1}{2} \\ \frac{1}{2}\mathbf{g}^T\mathbf{g} - \frac{1}{2} \\ \frac{1}{2}\mathbf{h}^T\mathbf{h} - \frac{1}{2} \\ \mathbf{f}^T\mathbf{g} \\ \mathbf{f}^T\mathbf{h} \\ \mathbf{g}^T\mathbf{h} \end{bmatrix} = \mathbf{0} \tag{4.1.4}$$

Let $\mathbf{q}_0$ satisfy $\mathbf{\Phi}(\mathbf{q}_0) = \mathbf{0}$. The Jacobian of the constraint function $\mathbf{\Phi}(\mathbf{q})$ of Eq. 4.1.4 is the 6×9 matrix

$$\mathbf{\Phi}_{\mathbf{q}} = \begin{bmatrix} \mathbf{f}^T & \mathbf{0} & \mathbf{0} \\ \mathbf{0} & \mathbf{g}^T & \mathbf{0} \\ \mathbf{0} & \mathbf{0} & \mathbf{h}^T \\ \mathbf{g}^T & \mathbf{f}^T & \mathbf{0} \\ \mathbf{h}^T & \mathbf{0} & \mathbf{f}^T \\ \mathbf{0} & \mathbf{h}^T & \mathbf{g}^T \end{bmatrix} \tag{4.1.5}$$

The rows of $\mathbf{\Phi}_{\mathbf{q}}$ are linearly independent if and only if the columns of $\mathbf{\Phi}_{\mathbf{q}}^T$ are linearly independent; that is, if and only if

$$\mathbf{\Phi}_{\mathbf{q}}^T\boldsymbol{\alpha} = \mathbf{0} \tag{4.1.6}$$

implies that $\boldsymbol{\alpha} = \mathbf{0}$. Expanding Eq. 4.1.6 yields

$$\boldsymbol{\Phi}_{\mathbf{q}}^T\boldsymbol{\alpha} = \begin{bmatrix} \mathbf{f} & \mathbf{0} & \mathbf{0} & \mathbf{g} & \mathbf{h} & \mathbf{0} \\ \mathbf{0} & \mathbf{g} & \mathbf{0} & \mathbf{f} & \mathbf{0} & \mathbf{h} \\ \mathbf{0} & \mathbf{0} & \mathbf{h} & \mathbf{0} & \mathbf{f} & \mathbf{g} \end{bmatrix} \begin{bmatrix} \alpha_1 \\ \alpha_2 \\ \alpha_3 \\ \alpha_4 \\ \alpha_5 \\ \alpha_6 \end{bmatrix} = \mathbf{0} \tag{4.1.7}$$

Reordering variables and expanding each of the three rows of Eq. 4.1.7, this yields

$$\begin{aligned} [\mathbf{f} \quad \mathbf{g} \quad \mathbf{h}] \begin{bmatrix} \alpha_1 \\ \alpha_4 \\ \alpha_5 \end{bmatrix} &= \mathbf{A} \begin{bmatrix} \alpha_1 \\ \alpha_4 \\ \alpha_5 \end{bmatrix} = \mathbf{0} \\ [\mathbf{f} \quad \mathbf{g} \quad \mathbf{h}] \begin{bmatrix} \alpha_4 \\ \alpha_2 \\ \alpha_6 \end{bmatrix} &= \mathbf{A} \begin{bmatrix} \alpha_4 \\ \alpha_2 \\ \alpha_6 \end{bmatrix} = \mathbf{0} \\ [\mathbf{f} \quad \mathbf{g} \quad \mathbf{h}] \begin{bmatrix} \alpha_5 \\ \alpha_6 \\ \alpha_3 \end{bmatrix} &= \mathbf{A} \begin{bmatrix} \alpha_5 \\ \alpha_6 \\ \alpha_3 \end{bmatrix} = \mathbf{0} \end{aligned} \tag{4.1.8}$$

Since $|\mathbf{A}^T\mathbf{A}| = |\mathbf{I}| = 1 = |\mathbf{A}^T||\mathbf{A}| = |\mathbf{A}||\mathbf{A}| = |\mathbf{A}|^2$, $|\mathbf{A}| = 1$ and $\mathbf{A}$ is nonsingular. Thus Eqs. 4.1.8 imply that $\boldsymbol{\alpha} = \mathbf{0}$. Hence $\boldsymbol{\Phi}_{\mathbf{q}}$ has full row rank and there is a 6×6 nonsingular submatrix of $\boldsymbol{\Phi}_{\mathbf{q}}$. Defining the variables in $\mathbf{q}$ that correspond to columns of the nonsingular submatrix of $\boldsymbol{\Phi}_{\mathbf{q}}$ as dependent, by the implicit function theorem these six variables can be determined as functions of the remaining three variables, uniquely in some neighborhood of $\mathbf{q}_0$. Thus there are three degrees of rotational freedom of a rigid body in space.

While the nine direction cosines, subject to the constraints of Eq. 4.1.2, could be adopted as generalized coordinates that define orientation of the x'-y'-z' frame, this is neither practical nor convenient. Thus other orientation generalized coordinates are sought.

4.2 VELOCITY, ACCELERATION, AND VIRTUAL DISPLACEMENT OF A POINT FIXED IN A MOVING REFERENCE FRAME

Consider a typical point P that is fixed in the body of Fig. 4.1.5; that is, it is fixed in the x'-y'-z' frame, as shown in Fig. 4.2.1. The vector that locates P in the x-y-z reference frame is given by Eq. 4.1.1 as

$$\mathbf{r}^P = \mathbf{r} + \mathbf{A}\mathbf{s}'^P \tag{4.2.1}$$

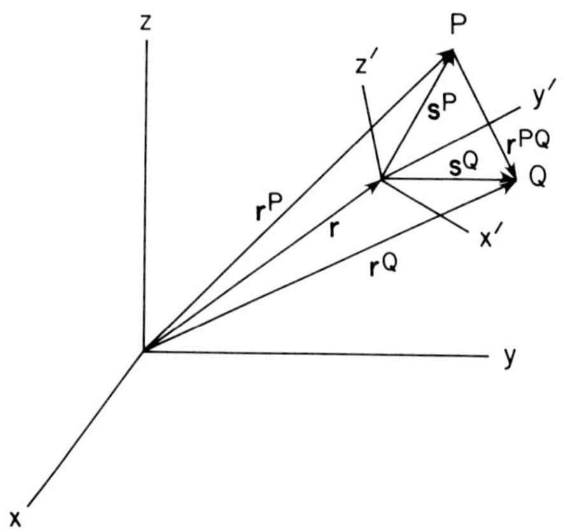

Figure 4.2.1 Points P and Q fixed in x'-y'-z' frame.

where $\mathbf{s}'^P$ is the constant vector of coordinates of P in the x'-y'-z' frame and $\mathbf{A}$ is the direction cosine matrix for the x'-y'-z' frame, relative to the x-y-z frame.

➧ **Example 4.2.1:** Points P and Q are fixed in a rigid body, as shown in Fig. 4.2.1. The points are located in the body-fixed x'-y'-z' frame by constant vectors $\mathbf{s}'^P$ and $\mathbf{s}'^Q$ and in the x-y-z reference frame by Eq. 4.2.1 and

$$\mathbf{r}^Q = \mathbf{r} + \mathbf{A}\mathbf{s}'^Q$$

Note that the locations of both points depend on $\mathbf{r}$ and $\mathbf{A}$, which locate and orient the x'-y'-z' frame, hence the body. Since the body is rigid, the distance between P and Q should be constant, regardless of the values of $\mathbf{r}$ and $\mathbf{A}$. To see that this is the case, form the vector from P to Q.

$$\begin{aligned}\mathbf{r}^{PQ} &= \mathbf{r}^Q - \mathbf{r}^P \\ &= \mathbf{r} + \mathbf{A}\mathbf{s}'^Q - (\mathbf{r} + \mathbf{A}\mathbf{s}'^P) \\ &= \mathbf{A}(\mathbf{s}'^Q - \mathbf{s}'^P)\end{aligned}$$

The length of $\mathbf{r}^{PQ}$ is

$$\begin{aligned}|\mathbf{r}^{PQ}| &= [\mathbf{r}^{PQ^T}\mathbf{r}^{PQ}]^{1/2} \\ &= [(\mathbf{s}'^Q - \mathbf{s}'^P)^T\mathbf{A}^T\mathbf{A}(\mathbf{s}'^Q - \mathbf{s}'^P)]^{1/2} \\ &= [(\mathbf{s}'^Q - \mathbf{s}'^P)^T(\mathbf{s}'^Q - \mathbf{s}'^P)]^{1/2}\end{aligned}$$

which is constant, independent of $\mathbf{r}$ and $\mathbf{A}$.

Velocity, Acceleration, and Angular Velocity

➧ **Example 4.2.2:** Consider the flywheel of Example 2.2.1 (Fig. 2.2.1) that rotates about the z axis with angular velocity ω. Attach a body-fixed x'-y'-z' frame to the flywheel with its z' axis along the axis of rotation and with the x' and y' axes coincident with the x and y axes at $t = 0$. Thus the x'-y'-z' frame rotates counterclockwise about the z axis by angle ωt. In the notation of Euler's theorem, the orientation axis is defined by $\mathbf{u} = \mathbf{k}$ and $\chi = \omega t$. Thus, from Eq. 2.5.46,

$$\mathbf{A} = \begin{bmatrix} 2\cos^2\frac{\omega t}{2} - 1 & -2\cos\frac{\omega t}{2}\sin\frac{\omega t}{2} & 0 \\ 2\cos\frac{\omega t}{2}\sin\frac{\omega t}{2} & 2\cos^2\frac{\omega t}{2} - 1 & 0 \\ 0 & 0 & 2\cos^2\frac{\omega t}{2} - 1 + 2\sin^2\frac{\omega t}{2} \end{bmatrix}$$

$$= \begin{bmatrix} \cos\omega t & -\sin\omega t & 0 \\ \sin\omega t & \cos\omega t & 0 \\ 0 & 0 & 1 \end{bmatrix}$$

Thus for any point P on the flywheel,

$$\mathbf{r}^P = \mathbf{s}^P = \mathbf{A}\mathbf{s}'^P = \begin{bmatrix} s_{x'}^P \cos\omega t - s_{y'}^P \sin\omega t \\ s_{x'}^P \sin\omega t + s_{y'}^P \cos\omega t \\ s_{z'}^P \end{bmatrix}$$

Differentiating both sides of this equation with respect to time, the velocity of point P is

$$\dot{\mathbf{r}}^P = \omega \begin{bmatrix} -s_{x'}^P \sin\omega t - s_{y'}^P \cos\omega t \\ s_{x'}^P \cos\omega t - s_{y'}^P \sin\omega t \\ 0 \end{bmatrix}$$

A direct multiplication verifies that

$$\dot{\mathbf{r}}^P = \omega\tilde{\mathbf{k}}\mathbf{s}^P$$

In geometric vector form, this is

$$\vec{v}^P = \vec{\omega} \times \vec{s}^P$$

where the angular velocity is $\vec{\omega} = \omega\vec{k}$. Thus direct manipulation with the orientation matrix $\mathbf{A}$ yields a familiar result for velocity of points that are fixed in a body that rotates about an *axis fixed in space*, in this case the z axis.

Since the body-fixed x'-y'-z' frame translates and rotates relative to the x-y-z frame, the vector $\mathbf{r}$ and transformation matrix $\mathbf{A}$ are functions of time. Differentiation rules of Section 2.2 can be used to obtain the time derivative of $\mathbf{r}^P$ as

$$\dot{\mathbf{r}}^P = \dot{\mathbf{r}} + \dot{\mathbf{A}}\mathbf{s}'^P \tag{4.2.2}$$

since $\dot{\mathbf{s}}'^P = \mathbf{0}$ because the vector $\mathbf{s}'^P$ defined in the x'-y'-z' frame is constant. Using Eq. 2.5.15, this result may be rewritten as

$$\dot{\mathbf{r}}^P = \dot{\mathbf{r}} + \dot{\mathbf{A}}\mathbf{A}^T\mathbf{s}^P \tag{4.2.3}$$

In order to interpret the second term on the right side of Eq. 4.2.3, it is helpful to derive an identity that involves the transformation matrix $\mathbf{A}$. Applying Eq. 2.2.9 to differentiate both sides of $\mathbf{A}\mathbf{A}^T = \mathbf{I}$ with respect to time yields

$$\dot{\mathbf{A}}\mathbf{A}^T + \mathbf{A}\dot{\mathbf{A}}^T = \mathbf{0}$$

Thus

$$(\dot{\mathbf{A}}\mathbf{A}^T)^T = \mathbf{A}\dot{\mathbf{A}}^T = -\dot{\mathbf{A}}\mathbf{A}^T \tag{4.2.4}$$

and $\dot{\mathbf{A}}\mathbf{A}^T$ is a 3×3 skew-symmetric matrix. Thus there exists a vector $\boldsymbol{\omega}$, such that

$$\tilde{\boldsymbol{\omega}} = \dot{\mathbf{A}}\mathbf{A}^T \tag{4.2.5}$$

which is the definition of $\boldsymbol{\omega}$ in terms of components of the transformation matrix $\mathbf{A}$ and its time derivative $\dot{\mathbf{A}}$; that is, according to Eq. 2.5.21, ω_x is the 3-2 entry in $\dot{\mathbf{A}}\mathbf{A}^T$, ω_y is the 1-3 entry, and ω_z is the 2-1 entry in the same matrix.

Using Eq. 4.2.5 in Eq. 4.2.3 yields

$$\dot{\mathbf{r}}^P = \dot{\mathbf{r}} + \tilde{\boldsymbol{\omega}}\mathbf{s}^P \tag{4.2.6}$$

If $\mathbf{r} = \dot{\mathbf{r}} = 0$, this equation defines the geometry shown in Fig. 4.2.2. In geometric vector notation, Eq. 4.2.6 is the familiar relationship

$$\dot{\vec{r}}^P = \dot{\vec{r}} + \vec{\omega} \times \vec{s}^P$$

where $\vec{\omega}$ is the angular velocity of the x'-y'-z' frame, relative to the x-y-x reference frame. Thus the vector $\boldsymbol{\omega}$ defined in Eq. 4.2.5 is the *angular velocity* of the x'-y'-z' frame, represented in the x-y-z frame. Taking the vector product in the x'-y'-z' frame and transforming to the x-y-z frame yields (Prob. 4.2.1)

$$\dot{\mathbf{r}}^P = \dot{\mathbf{r}} + \mathbf{A}\tilde{\boldsymbol{\omega}}'\mathbf{s}'^P \tag{4.2.7}$$

Thus angular velocity is a vector quantity.

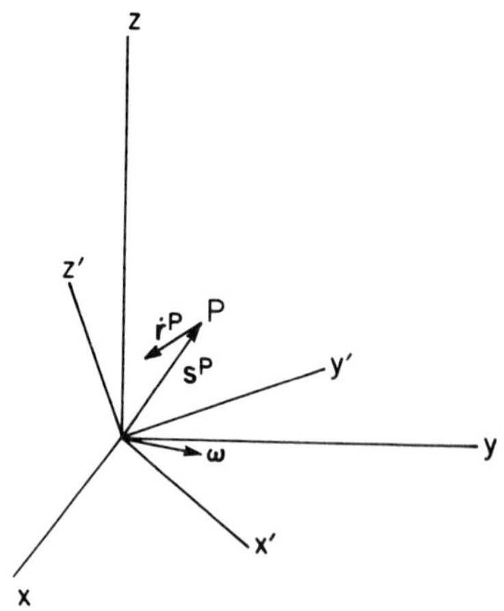

Figure 4.2.2 Angular velocity of x'-y'-z' frame.

♦ **Example 4.2.3:** For the transformation matrix **A** of Example 4.2.2,

$$\dot{\mathbf{A}}\mathbf{A}^T = \begin{bmatrix} 0 & -\omega & 0 \\ \omega & 0 & 0 \\ 0 & 0 & 0 \end{bmatrix}$$

Thus

$$\boldsymbol{\omega} = \begin{bmatrix} 0 \\ 0 \\ \omega \end{bmatrix} = \omega\mathbf{k}$$

which agrees with the result obtained directly in Example 4.2.2.

While the introduction of angular velocity $\boldsymbol{\omega}$ using Eq. 4.2.5 may seem to be formal and mathematical, it is correct and general. In much of the literature on kinematics and dynamics, Eq. 2.5.46 is used in an argument that is similar to that used in Example 4.2.2, with **u** constant. While this was indeed the case in Example 4.2.2, in general motion of a rigid body $\mathbf{u} = \mathbf{u}(t)$ is a function of time. Such physically based derivations may thus yield the correct result, but with invalid reasoning.

Multiplying both sides of Eq. 4.2.5 on the right by **A** yields the useful relationship

$$\dot{\mathbf{A}} = \tilde{\boldsymbol{\omega}}\mathbf{A} \tag{4.2.8}$$

Further, using Eq. 2.5.31 with $\mathbf{s} = \boldsymbol{\omega}$, Eq. 4.2.8 yields

$$\begin{aligned} \dot{\mathbf{A}} &= \mathbf{A}\tilde{\boldsymbol{\omega}}' \\ \tilde{\boldsymbol{\omega}}' &= \mathbf{A}^T\dot{\mathbf{A}} \end{aligned} \tag{4.2.9}$$

Equations 4.2.8 and 4.2.9 provide frequently used relationships between the time derivative of the transformation matrix **A** and the angular velocity $\boldsymbol{\omega}$ of the x'-y'-z' frame.

Both sides of Eq. 4.2.2 may be differentiated with respect to time to obtain the *acceleration equation*

$$\ddot{\mathbf{r}}^P = \ddot{\mathbf{r}} + \ddot{\mathbf{A}}\mathbf{s}'^P \tag{4.2.10}$$

Using the differentiation formula of Eq. 2.2.9 and Eqs. 4.2.5 and 4.2.8, the following relationships for the second time derivative of the transformation matrix **A** are obtained (Prob. 4.2.2):

$$\ddot{\mathbf{A}} = \dot{\tilde{\boldsymbol{\omega}}}\mathbf{A} + \tilde{\boldsymbol{\omega}}\dot{\mathbf{A}} = \dot{\tilde{\boldsymbol{\omega}}}\mathbf{A} + \tilde{\boldsymbol{\omega}}\tilde{\boldsymbol{\omega}}\mathbf{A} \tag{4.2.11}$$

Similarly, differentiating both sides of the first of Eqs. 4.2.9 and using Eq. 4.2.9 yields

$$\ddot{\mathbf{A}} = \mathbf{A}\dot{\tilde{\boldsymbol{\omega}}}' + \mathbf{A}\tilde{\boldsymbol{\omega}}'\tilde{\boldsymbol{\omega}}' \tag{4.2.12}$$

Differentiating both sides of $\boldsymbol{\omega} = \mathbf{A}\boldsymbol{\omega}'$ and using Eq. 4.2.9 yields

$$\dot{\boldsymbol{\omega}} = \mathbf{A}\dot{\boldsymbol{\omega}}' + \dot{\mathbf{A}}\boldsymbol{\omega}' = \mathbf{A}\dot{\boldsymbol{\omega}}' + \mathbf{A}\tilde{\boldsymbol{\omega}}'\boldsymbol{\omega}' = \mathbf{A}\dot{\boldsymbol{\omega}}' \tag{4.2.13}$$

so *angular acceleration* $\dot{\boldsymbol{\omega}}$ is a vector quantity.

♦ **Example 4.2.4:** Using the results of Examples 4.2.2 and 4.2.3 and Eq. 4.2.7, the velocity of a point $\mathbf{s}'^P = [1, 1, 1]$ in the x'-y'-z' frame is

$$\dot{\mathbf{r}}^P = \mathbf{A}\tilde{\boldsymbol{\omega}}'\mathbf{s}'^P = \omega[-\cos\omega t - \sin\omega t, \cos\omega t - \sin\omega t, 0]^T$$

where $\dot{\mathbf{r}} = \mathbf{0}$. The acceleration of point P, from Eqs. 4.2.10 and 4.2.11, is

$$\ddot{\mathbf{r}}^P = \ddot{\mathbf{A}}\mathbf{s}'^P = \omega^2 \begin{bmatrix} \sin\omega t - \cos\omega t \\ -\cos\omega t - \sin\omega t \\ 0 \end{bmatrix}$$

The derivative relations derived here serve as the foundation for kinematic analysis of rigid bodies in space. The reader is encouraged to become proficient in their use, in preparation for dynamic analysis that follows (Probs. 4.2.3 and 4.2.4).

Virtual Displacements and Rotations

A *virtual displacement* $\delta\mathbf{r}^P$ in the position of a point P is defined as an *infinitesimal displacement* (see Section 2.8), or arbitrarily small displacement of the point. If point P is attached to a rigid body, or equivalently if it is fixed in an x'-y'-z' frame, a virtual displacement of point P may be represented in terms of the virtual displacement of the origin of the frame and the variation in its direction cosines.

For a point P that is fixed in the x'-y'-z' frame, the vector $\mathbf{r}^P$ in Eq. 4.2.1 depends on $\mathbf{r}$ and $\mathbf{A}$. If the frame is perturbed slightly, $\mathbf{r}$ changes to $\mathbf{r} + \delta\mathbf{r} + \mathbf{0}(\delta\mathbf{r}^2)$ and $\mathbf{A}$ changes to $\mathbf{A} + \delta\mathbf{A} + \mathbf{0}(\delta\mathbf{A}^2)$. Since the direction cosine matrix must always be orthogonal, it is required that

$$(\mathbf{A} + \delta\mathbf{A} + \mathbf{0}(\delta\mathbf{A}^2))(\mathbf{A} + \delta\mathbf{A} + \mathbf{0}(\delta\mathbf{A}^2))^T$$
$$= \mathbf{A}\mathbf{A}^T + \mathbf{A}\,\delta\mathbf{A}^T + \delta\mathbf{A}\mathbf{A}^T + \delta\mathbf{A}\,\delta\mathbf{A}^T + \mathbf{0}(\delta\mathbf{A}^2) = \mathbf{I}$$

Since $\mathbf{A}$ is orthogonal, this is

$$\mathbf{A}\,\delta\mathbf{A}^T + \delta\mathbf{A}\mathbf{A}^T = \mathbf{0}(\delta\mathbf{A}^2)$$

Equating linear terms in $\delta\mathbf{A}$ on both sides (infinitesimals), this is

$$\mathbf{A}\,\delta\mathbf{A}^T + \delta\mathbf{A}\mathbf{A}^T = \mathbf{0}$$

This relation may also be obtained by taking the differential of both sides of the identity $\mathbf{A}^T\mathbf{A} = \mathbf{I}$ (Prob. 4.2.5). Using this result, observe that

$$(\delta\mathbf{A}\mathbf{A}^T)^T = \mathbf{A}\,\delta\mathbf{A}^T = -\delta\mathbf{A}\mathbf{A}^T \tag{4.2.14}$$

so the matrix $\delta\mathbf{A}\mathbf{A}^T$ is skew-symmetric and can be represented as

$$\delta\mathbf{A}\mathbf{A}^T = \delta\tilde{\boldsymbol{\pi}} \tag{4.2.15}$$

where the vector $\delta\boldsymbol{\pi}$ depends on both the components of $\mathbf{A}$ and $\delta\mathbf{A}$ (Prob. 4.2.6). Multiplying both sides of Eq. 4.2.15 on the right by $\mathbf{A}$ yields

$$\delta\mathbf{A} = \delta\tilde{\boldsymbol{\pi}}\mathbf{A} \tag{4.2.16}$$

To use the foregoing results, take the variation of both sides of Eq. 4.2.1, to obtain (Prob. 4.2.7)

$$\delta\mathbf{r}^P = \delta\mathbf{r} + \delta\mathbf{A}\mathbf{s}'^P \tag{4.2.17}$$

Using Eq. 4.2.16, this is

$$\begin{aligned}\delta\mathbf{r}^P &= \delta\mathbf{r} + \delta\tilde{\boldsymbol{\pi}}\mathbf{A}\mathbf{s}'^P \\ &= \delta\mathbf{r} + \delta\tilde{\boldsymbol{\pi}}\mathbf{s}^P\end{aligned} \tag{4.2.18}$$

Thus $\delta\boldsymbol{\pi}$ plays the role of a rotation, called a *virtual rotation* of the x'-y'-z' frame, relative to the x-y-x reference frame and represented in the latter. Using $\delta\boldsymbol{\pi} = \mathbf{A}\,\delta\boldsymbol{\pi}'$ and Eq. 2.5.31 in Eq. 4.2.16 gives

$$\begin{aligned}\delta\mathbf{A} &= \mathbf{A}\,\delta\tilde{\boldsymbol{\pi}}' \\ \delta\tilde{\boldsymbol{\pi}}' &= \mathbf{A}^T\,\delta\mathbf{A}\end{aligned} \tag{4.2.19}$$

Using this relation in Eq. 4.2.17 yields

$$\delta\mathbf{r}^P = \delta\mathbf{r} + \mathbf{A}\,\delta\tilde{\boldsymbol{\pi}}'\mathbf{s}'^P \tag{4.2.20}$$

Equation 4.2.18 may be written in geometric vector notation as

$$\delta\vec{r}^P = \delta\vec{r} + \delta\vec{\pi} \times \vec{s}^P$$

which verifies that virtual rotation is a vector quantity. This fact is very important. Even though large rotation cannot be represented as a vector quantity (see Section 4.1), virtual rotation and angular velocity are vector quantities.

For a vector $\mathbf{s}$ that is fixed in the x'-y'-z' frame, $\mathbf{s}'$ is constant and

$$\begin{aligned}\delta\mathbf{s} &= \delta\mathbf{A}\mathbf{s}' \\ &= \mathbf{A}\,\delta\tilde{\boldsymbol{\pi}}'\mathbf{s}'\end{aligned} \tag{4.2.21}$$

Using Eq. 2.5.22, this may be written as

$$\delta\mathbf{s} = -\mathbf{A}\tilde{\mathbf{s}}'\,\delta\boldsymbol{\pi}' \tag{4.2.22}$$

The δ-notation associated with virtual displacements and virtual rotations may be interpreted as the *partial differential operator* of calculus, with time held fixed. This is the case, since vectors in the x'-y'-z' frame are functions of whatever set of parameters is used to define position and orientation of the frame. If a vector $\mathbf{h}(\mathbf{q})$ depends on a variable $\mathbf{q}$ in a differentiable way, the *differential* of the vector is $\delta\mathbf{h} = \mathbf{h}_\mathbf{q}\,\delta\mathbf{q}$, where $\delta\mathbf{q}$ is a *differential* or variation of $\mathbf{q}$. Even before the variable $\mathbf{q}$ is defined, the differential $\delta\mathbf{h}$ can be defined and used in analysis, since the linear differential operator obeys the same rules of calculus as the partial derivative operator, with time held fixed.

The differential of the sum of two vectors is the sum of their differentials; that is,

$$\delta(\mathbf{g} + \mathbf{h}) = \delta\mathbf{g} + \delta\mathbf{h} \tag{4.2.23}$$

where the differential δ operates on only the quantity to its immediate right. The *product rule for differentials* is analogous to Eq. 2.2.13; that is,

$$\delta(\mathbf{g}^T\mathbf{h}) = \mathbf{h}^T\,\delta\mathbf{g} + \mathbf{g}^T\,\delta\mathbf{h} \tag{4.2.24}$$

Since the vector product ($\tilde{\ }$) operator is linear (see Eq. 2.5.28), the δ and ($\tilde{\ }$) operators may be interchanged in order of application. Thus

$$\begin{aligned}\delta(\tilde{\mathbf{g}}\mathbf{h}) &= \delta\tilde{\mathbf{g}}\mathbf{h} + \tilde{\mathbf{g}}\,\delta\mathbf{h}\\ &= -\tilde{\mathbf{h}}\,\delta\mathbf{g} + \tilde{\mathbf{g}}\,\delta\mathbf{h}\end{aligned} \tag{4.2.25}$$

Finally, the differential of a triple product may be expanded, using the product rule of differential calculus, to obtain

$$\begin{aligned}\delta(\mathbf{g}^T\mathbf{B}\mathbf{h}) &= \delta\mathbf{g}^T\,\mathbf{B}\mathbf{h} + \mathbf{g}^T\,\delta\mathbf{B}\mathbf{h} + \mathbf{g}^T\mathbf{B}\,\delta\mathbf{h}\\ &= \mathbf{h}^T\mathbf{B}^T\,\delta\mathbf{g} + \mathbf{g}^T\,\delta\mathbf{B}\mathbf{h} + \mathbf{g}^T\mathbf{B}\,\delta\mathbf{h}\end{aligned} \tag{4.2.26}$$

where $\delta\mathbf{g}^T\mathbf{B}\mathbf{h} = \mathbf{h}^T\mathbf{B}^T\,\delta\mathbf{g}$, since the transpose of a scalar is the same scalar.

These and many additional identities that involve virtual displacements, or variations, in position and orientation of reference frames will be used to advantage in analysis of the dynamics of mechanical systems. Of particular importance in analytical dynamics is the powerful coupling between differential calculus and vector analysis. If the abundant manipulation rules of calculus and vector analysis are used to advantage, geometrically clear results can be obtained and understood.

♦ **Example 4.2.5:** To illustrate the power of combining the vector properties of virtual displacements and rotations and differential calculus, consider a commonly used constraint between two vectors $\mathbf{s}_i'$ and $\mathbf{s}_j'$ that are fixed in their respective x_i'-y_i'-z_i' and x_j'-y_j'-z_j' frames, namely that they are orthogonal; that is,

$$\Phi = \mathbf{s}_i^T\mathbf{s}_j = 0 \tag{4.2.27}$$

Taking the differential of both sides of this equation yields

$$\delta\Phi = \mathbf{s}_j^T\,\delta\mathbf{s}_i + \mathbf{s}_i^T\,\delta\mathbf{s}_j = 0 \tag{4.2.28}$$

Substituting for the differentials of the two vectors from Eq. 4.2.22, the differential of Eq. 4.2.28 is obtained in terms of virtual rotations as

$$\delta\Phi = -(\mathbf{s}_j'^T\mathbf{A}_j^T\mathbf{A}_i\tilde{\mathbf{s}}_i'\,\delta\boldsymbol{\pi}_i' + \mathbf{s}_i'^T\mathbf{A}_i^T\mathbf{A}_j\tilde{\mathbf{s}}_j'\,\delta\boldsymbol{\pi}_j') = 0 \tag{4.2.29}$$

This and related differential constraint expressions, or linearized constraints, are of great value in formulating the equations of motion of constrained dynamic systems.

Example 4.2.6: As a second illustration of the power of vector properties of virtual rotations, consider the reference frames shown in Fig. 2.5.7 (Example 2.5.3), which are constrained to rotate about their common $\mathbf{h}_i$ and $\mathbf{h}_j$ unit vectors. Since virtual rotations $\delta\boldsymbol{\pi}_i$ and $\delta\boldsymbol{\pi}_j$ are vector quantities, the infinitesimal relative virtual rotation is

$$\delta\theta\mathbf{h}_i = \delta\boldsymbol{\pi}_j - \delta\boldsymbol{\pi}_i$$

Taking the scalar product of both sides with $\mathbf{h}_i$ and writing vectors in their respective body reference frames yields

$$\begin{aligned}\delta\theta &= \mathbf{h}_i^T(\delta\boldsymbol{\pi}_j - \delta\boldsymbol{\pi}_i) \\ &= \mathbf{h}_i'^T(\mathbf{A}_i^T\mathbf{A}_j\delta\boldsymbol{\pi}_j' - \delta\boldsymbol{\pi}_i')\end{aligned} \tag{4.2.30}$$

Similarly, the relative angular velocity about the axis defined by $\mathbf{h}_i$ and $\mathbf{h}_j$ is

$$\dot{\theta} = \mathbf{h}_i'^T(\mathbf{A}_i^T\mathbf{A}_j\boldsymbol{\omega}_j' - \boldsymbol{\omega}_i') \tag{4.2.31}$$

Example 4.2.7: To further illustrate the power of coupling differential and vector calculus, consider a set of independent generalized coordinates, denoted by a vector $\mathbf{q}$, that defines orientation of bodies that make up a mechanical system. Such generalized coordinates are defined in Sections 4.3 and 4.4, where relationships between variations in generalized coordinates and virtual rotations are established in the form

$$\begin{aligned}\delta\boldsymbol{\pi}_i' &= \mathbf{B}_i\delta\mathbf{q} \\ \delta\boldsymbol{\pi}_j' &= \mathbf{B}_j\delta\mathbf{q}\end{aligned} \tag{4.2.32}$$

where $\delta\mathbf{q}$ is a variation in the generalized coordinates and $\mathbf{B}_i$ and $\mathbf{B}_j$ are matrices that depend on the generalized coordinate vector $\mathbf{q}$. Once the relationship of Eq. 4.2.32 is established, it may be substituted into Eq. 4.2.29, which was derived using vector relationships, to obtain the differential of Φ as

$$\delta\Phi = -[\mathbf{s}_j'^T\mathbf{A}_j^T\mathbf{A}_i\tilde{\mathbf{s}}_i'\mathbf{B}_i + \mathbf{s}_i'^T\mathbf{A}_i^T\mathbf{A}_j\tilde{\mathbf{s}}_j'\mathbf{B}_j]\,\delta\mathbf{q} \tag{4.2.33}$$

Since the coefficient of the differential $\delta\mathbf{q}$ is $\partial\Phi/\partial\mathbf{q}$ if components of $\mathbf{q}$ are independent,

$$\frac{\partial\Phi}{\partial\mathbf{q}} \equiv -(\mathbf{s}_j'^T\mathbf{A}_j^T\mathbf{A}_i\tilde{\mathbf{s}}_i'\mathbf{B}_i + \mathbf{s}_i'^T\mathbf{A}_i^T\mathbf{A}_j\tilde{\mathbf{s}}_j'\mathbf{B}_j) \tag{4.2.34}$$

Recall this is valid only if $\mathbf{q}$ are independent generalized coordinates.

Relative Rotation Relations

Examples 4.2.6 and 4.2.7 illustrate some special vector properties of relative rotation of a pair of reference frames. To be more general, consider the transformation relation from an x_j'-y_j'-z_j' frame to an x_i'-y_i'-z_i' frame, given by Eq. 2.5.35, repeated here as

$$\mathbf{A}_{ij} = \mathbf{A}_i^T\mathbf{A}_j \tag{4.2.35}$$

where $\mathbf{A}_i$ and $\mathbf{A}_j$ are transformation matrices from the x_i'-y_i'-z_i' and x_j'-y_j'-z_j' frames to the global x-y-z reference frame, respectively. Taking the time derivative of both sides of Eq. 4.2.35 and using Eq. 4.2.8 gives

$$\begin{aligned}\dot{\mathbf{A}}_{ij} &= \dot{\mathbf{A}}_i^T\mathbf{A}_j + \mathbf{A}_i^T\dot{\mathbf{A}}_j \\ &= (\tilde{\boldsymbol{\omega}}_i\mathbf{A}_i)^T\mathbf{A}_j + \mathbf{A}_i^T\tilde{\boldsymbol{\omega}}_j\mathbf{A}_j \\ &= \mathbf{A}_i^T(\tilde{\boldsymbol{\omega}}_j - \tilde{\boldsymbol{\omega}}_i)\mathbf{A}_j \\ &= \mathbf{A}_i^T\tilde{\boldsymbol{\omega}}_{ij}\mathbf{A}_j\end{aligned} \tag{4.2.36}$$

where the *relative angular velocity* $\boldsymbol{\omega}_{ij}$ of body j with respect to body i is defined as

$$\boldsymbol{\omega}_{ij} \equiv \boldsymbol{\omega}_j - \boldsymbol{\omega}_i \tag{4.2.37}$$

Using orthogonality of $\mathbf{A}_i$, transforming the relative angular velocity to the x_i'-y_i'-z_i' frame, and using Eq. 2.5.31,

$$\begin{aligned}\dot{\mathbf{A}}_{ij} &= \mathbf{A}_i^T\tilde{\boldsymbol{\omega}}_{ij}\mathbf{A}_j \\ &= \mathbf{A}_i^T\mathbf{A}_i\tilde{\boldsymbol{\omega}}_{ij}'\mathbf{A}_i^T\mathbf{A}_j \\ &= \tilde{\boldsymbol{\omega}}_{ij}'\mathbf{A}_{ij}\end{aligned} \tag{4.2.38}$$

where $\boldsymbol{\omega}_{ij}'$ is the angular velocity of the x_j'-y_j'-z_j' frame relative to the x_i'-y_i'-z_i' frame, represented in the x_i'-y_i'-z_i' frame; that is,

$$\boldsymbol{\omega}_{ij}' \equiv (\boldsymbol{\omega}_{ij})_i' \tag{4.2.39}$$

Taking the time derivative of both sides of Eq. 4.2.38 yields the second derivative of the transformation matrix as

$$\ddot{\mathbf{A}}_{ij} = \dot{\tilde{\boldsymbol{\omega}}}_{ij}'\mathbf{A}_{ij} + \tilde{\boldsymbol{\omega}}_{ij}'\tilde{\boldsymbol{\omega}}_{ij}'\mathbf{A}_{ij} \tag{4.2.40}$$

which corresponds to Eq. 4.2.11, since $\mathbf{A}_{ij}$ is a transformation to the x_i'-y_i'-z_i' frame.

Analogous to the relative angular velocity relationship of Eq. 4.2.36, the differential of both sides of Eq. 4.2.35 may be taken and simplified, using Eq. 4.2.16, to obtain

$$\begin{aligned}\delta\mathbf{A}_{ij} &= \delta\mathbf{A}_i^T\mathbf{A}_j + \mathbf{A}_i^T\delta\mathbf{A}_j \\ &= \mathbf{A}_i^T(\delta\tilde{\boldsymbol{\omega}}_j - \delta\tilde{\boldsymbol{\omega}}_i)\mathbf{A}_j \\ &= \mathbf{A}_i^T\delta\tilde{\boldsymbol{\pi}}_{ij}\mathbf{A}_j\end{aligned} \tag{4.2.41}$$

where the *relative virtual rotation* is defined as

$$\delta\boldsymbol{\pi}_{ij} \equiv \delta\boldsymbol{\pi}_j - \delta\boldsymbol{\pi}_i \tag{4.2.42}$$

Carrying out the same manipulation that leads to Eq. 4.2.38 yields

$$\begin{aligned}\delta\mathbf{A}_{ij} &= \mathbf{A}_i^T\delta\tilde{\boldsymbol{\pi}}_{ij}\mathbf{A}_j \\ &= \delta\tilde{\boldsymbol{\pi}}_{ij}'\mathbf{A}_{ij}\end{aligned} \tag{4.2.43}$$

where $\delta\boldsymbol{\pi}'_{ij}$ is the virtual rotation of the x'_j-y'_j-z'_j frame relative to the x'_i-y'_i-z'_i frame, represented in the x'_i-y'_i-z'_i frame; that is,

$$\delta\boldsymbol{\pi}'_{ij} = (\delta\boldsymbol{\pi}_{ij})'_i \tag{4.2.44}$$

As an illustration of use of the foregoing relations, consider the pair of reference frames that move in a global x-y-z frame, as shown in Fig. 4.2.3. The x'_i-y'_i-z'_i frame moves relative to the global frame, as specified by the vector $\mathbf{r}_i$ and transformation matrix $\mathbf{A}_i$, and the x'_j-y'_j-z'_j frame moves relative to the x'_i-y'_i-z'_i frame, as specified by the vector $\mathbf{r}_{ij}$ and the relative rotation transformation matrix $\mathbf{A}_{ij}$.

The position of a point P that is fixed in the x'_j-y'_j-z'_j frame is

$$\begin{aligned}\mathbf{r}^P &= \mathbf{r}_i + \mathbf{r}_{ij} + \mathbf{s}_j^P \\ &= \mathbf{r}_i + \mathbf{A}_i\mathbf{r}'_{ij} + \mathbf{A}_i\mathbf{A}_{ij}\mathbf{s}'^P_j\end{aligned} \tag{4.2.45}$$

where $\mathbf{r}'_{ij} \equiv (\mathbf{r}_{ij})'_i$ is the representation of $\mathbf{r}_{ij}$ in the x'_i-y'_i-z'_i frame. Taking the time derivative of both sides of Eq. 4.2.45, using Eq. 4.2.38, yields

$$\dot{\mathbf{r}}^P = \dot{\mathbf{r}}_i + \mathbf{A}_i\tilde{\boldsymbol{\omega}}'_i(\mathbf{r}'_{ij} + \mathbf{A}_{ij}\mathbf{s}'^P_j) + \mathbf{A}_i(\dot{\mathbf{r}}'_{ij} + \tilde{\boldsymbol{\omega}}'_{ij}\mathbf{A}_{ij}\mathbf{s}'^P_j) \tag{4.2.46}$$

Taking the time derivative of both sides of Eq. 4.2.46 yields

$$\begin{aligned}\ddot{\mathbf{r}}^P = \ddot{\mathbf{r}}_i &+ \mathbf{A}_i\dot{\tilde{\boldsymbol{\omega}}}'_i(\mathbf{r}'_{ij} + \mathbf{A}_{ij}\mathbf{s}'^P_j) + \mathbf{A}_i(\ddot{\mathbf{r}}'_{ij} + \dot{\tilde{\boldsymbol{\omega}}}'_{ij}\mathbf{A}_{ij}\mathbf{s}'^P_j) \\ &+ 2\mathbf{A}_i\tilde{\boldsymbol{\omega}}'_i(\dot{\mathbf{r}}'_{ij} + \tilde{\boldsymbol{\omega}}'_{ij}\mathbf{A}_{ij}\mathbf{s}'^P_j) + \mathbf{A}_i\tilde{\boldsymbol{\omega}}'_i\tilde{\boldsymbol{\omega}}'_i(\mathbf{r}'_{ij} + \mathbf{A}_{ij}\mathbf{s}'^P_j) \\ &+ \mathbf{A}_i\tilde{\boldsymbol{\omega}}'_{ij}\tilde{\boldsymbol{\omega}}'_{ij}\mathbf{A}_{ij}\mathbf{s}'^P_j\end{aligned} \tag{4.2.47}$$

where the last three terms on the right are quadratic expressions in velocities, which invariably arise in calculating accelerations in geometrically nonlinear kinematics.

Similarly, taking the differential of both sides of Eq. 4.2.45, using Eq. 4.2.43, yields

$$\delta\mathbf{r}^P = \delta\mathbf{r}_i + \mathbf{A}_i\,\delta\tilde{\boldsymbol{\pi}}'_i(\mathbf{r}'_{ij} + \mathbf{A}_{ij}\mathbf{s}'^P_j) + \mathbf{A}_i(\delta\mathbf{r}'_{ij} + \delta\tilde{\boldsymbol{\pi}}'_{ij}\mathbf{A}_{ij}\mathbf{s}'^P_j) \tag{4.2.48}$$

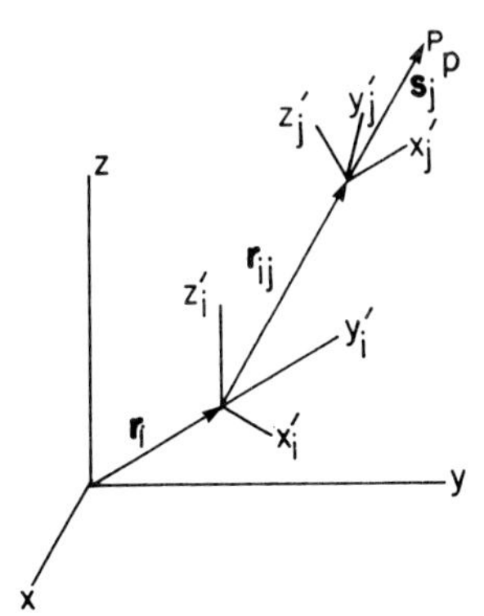

Figure 4.2.3 Relative motion of moving frames.

♦ **Example 4.2.8:** As an illustration of the use of matrix relations such as Eq. 4.2.48, virtual displacements of the three degree of freedom mechanism of Example 2.6.3 may be written in terms of a variation $\delta\mathbf{q}$ in the generalized coordinates $\mathbf{q} = [\phi, \alpha, \theta]^T$. By inspection in Fig. 2.6.3(b),

$$\begin{aligned}
\mathbf{r}_1 &= 0 \\
\mathbf{r}'_{12} &= q_2\mathbf{i}'_1 \\
\boldsymbol{\omega}'_1 &= \dot{q}_1\mathbf{k}'_1 \\
\boldsymbol{\omega}'_{12} &= \dot{q}_3\mathbf{i}'_1 \\
\delta\boldsymbol{\pi}'_1 &= \delta q_1\mathbf{k}'_1 \\
\delta\boldsymbol{\pi}'_{12} &= \delta q_3\mathbf{i}'_1
\end{aligned}$$

Substituting these results into Eq. 4.2.48 for the virtual displacement of point P yields

$$\begin{aligned}
\delta\mathbf{r}^P &= \delta q_1\mathbf{A}_1(q_2\mathbf{j}'_1 + \tilde{\mathbf{k}}'_1\mathbf{A}_{12}\mathbf{s}'^P) + \delta q_2\mathbf{A}_1\mathbf{i}'_1 + \delta q_3\mathbf{A}_1\tilde{\mathbf{i}}'_1\mathbf{A}_{12}\mathbf{s}'^P \\
&= \begin{bmatrix} -(q_2 + 1)\sin q_1 + \cos q_1(\sin q_3 - \cos q_3) & \cos q_1 & \sin q_1(\cos q_3 + \sin q_3) \\ (q_2 + 1)\cos q_1 + \sin q_1(\sin q_3 - \cos q_3) & \sin q_1 & -\cos q_1(\cos q_3 + \sin q_3) \\ 0 & 0 & (\cos q_3 - \sin q_3) \end{bmatrix} \begin{bmatrix} \delta q_1 \\ \delta q_2 \\ \delta q_3 \end{bmatrix}
\end{aligned}$$

which agrees with the result calculated by direct differentiation in Example 2.6.3. Similarly, the velocity and acceleration of point P may be obtained using Eqs. 4.2.46 and 4.2.47 (Prob. 4.2.8).

4.3 EULER PARAMETER ORIENTATION GENERALIZED COORDINATES

As shown in Section 2.5, the position and orientation of a rigid body in space can be defined by locating the origin of a body-fixed x'-y'-z' frame and an orthogonal direction cosine matrix that defines the orientation of the x'-y'-z' frame. To characterize orientation of the body analytically, a set of orientation generalized coordinates must be defined. Euler parameters are presented in this section, with the more conventional Euler angles presented in Section 4.4.

Recall from Euler's theorem of Section 2.5 that any orientation of a body can be achieved by a rotation χ about some unit vector $\mathbf{u}$, as shown in Fig. 4.3.1. Define

$$\begin{aligned}
e_0 &= \cos\frac{\chi}{2} \\
\mathbf{e} &= \mathbf{u}\sin\frac{\chi}{2}
\end{aligned} \tag{4.3.1}$$

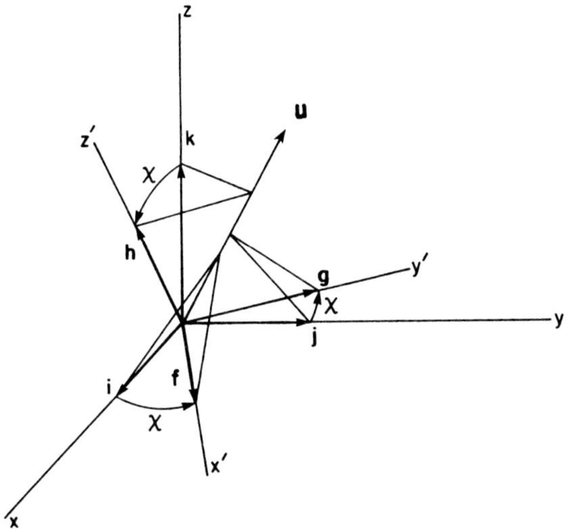

Figure 4.3.1 Euler rotation of reference frames.

The four quantities, e_0, e_1, e_2, and e_3, are called *Euler parameters*. For convenience, denote the vector of Euler parameters as

$$\mathbf{p} = [e_0, e^T]^T = [e_0, e_1, e_2, e_3]^T \tag{4.3.2}$$

Euler parameters are not independent, since

$$e_0^2 + \mathbf{e}^T\mathbf{e} = \cos\frac{2\chi}{2} + \mathbf{u}^T\mathbf{u}\sin\frac{2\chi}{2} = 1$$

That is,

$$\mathbf{p}^T\mathbf{p} = e_0^2 + e_1^2 + e_2^2 + e_3^2 = 1 \tag{4.3.3}$$

which is called the *Euler parameter normalization constraint.*

Since Eq. 4.3.3 implies that $\mathbf{p} \neq \mathbf{0}$, at least one component, say e_k, of $\mathbf{p}$ is not zero. Since the Jacobian of Eq. 4.3.3 is $2[e_0, e_1, e_2, e_3]$ and $e_k \neq 0$, the Jacobian has full row rank. Thus Eq. 4.3.3 determines e_k as a function of the remaining three components of $\mathbf{p}$, which are the three degrees of rotational freedom, at least locally. The Euler parameters thus serve as four *orientation generalized coordinates*. Euler parameters are not independent, however, since they must satisfy the normalization constraint of Eq. 4.3.3.

To see that use of four Euler parameters that must satisfy a normalization constraint is not unnatural, consider the elementary problem of orientation of the x'-y' frame in the x-y plane in Example 2.5.2. If $q_1 = \sin\phi$ and $q_2 = \cos\phi$ were chosen as orientation generalized coordinates, the planar rotation matrix of Eq. 2.5.20 is

$$\mathbf{A} = \begin{bmatrix} q_2 & -q_1 \\ q_1 & q_2 \end{bmatrix}$$

However, analogous to Eq. 4.3.3, $\mathbf{q} = [q_1, q_2]^T$ must satisfy the normalization constraint

$$q_1^2 + q_2^2 = \sin^2\phi + \cos^2\phi = 1$$

This approach, even though valid, is not used in planar kinematics, since ϕ is an independent orientation generalized coordinate in the plane. However, in three dimensional space, any set of three orientation generalized coordinates must have a singular configuration [10]. As shown in the following, Euler parameters avoid this difficulty. The penalty, however, is that four generalized coordinates that satisfy one normalization constraint must be dealt with.

Expanding Eq. 2.5.46 for the orthogonal direction cosine matrix $\mathbf{A}$, as a function of Euler parameters yields

$$\begin{aligned}\mathbf{A} &= (2e_0^2 - 1)\mathbf{I} + 2(\mathbf{e}\mathbf{e}^T + e_0\tilde{\mathbf{e}}) \\ &= 2\begin{bmatrix} e_0^2 + e_1^2 - \frac{1}{2} & e_1e_2 - e_0e_3 & e_1e_3 + e_0e_2 \\ e_1e_2 + e_0e_3 & e_0^2 + e_2^2 - \frac{1}{2} & e_2e_3 - e_0e_1 \\ e_1e_3 - e_0e_2 & e_2e_3 + e_0e_1 & e_0^2 + e_3^2 - \frac{1}{2} \end{bmatrix}\end{aligned} \tag{4.3.4}$$

It is possible to derive explicit formulas for Euler parameters in terms of elements of the direction cosine matrix $\mathbf{A}$ of Eq. 4.3.4. Assume that nine direction cosines of an orthogonal transformation matrix $\mathbf{A}$ are given. The *trace* of matrix $\mathbf{A}$, denoted by tr $\mathbf{A}$, is defined as

$$\text{tr}\,\mathbf{A} = a_{11} + a_{22} + a_{33} \tag{4.3.5}$$

From Eq. 4.3.4,

$$\text{tr}\,\mathbf{A} = 2(3e_0^2 + e_1^2 + e_2^2 + e_3^2) - 3 = 4e_0^2 - 1 \tag{4.3.6}$$

Thus

$$e_0^2 = \frac{\text{tr}\,\mathbf{A} - 1}{4} \tag{4.3.7}$$

Substitution of Eq. 4.3.7 into the diagonal elements of $\mathbf{A}$ in Eq. 4.3.4 results in

$$\begin{aligned}a_{11} &= 2(e_0^2 + e_1^2) - 1 \\ &= 2\left(\frac{\text{tr}\,\mathbf{A} + 1}{4} + e_1^2\right) - 1\end{aligned}$$

Thus

$$e_1^2 = \frac{1 + 2a_{11} - \text{tr}\,\mathbf{A}}{4} \tag{4.3.8}$$

Similarly,

$$e_2^2 = \frac{1 + 2a_{22} - \text{tr}\,\mathbf{A}}{4} \tag{4.3.9}$$

$$e_3^2 = \frac{1 + 2a_{33} - \text{tr}\,\mathbf{A}}{4} \tag{4.3.10}$$

It is interesting, and computationally important, to note that Eqs. 4.3.8 to 4.3.10 determine magnitudes of the Euler parameters in terms of only the diagonal elements of the direction cosine matrix **A**. To find algebraic signs of the Euler parameters, off-diagonal terms of **A** must be used.

Equation 4.3.3 shows that at least one Euler parameter must be nonzero. Consider first the case $e_0 \neq 0$. The sign of e_0 may be selected as positive or negative. Subtracting symmetrically placed off-diagonal terms of matrix **A** in Eq. 4.3.4 yields

$$\begin{aligned} a_{32} - a_{23} &= 4e_0e_1 \\ a_{13} - a_{31} &= 4e_0e_2 \\ a_{21} - a_{12} &= 4e_0e_3 \end{aligned}$$

or

$$\begin{aligned} e_1 &= \frac{a_{32} - a_{23}}{4e_0} \\ e_2 &= \frac{a_{13} - a_{31}}{4e_0} \\ e_3 &= \frac{a_{21} - a_{12}}{4e_0} \end{aligned} \tag{4.3.11}$$

If e_0 calculated from Eq. 4.3.7 is nonzero and its sign is selected, Eq. 4.3.11 can be used to determine e_1, e_2, and e_3. Suppose that for the selected sign of e_0 and the computed values of e_1, e_2, and e_3, the angle of rotation and the axis of rotation are determined to be χ and **e**, respectively. If the sign of e_0 is inverted, the signs of e_1, e_2, and e_3 are inverted also. Changing the signs of all four parameters does not influence the transformation matrix **A**, since variable elements in the matrix of Eq. 4.3.4 are quadratic in these variables.

Either sign may be selected for $e_0 \neq 0$ to define the physical orientation of the body. To see this, denote $a > 0$ as the magnitude of e_0 from Eq. 4.3.7 and denote $\cos(\chi^+/2) = e_0^+ = a$ and $\cos(\chi^-/2) = e_0^- = -a$, where $0 < \chi^+/2 \leq \pi/2$

and $\pi/2 \leq \chi^-/2 < \pi$. Since the cosine function is antisymmetric about $\pi/2$ [i.e., $\cos(\pi/2 + \theta) = -\cos(\pi/2 - \theta)$],

$$\frac{\chi^-}{2} - \frac{\pi}{2} = \frac{\pi}{2} - \frac{\chi^+}{2}$$

or

$$\chi^- = 2\pi - \chi^+ \tag{4.3.12}$$

Thus $\sin(\chi^-/2) = \sin(\chi^+/2)$. From Eq. 4.3.11, $\mathbf{e}^- = -\mathbf{e}^+$. The unit vector $\mathbf{u}$ in Eq. 4.3.1 defines the positive sense along the orientation axis, for implementation of the rotation χ by the right-hand rule. Using the above and Eq. 4.3.1, $\mathbf{u}^- = -\mathbf{u}^+$, as shown in Fig. 4.3.2. Thus if χ^- of Eq. 4.3.12 is selected, a rotation of $2\pi - \chi^+$ is implemented counterclockwise about the $\mathbf{u}^-$ axis, or equivalently clockwise about the $\mathbf{u}^+$ axis, as shown in Fig. 4.3.2, and precisely the same orientation is achieved.

If e_0^2 in Eq. 4.3.7 is zero, then Eq. 4.3.1 shows that $\chi/2 = k\pi/2$, or $\chi = k\pi$, for k an odd integer. To find the algebraic signs of e_1, e_2, and e_3, symmetrically placed off-diagonal terms of the matrix $\mathbf{A}$ in Eq. 4.3.4 may be added to obtain

$$\begin{aligned} a_{21} + a_{12} &= 4e_1e_2 \\ a_{31} + a_{13} &= 4e_1e_3 \\ a_{32} + a_{23} &= 4e_2e_3 \end{aligned} \tag{4.3.13}$$

One of the three remaining Euler parameters e_1, e_2, or e_3 from Eqs. 4.3.8 to 4.3.10 must be nonzero and its sign may be selected as positive or negative. Then Eq.

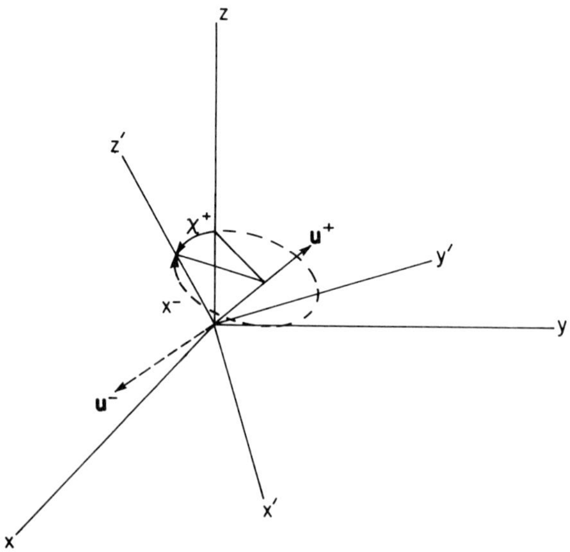

Figure 4.3.2 Alternative values of Euler parameters.

4.3.13 can be used to determine the magnitude and the sign of the other two Euler parameters.

Even though two distinct values of Euler parameters, namely $\mathbf{p}^+$ and $\mathbf{p}^-$ with $\mathbf{p}^- = -\mathbf{p}^+$, specify the same orientation, the mapping from $\mathbf{A}(\mathbf{p})$ to $\mathbf{p}$ is locally one-to-one. The two alternative values of $\mathbf{p}$ are never close together, because

$$(\mathbf{p}^+ - \mathbf{p}^-)^T(\mathbf{p}^+ - \mathbf{p}^-) = 4\mathbf{p}^{+T}\mathbf{p}^+ = 4$$

Thus for neighboring orientations that are specified by $\mathbf{A}_1$ and $\mathbf{A}_2$, $\mathbf{p}_1$ and $\mathbf{p}_2$ can be selected to be close to one another. This observation is important in kinematic and dynamic analysis, since the objective is to solve for $\mathbf{p}(t)$ to specify the time history of orientation of a body. Thus continuity of $\mathbf{p}(t)$ may be used to determine a unique time history $\mathbf{p}(t)$, once the sign of $\mathbf{p}(0)$ is selected.

Properties of Euler Parameters

Important relations between Euler parameters, their time derivatives, their variations, and the associated transformation matrices, angular velocities, and virtual rotations are now derived. Derivation of some of the identities is given in the text. To avoid lengthy expansions in the text, references are given to problems at the end of the chapter, where calculations are outlined and the reader is invited to fill in the details.

A pair of 3×4 matrices $\mathbf{E}$ and $\mathbf{G}$ is defined as

$$\mathbf{E} \equiv [-\mathbf{e}, \tilde{\mathbf{e}} + e_0\mathbf{I}] = \begin{bmatrix} -e_1 & e_0 & -e_3 & e_2 \\ -e_2 & e_3 & e_0 & -e_1 \\ -e_3 & -e_2 & e_1 & e_0 \end{bmatrix} \tag{4.3.14}$$

$$\mathbf{G} \equiv [-\mathbf{e}, -\tilde{\mathbf{e}} + e_0\mathbf{I}] = \begin{bmatrix} -e_1 & e_0 & e_3 & -e_2 \\ -e_2 & -e_3 & e_0 & e_1 \\ -e_3 & e_2 & -e_1 & e_0 \end{bmatrix} \tag{4.3.15}$$

It is observed that each row of E is orthogonal to $\mathbf{p}$; that is,

$$\begin{aligned} \mathbf{Ep} &= [-\mathbf{e}, \tilde{\mathbf{e}} + e_0\mathbf{I}]\begin{bmatrix} e_0 \\ \mathbf{e} \end{bmatrix} \\ &= -e_0\mathbf{e} + \tilde{\mathbf{e}}\mathbf{e} + e_0\mathbf{e} = \mathbf{0} \end{aligned} \tag{4.3.16}$$

where Eq. 2.5.24 has been used. Similarly (Prob. 4.3.2),

$$\mathbf{Gp} = \mathbf{0} \tag{4.3.17}$$

A direct calculation shows that the rows of $\mathbf{E}$ (Prob. 4.3.3) and the rows of $\mathbf{G}$ (Prob. 4.3.4) are orthonormal; that is,

$$\mathbf{EE}^T = \mathbf{GG}^T = \mathbf{I} \tag{4.3.18}$$

However, $\mathbf{E}^T\mathbf{E}$ is a 4×4 matrix of the form

$$\mathbf{E}^T\mathbf{E} = \begin{bmatrix} 1 - e_0^2 & -e_0\mathbf{e}^T \\ -e_0\mathbf{e} & -\mathbf{e}\mathbf{e}^T + \mathbf{I}_3 \end{bmatrix} = \mathbf{I}_4 - \mathbf{p}\mathbf{p}^T \tag{4.3.19}$$

where $\mathbf{I}_4$ is the 4×4 identity matrix. Similarly,

$$\mathbf{G}^T\mathbf{G} = \mathbf{I}_4 - \mathbf{p}\mathbf{p}^T \tag{4.3.20}$$

The key relationship involving these matrices is found by evaluating the matrix product

$$\begin{aligned} \mathbf{E}\mathbf{G}^T &= [-\mathbf{e}, \tilde{\mathbf{e}} + e_0\mathbf{I}]\begin{bmatrix} \mathbf{e}^T \\ \tilde{\mathbf{e}} + e_0\mathbf{I} \end{bmatrix} \\ &= \mathbf{e}\mathbf{e}^T + (\tilde{\mathbf{e}} + e_0\mathbf{I})(\tilde{\mathbf{e}} + e_0\mathbf{I}) \\ &= (2e_0^2 - 1)\mathbf{I} + 2(\mathbf{e}\mathbf{e}^T + e_0\tilde{\mathbf{e}}) \end{aligned} \tag{4.3.21}$$

Comparing Eq. 4.3.21 with the transformation matrix $\mathbf{A}$ of Eq. 4.3.4 reveals that

$$\mathbf{A} = \mathbf{E}\mathbf{G}^T \tag{4.3.22}$$

Thus the transformation matrix $\mathbf{A}$, with its quadratic terms in Euler parameters, is the product of two matrices whose terms are linear in the Euler parameters.

Time Derivatives of Euler Parameters

Taking the first time derivative of both sides of Eq. 4.3.3 yields

$$\dot{\mathbf{p}}^T\mathbf{p} + \mathbf{p}^T\dot{\mathbf{p}} = 2\dot{\mathbf{p}}^T\mathbf{p} = 2\mathbf{p}^T\dot{\mathbf{p}} = 0$$

or

$$\mathbf{p}^T\dot{\mathbf{p}} = \dot{\mathbf{p}}^T\mathbf{p} = 0 \tag{4.3.23}$$

Similarly, taking the first time derivatives of both sides of Eqs. 4.3.16 and 4.3.17 yields

$$\mathbf{E}\dot{\mathbf{p}} = -\dot{\mathbf{E}}\mathbf{p} \tag{4.3.24}$$

$$\mathbf{G}\dot{\mathbf{p}} = -\dot{\mathbf{G}}\mathbf{p} \tag{4.3.25}$$

Direct expansion, using Eqs. 4.3.14 and 4.3.15, shows (Prob. 4.3.6) that

$$\mathbf{E}\dot{\mathbf{G}}^T = \dot{\mathbf{E}}\mathbf{G}^T \tag{4.3.26}$$

Taking the time derivative of both sides of Eq. 4.3.22 and using Eq. 4.3.26 gives

$$\dot{\mathbf{A}} = \dot{\mathbf{E}}\mathbf{G}^T + \mathbf{E}\dot{\mathbf{G}}^T = 2\mathbf{E}\dot{\mathbf{G}}^T \tag{4.3.27}$$

The product $\mathbf{G}\dot{\mathbf{p}}$ can be expanded as

$$\mathbf{G}\dot{\mathbf{p}} = [-\mathbf{e}, -\tilde{\mathbf{e}} + e_0\mathbf{I}]\begin{bmatrix} \dot{e}_0 \\ \dot{\mathbf{e}} \end{bmatrix} = -\dot{e}_0\mathbf{e} - \tilde{\mathbf{e}}\dot{\mathbf{e}} + e_0\dot{\mathbf{e}}$$

Transforming both sides of this equation to skew-symmetric matrices, by the vector product (˜) operation, yields

$$\begin{aligned}
\widetilde{\mathbf{G}\dot{\mathbf{p}}} &= -\dot{e}_0\tilde{\mathbf{e}} - \widetilde{\tilde{\mathbf{e}}\dot{\mathbf{e}}} + e_0\tilde{\dot{\mathbf{e}}} \\
&= -\dot{e}_0\tilde{\mathbf{e}} - \tilde{\mathbf{e}}\tilde{\dot{\mathbf{e}}} + \tilde{\dot{\mathbf{e}}}\tilde{\mathbf{e}} + e_0\tilde{\dot{\mathbf{e}}} \\
&= -\dot{e}_0\tilde{\mathbf{e}} - \tilde{\mathbf{e}}\tilde{\dot{\mathbf{e}}} + \mathbf{e}\dot{\mathbf{e}}^T - \dot{\mathbf{e}}^T\mathbf{e}\mathbf{I} + e_0\tilde{\dot{\mathbf{e}}} \\
&= -\dot{e}_0\tilde{\mathbf{e}} - \tilde{\mathbf{e}}\tilde{\dot{\mathbf{e}}} + \mathbf{e}\dot{\mathbf{e}}^T + e_0\dot{e}_0\mathbf{I} + e_0\tilde{\dot{\mathbf{e}}} \\
&= [-\mathbf{e}, \tilde{\mathbf{e}} + e_0\mathbf{I}]\begin{bmatrix} -\dot{\mathbf{e}}^T \\ \tilde{\dot{\mathbf{e}}} + \dot{e}_0\mathbf{I} \end{bmatrix} \\
&= \mathbf{G}\dot{\mathbf{G}}^T
\end{aligned} \tag{4.3.28}$$

where Eqs. 2.5.22 and 2.5.23 and the identity $\mathbf{p}^T\dot{\mathbf{p}} = e_0\dot{e}_0 + \mathbf{e}^T\dot{\mathbf{e}} = 0$ have been used. In a similar manner (Prob. 4.3.5), it can be shown that

$$\widetilde{\mathbf{E}\dot{\mathbf{p}}} = -\mathbf{E}\dot{\mathbf{E}}^T$$

Euler Parameter Time Derivative and Angular Velocity Relations

Identities between the time derivative of Euler parameters and angular velocity vectors $\boldsymbol{\omega}$ and $\boldsymbol{\omega}'$ are needed in kinematic and dynamic analysis. Using Eqs. 4.3.22 and 4.3.27, Eq. 4.2.9 becomes

$$\tilde{\boldsymbol{\omega}}' = \mathbf{A}^T\dot{\mathbf{A}} = 2\mathbf{G}\mathbf{E}^T\mathbf{E}\dot{\mathbf{G}}^T$$

which, upon application of Eqs. 4.3.19 and 4.3.17, yields

$$\tilde{\boldsymbol{\omega}}' = 2\mathbf{G}\dot{\mathbf{G}}^T \tag{4.3.29}$$

Substituting Eq. 4.3.28 into Eq. 4.3.29 gives

$$\tilde{\boldsymbol{\omega}}' = 2\widetilde{\mathbf{G}\dot{\mathbf{p}}}$$

or

$$\boldsymbol{\omega}' = 2\mathbf{G}\dot{\mathbf{p}} \tag{4.3.30}$$

Multiplying both sides of Eq. 4.3.30 on the left by $\mathbf{G}^T$ yields

$$\mathbf{G}^T\boldsymbol{\omega}' = 2\mathbf{G}^T\mathbf{G}\dot{\mathbf{p}}$$

which, upon application of Eqs. 4.3.20 and 4.3.23, results in the inverse transformation

$$\dot{\mathbf{p}} = \tfrac{1}{2}\mathbf{G}^T\boldsymbol{\omega}' \tag{4.3.31}$$

Using Eqs. 4.3.22 and 4.3.30, angular velocity can be written in the x-y-z reference frame as

$$\boldsymbol{\omega} = \mathbf{A}\boldsymbol{\omega}' = 2\mathbf{E}\mathbf{G}^T\mathbf{G}\dot{\mathbf{p}}$$

Application of Eqs. 4.3.20 and 4.3.23 yields

$$\boldsymbol{\omega} = 2\mathbf{E}\dot{\mathbf{p}} \tag{4.3.32}$$

The inverse transformation of Eq. 4.3.32 is (Prob. 4.3.7)

$$\dot{\mathbf{p}} = \tfrac{1}{2}\mathbf{E}^T\boldsymbol{\omega} \tag{4.3.33}$$

Relationships between Euler Parameter Variations and Virtual Rotations

It was shown in Section 4.2 that virtual displacements and virtual rotations of a reference frame are related to variations in the generalized coordinates that are used to define position and orientation. The purpose of this subsection is to develop relationships between variations in Euler parameters and virtual rotations of the frame whose orientation is defined by the Euler parameters. The principal tool in this development is the differential operator of calculus, which is a linear operator that satisfies the same rules of manipulation as the time derivative operator, to develop relationships between the Euler parameter time derivative and angular velocity.

To be more specific, taking the differential, or variation, of both sides of Eq. 4.3.3 yields

$$\mathbf{p}^T\,\delta\mathbf{p} = 0 \tag{4.3.34}$$

Since Euler parameter relations do not involve time explicitly, the differential operator satisfies the same rules as the time derivative operator, so it is not surprising that Eq. 4.3.34 is of exactly the same form as Eq. 4.3.23, with time derivative replaced by differential. Similarly, taking the differential of both sides of Eq. 4.3.22 yields

$$\delta\mathbf{A} = \delta\mathbf{E}\mathbf{G}^T + \mathbf{E}\,\delta\mathbf{G}^T \tag{4.3.35}$$

which is analogous to the time derivative of the transformation matrix given in Eq. 4.3.27.

Replacing the time derivative operator by the differential operator, all Euler parameter relations that involve time derivatives can be replaced by analogous expressions in terms of differentials. Rather than repeating all these equations and identities, they will simply be used to obtain the desired relationships between variations in Euler parameters and virtual rotations.

Recall the second of Eqs. 4.2.19,

$$\delta\tilde{\boldsymbol{\pi}}' = \mathbf{A}^T\,\delta\mathbf{A} \tag{4.3.36}$$

Substituting for $\delta\mathbf{A}$ from Eq. 4.3.35 and manipulating in exactly the same manner as in the derivation of Eq. 4.3.30 yields (Prob. 4.3.8)

$$\delta\boldsymbol{\pi}' = 2\mathbf{G}\,\delta\mathbf{p} \tag{4.3.37}$$

which provides the desired relationship between a variation in Euler parameters and a virtual rotation of the body-fixed frame, represented in the latter frame. Analogous to Eq. 4.3.32 (Prob. 4.3.9),

$$\delta\boldsymbol{\pi} = 2\mathbf{E}\,\delta\mathbf{p} \tag{4.3.38}$$

which relates a variation in Euler parameters to the virtual rotation vector in global coordinates.

Just as in Eq. 4.3.31, the inverse relation of Eq. 4.3.37 is (Prob. 4.3.10)

$$\delta\mathbf{p} = \tfrac{1}{2}\mathbf{G}^T\,\delta\boldsymbol{\pi}' \tag{4.3.39}$$

Finally, analogous to Eq. 4.3.33 (Prob. 4.3.11),

$$\delta\mathbf{p} = \tfrac{1}{2}\mathbf{E}^T\,\delta\boldsymbol{\pi} \tag{4.3.40}$$

These relationships provide transformations between virtual rotations and generalized coordinate variations that are written symbolically in Eq. 4.2.32. In the case of Euler parameters, the matrix $\mathbf{B}_i$ in Eq. 4.2.32 is simply $2\mathbf{G}_i$. The differential and partial derivative calculations that are written symbolically in Eq. 4.2.33 can now be written explicitly in terms of Euler parameters. These results will be used extensively in formulating the Jacobian matrix of constraint equations for kinematic analysis.

4.4 EULER ANGLE ORIENTATION GENERALIZED COORDINATES

Introduction of four Euler parameter orientation coordinates that must satisfy a normalization constraint in Section 4.3 may seem a bit peculiar. It was shown that one of the Euler parameters could be determined as a function of the remaining three, from the normalization constraint, and that orientation may thus be defined by a system of three parameters. The fact that this reduction can uniquely be carried out locally shows that there are indeed three rotational degrees of freedom for orientation of bodies in space. An attractive feature of Euler parameters is their one-to-one association with orientation. No single set of three independent orientation parameters has ever been found, and arguments have been made that none exist that are both independent and have no singular configurations [10].

The most common conventional form of orientation parameters, *Euler angles*, is developed in this section to define orientation of bodies in space. Three sequential rotation angles are used to determine orientation. Euler angles suffer from having a singular orientation that must be avoided in practice if computational difficulties are to be circumvented. Nevertheless, Euler angles and a number of related orientation angles [9, 10] have attractive features for applications in which singular configurations are naturally avoided. Euler angles have been extensively used in the classical literature. While the multibody dynamics community is turning to Euler parameters, there is adequate historical virtue associated with Euler angles to justify their treatment in this section.

The sequence of three rotations shown in Fig. 4.4.1 defines a general orientation of the x'-y'-z' frame, relative to the x-y-z reference frame. With the moving frame initially coincident with the x-y-z reference frame, rotation by angle ϕ about the positive z axis is implemented to form the x'''-y'''-z''' frame shown in Fig. 4.4.1. Next, rotation by angle θ about the positive x''' axis is implemented, to obtain the x''-y''-z'' frame shown in Fig. 4.4.1. Finally, rotation by angle ψ about the positive z'' axis is implemented to obtain the desired x'-y'-z' frame, in its general orientation shown in Fig. 4.4.1.

Three questions naturally arise upon introduction of such a sequence of orientation generalized coordinates:

1. Must the ϕ-θ-ψ order of rotations be observed?
2. Can every orientation of the x'-y'-z' frame be described by these three angles?
3. If the answer to question 2 is yes, is the mapping from the direction cosine matrix **A** to Euler angles nonsingular?

The answer to question 1 is clearly yes, as shown by the geometric analysis in Section 4.1, where it is shown that interchanging orders of rotation drastically influences the resulting orientation. Thus the order of Euler angle rotations must be adhered to strictly.

Before questions 2 and 3 can be answered, it is most natural to develop an analytical expression for the direction cosine matrix as a function of Euler angles. To this end, each of the rotations is considered separately. The first rotation is shown in Fig. 4.4.2. Since this is precisely the rotation defined in Example 2.5.2, the results of calculations carried out in that example yield the direction cosine

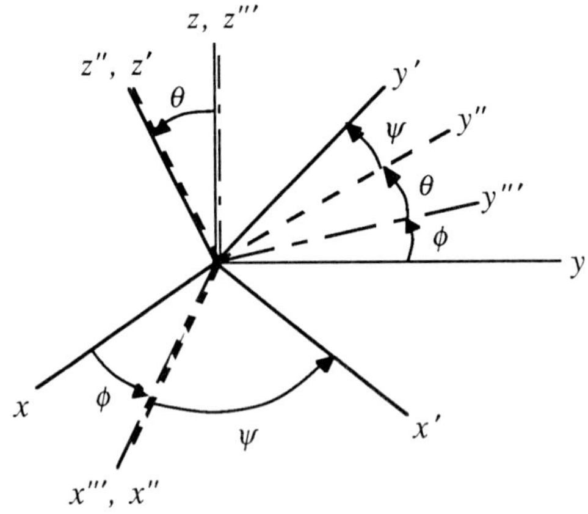

Figure 4.4.1 Euler angle definition.

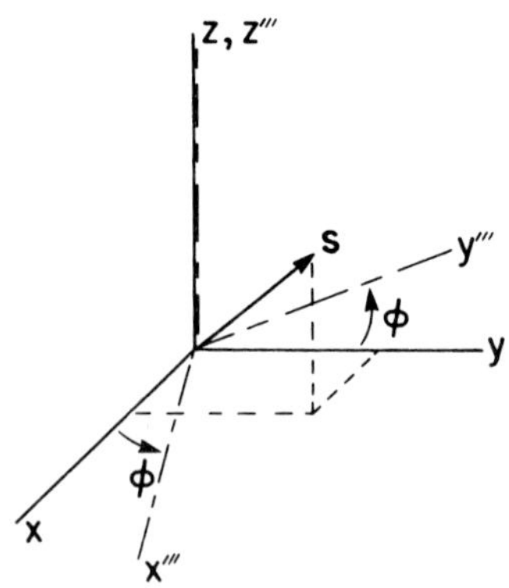

Figure 4.4.2 First Euler angle rotation.

transformation

$$\mathbf{s} = \begin{bmatrix} c_\phi & -s_\phi & 0 \\ s_\phi & c_\phi & 0 \\ 0 & 0 & 1 \end{bmatrix} \mathbf{s}''' \equiv \mathbf{B}\mathbf{s}''' \tag{4.4.1}$$

where $s_\phi = \sin \phi$ and $c_\phi = \cos \phi$.

The second Euler angle rotation is shown in Fig. 4.4.3. Much as in the case of the first rotation treated in Example 2.5.2, the direction cosine transformation is (Prob. 4.4.1)

$$\mathbf{s}''' = \begin{bmatrix} 1 & 0 & 0 \\ 0 & c_\theta & -s_\theta \\ 0 & s_\theta & c_\theta \end{bmatrix} \mathbf{s}'' \equiv \mathbf{C}\mathbf{s}'' \tag{4.4.2}$$

where $s_\theta = \sin \theta$ and $c_\theta = \cos \theta$.

Finally, the third Euler angle rotation is shown in Fig. 4.4.4. Since it is identical in form to the first, the direction cosine transformation relation is obtained immediately as

$$\mathbf{s}'' = \begin{bmatrix} c_\psi & -s_\psi & 0 \\ s_\psi & c_\psi & 0 \\ 0 & 0 & 1 \end{bmatrix} \mathbf{s}' \equiv \mathbf{D}\mathbf{s}' \tag{4.4.3}$$

where $s_\psi = \sin \psi$ and $c_\psi = \cos \psi$.

Substituting the result of Eq. 4.4.3 into Eq. 4.4.2 and subsequently into Eq. 4.4.1, the composite transformation from the x'-y'-z' frame to the x-y-z reference frame is

$$\mathbf{s} = \mathbf{BCD}\mathbf{s}' \equiv \mathbf{A}\mathbf{s}' \tag{4.4.4}$$

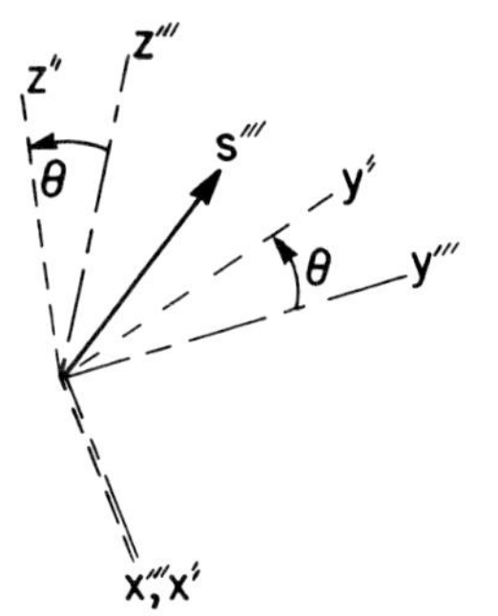

Figure 4.4.3 Second Euler angle rotation.

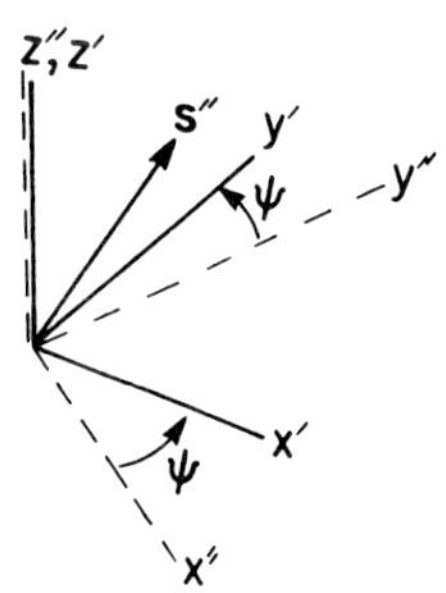

Figure 4.4.4 Third Euler angle rotation.

where the direction cosine transformation matrix **A** is

$$\mathbf{A} = \mathbf{BCD} \tag{4.4.5}$$

Substituting for **B**, **C**, and **D** from Eqs. 4.4.1 through 4.4.3 into Eq. 4.4.5 and carrying out the expansion (Prob. 4.4.2),

$$\mathbf{A} = \begin{bmatrix} c_\psi c_\phi - c_\theta s_\phi s_\psi & -s_\psi c_\phi - c_\theta s_\phi c_\psi & s_\theta s_\phi \\ c_\psi s_\phi + c_\theta c_\phi s_\psi & -s_\psi s_\phi + c_\theta c_\phi c_\psi & -s_\theta c_\phi \\ s_\psi s_\phi & c_\psi s_\theta & c_\theta \end{bmatrix} \tag{4.4.6}$$

Equation 4.4.4 provides the desired direction cosine transformation, where **A** is defined by Eq. 4.4.6.

In order to resolve question 3, concerning singularity of the transformation from **A** to Euler angles, consider the transformation sequence shown in Fig. 4.4.5 with $\theta = 0$. In this special case, the first and third rotation angles, ϕ and ψ, are about the same axis. Therefore, for any angle α, ϕ can be replaced by $\phi + \alpha$ and ψ can be replaced by $\psi - \alpha$ to yield the same orientation transformation matrix. Thus the Euler angles are not uniquely determined, even locally, by the direction cosine transformation matrix when $\theta = 0$. Thus the transformation from **A** to Euler angles is singular when $\theta = 0$ or any multiple of π.

While the details are not presented here, it may be shown that for an arbitrary orthogonal transformation **A**, Eq. 4.4.6 determines at least one set of Euler angles that yield this reference frame. This resolves question 2 and indicates that every orientation can be represented by Euler angles, however not uniquely when $\theta = 0$.

Since angular velocity is of concern in kinematic and dynamic analysis, it is desired to obtain relationships between time derivatives of Euler angles and an-

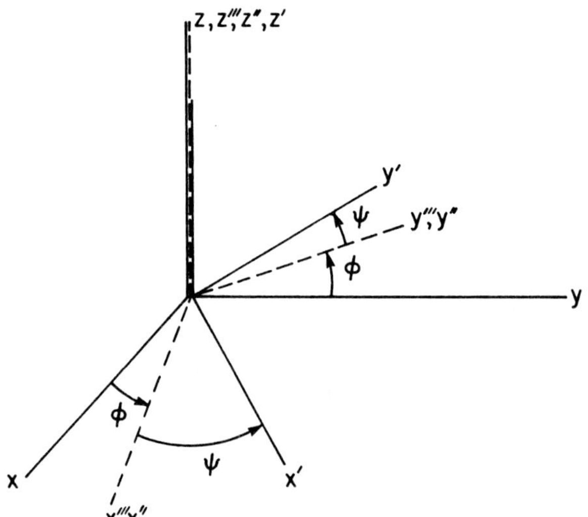

Figure 4.4.5 Euler angle rotations with $\theta = 0$.

gular velocity. First, the time derivative of the transformation matrix $\mathbf{A}$ is obtained by taking the time derivative of both sides of Eq. 4.4.5 and using the product rule of differentiation,

$$\dot{\mathbf{A}} = \dot{\mathbf{B}}\mathbf{C}\mathbf{D} + \mathbf{B}\dot{\mathbf{C}}\mathbf{D} + \mathbf{B}\mathbf{C}\dot{\mathbf{D}} \tag{4.4.7}$$

Equation 4.2.9, using Eqs. 4.4.5 and 4.4.8, yields

$$\begin{aligned} \tilde{\boldsymbol{\omega}}' &= \mathbf{A}^T\dot{\mathbf{A}} \\ &= \mathbf{D}^T\mathbf{C}^T\mathbf{B}^T(\dot{\mathbf{B}}\mathbf{C}\mathbf{D} + \mathbf{B}\dot{\mathbf{C}}\mathbf{D} + \mathbf{B}\mathbf{C}\dot{\mathbf{D}}) \\ &= \mathbf{D}^T\mathbf{C}^T\mathbf{B}^T\dot{\mathbf{B}}\mathbf{C}\mathbf{D} + \mathbf{D}^T\mathbf{C}^T\dot{\mathbf{C}}\mathbf{C}\mathbf{D} + \mathbf{D}^T\dot{\mathbf{D}} \end{aligned} \tag{4.4.8}$$

where simplifications have been made, using the fact that $\mathbf{B}$, $\mathbf{C}$, and $\mathbf{D}$ are orthogonal matrices. Beginning with the third term on the right of Eq. 4.4.9, direct calculation yields

$$\begin{aligned} \mathbf{D}^T\dot{\mathbf{D}} &= \begin{bmatrix} c_\psi & s_\psi & 0 \\ -s_\psi & c_\psi & 0 \\ 0 & 0 & 1 \end{bmatrix} \begin{bmatrix} -s_\psi & -c_\psi & 0 \\ c_\psi & -s_\psi & 0 \\ 0 & 0 & 0 \end{bmatrix} \dot{\psi} \\ &= \begin{bmatrix} 0 & -1 & 0 \\ 1 & 0 & 0 \\ 0 & 0 & 0 \end{bmatrix} \dot{\psi} \end{aligned} \tag{4.4.9}$$

Similarly (Prob. 4.4.3),

$$\begin{aligned} \mathbf{D}^T\mathbf{C}^T\dot{\mathbf{C}}\mathbf{D} &= \mathbf{D}^T \begin{bmatrix} 0 & 0 & 0 \\ 0 & 0 & -1 \\ 0 & 1 & 0 \end{bmatrix} \mathbf{D}\dot{\theta} \\ &= \begin{bmatrix} 0 & 0 & -s_\psi \\ 0 & 0 & -c_\psi \\ s_\psi & c_\psi & 0 \end{bmatrix} \dot{\theta} \end{aligned} \tag{4.4.10}$$

Finally, carrying out a somewhat longer expansion (Prob. 4.4.4) yields

$$\mathbf{D}^T\mathbf{C}^T\mathbf{B}^T\dot{\mathbf{B}}\mathbf{C}\mathbf{D} = \begin{bmatrix} 0 & -c_\theta & c_\psi s_\theta \\ c_\theta & 0 & -s_\psi s_\theta \\ -s_\theta c_\psi & s_\theta s_\psi & 0 \end{bmatrix} \dot{\phi} \tag{4.4.11}$$

Substituting terms from Eqs. 4.4.9 to 4.4.11 into Eq. 4.4.8 yields a 3×3 matrix for $\tilde{\boldsymbol{\omega}}'$. Using the definition of Eq. 2.5.21, components of angular velocity $\boldsymbol{\omega}'$ are identified as (Prob. 4.4.5)

$$\boldsymbol{\omega}' = \begin{bmatrix} 0 \\ 0 \\ 1 \end{bmatrix} \dot{\psi} + \begin{bmatrix} \cos\psi \\ -\sin\psi \\ 0 \end{bmatrix} \dot{\theta} + \begin{bmatrix} \sin\psi\sin\theta \\ \cos\psi\sin\theta \\ \cos\theta \end{bmatrix} \dot{\phi} \tag{4.4.12}$$

Equation 4.4.12 could be obtained by defining $\dot{\phi}$, $\dot{\theta}$, and $\dot{\psi}$ as angular velocities about the associated axes of rotation, transforming each of these components into the final reference frame, and summing to obtain Eq. 4.4.12 (Prob. 4.4.6). This approach may be conveniently used to obtain the component form of the angular velocity vector in the x-y-z reference frame as (Prob. 4.4.7)

$$\boldsymbol{\omega} = \begin{bmatrix} \sin\theta\sin\phi \\ -\sin\theta\cos\phi \\ \cos\theta \end{bmatrix}\dot{\psi} + \begin{bmatrix} \cos\phi \\ \sin\phi \\ 0 \end{bmatrix}\dot{\theta} + \begin{bmatrix} 0 \\ 0 \\ 1 \end{bmatrix}\dot{\phi} \tag{4.4.13}$$

To determine $\dot{\phi}$, $\dot{\theta}$, and $\dot{\psi}$ once $\boldsymbol{\omega}'$ is known, Eq. 4.4.12 may be viewed as a matrix equation of the form

$$\begin{bmatrix} \sin\psi\sin\theta & \cos\psi & 0 \\ \cos\psi\sin\theta & -\sin\psi & 0 \\ \cos\theta & 0 & 1 \end{bmatrix}\begin{bmatrix} \dot{\phi} \\ \dot{\theta} \\ \dot{\psi} \end{bmatrix} = \boldsymbol{\omega}' \tag{4.4.14}$$

To determine whether a solution exists, the determinant of the coefficient matrix on the left may be evaluated and found to be $-\sin\theta$. Therefore, there exists a unique solution for Euler angle time derivatives, for any angular velocity $\boldsymbol{\omega}'$, except when θ is equal to an integral multiple of π, including zero. In this singular case, not every angular velocity can be represented. For those that can be represented, an infinite family of Euler angle time derivatives results. For computational purposes, Eq. 4.4.14 may be solved to obtain

$$\begin{aligned} \dot{\phi} &= \frac{\omega_{y'}\cos\psi + \omega_{x'}\sin\psi}{\sin\theta} \\ \dot{\theta} &= \omega_{x'}\cos\psi - \omega_{y'}\sin\psi \\ \dot{\psi} &= \omega_{z'} - \frac{\cos\theta(\omega_{y'}\cos\psi + \omega_{x'}\sin\psi)}{\sin\theta} \end{aligned} \tag{4.4.15}$$

which shows clearly that a singular case occurs when θ is equal to an integral multiple of π.

Equation 4.4.14 may be interpreted as a definition of the components of body-fixed angular velocity $\boldsymbol{\omega}'$, as a differential form in ϕ, θ, ψ. It is of interest to know whether angular velocity can be obtained as the time derivative of some orientation quantity (i.e., if the components of $\boldsymbol{\omega}'$ are time derivatives of functions of ϕ, θ, and ψ). To see that they are not, consider the z' component of $\boldsymbol{\omega}'$. From Eq. 4.4.14,

$$\omega_{z'} = \cos\theta\,\dot{\phi} + 0\dot{\theta} + \dot{\psi} \tag{4.4.16}$$

The conditions of Eq. 2.7.4 for exactness of a differential form must be tested. Consider first the coefficients of $\dot{\phi}$ and $\dot{\theta}$ in Eq. 4.4.17. The derivatives of the coefficient of $\dot{\phi}$ with respect to θ and of the coefficient of $\dot{\theta}$ with respect to ϕ are

$$\frac{\partial \cos\theta}{\partial\theta} = -\sin\theta \neq 0 = \frac{\partial 0}{\partial\phi}$$

which demonstrates that ω_z' in Eq. 4.4.16 is not a total differential. Therefore, angular velocity cannot be obtained as the time derivative of any function of the Euler angles. This nonintegrability of angular velocity is associated with the complexity of defining angular orientation. The reader is cautioned that angular velocity cannot be integrated to obtain orientation variables. Similarly, while the virtual rotation $\delta\boldsymbol{\pi}'$ can be used as an infinitesimal rotation quantity, it is not the differential of any function that defines orientation.

PROBLEMS

Section 4.2

4.2.1 Show that Eq. 4.2.7 is correct.

4.2.2 Show that Eqs. 4.2.11 and 4.2.12 are correct.

4.2.3 Use Eqs. 4.2.10 to 4.2.12 to show that

$$\begin{aligned}\ddot{\mathbf{r}}^P &= \ddot{\mathbf{r}} - \tilde{\mathbf{s}}^P\dot{\boldsymbol{\omega}} - \tilde{\boldsymbol{\omega}}\tilde{\mathbf{s}}^P\boldsymbol{\omega} \\ &= \ddot{\mathbf{r}} - \mathbf{A}\tilde{\mathbf{s}}'^P\dot{\boldsymbol{\omega}}' - \mathbf{A}\tilde{\boldsymbol{\omega}}'\tilde{\mathbf{s}}'^P\boldsymbol{\omega}'\end{aligned}$$

4.2.4 Write the vector $\mathbf{r}^P$ for the point $\mathbf{s}'^P$ of Example 4.2.2, with $\mathbf{r} = \mathbf{0}$, explicitly as a function of ϕ. Differentiate the result to obtain $\dot{\mathbf{r}}^P$ and $\ddot{\mathbf{r}}^P$ and compare with the results presented in Examples 4.2.2 and 4.2.3.

4.2.5 If matrices $\mathbf{A}_{m\times p}$ and $\mathbf{B}_{p\times n}$ depend on a scalar variable q, then $\delta\mathbf{A} = (\partial a_{ij}/\partial q)\delta q$ and $\delta\mathbf{B} = (\partial b_{ij}/\partial q)\delta q$ are linear in δq. Use this result to show that

$$\delta(\mathbf{AB}) = (\delta\mathbf{A})\mathbf{B} + \mathbf{A}(\delta\mathbf{B})$$

In particular, show that this result and $\mathbf{A}^T\mathbf{A} = \mathbf{I}$ lead to the equation above Eq. 4.2.14.

4.2.6 Use Eq. 4.2.15 to show that the virtual rotation in Example 2.5.2 is $\delta\boldsymbol{\pi} = \delta\boldsymbol{\pi}' = [0, 0, \delta\phi]^T$. (*Hint*: Follow the calculation carried out in Example 4.2.2.)

4.2.7 Use the result of Problem 4.2.5 and the fact that $\mathbf{s}'^P$ is constant to obtain Eq. 4.2.17.

4.2.8 Write $\dot{\mathbf{r}}^P$ and $\ddot{\mathbf{r}}^P$ in Example 4.2.7, using Eq. 4.2.47, and compare with the result obtained in Example 2.3.3.

Section 4.3

4.3.1 Use the relations of Eqs. 4.3.6 to 4.3.10 with the matrix $\mathbf{A}$ derived in Example 2.5.2 to recover the geometry of the rotation of Fig. 2.5.5.

4.3.2 Show that each row of $\mathbf{G}$ is orthogonal to $\mathbf{p}$ (Eq. 4.3.16).

4.3.3 Show that the rows of $\mathbf{E}$ are orthogonal unit vectors (Eq. 4.3.18),

(a) Using the expanded form of matrix $\mathbf{E}$ in Eq. 4.3.14.
(b) Using the compact form of matrix $\mathbf{E}$ in Eq. 4.3.14 and evaluating the matrix product

$$\mathbf{E}\mathbf{E}^T = [-\mathbf{e}, \tilde{\mathbf{e}} + e_0\mathbf{I}]\begin{bmatrix} -\mathbf{e}^T \\ -\tilde{\mathbf{e}} + e_0\mathbf{I} \end{bmatrix}$$

4.3.4 Repeat Problem 4.3.3 for matrix $\mathbf{G}$.

4.3.5 Follow the steps in Eq. 4.3.28 to show that $\mathbf{E}\dot{\mathbf{p}} = -\dot{\mathbf{E}}\mathbf{E}^T$.

4.3.6 Use the definitions of Eqs. 4.3.14 and 4.3.15 to expand the products $\dot{\mathbf{E}}\mathbf{G}^T$ and $\mathbf{E}\dot{\mathbf{G}}^T$ and show that they are equal.

4.3.7 Show that the inverse transformation of Eq. 4.3.33 is valid.

4.3.8 Carry out the detailed derivation of Eq. 4.3.37, following the procedure outlined in the text.

4.3.9 Carry out the detailed derivation of Eq. 4.3.38, following the procedure outlined in the text.

4.3.10 Carry out the detailed derivation of Eq. 4.3.39, following the procedure outlined in the text.

4.3.11 Carry out the detailed derivation of Eq. 4.3.40, following the procedure outlined in the text.

Section 4.4

4.4.1 Derive the transformation matrix of Eq. 4.4.2 from the x''-y''-z'' frame to the x'''-y'''-z''' frame of Fig. 4.4.3.

4.4.2 Carry out the expansion of Eq. 4.4.5, using Eqs. 4.4.1 to 4.4.3, to obtain Eq. 4.4.6.

4.4.3 Carry out the expansions in Eq. 4.4.10.

4.4.4 Carry out the expansions in Eq. 4.4.11.

4.4.5 Add terms in Eq. 4.4.8, using Eqs. 4.4.9 to 4.4.11 to obtain $\tilde{\boldsymbol{\omega}}'$. Use the definition of Eq. 2.5.21 to identify components of $\tilde{\boldsymbol{\omega}}'$ and obtain Eq. 4.4.12.

4.4.6 Add the components of angular velocity shown in the x'-y'-z' frame of Fig. P4.4.6 to obtain angular velocity in that frame given in Eq. 4.4.12.

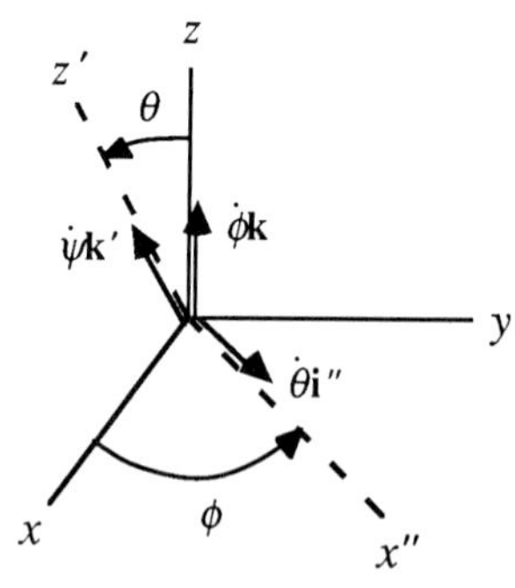

Figure P4.4.6

4.4.7 Add the components of angular velocity shown in Problem 4.4.6 in the x-y-z frame to obtain the angular velocity in that frame given in Eq. 4.4.13.

SUMMARY OF KEY FORMULAS

Velocity, Acceleration, and Angular Velocity

$$\dot{\mathbf{r}}^P = \dot{\mathbf{r}} + \dot{\mathbf{A}}\mathbf{s}'^P = \dot{\mathbf{r}} + \tilde{\boldsymbol{\omega}}\mathbf{s}^P = \dot{\mathbf{r}} + \mathbf{A}\tilde{\boldsymbol{\omega}}'\mathbf{s}'^P \quad (4.2.2,6,7)$$

$$\dot{\mathbf{A}} = \tilde{\boldsymbol{\omega}}\mathbf{A} = \mathbf{A}\tilde{\boldsymbol{\omega}}', \qquad \tilde{\boldsymbol{\omega}} = \dot{\mathbf{A}}\mathbf{A}^T, \qquad \tilde{\boldsymbol{\omega}}' = \mathbf{A}^T\dot{\mathbf{A}} \quad (4.2.8,9)$$

$$\ddot{\mathbf{r}}^P = \ddot{\mathbf{r}} + \ddot{\mathbf{A}}\mathbf{s}'^P, \qquad \ddot{\mathbf{A}} = \dot{\tilde{\boldsymbol{\omega}}}\mathbf{A} + \tilde{\boldsymbol{\omega}}\tilde{\boldsymbol{\omega}}\mathbf{A} = \mathbf{A}\dot{\tilde{\boldsymbol{\omega}}}' + \mathbf{A}\tilde{\boldsymbol{\omega}}'\tilde{\boldsymbol{\omega}}' \quad (4.2.10,11,12)$$

$$\dot{\mathbf{A}}_{ij} = \mathbf{A}_i^T\tilde{\boldsymbol{\omega}}_{ij}\mathbf{A}_j, \qquad \boldsymbol{\omega}_{ij} = \boldsymbol{\omega}_j - \boldsymbol{\omega}_i \quad (4.2.36,37)$$

Virtual Displacements and Rotations

$$\delta\mathbf{r}^P = \delta\mathbf{r} + \delta\mathbf{A}\mathbf{s}'^P = \delta\mathbf{r} + \delta\tilde{\boldsymbol{\pi}}\mathbf{s}^P = \delta\mathbf{r} + \mathbf{A}\,\delta\tilde{\boldsymbol{\pi}}'\mathbf{s}'^P \quad (4.2.17,18,20)$$

$$\delta\tilde{\boldsymbol{\pi}} = \delta\mathbf{A}\mathbf{A}^T, \qquad \delta\mathbf{A} = \delta\tilde{\boldsymbol{\pi}}\mathbf{A} = \mathbf{A}\delta\tilde{\boldsymbol{\pi}}', \qquad \delta\tilde{\boldsymbol{\pi}}' = \mathbf{A}^T\,\delta\mathbf{A} \quad (4.2.15,16,19)$$

$$\delta(\mathbf{g}^T\mathbf{h}) = \mathbf{h}^T\,\delta\mathbf{g} + \mathbf{g}^T\,\delta\mathbf{h}, \qquad \delta(\tilde{\mathbf{g}}\mathbf{h}) = \tilde{\mathbf{g}}\,\delta\mathbf{h} - \tilde{\mathbf{h}}\,\delta\mathbf{g} \quad (4.2.24,25)$$

$$\delta\mathbf{A}_{ij} = \mathbf{A}_i^T\,\delta\tilde{\boldsymbol{\pi}}_{ij}\mathbf{A}_j, \qquad \delta\boldsymbol{\pi}_{ij} = \delta\boldsymbol{\pi}_j - \delta\boldsymbol{\pi}_i \quad (4.2.41,42)$$

Euler Parameter Definitions

$$e_0 = \cos\frac{\chi}{2}\;\mathbf{e} = \mathbf{u}\sin\frac{\chi}{2}, \qquad \mathbf{A} = (2e_0^2 - 1)\mathbf{I} + 2(\mathbf{e}\mathbf{e}^T + e_0\tilde{\mathbf{e}}) \quad (4.3.1,4)$$

$$\mathbf{p} = [e_0, \mathbf{e}^T]^T = [e_0, e_1, e_2, e_3]^T, \qquad \mathbf{p}^T\mathbf{p} = 1 \quad (4.3.2,3)$$

Euler Parameter Relations

$$\mathbf{E} = [-\mathbf{e}, \tilde{\mathbf{e}} + e_0\mathbf{I}], \qquad \mathbf{G} = [-\mathbf{e}, -\tilde{\mathbf{e}} + e_0\mathbf{I}] \quad (4.3.14,15)$$

$$\mathbf{E}\mathbf{E}^T = \mathbf{G}\mathbf{G}^T = \mathbf{I}, \qquad \mathbf{E}^T\mathbf{E} = \mathbf{G}^T\mathbf{G} = \mathbf{I}_4 - \mathbf{p}\mathbf{p}^T, \qquad \mathbf{A} = \mathbf{E}\mathbf{G}^T \quad (4.3.18,19,20,22)$$

$$\boldsymbol{\omega}' = 2\mathbf{G}\dot{\mathbf{p}}, \qquad \boldsymbol{\omega} = 2\mathbf{E}\dot{\mathbf{p}}, \qquad \dot{\mathbf{p}} = \tfrac{1}{2}\mathbf{G}^T\boldsymbol{\omega}' = \tfrac{1}{2}\mathbf{E}^T\boldsymbol{\omega} \quad (4.3.30\text{–}33)$$

$$\delta\boldsymbol{\pi}' = 2\mathbf{G}\,\delta\mathbf{p}, \qquad \delta\boldsymbol{\pi} = 2\mathbf{E}\,\delta\mathbf{p}, \qquad \delta\mathbf{p} = \tfrac{1}{2}\mathbf{G}^T\,\delta\boldsymbol{\pi}' = \tfrac{1}{2}\mathbf{E}^T\,\delta\boldsymbol{\pi} \quad (4.3.37\text{–}40)$$

5

Dynamics of a Rigid Body

As shown in Section 3.2, a rigid body may be modeled as a collection of particles, all of which are fixed in a reference frame that moves with the body. While such a discrete model of rigid body could be used in some applications, a far more general formulation is obtained by considering a rigid body as a continuum of distributed mass with fixed shape and a body-fixed reference frame that characterizes the position and orientation of the body in space, as developed in Chapter 4. The equations of motion of a single rigid body are derived and analyzed in this chapter using the kinematics formulation developed in Chapter 4, a model of a rigid body in which differential masses in the body are constrained to have no motion relative to one another, and using Newton's equations of motion for particles with workless constraints. The variational formulation of the equations of motion is shown to yield, effectively and generally, the equations of motion of a rigid body in space.

5.1 EQUATIONS OF MOTION OF A RIGID BODY

Consider the rigid body shown in Fig. 5.1.1, which is located in space by a vector $\mathbf{r}$ to the origin of the body-fixed x'-y'-z' frame and a set of generalized coordinates that define orientation of the x'-y'-z' frame, relative to the inertial x-y-z frame. A differential mass $dm(P)$ at a typical point P in the body is located by the vector $\mathbf{s}^P = \mathbf{A}\mathbf{s}'^P$, where $\mathbf{s}'^P$ is a constant vector in the x'-y'-z' frame. Forces that act

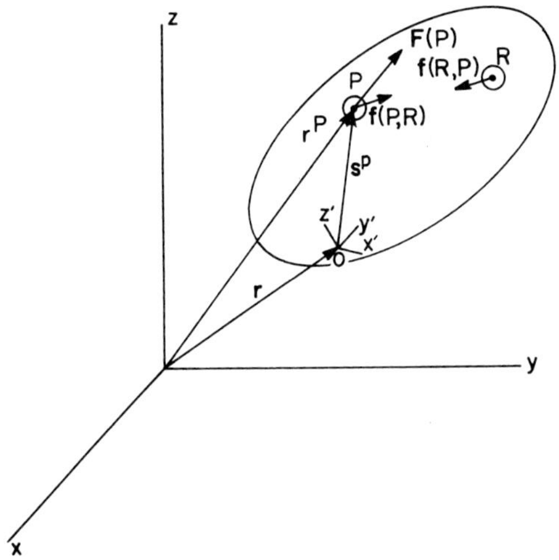

Figure 5.1.1 Forces acting on rigid body in space.

on this differential element of mass include *external force* $\mathbf{F}(P)$ per unit mass at point P and *internal force* $\mathbf{f}(P, R)$ per unit masses at points P and R, as shown in Fig. 5.1.1.

Newton's equation of motion for differential mass $dm(P)$ is

$$\ddot{\mathbf{r}}^P\, dm(P) - \mathbf{F}(P)\, dm(P) - \left[\int_m \mathbf{f}(P, R)\, dm(R)\right] dm(P) = \mathbf{0} \tag{5.1.1}$$

where integration of the internal force $\mathbf{f}(P, R)$ is taken over the entire body.

Let $\delta\mathbf{r}^P$ denote a *virtual displacement* of point P; that is, an infinitesimal variation in the location of point P that is consistent with constraints on the motion of point P. Multiplying both sides of Eq. 5.1.1 on the left by $\delta\mathbf{r}^{P^T}$ and integrating over the total mass of the body yields

$$\int_m \delta\mathbf{r}^{P^T}\ddot{\mathbf{r}}^P\, dm(P) - \int_m \delta\mathbf{r}^{P^T}\mathbf{F}(P)\, dm(P) - \int_m\int_m \delta\mathbf{r}^{P^T}\mathbf{f}(P, R)\, dm(R)\, dm(P) = 0 \tag{5.1.2}$$

This is called the *variational equation of motion* of a rigid body.

Manipulation of the double integral that appears in Eq. 5.1.2 shows that (Prob. 5.1.1)

$$\begin{aligned} \int_m\int_m \delta\mathbf{r}^{P^T}\mathbf{f}(P, R)\, dm(R)\, dm(P) &= \frac{1}{2}\int_m\int_m \delta\mathbf{r}^{P^T}\mathbf{f}(P, R)\, dm(R)\, dm(P) \\ &\quad + \frac{1}{2}\int_m\int_m \delta\mathbf{r}^{R^T}\mathbf{f}(R,P)\, dm(P)\, dm(R) \\ &= \frac{1}{2}\int_m\int_m (\delta\mathbf{r}^P - \delta\mathbf{r}^R)^T\mathbf{f}(P, R)\, dm(P)\, dm(R) \end{aligned} \tag{5.1.3}$$

where the first equality simply rewrites the expression and interchanges dummy variables P and R of integration in the second integral. The second equality follows from interchanging of the order of integration and using Newton's third law (i.e., $\mathbf{f}(P, R) = -\mathbf{f}(R, P)$).

As a model of a rigid body, a distance constraint is imposed between every pair of points P and R; that is,

$$(\mathbf{r}^P - \mathbf{r}^R)^T(\mathbf{r}^P - \mathbf{r}^R) = C^{PR}$$

Taking the differential of both sides yields the relation

$$(\delta\mathbf{r}^P - \delta\mathbf{r}^R)^T(\mathbf{r}^P - \mathbf{r}^R) = 0 \tag{5.1.4}$$

Recall from Example 3.2.1 that the constraint reaction forces on a pair of particles that are connected by a distance constraint are equal, opposite, and collinear; that is,

$$\mathbf{f}(P, R) = k(\mathbf{r}^P - \mathbf{r}^R) \tag{5.1.5}$$

where k is a constant. From Eqs. 5.1.4 and 5.1.5, the right side of Eq. 5.1.3 is zero.

In general, collinearity of $\mathbf{f}(P, R)$ and $\mathbf{f}(R, P)$ (i.e., Eq. 5.1.5) is not implied by Newton's third law* [11]. Goldstein [9] calls $\mathbf{f}(P\ R) = -\mathbf{f}(R, P)$, with collinearity, the *strong form of Newton's third law*. Counterexamples that demonstrate noncollinearity of $\mathbf{f}(P, R)$ and $\mathbf{f}(R, P)$ include charged particles in an electric field, dipoles in a magnetic field, and particles in a nonuniform gravitational field. In the presence of such effects, the integral on the right of Eq. 5.1.3 is not zero. Its value must therefore be computed, based on the nature of the forces that act, and included as an applied force.

Note that the right side of Eq. 5.1.3 is the virtual work of the internal forces in the rigid body. In this book attention is focused on rigid bodies for which internal forces are workless. Thus, under this assumption, or if the strong form of Newton's third law holds,

$$\int_m \int_m \delta\mathbf{r}^{PT}\mathbf{f}(P, R)\, dm(P)\, dm(R) = 0 \tag{5.1.6}$$

* Collinearity of $\mathbf{f}(P, R)$ and $\mathbf{f}(R, P)$ in the model used here for a rigid body is a function of the model. While $\mathbf{f}(P, R)\, dm(R)\, dm(P) = -\mathbf{f}(R, P)\, dm(P)\, dm(R)$ follows from Newton's law of action–reaction, in general Newton said nothing about collinearity. This matter has been studied by many scholars and is elegantly summarized in Trusedell's essay "Whence the Law of Moment of Momentum?" in Ref. 11. The rigid-body model used here is adequate to represent internal gravitational attraction and effects of stress distributions (as a limit of an elastic body) with symmetric stress tensors. The model used here requires that if forces of interaction occur in a body due to electric or magnetic field effects, they must be accounted for as external forces. This model of a rigid body is adequate for commonly used engineering materials and machine dynamics applications. For more philosophical considerations, the reader may wish to consult Ref. 11 and references cited therein.

Using equations of motion for flexible bodies [11, 13], if the body is elastic and if its stiffness approaches infinity to form a rigid body, Eq. 5.1.6 holds. There is thus ample justification for Eq. 5.1.6, on both physical and mathematical grounds, for virtually all essentially rigid bodies that are encountered in engineering application.

Using the result of Eq. 5.1.6, Eq. 5.1.2 simplifies to

$$\int_m \delta\mathbf{r}^{P^T}\ddot{\mathbf{r}}^P\, dm(P) - \int_m \delta\mathbf{r}^{P^T}\mathbf{F}(P)\, dm(P) = 0 \tag{5.1.7}$$

which must hold for all virtual displacements $\delta\mathbf{r}^P$ that can occur in a rigid body. As in Chapter 3 for particles, this result is equivalent to the *principle of virtual work*. To take full advantage of Eq. 5.1.7, the virtual displacement of point P will be written in terms of a virtual displacement of the origin of the body-fixed x'-y'-z' frame and a virtual rotation of the frame.

From Eq. 4.2.20, a virtual displacement of point P is

$$\begin{aligned}\delta\mathbf{r}^P &= \delta\mathbf{r} + \mathbf{A}\,\delta\tilde{\boldsymbol{\pi}}'\mathbf{s}'^P \\ &= \delta\mathbf{r} - \mathbf{A}\tilde{\mathbf{s}}'^P\,\delta\boldsymbol{\pi}'\end{aligned} \tag{5.1.8}$$

Similarly, from Eqs. 4.2.10 and 4.2.12, the acceleration of point P may be written as

$$\begin{aligned}\ddot{\mathbf{r}}^P &= \ddot{\mathbf{r}} + \ddot{\mathbf{A}}\mathbf{s}'^P \\ &= \ddot{\mathbf{r}} + \mathbf{A}\dot{\tilde{\boldsymbol{\omega}}}'\mathbf{s}'^P + \mathbf{A}\tilde{\boldsymbol{\omega}}'\tilde{\boldsymbol{\omega}}'\mathbf{s}'^P\end{aligned} \tag{5.1.9}$$

Substituting Eqs. 5.1.8 and 5.1.9 into the variational equation of Eq. 5.1.7 yields

$$\begin{aligned}&\int_m (\delta\mathbf{r}^T + \delta\boldsymbol{\pi}'^T\tilde{\mathbf{s}}'^P\mathbf{A}^T)(\ddot{\mathbf{r}} + \mathbf{A}\dot{\tilde{\boldsymbol{\omega}}}'\mathbf{s}'^P + \mathbf{A}\tilde{\boldsymbol{\omega}}'\tilde{\boldsymbol{\omega}}'\mathbf{s}'^P)\, dm(P) \\ &\qquad - \int_m (\delta\mathbf{r}^T + \delta\boldsymbol{\pi}'^T\tilde{\mathbf{s}}'^P\mathbf{A}^T)\mathbf{F}(P)\, dm(P) = 0\end{aligned} \tag{5.1.10}$$

for arbitrary $\delta\mathbf{r}$ and $\delta\boldsymbol{\pi}'$. Expanding the integrals of Eq. 5.1.10 yields (Prob. 5.1.2)

$$\begin{aligned}&\delta\mathbf{r}^T\ddot{\mathbf{r}}\int_m dm(P) + \delta\mathbf{r}^T(\mathbf{A}\dot{\tilde{\boldsymbol{\omega}}}' + \mathbf{A}\tilde{\boldsymbol{\omega}}'\tilde{\boldsymbol{\omega}}')\int_m \mathbf{s}'^P\, dm(P) \\ &+ \delta\boldsymbol{\pi}'^T\left(\widetilde{\int_m \mathbf{s}'^P\, dm(P)}\right)\mathbf{A}^T\ddot{\mathbf{r}} + \delta\boldsymbol{\pi}'^T\int_m \tilde{\mathbf{s}}'^P\dot{\tilde{\boldsymbol{\omega}}}'\mathbf{s}'^P\, dm(P) \\ &\qquad + \delta\boldsymbol{\pi}'^T\int_m \tilde{\mathbf{s}}'^P\tilde{\boldsymbol{\omega}}'\tilde{\boldsymbol{\omega}}'\mathbf{s}'^P\, dm(P) - \delta\mathbf{r}^T\int_m \mathbf{F}(P)\, dm(P) \\ &\qquad\qquad - \delta\boldsymbol{\pi}'^T\int_m \tilde{\mathbf{s}}'^P\mathbf{F}'(P)\, dm(P) = 0\end{aligned} \tag{5.1.11}$$

for all $\delta\mathbf{r}$ and $\delta\boldsymbol{\pi}'$.

Equations of Motion with a Centroidal Body-Fixed Reference Frame

In order to simplify the form of Eq. 5.1.11, a body-fixed x'-y'-z' reference frame is selected with its origin at the *center of mass* of the body. It is thus called a *centroidal body-fixed frame*. The total mass is

$$m \equiv \int_m dm(P) \tag{5.1.12}$$

and by definition of the centroidal body-fixed frame,

$$\int_m \mathbf{s}'^P \, dm(P) = \mathbf{0} \tag{5.1.13}$$

The resultant *external force* that acts on the body is simply

$$\mathbf{F} \equiv \int_m \mathbf{F}(P) \, dm(P) \tag{5.1.14}$$

and the *external torque* with respect to the origin of the body-fixed x'-y'-z' frame, represented in that frame, is

$$\mathbf{n}' \equiv \int_m \tilde{\mathbf{s}}'^P \mathbf{F}'(P) \, dm(P) \tag{5.1.15}$$

The fourth integral in Eq. 5.1.11 may be written as

$$\begin{aligned} \int_m \tilde{\mathbf{s}}'^P \tilde{\dot{\boldsymbol{\omega}}}' \mathbf{s}'^P \, dm(P) &= -\left(\int_m \tilde{\mathbf{s}}'^P \tilde{\mathbf{s}}'^P \, dm(P) \right) \dot{\boldsymbol{\omega}}' \\ &\equiv \mathbf{J}' \dot{\boldsymbol{\omega}}' \end{aligned} \tag{5.1.16}$$

where the constant *inertia matrix* (or *inertia tensor*) $\mathbf{J}'$ with respect to the x'-y'-z' centroidal body reference frame is defined as

$$\begin{aligned} \mathbf{J}' &\equiv -\int_m \tilde{\mathbf{s}}'^P \tilde{\mathbf{s}}'^P \, dm(P) \\ &= \int_m \begin{bmatrix} y'^{P2} + z'^{P2} & -x'^P y'^P & -x'^P z'^P \\ -x'^P y'^P & x'^{P2} + z'^{P2} & -y'^P z'^P \\ -x'^P z'^P & -y'^P z'^P & x'^{P2} + y'^{P2} \end{bmatrix} dm(P) \end{aligned} \tag{5.1.17}$$

where each term in the matrix on the right is integrated separately. The diagonal elements in the matrix $\mathbf{J}'$ are called *moments of inertia* and the off-diagonal terms are called *products of inertia*. Since symmetrically placed off-diagonal elements are equal, $\mathbf{J}'$ is symmetric (i.e., $\mathbf{J}' = \mathbf{J}'^T$).

The integrand of the fifth integral in Eq. 5.1.11 may be expanded, using Eqs.

2.5.23 and A.3.15, as

$$\begin{aligned}\tilde{\mathbf{s}}'^P\tilde{\boldsymbol{\omega}}'\tilde{\boldsymbol{\omega}}'\mathbf{s}'^P &= -\tilde{\mathbf{s}}'^P\tilde{\boldsymbol{\omega}}'\tilde{\mathbf{s}}'^P\boldsymbol{\omega}' \\ &= -\tilde{\boldsymbol{\omega}}'\tilde{\mathbf{s}}'^P\tilde{\mathbf{s}}'^P\boldsymbol{\omega}' - \boldsymbol{\omega}'\mathbf{s}'^{P^T}\tilde{\mathbf{s}}'^P\boldsymbol{\omega}' - \mathbf{s}'^P\boldsymbol{\omega}'^T\tilde{\boldsymbol{\omega}}'\mathbf{s}'^P \\ &= -\tilde{\boldsymbol{\omega}}'\tilde{\mathbf{s}}'^P\tilde{\mathbf{s}}'^P\boldsymbol{\omega}'\end{aligned}$$

which yields the integral

$$\int_m \tilde{\mathbf{s}}'^P\tilde{\boldsymbol{\omega}}'\tilde{\boldsymbol{\omega}}'\mathbf{s}'^P\, dm(P) = \tilde{\boldsymbol{\omega}}'\mathbf{J}'\boldsymbol{\omega}' \tag{5.1.18}$$

Substituting Eqs. 5.1.12 to 5.1.18 into Eq. 5.1.11 yields the *variational Newton–Euler equation of motion* for the rigid body as

$$\delta\mathbf{r}^T[m\ddot{\mathbf{r}} - \mathbf{F}] + \delta\boldsymbol{\pi}'^T[\mathbf{J}'\dot{\boldsymbol{\omega}}' + \tilde{\boldsymbol{\omega}}'\mathbf{J}'\boldsymbol{\omega}' - \mathbf{n}'] = 0 \tag{5.1.19}$$

which must hold for all virtual displacements $\delta\mathbf{r}$ and virtual rotations $\delta\boldsymbol{\pi}'$ of the centroidal body-fixed x'-y'-z' frame.

For a rigid body that moves in space under the action of the resultant external centroidal force $\mathbf{F}$ and torque $\mathbf{n}'$, $\delta\mathbf{r}$ and $\delta\boldsymbol{\pi}'$ are arbitrary, so their coefficients in Eq. 5.1.19 must be zero. This yields the *Newton–Euler equations of motion* for a body that moves in space,

$$\begin{aligned} m\ddot{\mathbf{r}} &= \mathbf{F} \\ \mathbf{J}'\dot{\boldsymbol{\omega}}' &= \mathbf{n}' - \tilde{\boldsymbol{\omega}}'\mathbf{J}'\boldsymbol{\omega}'\end{aligned} \tag{5.1.20}$$

Recall that this form of the equations of motion is valid only if the body-fixed x'-y'-z' frame is centroidal.

The Newton–Euler equations of Eq. 5.1.20 may be thought of as differential equations in $\mathbf{r}$ and $\boldsymbol{\omega}'$. In some special cases, they may be integrated for these variables. Integration for orientation of the body is possible, however, only after a set of orientation generalized coordinates, such as Euler parameters or Euler angles, is introduced. This is required, since $\boldsymbol{\omega}'$ cannot be integrated directly.

Equations of Motion of a Rigid Body Subject to Workless Constraints

Consider kinematic constraints of the form

$$\boldsymbol{\Phi}(\mathbf{q}) = \mathbf{0} \tag{5.1.21}$$

in which constraint reaction forces act, so

$$\begin{aligned}\mathbf{F} &= \mathbf{F}^A + \mathbf{F}^C \\ \mathbf{n}' &= \mathbf{n}'^A + \mathbf{n}'^C\end{aligned} \tag{5.1.22}$$

where $\mathbf{F}^A$ and $\mathbf{n}'^A$ are resultant *applied force* and *applied torque*, respectively, which may include gravity, spring and damper forces between bodies, and any forces other than the *constraint force* and *constraint torque* $\mathbf{F}^C$ and $\mathbf{n}'^C$, respectively. If the constraints are frictionless, they are *workless constraints;* that is, their constraint force and torque do no work,

$$\delta W^C \equiv \delta\mathbf{r}^T\mathbf{F}^C + \delta\boldsymbol{\pi}'^T\mathbf{n}'^C = 0 \tag{5.1.23}$$

for all virtual displacements $\delta\mathbf{r}$ and virtual rotations $\delta\boldsymbol{\pi}'$ such that

$$\boldsymbol{\Phi}_{\mathbf{r}}\delta\mathbf{r} + \boldsymbol{\Phi}_{\boldsymbol{\pi}'}\delta\boldsymbol{\pi}' = \mathbf{0} \tag{5.1.24}$$

which are called *kinematically admissible virtual displacements and rotations.*

Equation 5.1.19 can be written, using Eq. 5.1.22, as

$$\delta\mathbf{r}^T[m\ddot{\mathbf{r}} - \mathbf{F}^A] + \delta\boldsymbol{\pi}'^T[\mathbf{J}'\dot{\boldsymbol{\omega}}' + \tilde{\boldsymbol{\omega}}'\mathbf{J}'\boldsymbol{\omega}' - \mathbf{n}'^A] - \delta\mathbf{r}^T\mathbf{F}^C - \delta\boldsymbol{\pi}'\mathbf{n}'^C = 0 \tag{5.1.25}$$

for all $\delta\mathbf{r}$ and $\delta\boldsymbol{\pi}'$. If virtual displacements and rotations in Eq. 5.1.25 are restricted to satisfy Eq. 5.1.24, then by Eqs. 5.1.23 and 5.1.25,

$$\delta\mathbf{r}^T[m\ddot{\mathbf{r}} - \mathbf{F}^A] + \delta\boldsymbol{\pi}'^T[\mathbf{J}'\dot{\boldsymbol{\omega}}' + \tilde{\boldsymbol{\omega}}'\mathbf{J}'\boldsymbol{\omega}' - \mathbf{n}'^A] = 0 \tag{5.1.26}$$

for all kinematically admissible $\delta\mathbf{r}$ and $\delta\boldsymbol{\pi}$ (i.e., for all $\delta\mathbf{r}$ and $\delta\boldsymbol{\pi}'$ that satisfy Eq. 5.1.24.

♦ **Example 5.1.1:** A body shown in Fig. 5.1.2 is constrained so that point P on its z' axis moves without friction on the surface of a unit sphere. This constraint is defined by the constraint equation

$$\Phi = (\mathbf{r} + \mathbf{A}\mathbf{h}')^T(\mathbf{r} + \mathbf{A}\mathbf{h}') - 1 = 0$$

where $\mathbf{h}'$ is the z'-axis unit vector, $\mathbf{h}' = [0, 0, 1]^T$. Taking the variation of both sides of this equation yields

$$\begin{aligned}\delta\Phi &= 2(\mathbf{r} + \mathbf{A}\mathbf{h}')^T(\delta\mathbf{r} + \delta\mathbf{A}\mathbf{h}') \\ &= 2(\mathbf{r} + \mathbf{A}\mathbf{h}')^T\,\delta\mathbf{r} + 2(\mathbf{r} + \mathbf{A}\mathbf{h}')^T\mathbf{A}\,\delta\tilde{\boldsymbol{\pi}}'\mathbf{h}' \\ &= 2(\mathbf{r} + \mathbf{A}\mathbf{h}')^T\,\delta\mathbf{r} - (2\mathbf{r}^T\mathbf{A}\tilde{\mathbf{h}}' + 2\mathbf{h}'^T\tilde{\mathbf{h}}')\,\delta\boldsymbol{\pi}' \\ &= 2(\mathbf{r} + \mathbf{A}\mathbf{h}')^T\,\delta\mathbf{r} - (2\mathbf{r}^T\mathbf{A}\tilde{\mathbf{h}}')\,\delta\boldsymbol{\pi}' = 0\end{aligned}$$

Since Eq. 5.1.26 must hold for all $\delta\mathbf{r}$ and $\delta\boldsymbol{\pi}'$ that satisfy the preceding equation, there exists a Lagrange multiplier λ such that

$$\delta\mathbf{r}^T[m\ddot{\mathbf{r}} - \mathbf{F}^A] + 2\lambda(\mathbf{r} + \mathbf{A}\mathbf{h}')] + \delta\boldsymbol{\pi}'^T[\mathbf{J}'\dot{\boldsymbol{\omega}}' + \tilde{\boldsymbol{\omega}}'\mathbf{J}'\boldsymbol{\omega}' - \mathbf{n}'^A + 2\lambda\tilde{\mathbf{h}}'\mathbf{A}^T\mathbf{r}] = 0$$

which must hold for all $\delta\mathbf{r}$ and $\delta\boldsymbol{\pi}'$. Thus the equations of motion are

$$\begin{aligned}m\ddot{\mathbf{r}} + 2(\mathbf{r} + \mathbf{A}\mathbf{h}')\lambda &= \mathbf{F}^A \\ \mathbf{J}'\dot{\boldsymbol{\omega}}' + 2\tilde{\mathbf{h}}'\mathbf{A}^T\mathbf{r}\lambda &= \mathbf{n}'^A - \tilde{\boldsymbol{\omega}}'\mathbf{J}'\boldsymbol{\omega}'\end{aligned}$$

and the constraint equation.

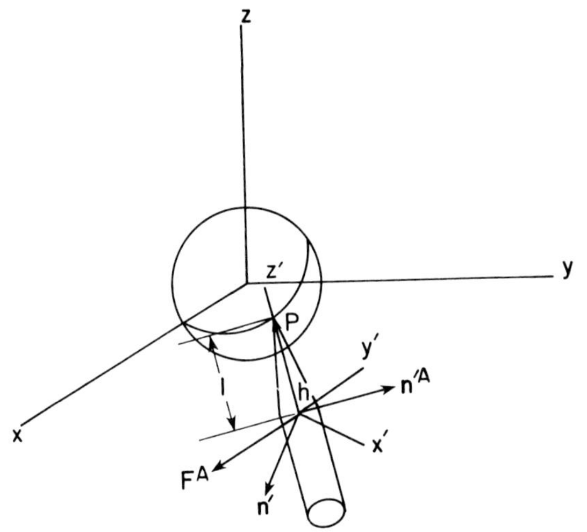

Figure 5.1.2 Pendulum with point on unit sphere.

Equations of Motion for a Rigid Body with One Point Fixed in Space

Consider the special case of a body, such as the top shown in Fig. 5.1.3, with the point O'' fixed in the inertial reference frame, without friction. Since point O'' will not generally be the center of mass of the body, to avoid confusion, define an x''-y''-z'' body-fixed frame with origin at point O''. Since point O'' is fixed in space, $\mathbf{r}$ is constant, its velocity and acceleration are zero, and kinematically admissible virtual displacements of point O'' are zero; that is,

$$\begin{aligned} \mathbf{r} &= \mathbf{c} \\ \dot{\mathbf{r}} &= \ddot{\mathbf{r}} = \mathbf{0} \\ \delta\mathbf{r} &= \mathbf{0} \end{aligned} \tag{5.1.27}$$

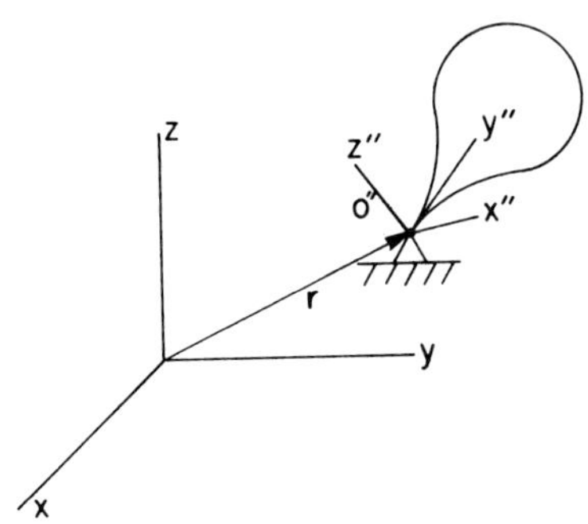

Figure 5.1.3 Body with point O'' fixed in space.

The general variational form of the equations of motion in a noncentroidal x''-y''-z'' frame of Eq. 5.1.11 can be written using the relations of Eq. 5.1.27. Using Eqs. 5.1.15 to 5.1.18, all of which are valid for the noncentroidal x''-y''-z'' frame, Eq. 5.1.11 reduces to

$$\delta\boldsymbol{\pi}''^T[\mathbf{J}''\dot{\boldsymbol{\omega}}'' + \tilde{\boldsymbol{\omega}}''\mathbf{J}''\boldsymbol{\omega}'' - \mathbf{n}''^A] = 0 \tag{5.1.28}$$

for all virtual rotations $\delta\boldsymbol{\pi}''$ that are consistent with any additional workless constraints that may be imposed on motion of the body.

If no orientational constraints are imposed on the body (e.g., if the body of Fig. 5.1.3 is free to move as a pendulum or a spinning top with a ball and socket constraint at point O''), then $\delta\boldsymbol{\pi}''$ of Eq. 5.1.28 is arbitrary and its coefficient must be zero, leading to the *Euler equations of motion*

$$\mathbf{J}''\dot{\boldsymbol{\omega}}'' = \mathbf{n}'' - \tilde{\boldsymbol{\omega}}''\mathbf{J}''\boldsymbol{\omega}'' \tag{5.1.29}$$

Since these equations are nonlinear in angular velocities that appear on the right side, a closed-form solution is generally not possible.

Example 5.1.2: As a special case of a body with a point fixed in space, consider the pendulum shown in Fig. 5.1.4 that is constrained to rotate about the inertial y axis. The body-fixed x''-y''-z'' frame is selected so that the y'' axis coincides with the inertial y axis and is the axis of rotation. By construction of the body-fixed x''-y''-z'' frame and definition of the angle ϕ of rotation defined in Fig. 5.1.4, the transformation from the x''-y''-z'' frame to the x-y-z frame is

$$\mathbf{s} = \mathbf{A}\mathbf{s}'' = \begin{bmatrix} \cos\phi & 0 & \sin\phi \\ 0 & 1 & 0 \\ -\sin\phi & 0 & \cos\phi \end{bmatrix} \mathbf{s}''$$

A direct calculation indicates that angular velocity, angular acceleration, and virtual rotation are about the y'' axis and are related to the angle ϕ by

$$\begin{aligned} \boldsymbol{\omega}'' &= [0, \dot{\phi}, 0]^T \\ \dot{\boldsymbol{\omega}}'' &= [0, \ddot{\phi}, 0]^T \\ \delta\boldsymbol{\pi}'' &= [0, \delta\phi, 0]^T \end{aligned} \tag{5.1.30}$$

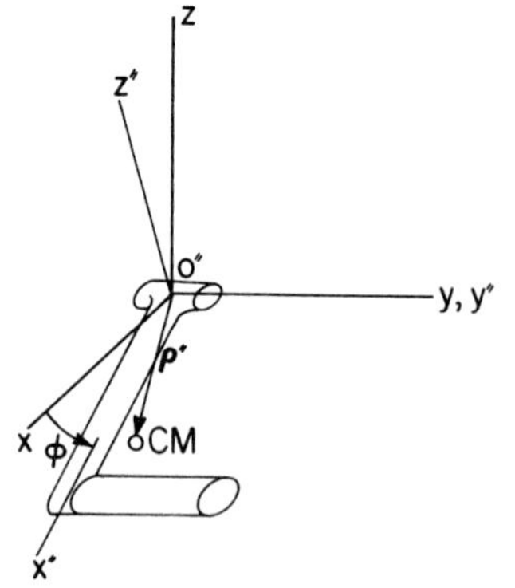

Figure 5.1.4 Pendulum rotating about y axis.

Substituting these relations into Eq. 5.1.28 yields the variational equation of motion of the pendulum as

$$[0,\ \delta\phi,\ 0]\left\{\begin{bmatrix} J_{x''y''}\ddot{\phi} & J_{z''y''}\dot{\phi}^2 \\ J_{y''y''}\ddot{\phi}+ & 0 \\ J_{z''y''}\ddot{\phi} & -J_{x''y''}\dot{\phi}^2 \end{bmatrix} - \begin{bmatrix} n^A_{x''} \\ n^A_{y''} \\ n^A_{z''} \end{bmatrix}\right\} = 0 \tag{5.1.31}$$

The gravitational force acts at the center of mass and is $\mathbf{F}^A = -mg\mathbf{k}$, in the x-y-z frame. In the x''-y''-z'' frame this is

$$\mathbf{F}''^A = \mathbf{A}^T\mathbf{F}^A = mg\begin{bmatrix} \sin\phi \\ 0 \\ -\cos\phi \end{bmatrix}$$

The torque $\mathbf{n}''^A$, from Eq. 5.1.15, is thus

$$\mathbf{n}''^A = \tilde{\boldsymbol{\rho}}''\mathbf{F}''^A = mg\begin{bmatrix} -\rho_{y''}\cos\phi \\ \rho_{z''}\sin\phi + \rho_{x''}\cos\phi \\ -\rho_{y''}\sin\phi \end{bmatrix}$$

Since $\delta\phi$ in Eq. 5.1.31 is arbitrary, its coefficient must be zero, yielding the differential equation of motion of the pendulum as

$$J_{y''y''}\ddot{\phi} = n^A_{y''} = mg(\rho_{z''}\sin\phi + \rho_{x''}\cos\phi) \tag{5.1.32}$$

Note that only the moment of inertia about the y'' axis of the body appears in the equation of motion. To determine reaction torques that act at the rotational joint at point O'' in Fig. 5.1.4, the general form of the constraint reaction torque may be written as

$$\mathbf{T}''^C = [T^C_{x''},\ 0,\ T^C_{z''}]^T \tag{5.1.33}$$

The y'' component of constraint reaction torque is zero, since the rotational joint permits free rotation of the body about that axis. Substituting $\mathbf{n}'' = \mathbf{n}''^A + \mathbf{T}''^C$ into Eq. 5.1.29 yields Eq. 5.1.32 and

$$\begin{aligned} T^C_{x''} &= J_{z''y''}\dot{\phi}^2 + J_{x''y''}\ddot{\phi} - n^A_{x''} \\ T^C_{z''} &= -J_{x''y''}\dot{\phi}^2 + J_{z''y''}\ddot{\phi} - n^A_{z''} \end{aligned} \tag{5.1.34}$$

Thus if Eq. 5.1.32 is solved for ϕ, $\dot{\phi}$, and $\ddot{\phi}$, reaction torques due to asymmetry of the pendulum are determined by Eq. 5.1.34. This result shows that products of inertia may have a substantial influence on forces that act in a system, if asymmetry is present.

Planar Equations of Motion of a Rigid Body

Consider a rigid body that is constrained to move in the x-y plane of an inertial reference frame, as shown in Fig. 5.1.5. Mass distribution in the body is symmetric with respect to the x-y plane, so it can be considered concentrated in the x-y plane. Such a body is called a *planar rigid body*, or *lamina*. The pendulum of Example 5.1.2 is thus not a planar rigid body, even though its body-fixed reference frame undergoes planar motion. A centroidal body-fixed frame is attached to the body,

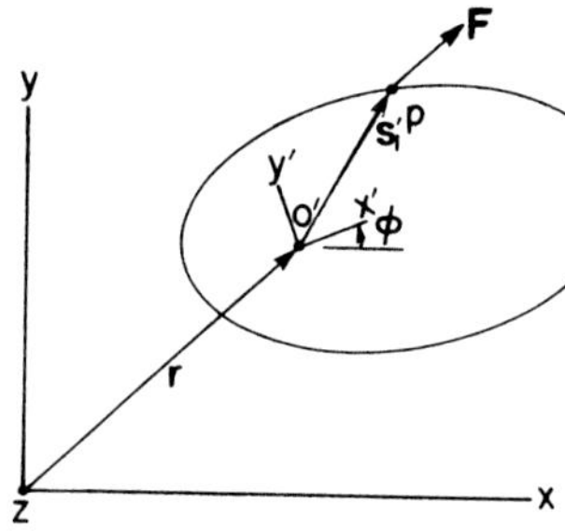

Figure 5.1.5 Body moving in x-y plane.

with its z' axis parallel to the inertial z axis. Orientation of the reference frame is then defined by a single angle ϕ of rotation, as shown in Fig. 5.1.5 and studied in Example 2.5.2.

For the planar case, position, virtual displacement, angular velocity, and virtual rotation are simply

$$\begin{aligned} \mathbf{r} &= [x, y, 0]^T \\ \delta\mathbf{r} &= [\delta x, \delta y, 0]^T \\ \boldsymbol{\omega}' &= [0, 0, \dot{\phi}]^T \\ \delta\boldsymbol{\pi}' &= [0, 0, \delta\phi]^T \end{aligned} \tag{5.1.35}$$

Substituting these quantities into the variational Newton–Euler equations of motion of Eq. 5.1.26 and eliminating zero multiplications yields

$$[\delta x, \delta y]^T \left\{ m \begin{bmatrix} \ddot{x} \\ \ddot{y} \end{bmatrix} - \begin{bmatrix} F_x^A \\ F_y^A \end{bmatrix} \right\} + \delta\phi[J_{z'z'}\ddot{\phi} - n_{z'}^A] = 0 \tag{5.1.36}$$

which must hold for all δx, δy, and $\delta\phi$ that are consistent with any workless constraints that act on the body.

To define a notation for planar dynamics that is consistent with that for spatial dynamics, let

$$\begin{aligned} \mathbf{r} &\equiv [x, y]^T \\ \mathbf{F}^A &\equiv [F_x^A, F_y^A]^T \\ J_0 &\equiv J_{z'z'} \end{aligned} \tag{5.1.37}$$

The inertia constant J_0 is called the *polar moment of inertia* of the body about a centroidal z' axis.

Since the z' axis is parallel to the global z axis, torque may be defined simply as n and written, using Eq. 5.1.15, as

$$\begin{aligned} n \equiv n_{z'} &= [-s_{y'}^P, s_{x'}^P, 0] \begin{bmatrix} F_{x'} \\ F_{y'} \\ F_{z'} \end{bmatrix} \\ &= s_{x'}^P F_{y'} - s_{y'}^P F_{x'} \end{aligned} \tag{5.1.38}$$

The planar transformation derived in Eq. 2.5.20 is written here as

$$\mathbf{s} = \mathbf{A}\mathbf{s}' = \begin{bmatrix} \cos\phi & -\sin\phi \\ \sin\phi & \cos\phi \end{bmatrix} \mathbf{s}' \tag{5.1.39}$$

It is trivial to show that $\mathbf{A}^T\mathbf{A} = \mathbf{I}$, so $\mathbf{A}$ is orthogonal and

$$\mathbf{s}' = \mathbf{A}^T\mathbf{s} \tag{5.1.40}$$

Using Eq. 5.1.40, the applied torque n^A of Eq. 5.1.38 may be written in terms of global components of applied force as

$$n^A = s_{x'}^P(-F_x^A \sin\phi + F_y^A \cos\phi) - s_{y'}^P(F_x^A \cos\phi + F_y^A \sin\phi) \tag{5.1.41}$$

Thus, if body-fixed components of $\mathbf{F}'^A$ are given, Eq. 5.1.38 yields the desired torque and if global components of $\mathbf{F}^A$ are given, Eq. 5.1.41 is used.

Using the streamlined planar notation of Eqs. 5.1.37 and 5.1.38, the *variational planar equation of motion* of Eq. 5.1.30 may be written as

$$\delta\mathbf{r}^T[m\ddot{\mathbf{r}} - \mathbf{F}^A] + \delta\phi[J_0\ddot{\phi} - n^A] = 0 \tag{5.1.42}$$

which must hold for all kinematically admissible virtual displacements $\delta\mathbf{r}$ and virtual rotations $\delta\phi$. Recall that this equation is valid only for a centroidal body-fixed x'-y' frame.

If a body is free to move in a plane without constraints, $\delta\mathbf{r}$ and $\delta\phi$ in Eq. 5.1.42 are arbitrary and their coefficients must be zero, leading to the *unconstrained equations of planar motion,*

$$\begin{aligned} m\ddot{\mathbf{r}} &= \mathbf{F}^A \\ J_0\ddot{\phi} &= n^A \end{aligned} \tag{5.1.43}$$

Example 5.1.3: A circular disk of radius R, mass m, and polar moment of inertia J_0 rolls without slipping on the inclined plane shown in Fig. 5.1.6. The unit

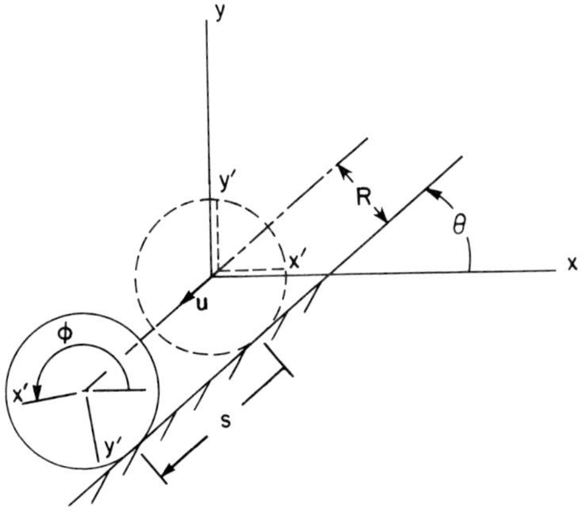

Figure 5.1.6 Rolling disk on incline plane.

vector **u** along the plane is

$$\mathbf{u} = [-\cos\theta,\ -\sin\theta]^T$$

Since the disk rolls without slipping,

$$s = R\phi$$

Thus

$$\mathbf{r} = R\phi\mathbf{u}$$

Since R and **u** are constant,

$$\dot{\mathbf{r}} = R\dot{\phi}\mathbf{u}$$

$$\ddot{\mathbf{r}} = R\ddot{\phi}\mathbf{u}$$

$$\delta\mathbf{r} = R\,\delta\phi\mathbf{u}$$

The variational planar equation of motion, under the action of gravitational force $\mathbf{F}^A = -mg\mathbf{j}$, is obtained by substituting the above into Eq. 5.1.42, noting that the forces of rolling contact are workless, to obtain

$$\delta\phi[mR^2\ddot{\phi} - mgR\sin\theta] + \delta\phi J_0\ddot{\phi} = 0$$

Since $\delta\phi$ is arbitrary,

$$(mR^2 + J_0)\ddot{\phi} = Rmg\sin\theta$$

The general solution of this differential equation is

$$\phi = \left[\frac{Rmg\sin\theta}{mR^2 + J_0}\right]\frac{t^2}{2} + C_0 + C_1 t$$

where C_0 and C_1 are to be determined by initial conditions on $\phi(t_0)$ and $\dot{\phi}(t_0)$ (e.g., if the disk is at rest with $\phi = 0$ at $t = 0$, then $C_0 = C_1 = 0$).

5.2 PROPERTIES OF CENTROID AND MOMENTS AND PRODUCTS OF INERTIA

Centroid

To find the centroid of a body, first select a body-fixed noncentroidal x''-y''-z'' frame with origin at O''. Let the centroid be designated as point O' and let a body-fixed x'-y'-z' frame have its origin at O', as shown in Fig. 5.2.1.

The vector $\boldsymbol{\rho}''$ in the x''-y''-z'' frame that locates O'' may be determined by writing $\mathbf{s}'^P = \mathbf{C}(\mathbf{s}''^P - \mathbf{e}'')$, where $\mathbf{C}$ is the orthogonal transformation matrix from the x''-y''-z'' frame to the x'-y'-z' frame (i.e., $\mathbf{s}' = \mathbf{C}\mathbf{s}''$), and using Eq. 5.1.13, yielding

$$\begin{aligned}\mathbf{0} &= \int_m \mathbf{s}'^P\,dm(P) = \int_m \mathbf{C}(\mathbf{s}''^P - \boldsymbol{\rho}'')\,dm(P)\\ &= \mathbf{C}\int_m \mathbf{s}''^P\,dm(P) - \mathbf{C}\boldsymbol{\rho}''\int_m dm(P)\\ &= \mathbf{C}\int_m \mathbf{s}''^P\,dm(P) - m\mathbf{C}\boldsymbol{\rho}''\end{aligned}$$

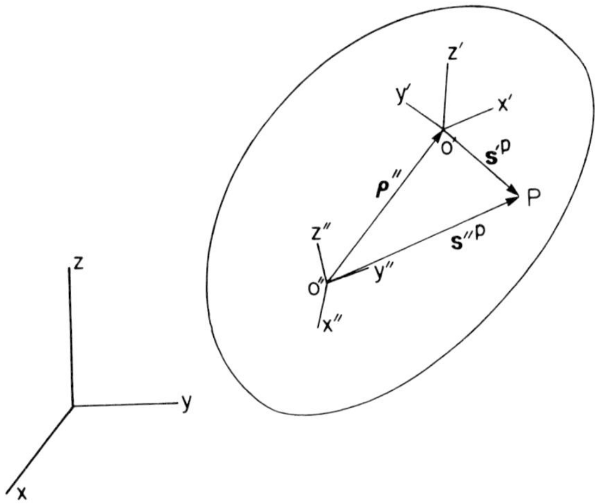

Figure 5.2.1 Location of centroid.

Multiplying both sides of this equation on the left by $\mathbf{C}^T$ yields

$$\boldsymbol{\rho}'' = \frac{1}{m}\int_m \mathbf{s}''^P \, dm(P) \tag{5.2.1}$$

Example 5.2.1: Consider the homogeneous body shown in Fig. 5.2.2. By Eq. 5.2.1, its centroid in the x''-y''-z'' frame is

$$\begin{aligned}
\boldsymbol{\rho}'' &= \frac{1}{m}\int_0^{10}\int_0^{10}\int_{9-0.2y''}^{11+0.2y''} \gamma[x'', y'', z'']^T \, dz''\, dy''\, dx'' \\
&= \frac{\gamma}{m}\int_0^{10}\int_0^{10} [x''(2 + 0.4y''), y''(2 + 0.4y''), 20 + 4y'']^T \, dy''\, dx'' \\
&= \frac{\gamma}{m}\int_0^{10} \left[40x'', \frac{700}{3}, 400\right]^T dx'' \\
&= \frac{\gamma}{m}\left[2000, \frac{7000}{3}, 4000\right]^T
\end{aligned}$$

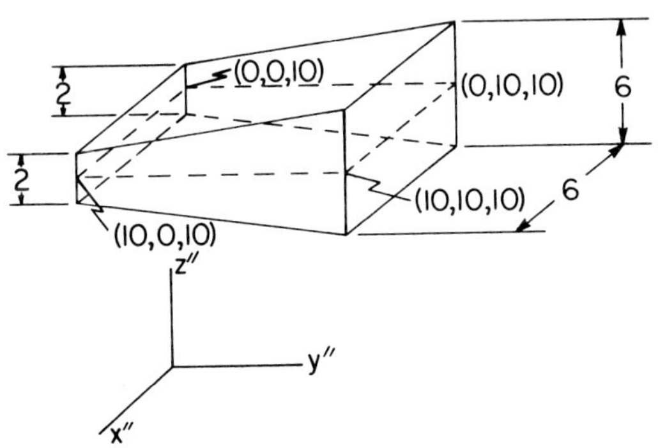

Figure 5.2.2 Homogeneous body.

where γ is mass density of the material. The mass of the body is $m = \gamma 400$. Thus

$$\boldsymbol{\rho}'' = \left[5, \frac{35}{6}, 10\right]^T$$

Consider the rigid body shown in Fig. 5.2.3, for which the location of points and mass distribution are symmetric about a *plane of symmetry* that has normal **n**. Select a body-fixed x''-y''-z'' frame so that its origin is on the plane of symmetry. Thus for points P^+ and P^- and associated differential masses that are symmetrically placed with respect to the plane of symmetry,

$$\begin{aligned} \mathbf{n}^T\mathbf{s}''^{P^-} &= -\mathbf{n}^T\mathbf{s}''^{P^+} \\ dm(P^-) &= dm(P^+) \end{aligned} \tag{5.2.2}$$

Taking the scalar product of both sides of Eq. 5.2.1 with **n** and using Eq. 5.2.2 to change variables of integration yields

$$\begin{aligned} \mathbf{n}^T\boldsymbol{\rho}'' &= \frac{1}{m}\int_m \mathbf{n}^T\mathbf{s}''^P\, dm(P) \\ &= \frac{1}{m}\int_{m^+} \mathbf{n}^T\mathbf{s}''^{P^+}\, dm(P^+) + \frac{1}{m}\int_{m^-} \mathbf{n}^T\mathbf{s}''^{P^-}\, dm(P^-) \\ &= \frac{1}{m}\int_{m^+} \mathbf{n}^T\mathbf{s}''^{P^+}\, dm(P^+) + \frac{1}{m}\int_{m^+} \mathbf{n}^T(-\mathbf{s}'')^{P^+}\, dm(P^+) \\ &= 0 \end{aligned} \tag{5.2.3}$$

Equation 5.2.3 shows that if a body has a plane of symmetry, the centroid lies in that plane. This agrees with the result of Example 5.2.1, in which the planes $x'' = 5$ and $z'' = 10$ are planes of symmetry.

If a body has an *axis of symmetry* (i.e., if every plane that contains the axis of symmetry is a plane of symmetry), the centroid lies on the axis of symmetry. Note, however, that if a body is geometrically symmetric but has asymmetric mass distribution, these helpful geometric results are no longer valid.

Transformation of Inertia Matrix

The inertia matrix with respect to the x''-y''-z'' frame in Fig. 5.2.1 is defined as

$$\mathbf{J}'' = -\int_m \tilde{\mathbf{s}}''^P\tilde{\mathbf{s}}''^P\, dm(P) \tag{5.2.4}$$

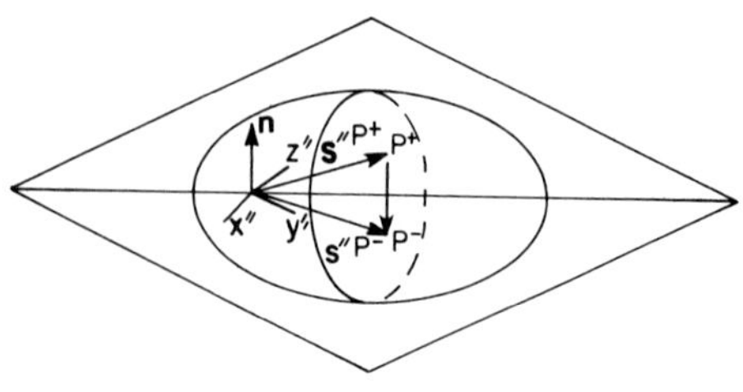

Figure 5.2.3 Body with plane of symmetry.

To obtain a relationship between $\mathbf{J}''$ and the inertia matrix $\mathbf{J}'$ with respect to an x'-y'-z' centroidal frame, as shown in Fig. 5.2.1, substitute $\mathbf{s}''^P = \boldsymbol{\rho}'' + \mathbf{C}^T\mathbf{s}'^P$, where $\mathbf{C}^T$ is the transformation from the x'-y'-z' frame to the x''-y''-z'' frame, into Eq. 5.2.4 to obtain

$$\mathbf{J}'' = -\int_m (\tilde{\boldsymbol{\rho}}'' + \widetilde{\mathbf{C}^T\mathbf{s}'^P})(\tilde{\boldsymbol{\rho}}'' + \widetilde{\mathbf{C}^T\mathbf{s}'^P})\, dm(P)$$

Using Eq. 2.5.3, in the form $\widetilde{\mathbf{C}^T\mathbf{s}'^P} = \mathbf{C}^T\tilde{\mathbf{s}}'^P\mathbf{C}$, and expanding terms, this becomes

$$\mathbf{J}'' = -m\tilde{\boldsymbol{\rho}}''\tilde{\boldsymbol{\rho}}'' - \tilde{\boldsymbol{\rho}}''\mathbf{C}^T \int_m \tilde{\mathbf{s}}'^P\, dm(P)\mathbf{C} - \mathbf{C}^T \int_m \tilde{\mathbf{s}}'^P\, dm(P)\mathbf{C}\tilde{\boldsymbol{\rho}}''$$
$$- \mathbf{C}^T \int_m \tilde{\mathbf{s}}'^P\tilde{\mathbf{s}}'^P\, dm(P)\mathbf{C} \qquad (5.2.5)$$

Since the x'-y'-z' frame is centroidal,

$$\int_m \tilde{\mathbf{s}}'^P\, dm(P) = \left(\widetilde{\int_m \mathbf{s}'^P\, dm(P)}\right) = \mathbf{0}$$

Using this result and Eq. 5.1.17, Eq. 5.2.3 becomes

$$\mathbf{J}'' = \mathbf{C}^T\mathbf{J}'\mathbf{C} - m\tilde{\boldsymbol{\rho}}''\tilde{\boldsymbol{\rho}}'' \qquad (5.2.6)$$

where $\mathbf{C}$ is the orthogonal transformation matrix from the x''-y''-z'' frame to the centroidal x'-y'-z' frame (i.e., $\mathbf{s}' = \mathbf{C}\mathbf{s}''$). Using Eq. 2.5.25, Eq. 5.2.6 may be written in the more conventional form [6, 9, 10, 12]

$$\mathbf{J}'' = \mathbf{C}^T\mathbf{J}'\mathbf{C} + m(\boldsymbol{\rho}''^T\boldsymbol{\rho}''\mathbf{I} - \boldsymbol{\rho}''\boldsymbol{\rho}''^T) \qquad (5.2.7)$$

As a special case, if $\mathbf{C} = \mathbf{I}$ (i.e., if the x'-y'-z' and x''-y''-z'' frames are parallel), moments of inertia on the diagonal of Eq. 5.2.7 are related by

$$\begin{aligned} J_{x''x''} &= J_{x'x'} + m(\rho_{y''}^2 + \rho_{z''}^2) \\ J_{y''y''} &= J_{y'y'} + m(\rho_{x''}^2 + \rho_{z''}^2) \\ J_{z''z''} &= J_{z'z'} + m(\rho_{x''}^2 + \rho_{y''}^2) \end{aligned} \qquad (5.2.8)$$

Thus moments of inertia with respect to the noncentroidal x''-y''-z'' frame are those with respect to the parallel centroidal x'-y'-z' frame plus the mass of the body times the square of the distances between the respective parallel axes. This is called the *parallel axis theorem*. The products of inertia for parallel axes are obtained from off-diagonal terms in Eq. 5.2.7 as

$$\begin{aligned} J_{x''y''} &= J_{x'y'} - m\rho_{x''}\rho_{y''} \\ J_{x''z''} &= J_{x'z'} - m\rho_{x''}\rho_{z''} \\ J_{y''z''} &= J_{y'z'} - m\rho_{y''}\rho_{z''} \end{aligned} \qquad (5.2.9)$$

Consider a special case in which one of the planes of a x''-y''-z'' frame is a plane of symmetry (i.e., it has both geometric and mass distribution symmetry).

If the x''-y'' plane is a plane of symmetry, then using the symmetry argument employed in deriving Eq. 5.2.3 (Prob. 5.2.2),

$$J_{z''x''} = J_{z''y''} = 0 \tag{5.2.10}$$

Similarly, if the x''-z'' plane is a plane of symmetry,

$$J_{x''y''} = J_{z''y''} = 0 \tag{5.2.11}$$

and if the y''-z'' plane is a plane of symmetry,

$$J_{x''y''} = J_{x''z''} = 0 \tag{5.2.12}$$

If any pair of coordinate planes of the x''-y''-z'' frame are planes of symmetry, then from Eqs. 5.2.10 to 5.2.12, all the products of inertia are zero (Prob. 5.2.3). This is a very helpful property, since bodies in mechanical systems often have two planes of symmetry. A common case of pairs of orthogonal planes of symmetry is a homogeneous body of revolution about one of the axes of a body-fixed reference frame.

Example 5.2.2: The moments and products of inertia of the body of Fig. 5.2.2 with respect to a centroidal x'-y'-z' frame that is parallel to the x''-y''-z'' frame can be calculated using the parallel axis theorem. Using the results of Example 5.2.1 in Eq. 5.2.8,

$$J_{x'x'} = J_{x''x''} - 400\gamma\left[\left(\frac{35}{6}\right)^2 + 100\right]$$

$$J_{y'y'} = J_{y''y''} - 400\gamma\,[25 + 100]$$

$$J_{z'z'} + J_{z''z''} - 400\gamma\left[25 + \left(\frac{35}{6}\right)^2\right]$$

which may be evaluated using the results of Problem 5.2.3. Since the x'-y' and z'-y' planes are planes of symmetry,

$$J_{x'y'} = J_{x'z'} = J_{y'z'} = 0$$

Principal Axes

Even if a body is not symmetric, it is possible to find a centroidal body-fixed frame with respect to which the products of inertia are all zero. Let the x''-y''-z'' and x'-y'-z' body-fixed frames have their origins at the centroid of a body. Then $\boldsymbol{\rho}'' = \mathbf{0}$, and presume that $\mathbf{J}''$ is known, in general as a full, symmetric matrix. Multiplying both sides of Eq. 5.2.7 on the left by $\mathbf{C}$ and on the right by $\mathbf{C}^T$ gives

$$\mathbf{J}' = \mathbf{C}\mathbf{J}''\mathbf{C}^T \tag{5.2.13}$$

The task is now to find an orthogonal transformation matrix $\mathbf{C}$ for which $\mathbf{J}'$ is diagonal.

From the definition of $\mathbf{J}''$ (Eq. 5.2.4), for any vector $\mathbf{a}$ in R^3,

$$\begin{aligned}\mathbf{a}^T\mathbf{J}''\mathbf{a} &= -\int_m \mathbf{a}^T\tilde{\mathbf{s}}''^P\tilde{\mathbf{s}}''^P\mathbf{a}\, dm \\ &= \int_m \mathbf{a}^T\tilde{\mathbf{s}}''^{PT}\tilde{\mathbf{s}}''^P\mathbf{a}\, dm \\ &= \int_m (\tilde{\mathbf{s}}''^P\mathbf{a})^T(\tilde{\mathbf{s}}''^P\mathbf{a})\, dm \geq 0\end{aligned} \tag{5.2.14}$$

Thus $\mathbf{J}''$ is *positive semidefinite* [2]. In fact, if mass density is nowhere zero in the body, then $\mathbf{a}^T\mathbf{J}''\mathbf{a} = 0$ implies that $\tilde{\mathbf{s}}''\mathbf{a} = \mathbf{0} = -\tilde{\mathbf{a}}\mathbf{s}''$, for all $\mathbf{s}''$ in the body. If there are three linearly independent vectors $\mathbf{s}_i''$, $i = 1, 2, 3$, to points in the body, then

$$\tilde{\mathbf{a}}[\mathbf{s}_1'', \mathbf{s}_2'', \mathbf{s}_3''] = \mathbf{0}$$

and since $[\mathbf{s}_1'', \mathbf{s}_2'', \mathbf{s}_3'']$ is nonsingular, $\tilde{\mathbf{a}} = \mathbf{0}$. Hence $\mathbf{a} = \mathbf{0}$. In this case, $\mathbf{J}''$ is *positive definite* [2]. This is the case for any homogeneous body whose volume is not zero.

The 3×3 positive semidefinite matrix $\mathbf{J}''$ has three orthonormal *eigenvectors* [2, 9, 10], denoted $\mathbf{f}''$, $\mathbf{g}''$, and $\mathbf{h}''$ in the x''-y''-z'' frame, with corresponding nonnegative eigenvalues ζ_i, $i = 1, 2, 3$, ordered such that $\zeta_1 \geq \zeta_2 \geq \zeta_3 \geq 0$, where

$$\begin{aligned}\mathbf{J}''\mathbf{f}'' &= \zeta_1\mathbf{f}'' \\ \mathbf{J}''\mathbf{g}'' &= \zeta_2\mathbf{g}'' \\ \mathbf{J}''\mathbf{h}'' &= \zeta_3\mathbf{h}''\end{aligned} \tag{5.2.15}$$

Since $-\mathbf{g}''$ and $-\mathbf{h}''$ are also eigenvectors, signs of the eigenvectors can be arranged so that $\mathbf{f}''$, $\mathbf{g}''$, and $\mathbf{h}''$ are unit coordinate vectors for a right-hand x'-y'-z' Cartesian frame. The transformation matrix $\mathbf{C}^T$ from the x'-y'-z' frame to the x''-y''-z'' frame is, by Eq. 2.5.13,

$$\mathbf{C}^T = [\mathbf{f}'', \mathbf{g}'', \mathbf{h}''] \tag{5.2.16}$$

Substituting $\mathbf{C}^T$ from Eq. 5.2.16 into Eq. 5.2.13 yields

$$\begin{aligned}\mathbf{J}' &= \mathbf{C}\mathbf{J}''\mathbf{C}^T \\ &= [\mathbf{f}'', \mathbf{g}'', \mathbf{h}'']^T\mathbf{J}''[\mathbf{f}'', \mathbf{g}'', \mathbf{h}''] \\ &= [\mathbf{f}'', \mathbf{g}'', \mathbf{h}'']^T[\mathbf{J}''\mathbf{f}'', \mathbf{J}''\mathbf{g}'', \mathbf{J}''\mathbf{h}''] \\ &= [\mathbf{f}'', \mathbf{g}'', \mathbf{h}'']^T[\zeta_1\mathbf{f}'', \zeta_2\mathbf{g}'', \zeta_3\mathbf{h}''] \\ &= \begin{bmatrix}\zeta_1 & 0 & 0 \\ 0 & \zeta_2 & 0 \\ 0 & 0 & \zeta_3\end{bmatrix}\end{aligned} \tag{5.2.17}$$

Thus, in the x'-y'-z' frame that is defined by $\mathbf{f}''$, $\mathbf{g}''$, and $\mathbf{h}''$, $J_{x'x'} = \xi_1$, $J_{y'y'} = \zeta_2$, $J_{z'z'} = \zeta_3$, and $J_{x'y'} = J_{x'y'} = J_{y'z'} = 0$. The axes of such a centroidal frame are called *principal axes* and the associated moments of inertia are called *principal moments of inertia*. By construction, the products of inertia with respect to principal axes are zero.

Example 5.2.3: The triangular plate shown in Fig. 5.2.4 has negligible thickness, so the mass is considered concentrated in the x-y plane, with mass density γ per unit area.

Evaluating the integrals in Eq. 5.1.17, in the x-y-z frame yields

$$J_{xx} = \gamma \int_0^1 \int_0^{1-x} y^2 \, dy \, dx$$

$$= \gamma \int_0^1 \frac{(1-x)^3}{3dx}$$

$$= \frac{\gamma}{12}$$

$$J_{yy} = J_{xx} = \frac{\gamma}{12}$$

$$J_{zz} = \gamma \int_0^1 \int_0^{1-x} (x^2 + y^2) \, dy \, dx$$

$$= \gamma \int_0^1 \left[x^2(1-x) + \frac{(1-x)^3}{3} \right] dx$$

$$= \frac{\gamma}{6}$$

$$J_{xy} = -\gamma \int_0^1 \int_0^{1-x} xy \, dy \, dx$$

$$= -\gamma \int_0^1 \frac{x(1-x)^2}{2dx}$$

$$= \frac{-\gamma}{24}$$

$$J_{xz} = J_{yz} = 0$$

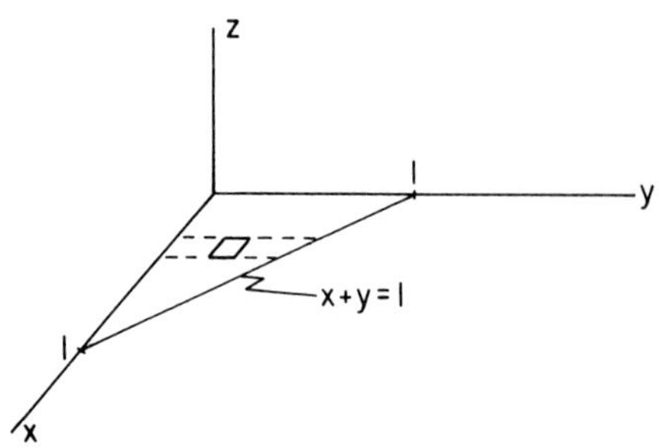

Figure 5.2.4 Triangular plate in x-y plane.

Thus the inertia matrix in the x-y-z frame is

$$\mathbf{J} = \frac{\gamma}{24}\begin{bmatrix} 2 & -1 & 0 \\ -1 & 2 & 0 \\ 0 & 0 & 4 \end{bmatrix}$$

The centroid is located by Eq. 5.2.1, with $m = \gamma/2$,

$$\begin{aligned}\boldsymbol{\rho} &= \frac{2}{\gamma}\int_0^1 \int_0^{1-x} [x, y, 0]^T \gamma \, dy \, dx \\ &= [\tfrac{1}{3}, \tfrac{1}{3}, 0]^T\end{aligned}$$

For a centroidal x''-y''-z'' frame that is parallel to the x-y-z noncentroidal frame shown in Fig. 5.2.5, the parallel axis theorem applies and, from Eqs. 5.2.8 and 5.2.9,

$$\begin{aligned}\mathbf{J}'' &= \mathbf{J} - m\begin{bmatrix} \rho_y^2 + \rho_z^2 & -\rho_x\rho_y & -\rho_x\rho_z \\ -\rho_x\rho_y & \rho_x^2 + \rho_z^2 & -\rho_y\rho_z \\ -\rho_x\rho_z & -\rho_y\rho_z & \rho_x^2 + \rho_y^2 \end{bmatrix} \\ &= \frac{\gamma}{72}\begin{bmatrix} 2 & 1 & 0 \\ 1 & 2 & 0 \\ 0 & 0 & 4 \end{bmatrix}\end{aligned}$$

The eigenvalues of $\mathbf{J}''$ are obtained by solving the characteristic equation [2],

$$\begin{aligned}|\mathbf{J}'' - \zeta\mathbf{I}| &= \left|\begin{bmatrix} \frac{\gamma}{36} - \zeta & \frac{\gamma}{72} & 0 \\ \frac{\gamma}{72} & \frac{\gamma}{36} - \zeta & 0 \\ 0 & 0 & \frac{\gamma}{18} - \zeta \end{bmatrix}\right| \\ &= \left(\frac{\gamma}{36} - \zeta\right)^2\left(\frac{\gamma}{18} - \zeta\right) - \left(\frac{\gamma}{18} - \zeta\right)\left(\frac{\gamma}{72}\right)^2 \\ &= \left[\left(\frac{\gamma}{36} - \zeta\right)^2 - \left(\frac{\gamma}{72}\right)^2\right]\left(\frac{\gamma}{18} - \zeta\right) = 0\end{aligned}$$

whose solutions are

$$\zeta_1 = \frac{\gamma}{18}, \qquad \zeta_2 = \frac{\gamma}{24}, \qquad \zeta_3 = \frac{\gamma}{72}$$

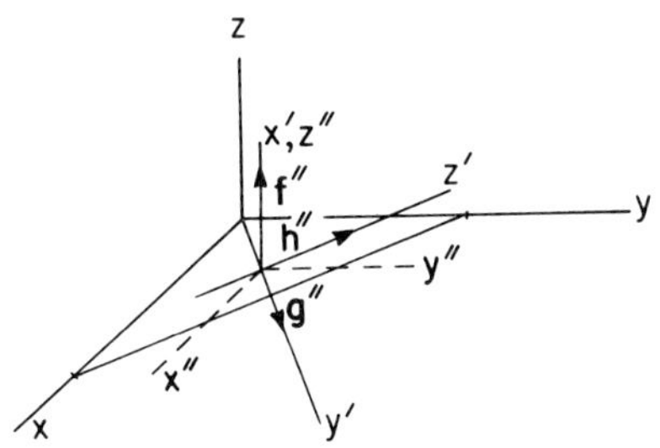

Figure 5.2.5 Principal axes for triangular plate.

Substituting these values of ζ into Eq. 5.2.15 yields the unit eigenvectors

$$\mathbf{f}'' = [0,\ 0,\ 1]^T$$

$$\mathbf{g}'' = \left[\frac{1}{\sqrt{2}},\ \frac{1}{\sqrt{2}},\ 0\right]^T$$

$$\mathbf{h}'' = \left[\frac{-1}{\sqrt{2}},\ \frac{1}{\sqrt{2}},\ 0\right]^T$$

Principal axes are shown in Fig. 5.2.5, in which the inertia matrix is

$$\mathbf{J}' = \begin{bmatrix} \frac{\gamma}{18} & 0 & 0 \\ 0 & \frac{\gamma}{24} & 0 \\ 0 & 0 & \frac{\gamma}{72} \end{bmatrix}$$

Inertia Properties of Composite Bodies

Components of machines are often made up of combinations of subcomponents that have standard shapes such as circles, disks, cylinders, spheres, rectangles, cubes, and so on, A typical example of such a *composite body* is shown in Fig. 5.2.6, in which all subcomponents or voids have shapes that are typical of those that result from standard manufacturing processes. The objective of this subsection is to develop expressions for inertia properties of the composite body, using easily calculated properties of individual subcomponents.

Let a noncentroidal x''-y''-z'' frame be fixed to the composite body in a convenient location (e.g, as shown in Fig. 5.2.6). Using this frame, the centroid of

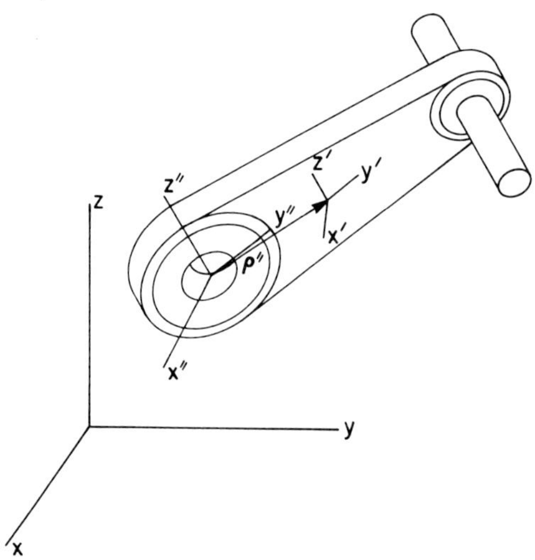

Figure 5.2.6 Composite body made up of subcomponents.

the composite body may be obtained using the definition of Eq. 5.2.1, employing the property that an integral over the composite body may be written as the sum of integrals over subcomponents m_i, to obtain

$$\begin{aligned} \boldsymbol{\rho}'' &= \frac{1}{m} \sum_{i=1}^{k} \int_{m_i} \mathbf{s}''^P \, dm_i(P) \\ &= \frac{1}{m} \sum_{i=1}^{k} m_i \boldsymbol{\rho}_i'' \end{aligned} \tag{5.2.18}$$

where $m = \sum_{i=1}^{k} m_i$. To use this result, centroids of subcomponents are first located in the x''-y''-z'' frame by vectors $\boldsymbol{\rho}_i''$, using mass and centroid location formation such as that found in Table 5.2.1 or from direct numerical calculation. Equation 5.2.18 is then used to locate the centroid of the composite body, in the x''-y''-z'' frame.

To calculate the inertia matrix $\mathbf{J}'$ with respect to the centroidal x'-y'-z' frame selected (e.g., as shown in Fig. 5.2.6), the defining equation of Eq. 5.1.17 may be written and its integral evaluated as a sum of integrals over subcomponents m_i that make up the composite body, to obtain

$$\begin{aligned} \mathbf{J}' &= -\int_{m} \tilde{\mathbf{s}}'^P \tilde{\mathbf{s}}'^P \, dm(P) \\ &= \sum_{i=1}^{k} \left[-\int_{m_i} \tilde{\mathbf{s}}_i'^P \tilde{\mathbf{s}}_i'^P \, dm_i(P) \right] \\ &= \sum_{i=1}^{k} \mathbf{J}_i' \end{aligned} \tag{5.2.19}$$

To use Eq. 5.2.19, the inertia matrix $\mathbf{J}_i'$ of subcomponent i in the x'-y'-z' frame must be calculated. To do this, a convenient x_i^*-y_i^*-z_i^* centroidal frame is defined for each subcomponent in the composite body. Denote by $\mathbf{J}_i^*$ the inertia matrix of subcomponent i, with respect to its selected centroidal frame and by $\boldsymbol{\rho}_i^{*\prime}$ the vector that locates the centroid of the subcomponent in the x'-y'-z' frame. Since the x'-y'-z' frame is not centroidal for the subcomponents, the inertia transformation of Eq. 5.2.7 must be used to calculate the inertia matrix of subcomponent i with respect to the composite body centroidal x'-y'-z' frame. Denoting by $\mathbf{C}_i$ the transformation matrix from the x'-y'-z' frame to the x_i^*-y_i^*-z_i^* frame, Eq. 5.2.7 yields

$$\mathbf{J}_i' = \mathbf{C}_i^T \mathbf{J}_i^* \mathbf{C}_i + m_i \left(\boldsymbol{\rho}_i^{*\prime T} \boldsymbol{\rho}_i^{*\prime} \mathbf{I} - \boldsymbol{\rho}_i^{*\prime} \boldsymbol{\rho}_i^{*\prime T} \right) \tag{5.2.20}$$

This result may be substituted into Eq. 5.2.19 to obtain the desired inertia matrix for the composite body. Inertia properties of a few homogeneous bodies, for use in such calculations, are provided in Table 5.2.1.

TABLE 5.2.1 INERTIA PROPERTIES OF SOME HOMOGENEOUS BODIES

Body	Mass and moment of inertia[a]
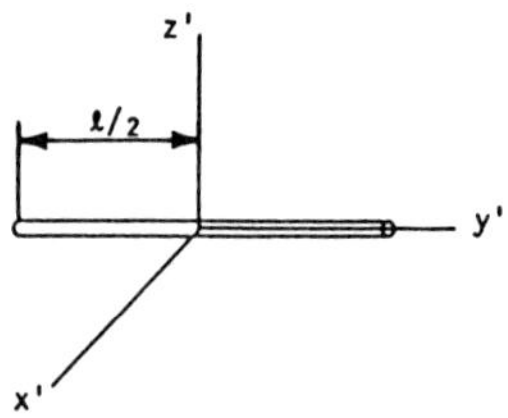 Thin Rod	$m = \gamma l A$ $J_{x'x'} = J_{z'z'} = \dfrac{m}{12} l^2$ $J_{y'y'} = 0$
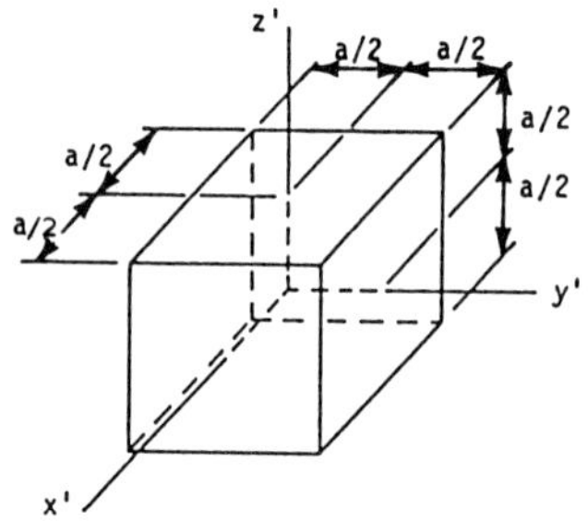 Cube	$m = \gamma a^3$ $J_{x'x'} = J_{y'y'} = J_{z'z'} = \frac{1}{6} m a^2$
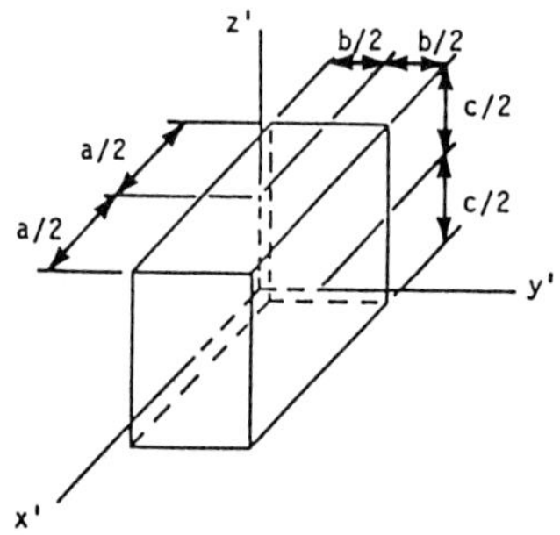 Rectangular Prism	$m = \gamma abc$ $J_{x'x'} = \frac{1}{12} m(b^2 + c^2)$ $J_{y'y'} = \frac{1}{12} m(a^2 + c^2)$ $J_{z'z'} = \frac{1}{12} m(a^2 + b^2)$
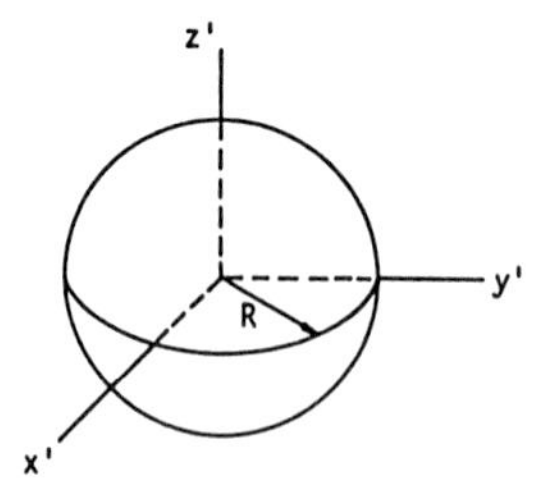 Sphere	$m = \frac{4}{3}\pi\gamma R^2$ $J_{x'x'} = J_{y'y'} = J_{z'z'} = \frac{2}{5} m R^2$

Body	Mass and moment of inertia[a]
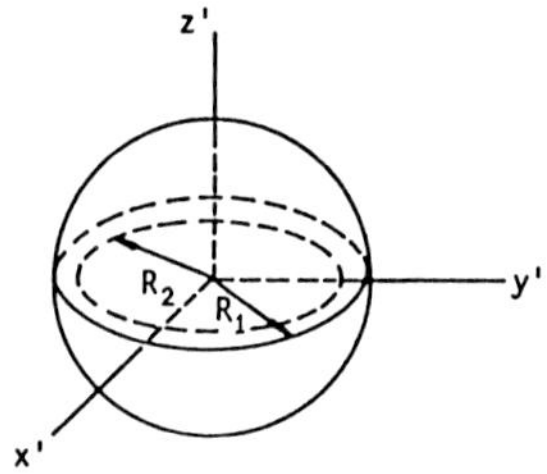 Hollow Sphere	$m = \frac{4}{3}\pi\gamma(R_1^2 - R_2^2)$ $J_{x'x'} = J_{y'y'} = J_{z'z'} = \frac{2}{5}m\,\frac{R_1^5 - R_2^5}{R_1^3\,R_2^3}$
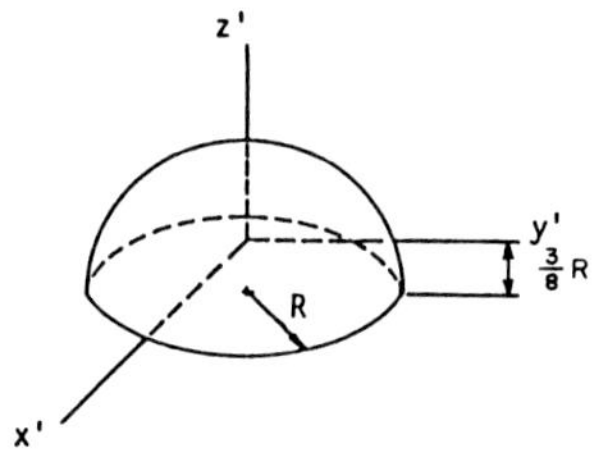 Hemisphere	$m = \frac{2}{3}\pi\gamma R^3$ $J_{x'x'} = J_{y'y'} = \frac{83}{320}mR^2$ $J_{z'z'} = \frac{2}{5}mR^2$
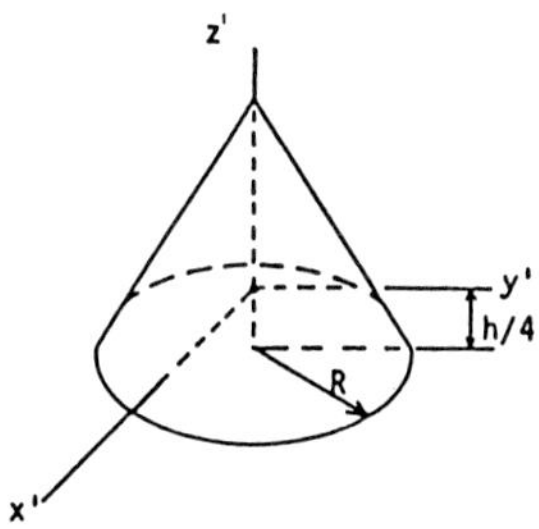 Right Circular Cone	$m = \frac{1}{3}\pi\gamma R^2 h$ $J_{x'x'} = J_{y'y'} = \frac{3}{80}m(4R^2 + h^2)$ $J_{z'z'} = \frac{3}{10}mR^2$
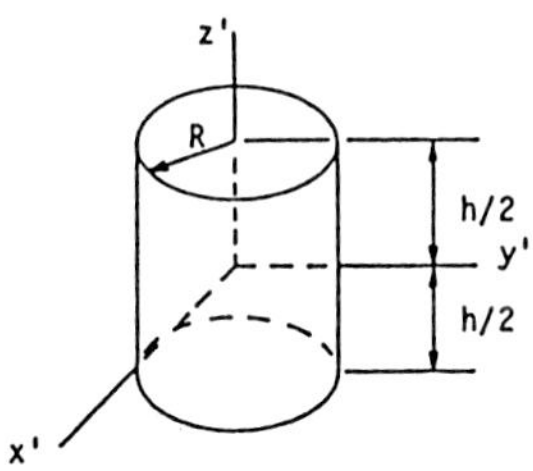 Right Circular Cylinder	$m = \pi\gamma R^2 h$ $J_{x'x'} = J_{y'y'} = \frac{1}{12}m(3R^2 + h^2)$ $J_{z'z'} = \frac{1}{2}mR^2$
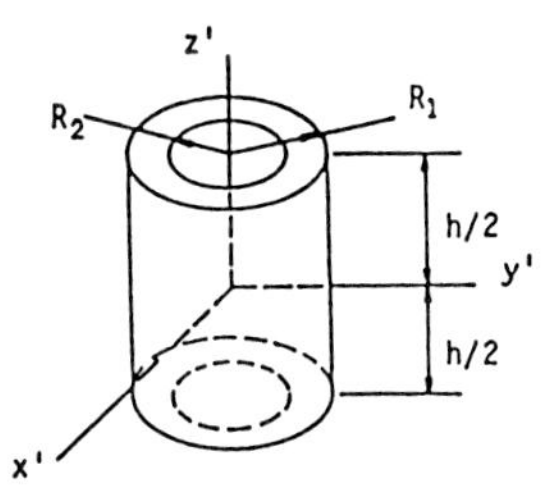 Hollow Right Circular Cylinder	$m = \pi\gamma h(R_1^2 - R_2^2)$ $J_{x'x'} = J_{y'y'} = \frac{1}{12}m(3R_1^2 + 3R_2^2 + h^2)$ $J_{z'z'} = \frac{1}{2}m(R_1^2 + R_2^2)$

[a] γ, mass density; A, cross-sectional area.

Example 5.2.4: The composite body of Fig. 5.2.7 is made up of a cylinder with hemispherical ends, all made of the same homogeneous material of density γ. Since all three coordinate planes are planes of symmetry, the centroid is at the origin of the x-y-z frame and all products of inertia with respect to this frame are zero. Using Eqs. 5.2.19 and 5.2.20, with $\mathbf{C}_i = \mathbf{I}$, and Table 5.2.1, the moments of inertia of the composite body are

$$J_{x'x'} = J_{z'z'} = \frac{\pi\gamma R^2 L}{12}(3R^2 + L^2) + 2\left[\frac{83\pi\gamma R^5}{480} + \frac{2\pi\gamma R^3}{3}\left(\frac{4L + 3R}{8}\right)^2\right]$$

$$J_{y'y'} = \frac{\pi\gamma R^4 L}{2} + 2\left(\frac{4\pi\gamma R^5}{15}\right)$$

If there are *voids* in a composite body, Eq. 5.2.19 may be used by assigning $-\mathbf{J}_i'$ as the inertia matrix of the void, where $\mathbf{J}_i'$ is the inertia matrix of a body that would occupy the void, with the same material density as the body from which it is removed.

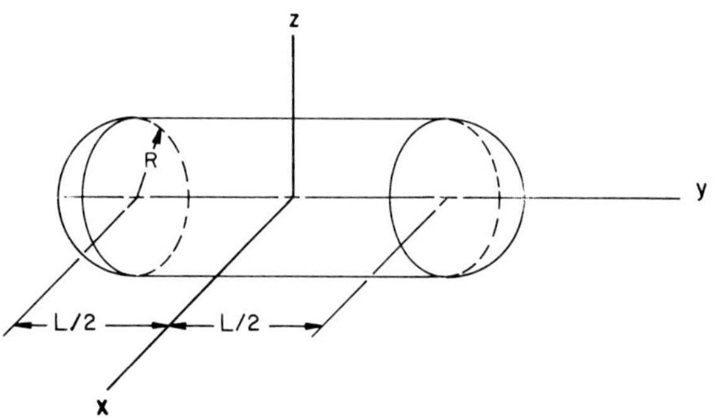

Figure 5.2.7 Cylinder with spherical ends.

Example 5.2.5: The composite body of Fig. 5.2.8 is formed by a rectangular bar with a hole of radius $R = c/2$ about the x axis. Since all three coordinate planes are planes of symmetry, the centroid is at the origin of the x-y-z frame and the products of inertia with respect to this frame are zero.

From Eq. 5.2.20, with $\mathbf{J}_1$ as the inertia matrix of the bar without a hole and $\mathbf{J}_2$ as the inertia matrix of a cylinder that would occupy the hole,

$$\mathbf{J} = \mathbf{J}_1 - \mathbf{J}_2$$

Using Table 5.2.1,

$$J_{x'x'} = \frac{\gamma abc}{12}(b^2 + c^2) - \frac{\pi\gamma c^4 a}{32}$$

$$J_{y'y'} = \frac{\gamma abc}{12}(a^2 + c^2) - \frac{\pi\gamma c^2 a}{48}\left(\frac{3c^2 + a^2}{4}\right)$$

$$J_{z'z'} = \frac{\gamma abc}{12}(a^2 + b^2) - \frac{\pi\gamma c^2 a}{48}\left(\frac{3c^2 + a^2}{4}\right)$$

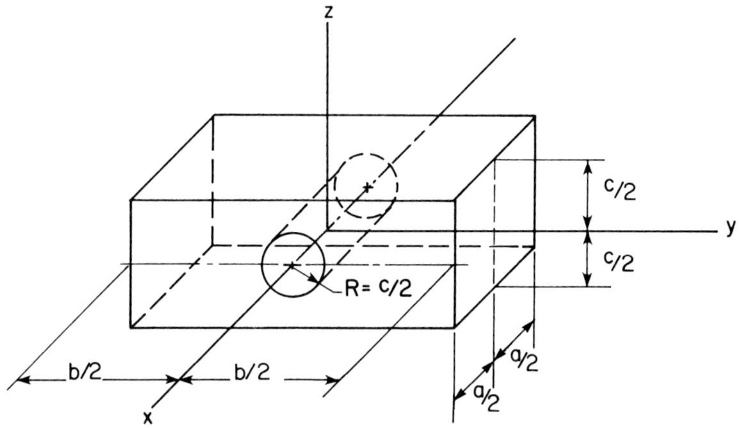

Figure 5.2.8 Rectangular solid with hole.

5.3 MOMENTUM AND KINETIC ENERGY OF RIGID BODIES

Linear Momentum and Impulse

The *linear momentum of a rigid body* is defined as the product of the mass of the body and the velocity vector of the center of mass:

$$\mathbf{P} = m\dot{\mathbf{r}} \tag{5.3.1}$$

For a body with constant mass, taking the time derivative of Eq. 5.3.1 and using the first of Eqs. 5.1.20, the translational equations of motion of Eq. 5.1.20 may be written as

$$\dot{\mathbf{P}} = m\ddot{\mathbf{r}} = \mathbf{F} \tag{5.3.2}$$

Note that if the resultant force $\mathbf{F}$ on the body is zero, its linear momentum remains constant. More generally, integrating both sides of Eq. 5.3.2 with respect to time from t_1 to t_2 yields

$$\int_{t_1}^{t_2} \dot{\mathbf{P}}\, dt = \int_{t_1}^{t_2} \mathbf{F}\, dt \equiv \overline{\mathbf{F}} \tag{5.3.3}$$

where $\overline{\mathbf{F}}$ is defined as the *linear impulse* of the force $\mathbf{F}$ from time t_1 to time t_2. From the fundamental theorem of calculus, the integral on the left of Eq. 5.3.3 is simply the change in linear momentum, leading to the *linear impulse–momentum relation*

$$\mathbf{P}(t_2) - \mathbf{P}(t_1) = \overline{\mathbf{F}} \tag{5.3.4}$$

Thus if the time integral of the resultant force that acts on a body over a period of time is used to calculate its linear impulse, the change in momentum of the body may be determined immediately.

Note that the linear impulse–momentum relation of Eq. 5.3.4 for rigid bodies is identical to that for particles derived in Section 1.4. The problem of impact of billiard balls studied in Example 1.4.1 could thus have been carried out in a rigid body setting.

Angular Momentum

The *angular momentum of a particle P* about the origin of an inertial x-y-z reference frame, is defined as the vector product of the vector from the origin of the reference frame to the particle P and the linear momentum of the particle. This definition can be extended to a rigid body by first calculating the angular momentum of a differential mass at point P in the body and subsequently integrating over all differential elements of mass in the body.

Using the geometry defined in Fig. 5.3.1, the *angular momentum of a rigid body* about the origin of the x-y-z inertial reference frame is defined and simplified as

$$
\begin{aligned}
\mathbf{H} &= \int_m \tilde{\mathbf{r}}^P \dot{\mathbf{r}}^P \, dm(P) \\
&= \int_m \widetilde{(\mathbf{r} + \mathbf{A}\mathbf{s}'^P)}(\dot{\mathbf{r}} + \mathbf{A}\tilde{\boldsymbol{\omega}}'\mathbf{s}'^P) \, dm(P) \\
&= m\tilde{\mathbf{r}}\dot{\mathbf{r}} + \mathbf{A} \int_m \tilde{\mathbf{s}}'^P \tilde{\boldsymbol{\omega}}'\mathbf{s}'^P \, dm(P) \\
&= m\tilde{\mathbf{r}}\dot{\mathbf{r}} + \mathbf{A}\mathbf{J}'\boldsymbol{\omega}'
\end{aligned}
\tag{5.3.5}
$$

where the fact that the x'-y'-z' frame is centroidal and the definition of the inertia matrix is used. Note that this relation may be interpreted as the angular momentum of the total mass m of the body, as if it were concentrated at the centroid, plus the body-fixed vector $\mathbf{J}'\boldsymbol{\omega}'$, transformed to the global reference frame. Thus $\mathbf{J}'\boldsymbol{\omega}'$ is the angular momentum of the body with respect to the centroidal body-fixed x'-y'-z' frame.

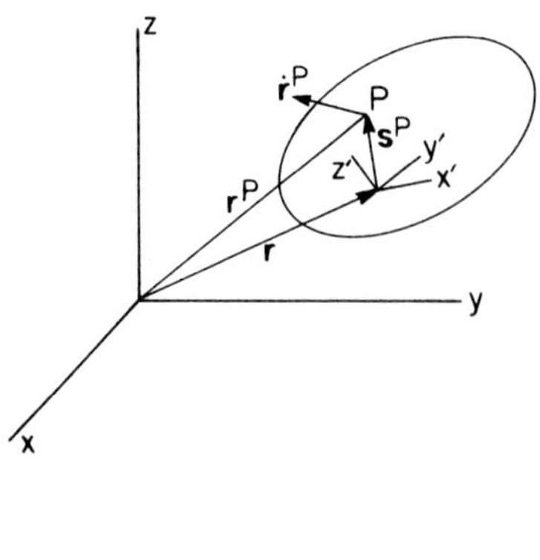

Figure 5.3.1 Rigid body for angular momentum calculation.

Taking the time derivative of angular momentum in Eq. 5.3.5 and simplifying yields

$$\begin{aligned}\dot{\mathbf{H}} &= m\dot{\tilde{\mathbf{r}}}\dot{\mathbf{r}} + m\tilde{\mathbf{r}}\ddot{\mathbf{r}} + \dot{\mathbf{A}}\mathbf{J}'\boldsymbol{\omega}' + \mathbf{A}\mathbf{J}'\dot{\boldsymbol{\omega}}' \\ &= m\tilde{\mathbf{r}}\ddot{\mathbf{r}} + \mathbf{A}(\mathbf{J}'\dot{\boldsymbol{\omega}}' + \tilde{\boldsymbol{\omega}}'\mathbf{J}'\boldsymbol{\omega}')\end{aligned} \tag{5.3.6}$$

Using the Newton–Euler equations of motion of Eq. 5.1.20, the right side of Eq. 5.3.6 may be written as

$$\dot{\mathbf{H}} = \tilde{r}\mathbf{F} + \mathbf{A}\mathbf{n}' \equiv \mathbf{N} \tag{5.3.7}$$

where $\mathbf{N}$ is the torque due to the resultant of all forces and torques that act on the rigid body, with respect to the inertial x-y-z reference frame.

The linear and angular momentum relations of Eqs. 5.3.2 and 5.3.7 may be regarded as equivalent to the Newton–Euler equations of motion of Eq. 5.1.20. Indeed, linear and angular momentum concepts have conventionally been used in the literature for development of the equations of motion of a rigid body [6, 8, 9, 10]. The more direct approach adopted in this book uses Newton's equations of motion for particles to arrive more directly at a concise set of equations of motion for rigid bodies, without the need for reliance on derived "principles" involving momentum.

As a final note regarding angular momentum, Eq. 5.3.7 may be integrated over an interval of time, beginning at t_1 and ending at t_2, to obtain

$$\int_{t_1}^{t_2} \dot{\mathbf{H}}\, dt = \int_{t_1}^{t_2} \mathbf{N}\, dt \equiv \overline{\mathbf{N}} \tag{5.3.8}$$

where $\overline{\mathbf{N}}$ is the *angular impulse* of torque $\mathbf{N}$ over the time interval t_1 to t_2. Using the fundamental theorem of calculus, the integral on the left of Eq. 5.3.8 is simply the change in angular momentum over the time interval, leading to the *angular impulse–momentum relation*

$$\mathbf{H}(t_2) - \mathbf{H}(t_1) = \overline{\mathbf{N}} \tag{5.3.9}$$

Certainly, if $\mathbf{N}$, hence $\overline{\mathbf{N}}$, is zero, then angular momentum does not change. In a more general case, if a known torque $\mathbf{N}(t)$ is applied to a body so that the angular impulse $\mathbf{N}$ can be computed, then Eq. 5.3.9 directly provides information regarding the change in angular momentum over the time interval of concern.

Example 5.3.1: Consider the problem of a baseball bat striking a ball, as shown in Fig. 5.3.2(a). If the velocity of the hands is small relative to the speed of the ball, the pendulum model of the system shown in Fig. 5.3.2(b) may yield useful information.

Let the angular velocity of the bat about the z axis prior to striking the ball be ω_0 and let the velocity of the ball be $-v_0$ in the y direction. The collision occurs at distance s along the x axis. The centroid of the bat is located at ρ along the x axis. The mass of the ball is m_2 and the mass and centroidal moment of inertia about the z axis of the bat are m_1 and $J_{z'}$. Finally, the coefficient of restitution between the bat and ball is e.

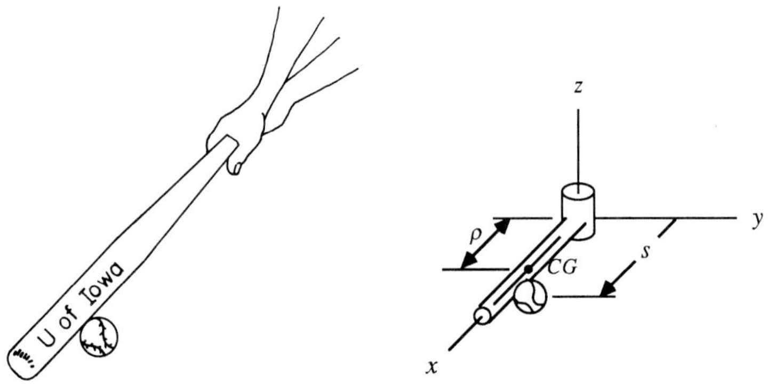

(a) (b)

Figure 5.3.2 Bat striking ball.

Modeling the force of contact between the ball and bat as $f(t)$, $t_1 \le t \le t_2$, in the y direction, Eq. 5.3.9 may be written for the bat and the ball, respectively, as

$$(m_1\rho^2 + J_{z'})(\omega(t_2) - \omega_0) = -\int_{t_1}^{t_2} sf(t)\, dt \equiv -s\bar{f} \tag{5.3.10}$$

$$m_2 s^2 \left(\frac{v(t_2)}{s} + \frac{v_0}{s}\right) = \int_{t_1}^{t_2} sf(t)\, dt \equiv s\bar{f} \tag{5.3.11}$$

Using the definition of coefficient of restitution yields

$$s\omega(t_2) - v(t_2) = -e(s\omega_0 + v_0) \tag{5.3.12}$$

Eliminating $\bar{f}$ from Eqs. 5.3.10 to 5.3.12 and solving for $\omega(t_2)$ and $v(t_2)$ yields

$$\omega(t_2) = (m_1\rho^2 + J_{z'} - em_2s^2)\omega_0 - (m_2s + em_2s)v_0\ m_1\rho^2 + J_{z'} + m_2s^2 \tag{5.3.13}$$

$$v(t_2) = s\omega(t_2) + e(s\omega_0 + v_0) \tag{5.3.14}$$

Let $h(t)$ be the reaction force at the hands (the origin of the x-y-z frame) on the bat, in the positive y direction. Applying the linear impulse–momentum relation of Eq. 5.3.4 to the bat yields

$$\begin{aligned} m_1[\rho\omega(t_2) - \rho\omega_0] &= -\int_{t_1}^{t_2} f(t)\, dt + \int_{t_1}^{t_2} h(t)\, dt \\ &\equiv -\bar{f} + \bar{h} \end{aligned} \tag{5.3.15}$$

Using Eq. 5.3.10 and 5.3.15 yields

$$\bar{h} = \frac{1}{s}[m_1\rho(s - \rho) - J_{z'}](\omega(t_2) - \omega_0) \tag{5.3.16}$$

Since $\bar{h}$ is the impulse of the force exerted by the hands on the bat, the hitter is most comfortable when $\bar{h} = 0$; that is, when he adjusts his swing so that the bat strikes the ball at

$$s = \rho + \frac{J_{z'}}{m_1\rho}$$

This is called the "sweet spot" on the bat. Note that if all the mass of the bat were concentrated at its centroid, $J_{z'} = 0$ and the sweet spot is at $s = \rho$ (i.e., particle striking particle), which is reasonable.

Kinetic Energy

The kinetic energy of a particle of mass m is $\frac{1}{2}mv^2$, where v is the magnitude of velocity of the particle. The kinetic energy of a differential mass located at point P, using the centroidal x'-y'-z' frame of Fig. 5.3.1, may be integrated over the body to obtain the *kinetic energy of a rigid body* as

$$\begin{aligned} T &\equiv \frac{1}{2}\int_m \dot{\mathbf{r}}^{P^T}\dot{\mathbf{r}}^P \, dm(P) \\ &= \frac{1}{2}\int_m (\dot{\mathbf{r}} + \mathbf{A}\tilde{\boldsymbol{\omega}}'\mathbf{s}'^P)^T(\dot{\mathbf{r}} + \mathbf{A}\tilde{\boldsymbol{\omega}}'\mathbf{s}'^P) \, dm(P) \\ &= \tfrac{1}{2}m\dot{\mathbf{r}}^T\dot{\mathbf{r}} - \tfrac{1}{2}\boldsymbol{\omega}'^T \int_m \tilde{\mathbf{s}}'^P\tilde{\mathbf{s}}'^P \, dm(P)\boldsymbol{\omega}' \\ &= \tfrac{1}{2}m\dot{\mathbf{r}}^T\dot{\mathbf{r}} + \tfrac{1}{2}\boldsymbol{\omega}'^T\mathbf{J}'\boldsymbol{\omega}' \end{aligned} \tag{5.3.17}$$

where the integrand has been expanded and simplified using identities of Chapter 2 and the centroidal inertia matrix has been employed. Note, as in the case of angular momentum, that the inertia matrix plays a natural role in characterizing kinetic energy of a rigid body. The first term in Eq. 5.3.17 might be viewed as the *translational kinetic energy* of the total mass of the body, as if it were concentrated at its center of mass. The second term is then the *rotational kinetic energy* of the body, due to angular velocity $\boldsymbol{\omega}'$.

The time rate of change of work done by external forces on a rigid body, equivalently the *power* associated with these forces, is

$$\dot{W} \equiv \dot{\mathbf{r}}^T\mathbf{F} + \boldsymbol{\omega}'^T\mathbf{n}' \tag{5.3.18}$$

Substituting for $\mathbf{F}$ and $\mathbf{n}'$ from the Newton–Euler equations of Eq. 5.1.20 into Eq. 5.3.18 yields

$$\begin{aligned} \dot{W} &= m\dot{\mathbf{r}}^T\ddot{\mathbf{r}} + \boldsymbol{\omega}'^T(\mathbf{J}'\dot{\boldsymbol{\omega}}' + \tilde{\boldsymbol{\omega}}'\mathbf{J}'\boldsymbol{\omega}') \\ &= m\dot{\mathbf{r}}^T\ddot{\mathbf{r}} + \boldsymbol{\omega}'^T\mathbf{J}'\dot{\boldsymbol{\omega}}' \end{aligned} \tag{5.3.19}$$

Since m and $\mathbf{J}'$ are constant, note that

$$\dot{W} = \frac{d}{dt}(\tfrac{1}{2}\dot{\mathbf{r}}^T\dot{\mathbf{r}} + \tfrac{1}{2}\boldsymbol{\omega}'^T\mathbf{J}'\boldsymbol{\omega}') = \dot{T} \tag{5.3.20}$$

Thus the time rate of change of work done by external forces is equal to the time rate of change of kinetic energy of the body.

If the force system that acts on a body is conservative, there exists a work function W that defines *potential energy* V, as in Section 3.6,

$$W = -V \tag{5.3.21}$$

Taking the time derivative of this relation and substituting it into Eq. 5.3.20 yields

$$\frac{d}{dt}(T + V) = 0 \tag{5.3.22}$$

Thus for conservative forces acting on a rigid body,

$$T + V = \text{constant} \tag{5.3.23}$$

This key relationship provides a *conservation of energy theorem* that can yield information regarding the solution, prior to constructing the explicit solution as a function of time.

Example 5.3.2: Since gravitational and workless constraint forces that act on the rolling disk of Example 5.1.3 are conservative, conservation of energy can be applied. The kinetic energy in this case is

$$T = \frac{m}{2} R^2\dot{\phi}^2 + \frac{J_0}{2}\dot{\phi}^2$$

and the potential energy is

$$V = -mgR\phi \sin\theta$$

Since $\phi(0) = \dot{\phi}(0) = 0$, $(T + V)(0) = 0$. Thus for all time,

$$T + V = \tfrac{1}{2}(mR^2 + J_0)\dot{\phi}^2 - mgR\,\phi \sin\theta = 0$$

and

$$\dot{\phi} = \left(\frac{2mgR\sin\theta}{mR^2 + J_0}\right)^{1/2} \phi^{1/2}$$

where the plus sign is selected, since the disk will roll down the incline in counterclockwise rotation. As often happens, conservation of energy in systems with one generalized coordinate yields a first order differential equation, or *first integral* of the second order equation of motion derived in Example 5.1.3.

Writing the first-order equation in differential form gives

$$\phi^{-1/2}\,d\phi = \left(\frac{2mgR\sin\theta}{mR^2 + J_0}\right)^{1/2} dt$$

which may be integrated to obtain

$$2\phi^{1/2} = \left(\frac{2mgR\sin\theta}{mR^2 - J_0}\right)^{1/2} t + a_1$$

where a_1 is a constant of integration. Since $\phi(0) = 0$, $a_1 = 0$. Squaring both sides yields

$$\phi = \frac{mgR\sin\theta}{2(mR^2 + J_0)}\,t^2$$

which is the solution that is obtained in Example 5.1.3 by integrating the equations of motion.

♦ **Example 5.3.3:** The pendulum of Example 5.1.2 is acted on by conservative gravitational and workless constraint forces and torques, so conservation of energy is applicable. Note first that

$$\boldsymbol{\rho} = \mathbf{A}\boldsymbol{\rho}'' = \begin{bmatrix} \rho_{x''} \cos \phi + \rho_{z''} \sin \phi \\ \rho_{y''} \\ -\rho_{x''} \sin \phi + \rho_{z''} \cos \phi \end{bmatrix}$$

and

$$\begin{aligned} \dot{\boldsymbol{\rho}} &= \tilde{\boldsymbol{\omega}}\boldsymbol{\rho} = \mathbf{A}\tilde{\boldsymbol{\omega}}''\boldsymbol{\rho}'' \\ &= \mathbf{A}\begin{bmatrix} \dot{\phi}\rho_{z''} \\ 0 \\ -\dot{\phi}\rho_{x''} \end{bmatrix} \\ &= \dot{\phi}\begin{bmatrix} \rho_{z''} \cos \phi - \rho_{x''} \sin \phi \\ 0 \\ -\rho_{z''} \sin \phi - \rho_{x''} \cos \phi \end{bmatrix} \end{aligned}$$

If a centroidal x'-y'-z' frame is selected parallel to the x''-y''-z'' frame, then $\boldsymbol{\omega}' = \boldsymbol{\omega}'' = [0, \dot{\phi}, 0]^T$. By the parallel axis theorem (i.e., Eq. 5.2.7 with $\mathbf{C} = \mathbf{I}$),

$$\mathbf{J}' = \mathbf{J}'' - m(\boldsymbol{\rho}''^T\boldsymbol{\rho}''\mathbf{I} - \boldsymbol{\rho}''\boldsymbol{\rho}''^T)$$

The kinetic energy is thus

$$\begin{aligned} T &= \frac{m}{2}\dot{\phi}^2(\rho_{z''}^2 + \rho_{x''}^2) + \tfrac{1}{2}\dot{\phi}^2 J_{y''y''} - \frac{m}{2}\dot{\phi}^2(\rho_{x''}^2 + \rho_{y''}^2 + \rho_{z''}^2 - \rho_{y''}^2) \\ &= \tfrac{1}{2}\dot{\phi}^2 J_{y''y''} \end{aligned}$$

The potential energy is just

$$V = mg\rho_z = mg(-\rho_{x''} \sin \phi + \rho_{z''} \cos \phi)$$

If $\phi(0) = \dot{\phi}(0) = 0$,

$$(T + V)(0) = mg\rho_{z''}$$

Thus for all t,

$$\begin{aligned} T + V &= \tfrac{1}{2}\dot{\phi}^2 J_{y''y''} + mg(-\rho_{x''} \sin \phi + \rho_{z''} \cos \phi) \\ &= mg\rho_{z''} \end{aligned}$$

Since $\dot{\phi} > 0$ for $t > 0$,

$$\dot{\phi} = \left(\frac{2mg}{J_{y''y''}}\right)^{1/2} [\rho_{x''} \sin \phi + \rho_{z''}(1 - \cos \phi)]^{1/2}$$

While this first order differential equation is not trivially integrable, it is a first integral of the rather complicated second order differential equation of motion of Eq. 5.1.26 that was derived in Example 5.1.2 (Prob. 5.3.1).

5.4 RIGID BODY EQUATIONS OF MOTION IN ORIENTATION GENERALIZED COORDINATES

As noted in Chapter 4, the angular velocity vector $\boldsymbol{\omega}'$ is not the time derivative of any vector. Thus while the Newton–Euler equations of motion of Eq. 5.1.20 may be solved for $\dot{\boldsymbol{\omega}}'$ and this variable integrated to obtain $\boldsymbol{\omega}'$, further direct integration is impossible. It remains to integrate kinematic differential equations that relate time derivatives of generalized coordinates to angular velocity, to obtain generalized coordinates. Two basically different approaches may be taken to accomplish this objective. The first is a mixed integration strategy in which the equations of motion are solved for $\dot{\boldsymbol{\omega}}'$, which is integrated to obtain $\boldsymbol{\omega}'$. Time derivatives of the generalized coordinates are obtained from $\boldsymbol{\omega}'$ and these are integrated to obtain orientation generalized coordinates. The second approach is to transform the equations of motion into second order differential equations in the generalized coordinates and to integrate these equations. Both approaches are outlined in this section.

Mixed Kinetic–Kinematic Equations

For an unconstrained body, the Euler orientation equations of motion are, from Eq. 5.1.20,

$$\mathbf{J}'\dot{\boldsymbol{\omega}}' = \mathbf{n}' - \tilde{\boldsymbol{\omega}}'\mathbf{J}'\boldsymbol{\omega}' \tag{5.4.1}$$

These equations may be viewed as first order differential equations in the angular velocity $\boldsymbol{\omega}'$.

If Euler parameters of Section 4.3 are used as orientation generalized coordinates, then Eq. 4.3.31 that determines $\dot{\mathbf{p}}$ in terms of $\boldsymbol{\omega}'$ and the normalization constraint of Eq. 4.3.3 form a system of differential-algebraic equations for the Euler parameters; i.e.,

$$\begin{aligned} \dot{\mathbf{p}} &= \tfrac{1}{2}\mathbf{G}^T\boldsymbol{\omega}' \\ \mathbf{p}^T\mathbf{p} - 1 &= 0 \end{aligned} \tag{5.4.2}$$

Equations 5.4.1 and 5.4.2 comprise a system of first order differential-algebraic equations for $\boldsymbol{\omega}'$ and $\mathbf{p}$. This system may be integrated numerically using methods outlined in Appendix D.

If Euler angles of Section 4.4 are used in place of Euler parameters, then time derivatives of Euler angles are determined from angular velocities by Eq. 4.4.16; that is,

$$\begin{bmatrix} \dot{\phi} \\ \dot{\theta} \\ \dot{\psi} \end{bmatrix} = \begin{bmatrix} \dfrac{(\omega_{y'}\cos\psi + \omega_{x'}\sin\psi)}{\sin\theta} \\ \omega_{x'}\cos\psi - \omega_{y'}\sin\psi \\ \omega_{z'} - \dfrac{\cos\theta(\omega_{y'}\cos\psi + \omega_{x'}\sin\psi)}{\sin\theta} \end{bmatrix} \tag{5.4.3}$$

Equations 5.4.1 and 5.4.3 may be viewed as a system of first order differential equations for $\boldsymbol{\omega}'$ and Euler angles ϕ, θ, and ψ. These differential equations may be integrated using methods of Appendix D. As is clear from Eq. 5.4.3, and observed earlier in general, if $\theta = n\pi$, a singularity occurs. This is due to the fundamental singularity associated with Euler angles. No such singularity occurs with Euler parameters, but at the cost of dealing with differential-algebraic equations.

Equations of Motion in Euler Parameters

To transform the variational Newton–Euler equations of Eq. 5.1.19 to Euler parameter form, first recall the angular velocity relation of Eq. 4.3.30,

$$\boldsymbol{\omega}' = 2\mathbf{G}\dot{\mathbf{p}} \tag{5.4.4}$$

Differentiating both sides of this equation yields

$$\dot{\boldsymbol{\omega}}' = 2\mathbf{G}\ddot{\mathbf{p}} + 2\dot{\mathbf{G}}\dot{\mathbf{p}}$$

Using the definition of Eq. 4.3.15 to write $\dot{\mathbf{G}}$ and carrying out a direct expansion yields

$$\begin{aligned}\dot{\mathbf{G}}\dot{\mathbf{p}} &= [-\dot{\mathbf{e}}, -\tilde{\dot{\mathbf{e}}} + \dot{e}_0\mathbf{I}]\begin{bmatrix}\dot{e}_0 \\ \dot{\mathbf{e}}\end{bmatrix} \\ &= -\dot{e}_0\dot{\mathbf{e}} - \tilde{\dot{\mathbf{e}}}\dot{\mathbf{e}} + \dot{e}_0\dot{\mathbf{e}} = \mathbf{0}\end{aligned}$$

Thus $\dot{\boldsymbol{\omega}}'$ may be written in terms of second derivatives of Euler parameters as

$$\dot{\boldsymbol{\omega}}' = 2\mathbf{G}\ddot{\mathbf{p}} \tag{5.4.5}$$

It is interesting to obtain an inverse relation for Eq. 5.4.5 (i.e., to solve for $\ddot{\mathbf{p}}$ in terms of $\dot{\boldsymbol{\omega}}'$). Differentiating Eq. 4.3.31

$$\begin{aligned}\ddot{\mathbf{p}} &= \tfrac{1}{2}\mathbf{G}^T\dot{\boldsymbol{\omega}}' + \tfrac{1}{2}\dot{\mathbf{G}}^T\boldsymbol{\omega}' \\ &= \tfrac{1}{2}\mathbf{G}^T\dot{\boldsymbol{\omega}}' + \dot{\mathbf{G}}^T\mathbf{G}\dot{\mathbf{p}}\end{aligned}$$

Differentiating both sides of Eq. 4.3.20 yields

$$\dot{\mathbf{G}}^T\mathbf{G} = -[\mathbf{G}^T\dot{\mathbf{G}} + \dot{\mathbf{p}}\mathbf{p}^T + \mathbf{p}\dot{\mathbf{p}}^T]$$

Thus

$$\begin{aligned}\dot{\mathbf{G}}^T\mathbf{G}\dot{\mathbf{p}} &= -\mathbf{G}^T\dot{\mathbf{G}}\dot{\mathbf{p}} - \dot{\mathbf{p}}\mathbf{p}^T\dot{\mathbf{p}} - \mathbf{p}\dot{\mathbf{p}}^T\dot{\mathbf{p}} \\ &= -(\dot{\mathbf{p}}^T\dot{\mathbf{p}})\mathbf{p} \\ &= -\tfrac{1}{4}(\boldsymbol{\omega}'^T\boldsymbol{\omega}')\mathbf{p}\end{aligned}$$

where $\dot{\mathbf{G}}\dot{\mathbf{p}} = 0$ and Eqs. 4.3.23 and 4.3.31 have been used. Finally,

$$\begin{aligned}\ddot{\mathbf{p}} &= \tfrac{1}{2}\mathbf{G}^T\dot{\boldsymbol{\omega}}' - (\dot{\mathbf{p}}^T\dot{\mathbf{p}})\mathbf{p} \\ &= \tfrac{1}{2}\mathbf{G}^T\dot{\boldsymbol{\omega}}' - \tfrac{1}{4}(\boldsymbol{\omega}'^T\boldsymbol{\omega}')\mathbf{p}\end{aligned} \tag{5.4.6}$$

Note that while $\mathbf{G}$ behaves similar to an orthogonal matrix in the relations between $\dot{\mathbf{p}}$ and $\boldsymbol{\omega}'$ in Eqs. 4.3.30 and 4.3.31, this property is lost in the relations between $\ddot{\mathbf{p}}$ and $\dot{\boldsymbol{\omega}}'$ in Eqs. 5.4.5 and 5.4.6.

Recall the relation between virtual rotation and Euler parameter variation in Eq. 4.3.37 (i.e., $\delta\boldsymbol{\pi}' = 2\mathbf{G}\,\delta\mathbf{p}$). Substituting Eqs. 5.4.4, 5.4.5, and 4.3.37 into Eq. 5.1.26 and using Eq. 4.3.28 yields

$$\delta\mathbf{r}^T[m\ddot{\mathbf{r}} - \mathbf{F}^A] + 2\delta\mathbf{p}^T[2\mathbf{G}^T\mathbf{J}'\mathbf{G}\ddot{\mathbf{p}} + 4\mathbf{G}^T\mathbf{G}\dot{\mathbf{G}}^T\mathbf{J}'\mathbf{G}\dot{\mathbf{p}} - \mathbf{G}^T\mathbf{n}'^A] = 0$$

Using Eqs. 4.3.20, 4.3.25, and 4.3.34, this becomes

$$\delta\mathbf{r}^T[m\ddot{\mathbf{r}} - \mathbf{F}^A] + \delta\mathbf{p}^T[4\mathbf{G}^T\mathbf{J}'\mathbf{G}\ddot{\mathbf{p}} - 8\dot{\mathbf{G}}^T\mathbf{J}'\dot{\mathbf{G}}\mathbf{p} - 2\mathbf{G}^T\mathbf{n}'^A] = 0 \tag{5.4.7}$$

which must hold for all $\delta\mathbf{r}$ and $\delta\mathbf{p}$ that satisfy kinematic constraints and the Euler parameter normalization constraint of Eq. 4.3.34,

$$\delta\mathbf{p}^T\mathbf{p} = 0 \tag{5.4.8}$$

This constraint may be treated by using the Lagrange multiplier theorem to introduce an *Euler parameter normalization Lagrange multiplier* β such that

$$\delta\mathbf{r}^T[m\ddot{\mathbf{r}} - \mathbf{F}^A] + \delta\mathbf{p}^T[4\mathbf{G}^T\mathbf{J}'\mathbf{G}\ddot{\mathbf{p}} - 8\dot{\mathbf{G}}^T\mathbf{J}'\dot{\mathbf{G}}\mathbf{p} - 2\mathbf{G}^T\mathbf{n}'^A + \beta\mathbf{p}] = 0 \tag{5.4.9}$$

which must hold for all $\delta\mathbf{r}$ and $\delta\mathbf{p}$ that satisfy kinematic constraints that act on the body.

If no kinematic constraints act on the body, $\delta\mathbf{r}$ and $\delta\mathbf{p}$ in Eq. 5.4.9 are arbitrary, hence their coefficients must be zero. This leads to the *Newton–Euler equations in Euler parameters* for a single body,

$$\begin{aligned} m\ddot{\mathbf{r}} &= \mathbf{F}^A \\ 4\mathbf{G}^T\mathbf{J}'\mathbf{G}\ddot{\mathbf{p}} &= 2\mathbf{G}^T\mathbf{n}'^A + 8\dot{\mathbf{G}}^T\mathbf{J}'\dot{\mathbf{G}}\mathbf{p} - \beta\mathbf{p} \end{aligned} \tag{5.4.10}$$

These equations must be solved, together with the Euler parameter normalization constraint

$$\mathbf{p}^T\mathbf{p} - 1 = 0 \tag{5.4.11}$$

Equations 5.4.10 and 5.4.11 thus form a system of differential-algebraic equations of motion for an unconstrained body, in terms of Euler parameters.

Equations of Motion in Euler Angles

The Euler angle time derivative-angular velocity relation of Eq. 4.4.15 may be written in the form

$$\boldsymbol{\omega}' = \begin{bmatrix} \sin\psi\sin\theta & \cos\psi & 0 \\ \cos\psi\sin\theta & -\sin\psi & 0 \\ \cos\theta & 0 & 1 \end{bmatrix} \begin{bmatrix} \dot{\phi} \\ \dot{\theta} \\ \dot{\psi} \end{bmatrix} \equiv \mathbf{B} \begin{bmatrix} \dot{\phi} \\ \dot{\theta} \\ \dot{\psi} \end{bmatrix} \tag{5.4.12}$$

and differentiated to obtain

$$\dot{\boldsymbol{\omega}}' = \mathbf{B}\begin{bmatrix}\ddot{\phi}\\ \ddot{\theta}\\ \ddot{\psi}\end{bmatrix} + \begin{bmatrix}\dot{\phi}(\dot{\psi}\cos\psi\sin\theta + \dot{\theta}\sin\psi\cos\theta) - \dot{\theta}\dot{\psi}\sin\psi\\ \dot{\phi}(-\dot{\psi}\sin\psi\sin\theta + \dot{\theta}\cos\psi\cos\theta) - \dot{\theta}\dot{\psi}\cos\psi\\ -\dot{\phi}\dot{\theta}\sin\theta\end{bmatrix}$$

$$\equiv \mathbf{B}\begin{bmatrix}\ddot{\phi}\\ \ddot{\theta}\\ \ddot{\psi}\end{bmatrix} + \mathbf{b} \tag{5.4.13}$$

Analogous to derivation of the angular velocity–Euler angle time derivative relation of Eq. 4.4.15, the relation between virtual rotation and Euler angle variation is

$$\delta\boldsymbol{\pi}' = \mathbf{B}\begin{bmatrix}\delta\phi\\ \delta\theta\\ \delta\psi\end{bmatrix} \tag{5.4.14}$$

Substituting Eqs. 5.4.12 to 5.4.14 into Eq. 5.1.26 yields the variational Newton–Euler equations of motion of a rigid body in terms of Euler angles,

$$\delta\mathbf{r}^T[m\ddot{\mathbf{r}} - \mathbf{F}^A] + [\delta\phi, \delta\theta, \delta\psi]\left\{\mathbf{B}^T\mathbf{J}'\mathbf{B}\begin{bmatrix}\ddot{\phi}\\ \ddot{\theta}\\ \ddot{\psi}\end{bmatrix} + \mathbf{B}^T\mathbf{J}'\mathbf{b}\right.$$

$$\left. + \mathbf{B}^T\widetilde{\left(\mathbf{B}\begin{bmatrix}\dot{\phi}\\ \dot{\theta}\\ \dot{\psi}\end{bmatrix}\right)}\mathbf{J}'\mathbf{B}\begin{bmatrix}\dot{\phi}\\ \dot{\theta}\\ \dot{\psi}\end{bmatrix} - \mathbf{B}^T\mathbf{n}'^A\right\} = 0 \tag{5.4.15}$$

for all $\delta\mathbf{r}$ and $[\delta\phi, \delta\theta, \delta\psi]^T$ that are consistent with constraints.

Consider the special case in which principal centroidal axes are selected, so that the inertia matrix $\mathbf{J}'$ is diagonal. In this case, Eqs. 5.4.12 and 5.4.13 may be substituted into the second of Eqs. 5.1.20 to obtain the Euler equations of motion in terms of Euler angles,

$$\begin{bmatrix}J_{x'x'}\sin\psi\sin\theta & J_{x'x'}\cos\psi & 0\\ J_{y'y'}\cos\psi\sin\theta & -J_{y'y'}\sin\psi & 0\\ J_{z'z'}\cos\theta & 0 & J_{z'z'}\end{bmatrix}\begin{bmatrix}\ddot{\phi}\\ \ddot{\theta}\\ \ddot{\psi}\end{bmatrix} = \mathbf{n}' - \mathbf{J}'\mathbf{b} - \widetilde{\left(\mathbf{B}\begin{bmatrix}\dot{\phi}\\ \dot{\theta}\\ \dot{\psi}\end{bmatrix}\right)}\mathbf{J}'\mathbf{B}\begin{bmatrix}\dot{\phi}\\ \dot{\theta}\\ \dot{\psi}\end{bmatrix} \tag{5.4.16}$$

Equation 5.4.16 is only moderately complicated, particularly as regards the quadratic nature of the right side of the equation in time derivatives of Euler angles. The more important complexity associated with Eq. 5.4.16 is the possibility that the coefficient matrix of the Euler angle second derivatives might become singular. A direct expansion of the determinant of this coefficient matrix shows that it is $-I_{x'x'}I_{y'y'}I_{z'z'} \sin \theta$. Thus if $\theta = n\pi$, for any integer n, Eq. 5.4.16 becomes singular and will lead to both analytical and numerical difficulties.

Lagrange's Equations of Motion of a Rigid Body

Since the right side of Eq. 3.6.8 involves a sum over as many particles as desired in modeling a system, as the number of particles approaches ∞, it may be expressed as an integral. Applying this argument to the model of a rigid body employed in Section 5.1, the *variational form of Lagrange's equations of motion for a rigid body* is, as in Eq. 3.6.9,

$$\delta \mathbf{q}^T \left[\frac{d}{dt}\left(\frac{\partial T}{\partial \dot{\mathbf{q}}}\right) - \left(\frac{\partial T}{\partial \mathbf{q}}\right)^T - \mathbf{Q}^A \right] = 0 \tag{5.4.17}$$

for all generalized coordinate variations $\delta \mathbf{q}$ that are consistent with constraints that act on the body. If the generalized coordinates are independent, then $\delta \mathbf{q}$ is arbitrary and, as in Eq. 3.6.11,

$$\frac{d}{dt}\left(\frac{\partial T}{\partial \dot{q}_j}\right) - \frac{\partial T}{\partial q_j} - Q_j^A = 0, \qquad j = 1, \ldots, nc \tag{5.4.18}$$

Equation 5.4.17 or 5.4.18 makes sense, of course, only if the kinetic energy and applied generalized force on the body can be explicitly written as functions of generalized coordinates and their time derivatives. Consider use of Euler angles for orientation, so the generalized coordinates are

$$\begin{aligned} \mathbf{q} &= [\mathbf{r}^T, \phi, \theta, \psi]^T \\ &\equiv [\mathbf{r}^T, \mathbf{u}^T]^T \end{aligned} \tag{5.4.19}$$

From Eq. 5.4.12,

$$\boldsymbol{\omega}' = \mathbf{B}\dot{\mathbf{u}} \tag{5.4.20}$$

Thus the kinetic energy of Eq. 5.3.17 is

$$T = \tfrac{1}{2} m \dot{\mathbf{r}}^T \dot{\mathbf{r}} + \tfrac{1}{2} \dot{\mathbf{u}}^T \mathbf{B}^T \mathbf{J}' \mathbf{B} \dot{\mathbf{u}} \tag{5.4.21}$$

Also, the virtual work of applied force $\mathbf{F}^A$ and torque $\mathbf{n}'^A$ is

$$\begin{aligned} \delta W &= \delta \mathbf{r}^T \mathbf{F}^A + \delta \boldsymbol{\pi}'^T \mathbf{n}'^A \\ &= \delta \mathbf{r}^T \mathbf{F}^A + \delta \mathbf{u}^T \mathbf{B}^T \mathbf{n}'^A \\ &= \delta \mathbf{q}^T \begin{bmatrix} \mathbf{F}^A \\ \mathbf{B}^T \mathbf{n}'^A \end{bmatrix} \\ &\equiv \delta \mathbf{q}^T \mathbf{Q}^A \end{aligned} \tag{5.4.22}$$

where Eq. 5.4.14 has been used.

Example 5.4.1: To illustrate the value of the Lagrange multiplier form of the equations of motion of a body, consider the *gyroscope* of Fig. 5.4.1. Due to symmetry of the rotor about its z' axis, its products of inertia are zero and $J_{x'x'} = J_{y'y'} \equiv I$ and $J_{z'z'} \equiv J$. Since the centroid of the rotor is fixed in space, the kinetic energy of Eq. 5.4.21 is

$$T = \tfrac{1}{2}[I(\dot{\phi}^2 \sin^2\theta + \dot{\theta}^2) + J(\dot{\psi} + \dot{\phi}\cos\theta)^2]$$

and the gravitational generalized force is zero. The only externally applied torque is about the x''' axis,

$$Q^A = [0, M_\theta, 0]^T$$

Since the three angles of rotation are independent, Eq. 5.4.18 yields

$$\frac{d}{dt}[I\dot{\phi}\sin^2\theta + J(\dot{\psi} + \dot{\phi}\cos\theta)\cos\theta] = 0$$

$$\frac{d}{dt}[I\dot{\theta}] - I\dot{\phi}^2 \sin\theta\cos\theta + J(\dot{\psi} + \dot{\phi}\cos\theta)\,\dot{\phi}\sin\theta = M_\theta$$

$$\frac{d}{dt}[J(\dot{\psi} + \dot{\phi}\cos\theta)] = 0$$

Thus ϕ and ψ are ignorable coordinates (see Section 3.6) that yield the first integrals

$$I\dot{\phi}\sin^2\theta + J(\dot{\psi} + \dot{\phi}\cos\theta)\cos\theta = C_1$$

$$J(\dot{\psi} + \dot{\phi}\cos\theta) = C_2$$

The second-order equation for θ is

$$I\ddot{\theta} - I\dot{\phi}^2\sin\theta\cos\theta + C_2\,\dot{\phi}\sin\theta = M_\theta$$

As noted in Section 3.6, one of the main values of Lagrange's equations of motion is to identify ignorable coordinates and to obtain first integrals that might not be found from the more complicated form of the equations of Eq. 5.4.16.

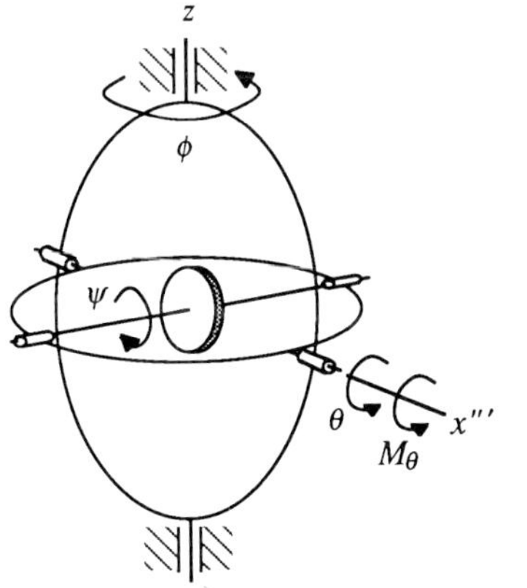

Figure 5.4.1 Gyroscope.

PROBLEMS

Section 5.1

5.1.1 Carry out the manipulations in detail to derive Eqs. 5.1.3 and 5.1.6.

5.1.2 Carry out the manipulations in detail to derive Eq. 5.1.11.

5.1.3 If the disk of Example 5.1.3 has $\phi = 0$ and $\dot{\phi} = -1$ (i.e., $\dot{\mathbf{r}} = -\mathbf{u}$) at $t = 0$, how far will it roll up the plane before it comes to rest?

5.1.4 Let the disk of Example 5.1.3 have a coefficient of friction of μ between itself and the plane. Since acceleration normal to the incline plane is zero, the normal force of contact is constant at mg cos θ. Thus if the disk is slipping, the magnitude of the friction force tangent to the plane is $\mu mg \cos \theta$.

(a) Write and solve the equation of motion of the disk, while slipping occurs.

(b) If the disk starts at $t = 0$ with $\phi = 0$ and $\dot{\phi} = -1$ [with $\dot{\mathbf{r}}(0) = 0$], when does slipping stop? How far does the disk roll up the plane before coming to rest?

5.1.5 Let the pendulum of Example 5.1.2 be permitted to rotate about and slide along the y axis of an inertial reference frame, as shown in Fig. P5.1.5. Write the equations of motion in terms of ϕ and s.

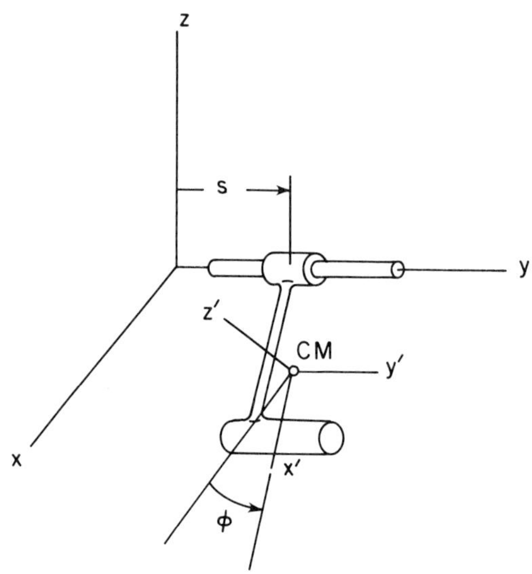

Figure P5.1.5

Section 5.2

5.2.1 Calculate moments of inertia of the body of Example 5.2.1 (Fig. 5.2.2) with respect to the x''-y''-z'' frame. Use these results, $\boldsymbol{\rho}''$ from Example 5.2.1, and the parallel axis transformation of Example 5.2.2 to evaluate moments of inertia of the body with respect to the centroidal x'-y'-z' frame.

5.2.2 Verify that Eqs. 5.2.10 to 5.2.12 are valid under the stated conditions.

5.2.3 Show that if any pair of coordinate planes are planes of symmetry for a body, all products of inertia with respect to this body-fixed reference frame are zero.

Section 5.3

5.3.1 Verify that the first order differential equation of motion derived in Example 5.3.3 is a first integral of Eq. 5.1.32. (*Hint:* Differentiate both sides of the first order differential equation and substitute for $\dot{\phi}$ on the right to obtain Eq. 5.1.32, thus showing that any solution of the first order equation satisfies Eq. 5.1.32.)

SUMMARY OF KEY FORMULAS

Newton–Euler Equations of Motion

$$m\ddot{\mathbf{r}} = \mathbf{F}, \qquad \mathbf{J}'\dot{\boldsymbol{\omega}}' = \mathbf{n}' - \tilde{\boldsymbol{\omega}}'\mathbf{J}'\boldsymbol{\omega}' \tag{5.1.20}$$

$$\delta\mathbf{r}^T[m\ddot{\mathbf{r}} - \mathbf{F}^A] + \delta\boldsymbol{\pi}'^T[\mathbf{J}'\dot{\boldsymbol{\omega}}' + \tilde{\boldsymbol{\omega}}'\mathbf{J}'\boldsymbol{\omega}' - \mathbf{n}'^A] = 0 \tag{5.1.26}$$

Planar Equations of Motion

$$\delta\mathbf{r}^T[m\ddot{\mathbf{r}} - \mathbf{F}^A] + \delta\phi[J_0\ddot{\phi} - n^A] = 0 \tag{5.1.42}$$

$$m\ddot{\mathbf{r}} = \mathbf{F}^A, \qquad J_0\ddot{\phi} = n^A \tag{5.1.43}$$

Inertia Properties

$$\int_m \mathbf{s}'^P\, dm(P) = \mathbf{0}, \qquad \boldsymbol{\rho}'' = \frac{1}{m}\int_m \mathbf{s}''^P\, dm(P) \tag{5.1.13, 5.2.1}$$

$$\mathbf{J}' = -\int_m \tilde{\mathbf{s}}'^P\tilde{\mathbf{s}}'^P\, dm(P), \qquad \mathbf{J}'' = -\int_m \tilde{\mathbf{s}}''^P\tilde{\mathbf{s}}''^P\, dm(P) \tag{5.1.17, 5.2.4}$$

$$\mathbf{J}'' = \mathbf{C}^T\mathbf{J}'\mathbf{C} + m(\boldsymbol{\rho}''^T\boldsymbol{\rho}''\mathbf{I} - \boldsymbol{\rho}''\boldsymbol{\rho}''^T), \qquad \mathbf{s}' = \mathbf{C}\mathbf{s}'' \tag{5.2.7}$$

$$\boldsymbol{\rho}'' = \frac{1}{m}\sum_{i=1}^{k} m_i\boldsymbol{\rho}_i'', \qquad \mathbf{J}' = \sum_{i=1}^{k}\mathbf{J}_i' \tag{5.2.18, 19}$$

Momentum and Kinetic Energy

$$\mathbf{P} \equiv m\dot{\mathbf{r}}, \qquad \dot{\mathbf{P}} = \mathbf{F} \tag{5.3.1, 2}$$

$$\mathbf{H} \equiv \int_m \tilde{\mathbf{r}}^P\dot{\mathbf{r}}^P\, dm(P) = m\tilde{\mathbf{r}}\dot{\mathbf{r}} + \mathbf{A}\mathbf{J}'\boldsymbol{\omega}' \tag{5.3.5}$$

$$\dot{\mathbf{H}} = \tilde{\mathbf{r}}\mathbf{F} + \mathbf{A}\mathbf{n}' \equiv \mathbf{N} \tag{5.3.7}$$

$$T \equiv \frac{1}{2}\int_m \dot{\mathbf{r}}^{PT}\dot{\mathbf{r}}^P\, dm(P) = \tfrac{1}{2}m\dot{\mathbf{r}}^T\dot{\mathbf{r}} + \tfrac{1}{2}\boldsymbol{\omega}'^T\mathbf{J}'\boldsymbol{\omega}' \tag{5.3.17}$$

$$T + V = \text{constant} \quad \text{(for conservative systems)} \tag{5.3.23}$$

6

An Introduction to Multibody Dynamics

Multibody systems consist of collections of rigid bodies that are constrained to move relative to one another by kinematic connections between bodies. A wide variety of distinctly different formulations for multibody dynamics exist, with fundamental differences in generalized coordinates that are used to describe the kinematics of a system. One extreme in this regard is use of a maximal set of absolute position and orientation coordinates for each body, such as those defined in Chapters 4 and 5, called *Cartesian generalized coordinates*. Kinematic constraints that act on and between bodies are represented as algebraic relationships, in the case of holonomic constraints, and differential relationships in the case of nonholonomic constraints and differential constraints for which an integral cannot be found. The Cartesian formulation is characterized by a large number of variables, a large number of kinematic constraint equations, and a simple form of the equations of motion. This formulation requires computer generation and solution of the equations of motion, since very large numbers of equations in many variables are involved. The basic ideas that support this approach are presented and illustrated with simple examples in Sections 6.1 and 6.2. A more detailed development of this approach, numerical methods for its implementation, and illustration with large scale applications may be found in Ref. 12.

In contrast to Cartesian coordinates, *relative coordinates* in joints between bodies may be used to define position and orientation of bodies relative to one another, yielding a minimal set of generalized coordinates. Kinematic restrictions on position, velocity, and acceleration of a system and the equations of motion in relative coordinates are minimal in number, but they are of substantial analytical complexity. The basic ideas of this approach are introduced in Sections 6.3 to 6.7 and illustrated through application to relatively elementary mechanical systems. More comprehensive developments of this and related formulations are presented and illustrated in Refs. 10, 13, and 19.

6.1 CARTESIAN COORDINATE FORMULATION FOR PLANAR SYSTEMS

A system of interconnected bodies that are symmetric with respect to and move only in an inertial plane is called a *planar system*. To specify the configuration or state of a planar system, it is first necessary to define generalized coordinates that locate and orient each body in the plane. Let the x-y frame shown in Fig. 6.1.1 be an inertial reference frame and define an x_i'-y_i' *body-fixed reference frame* embedded in body i, with its origin O_i at the center of mass of the body.

Any point on body i can be located in the inertial plane by locating it in the x_i'-y_i' frame and specifying the global coordinates $\mathbf{r}_i = [x_i, y_i]^T \equiv [x, y]_i^T$ of point O_i and the angle ϕ_i of rotation of this frame relative to the x-y frame. Let P be a point on body i. The vector that locates this point in the body-fixed x_i'-y_i' frame is the constant vector $\mathbf{s}_i' = [x'^P, y'^P]_i^T$, where the superscript P on $\mathbf{s}_i'$ is suppressed to simplify notation. Point P may be located in the x-y frame by $\mathbf{r}^P = [x^P, y^P]^T$. The relation between the local and global coordinates of point P is given by Eq. 4.2.1 as

$$\mathbf{r}^P = \mathbf{r}_i + \mathbf{A}_i\mathbf{s}_i' \tag{6.1.1}$$

where the orthogonal matrix $\mathbf{A}_i$ for a planar body is given by 5.1.39 as

$$\mathbf{A}_i = \begin{bmatrix} \cos\phi_i & -\sin\phi_i \\ \sin\phi_i & \cos\phi_i \end{bmatrix} \tag{6.1.2}$$

Equation 6.1.1 uniquely defines the position of any point P on body i, in terms of its x_i'-y_i' body-fixed coordinates $\mathbf{s}_i'$ and the variables x_i, y_i, and ϕ_i that define the position and orientation of the x_i'-y_i' frame. Thus x_i, y_i, and ϕ_i serve

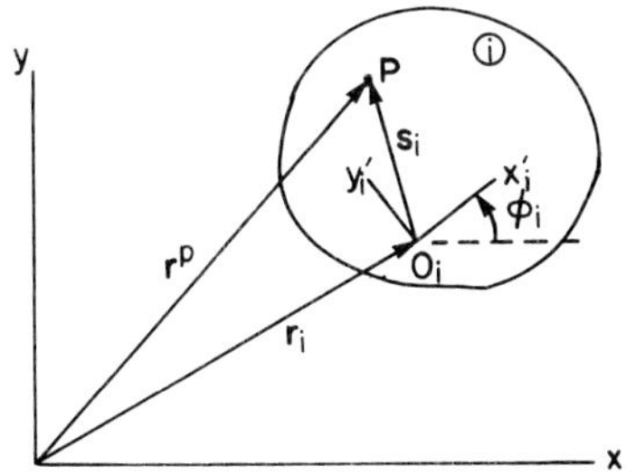

Figure 6.1.1 Location of point P on body i

as *planar Cartesian generalized coordinates* for body i, denoted by the vector

$$\mathbf{q}_i = [x_i, y_i, \phi_i]^T \equiv [x, y, \phi]_i^T \tag{6.1.3}$$

For a planar system with *nb* rigid bodies, the composite set of *Cartesian generalized coordinates* is the $3 \times nb$ vector

$$\begin{aligned} \mathbf{q} &= [\mathbf{q}_1^T, \mathbf{q}_2^T, \ldots, \mathbf{q}_{nb}^T]^T \\ &= [x_1, y_1, \phi_1, x_2, y_2, \phi_2, \ldots, x_{nb}, y_{nb}, \phi_{nb}]^T \end{aligned} \tag{6.1.4}$$

One of the attractive features of using Cartesian generalized coordinates is the ease of collecting data that are needed to describe bodies that make up a mechanical system. Points on bodies, such as point P in Fig. 6.1.1, locate centers of rotational bearings; attachment points of springs, dampers, and hydraulic actuators; and other physical characteristics in body-fixed x_i'-y_i' frames (drafting board frames).

Constraints between a Body and the Global Frame

An *absolute position constraint* on a point P that is fixed in body i in the x or y direction is shown physically for an x constraint on the piston of an engine in Fig. 6.1.2(a) and in general in Fig. 6.1.2(b). The following equations define the geometry of absolute x and y constraints, using Eq. 6.1.1:

$$\begin{aligned} \Phi^{axi} &\equiv x^P - C_1 \\ &= x_i + x_i'^P \cos\phi_i - y_i'^P \sin\phi_i - C_1 = 0 \end{aligned} \tag{6.1.5}$$

$$\begin{aligned} \Phi^{ayi} &\equiv y^P - C_2 \\ &= y_i + x_i'^P \sin\phi_i + y_i'^P \cos\phi_i - C_2 = 0 \end{aligned} \tag{6.1.6}$$

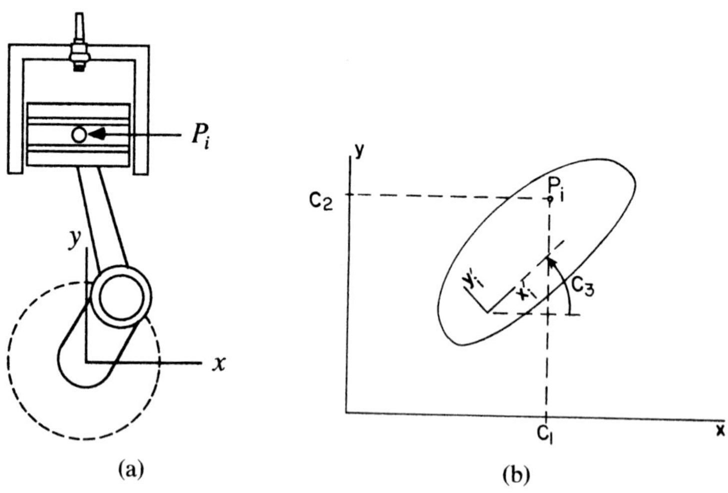

Figure 6.1.2 Constraints on absolute coordinates of point P_i and angular orientation.

where C_1 and C_2 are given constants and $x_i'^P$ and $y_i'^P$ are the body-fixed coordinates of point P.

For an *absolute rotation constraint* on body i, such as the restriction on rotation of the piston in Fig. 6.1.2(a) or the more general restriction shown in Fig. 6.1.2(b), the following constraint equation defines the geometry of the joint:

$$\Phi^{a\phi i} \equiv \phi_i - C_3 = 0 \tag{6.1.7}$$

where C_3 is a given constant.

Jacobians of these constraint functions are obtained, by direct differentiation, as

$$\begin{aligned} \Phi^{axi}_{\mathbf{q}_i} &= [1,\ 0,\ -x_i'^P \sin \phi_i - y_i'^P \cos \phi_i] \\ \Phi^{ayi}_{q_i} &= [0,\ 1,\ x_i'^P \cos \phi_i - y_i'^P \sin \phi_i] \\ \Phi^{a\phi i}_{\mathbf{q}_i} &= [0,\ 0,\ 1] \end{aligned} \tag{6.1.8}$$

♦ **Example 6.1.1:** Figure 6.1.3 shows a pendulum with its pivot pin P located at $(C_1, C_2) = (1, 1)$ in the x-y plane. With the origin of the x_1'-y_1' system at the center of the pendulum, whose length is 2 units, $x_1'^P = 1$ and $y_1'^P = 0$. Equations 6.1.5 and 6.1.6 thus describe kinematics of the pendulum, that is,

$$x_1 + \cos \phi_1 - 1 = 0$$

$$y_1 + \sin \phi_1 - 1 = 0$$

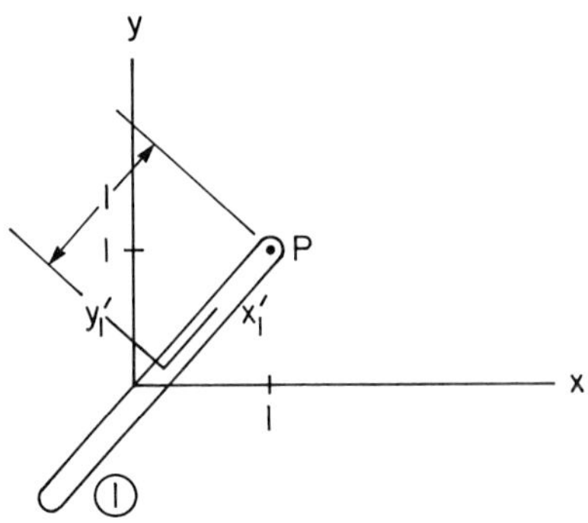

Figure 6.1.3 Pendulum with absolute constraints.

♦ **Example 6.1.2:** As a second example, Fig. 6.1.4 shows a joint that allows body 1 to slide along the x axis, with the body-fixed x_1' axis and the x axis coinciding (e.g., the cutting head of a shaper). This kinematic constraint between the slider and ground can be defined by Eq. 6.1.6, with $x_1'^P = y_1'^P = 0$ and $C_2 = 0$, and by Eq. 6.1.7, with $C_3 = 0$; that is,

$$y_1 = 0$$
$$\phi_1 = 0$$

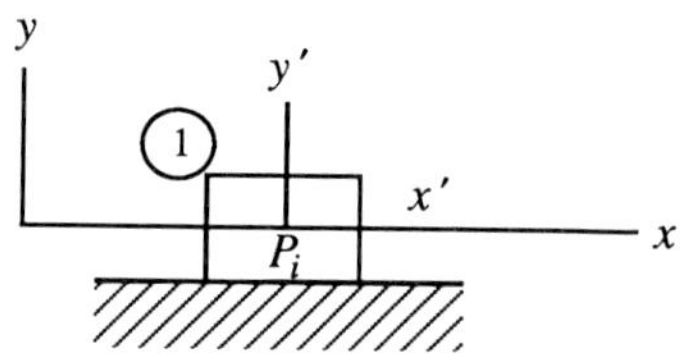

Figure 6.1.4 Slider along x axis.

Revolute and Translational Joints between Bodies

A *revolute joint* allows relative rotation about a point P that is common to bodies i and j, as shown in Fig. 6.1.5(b). Physically, such a joint is a rotational bearing between the bodies; for example, the three rotational bearings shown in Fig. 6.1.5(a) involving the crankshaft, connecting rod, and piston of an engine. A revolute joint is defined by locating point P on body i by $\mathbf{s}_i'$ and on body j by $\mathbf{s}_j'$, in their respective body-fixed x'-y' frames, as shown in Fig. 6.1.5(b).

Constraint equations that define the geometry of a revolute joint are obtained by requiring that point P on bodies i and j coincide; that is

$$\mathbf{\Phi}^{r(i,j)} = \mathbf{r}_j + \mathbf{A}_j\mathbf{s}_j' - \mathbf{r}_i - \mathbf{A}_i\mathbf{s}_i' = \mathbf{0} \tag{6.1.9}$$

which is in fact a pair of constraint equations.

To simplify Jacobian expressions, let

$$\frac{d}{d\phi}\mathbf{A}(\phi) = \begin{bmatrix} -\sin\phi & -\cos\phi \\ \cos\phi & -\sin\phi \end{bmatrix} \equiv \mathbf{B}(\phi) \tag{6.1.10}$$

With this notation, the Jacobians of the constraint of Eq. 6.1.9 with respect to $\mathbf{q}_i$ and $\mathbf{q}_j$ are (Prob. 6.1.2)

$$\begin{aligned} \mathbf{\Phi}_{\mathbf{q}_i}^{r(i,j)} &= -[\mathbf{I}_2,\ \mathbf{B}_i\mathbf{s}_i'] \\ \mathbf{\Phi}_{\mathbf{q}_j}^{r(i,j)} &= [\mathbf{I}_2,\ \mathbf{B}_j\mathbf{s}_j'] \end{aligned} \tag{6.1.11}$$

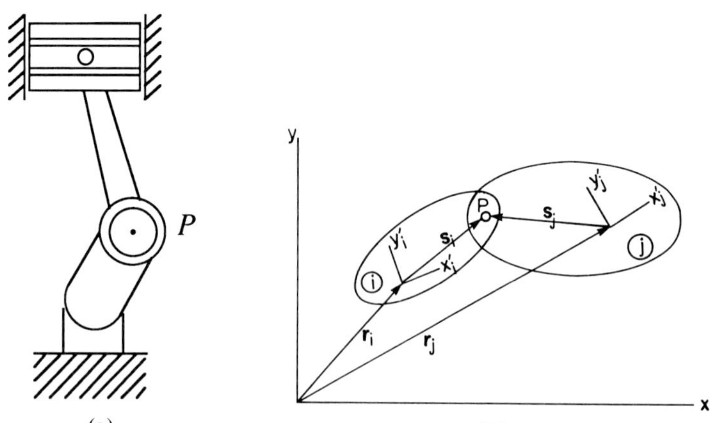

(a) (b)

Figure 6.1.5 Revolute joint.

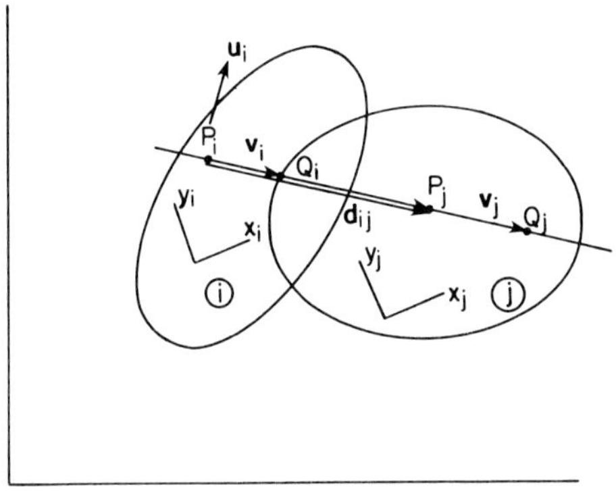

Figure 6.1.6 Translational joint.

A *translational joint* allows relative translation of a pair of bodies along a common axis, but no relative rotation of the bodies. Physically, such a joint may be defined as a straight block (or key) on one body that fits precisely in a straight slot (or keyway) on the second body and allows relative translation along their common centerlines. The piston in the cylinder of an engine in the planar model of Fig. 6.1.2(a) is an example of a translational joint. For the translational joint shown schematically in Fig. 6.1.6, let points P_i, Q_i, P_j, and Q_j lie on a line that is on the path of relative translation between bodies i and j. Points P_i and Q_i are located on body i and P_j and Q_j are on body j. The vector $\mathbf{v}_i$ is embedded in body i and connects points P_i and Q_i and vector $\mathbf{v}_j$ is embedded in body j and connects points P_j and Q_j. The vector $\mathbf{d}_{ij}$ that connects points P_i and P_j has variable length. An analytical condition that defines the geometry of the translational joint is that vectors $\mathbf{v}_i$ and $\mathbf{v}_j$ must remain collinear at all times.

In order to formulate analytical conditions that define the translational joint, a vector $\mathbf{u}_i$ that is perpendicular to $\mathbf{v}_i$, as shown in Fig. 6.1.6, is

$$\mathbf{u}_i' = \begin{bmatrix} -v_{y_i}' \\ v_{x_i}' \end{bmatrix} = \begin{bmatrix} 0 & -1 \\ 1 & 0 \end{bmatrix} \mathbf{v}_i' \tag{6.1.12}$$

The vector $\mathbf{d}_{ij}$ from P_i to P_j is

$$\mathbf{d}_{ij} = \mathbf{r}_j + \mathbf{A}_j\mathbf{s}_j' - \mathbf{r}_i - \mathbf{A}_i\mathbf{s}_i' \tag{6.1.13}$$

In terms of these vectors, a necessary and sufficient condition that $\mathbf{v}_i$ and $\mathbf{v}_j$ are collinear is that $\mathbf{d}_{ij}$ and $\mathbf{v}_j$ are orthogonal to $\mathbf{u}_i$. This pair of equations of constraint that define the geometry of the translational joint are

$$\mathbf{\Phi}^{t(i,j)} = \begin{bmatrix} \mathbf{u}_i^T\mathbf{d}_{ij} \\ \mathbf{u}_i^T\mathbf{v}_j \end{bmatrix} = \mathbf{0} \tag{6.1.14}$$

Note that even when $\mathbf{d}_{ij} = \mathbf{0}$, the second of Eqs. 6.1.14 requires that $\mathbf{v}_i$ and $\mathbf{v}_j$ are parallel. Since they have a point in common when $\mathbf{d}_{ij} = \mathbf{0}$, they are in fact collinear.

For later use in formulation of the equations of motion, the Jacobian matrices of this constraint function with respect to $\mathbf{q}_i$ and $\mathbf{q}_j$ are expanded as (Prob. 6.1.3)

$$\Phi^{t(i,j)}_{\mathbf{q}_i} = \begin{bmatrix} -\mathbf{u}_i'^T\mathbf{A}_i^T & \mathbf{u}_i'^T\mathbf{B}_i^T(\mathbf{r}_j + \mathbf{A}_j\mathbf{s}_j' - \mathbf{r}_i) \\ \mathbf{0} & \mathbf{u}_i'^T\mathbf{B}_i^T\mathbf{A}_j\mathbf{v}_j' \end{bmatrix}$$
$$\Phi^{t(i,j)}_{\mathbf{q}_j} = \begin{bmatrix} \mathbf{u}_i'^T\mathbf{A}_i^T & \mathbf{u}_i'^T\mathbf{A}_i^T\mathbf{B}_j\mathbf{s}_j' \\ \mathbf{0} & \mathbf{u}_i'^T\mathbf{A}_i^T\mathbf{B}_j\mathbf{v}_j' \end{bmatrix} \tag{6.1.15}$$

♦ **Example 6.1.3:** Alternative kinematic models of a slider–crank mechanism may be formulated using different combinations of bodies, joints between bodies, and absolute constraints.

A conventional four body model of the slider–crank mechanism is shown in Fig. 6.1.7. Three revolute joints and one translational joint define the kinematic constraints between the three moving bodies and body 4 is rigidly attached to ground. The constraint equations are defined as follows:

Absolute constraints of Eqs. 6.1.5 to 6.1.7 on body 4 (ground) are

$$x_4'^P = y_4'^P = 0 \quad \text{and} \quad C_1 = C_2 = C_3 = 0$$

Revolute joints of Eq. 6.1.9 between

bodies 1 and 4: $x_1'^P = y_1'^P = x_4'^P = y_4'^P = 0$

bodies 1 and 2: $x_1'^P = a_1,\quad y_1'^P = y_2'^P = 0,\quad x_2'^P = \dfrac{-a_2}{2}$

bodies 2 and 3: $x_2'^P = \dfrac{a_2}{2},\quad y_2'^P = y_3'^P = x_3'^P = 0$

Translational joint of Eq. 6.1.14 between bodies 3 and 4:

$$x_3'^P = -1,\quad y_3'^P = y_3'^Q = 0,\quad x_3'^Q = 1,\quad x_4'^P = -1,\quad y_4'^P = y_4'^Q = 0,\quad x_4'^Q = 1$$

This model has 12 generalized coordinates and 11 independent constraint equations, hence one degree of freedom (Prob. 6.1.4).

A two body model of the slider–crank mechanism is shown in Fig. 6.1.8. The crank and connecting rod are bodies, connected by a revolute joint. The coupling with ground is represented by absolute constraints. The constraint equations for this model are defined as follows:

Absolute constraints:

Eqs. 6.1.5 and 6.1.6 on body 1: $x_1'^P = y_1'^P = 0$ and $C_1 = C_2 = 0$

Eq. 6.1.6 on body 2: $x_2'^P = \dfrac{a_2}{2},\ y_2'^P = 0$, and $C_2 = 0$

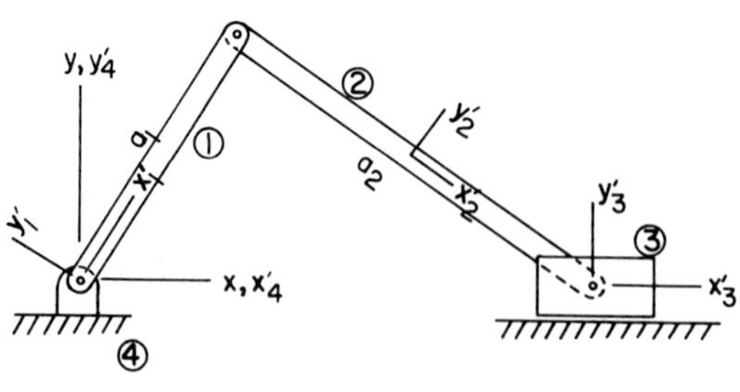

Figure 6.1.7 Four body model of slider–crank.

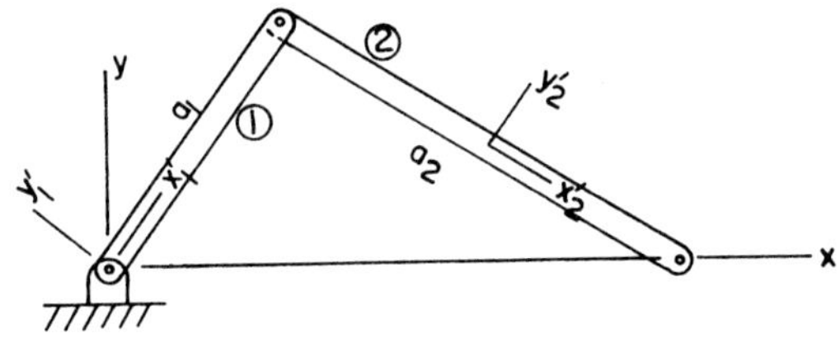

Figure 6.1.8 Two body model of slider–crank.

Revolute joint of Eq. 6.1.9 between bodies 1 and 2:

$$x_1'^P = l_1, \quad y_1'^P = y_2'^P = 0, \quad x_2'^P = \frac{-a_2}{2}$$

This model has six generalized coordinates and five independent constraint equations, hence one degree of freedom.

Position, Velocity, and Acceleration Analysis

Kinematic constraints define permissible motion of a multibody system. For holonomic constraints, they may be written in matrix form as

$$\mathbf{\Phi}(\mathbf{q}) \equiv [\Phi_1(\mathbf{q}), \ldots, \Phi_{nh}(\mathbf{q})]^T = \mathbf{0} \tag{6.1.16}$$

where $\mathbf{q}$ is the vector of $nc = 3nb$ generalized coordinates for the complete system and nh is the number of holonomic constraint equations.

A crucial ingredient in formulating well-posed dynamics models is to assure that the *constraint Jacobian*

$$\mathbf{\Phi_q}(\mathbf{q}) = \left[\frac{\partial \Phi_i(\mathbf{q})}{\partial q_j}\right]_{nh \times nc} \tag{6.1.17}$$

has full row rank (i.e., assure that the constraints are independent). This matrix is formed using Jacobians of the individual constraints (i.e., Eqs. 6.1.8, 6.1.11, and 6.1.15).

Example 6.1.4: To illustrate kinematic analysis, the elementary two body model of a slider–crank mechanism from Example 6.1.3 is analyzed. The mechanism model, with its body-fixed frames, is shown in Fig. 6.1.9. The analytical form of absolute constraints that define the pivot at the left of body 1 and the slider at the right of body 2 are simply $x_1 = y_1 = y_2 = 0$. Using these conditions to suppress the associated generalized coordinates, a reduced generalized coordinate vector $\mathbf{q} = [\phi_1, x_2, \phi_2]^T$ is obtained and the revolute joint constraint equations between bodies 1 and 2 reduce to

$$\mathbf{\Phi}(\mathbf{q}) = \begin{bmatrix} x_2 - \cos\phi_2 - a\cos\phi_1 \\ -\sin\phi_2 - a\sin\phi_1 \end{bmatrix} = \mathbf{0} \tag{6.1.18}$$

The Jacobian matrix for this set of equations is

$$\mathbf{\Phi_q} = \begin{bmatrix} a\sin\phi_1 & 1 & \sin\phi_2 \\ -a\cos\phi_1 & 0 & -\cos\phi_2 \end{bmatrix} \tag{6.1.19}$$

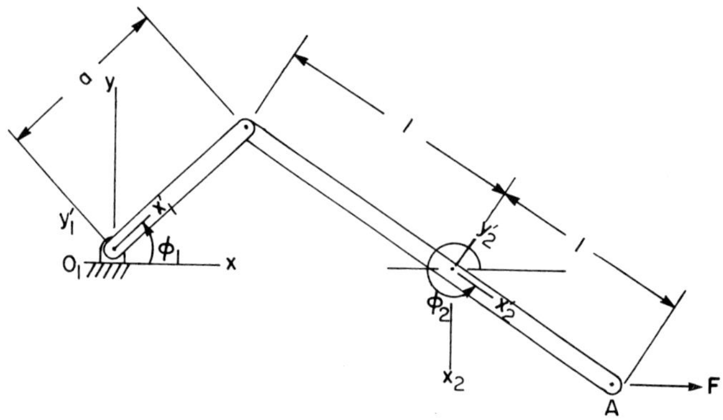

Figure 6.1.9 Elementary model of slider–crank.

Note that this Jacobian is row rank deficient only if $\cos\phi_1 = \cos\phi_2 = 0$. This can occur only if $a = 1$, and even then only when both bodies are vertical.

Since Eq. 6.1.16 must hold for all time, both sides may be differentiated with respect to time to obtain the *velocity equation*

$$\mathbf{\Phi_q \dot{q}} = \mathbf{0} \tag{6.1.20}$$

Just as the velocity equation of Eq. 6.1.20 is obtained by differentiating the kinematic equations of Eq. 6.1.16, both sides of Eq. 6.1.20 can be differentiated to obtain the *acceleration equation*

$$\mathbf{\Phi_q \ddot{q}} = -(\mathbf{\Phi_q \dot{q}})_\mathbf{q}\mathbf{\dot{q}} \equiv \boldsymbol{\gamma} \tag{6.1.21}$$

Note that in both the velocity and acceleration equations, the constraint Jacobian plays a dominant role.

Direct calculations show that $\boldsymbol{\gamma}$ of Eq. 6.1.21 is zero for the absolute position and rotation constraints. For the revolute and translational joint constraints of Eqs. 6.1.9 and 6.1.14 (Prob. 6.1.8),

$$\begin{aligned}\boldsymbol{\gamma}^{r(i,j)} &= \mathbf{A}_j\mathbf{s}_j'\dot{\phi}_j^2 - \mathbf{A}_i\mathbf{s}_i'\dot{\phi}_i^2 \\ \boldsymbol{\gamma}^{t(i,j)} &= \begin{bmatrix} -2\mathbf{u}_i'^T\mathbf{B}_i^T(\dot{\mathbf{r}}_j - \dot{\mathbf{r}}_i)\dot{\phi}_i + \mathbf{u}^{i^T}\mathbf{A}_i^T(\mathbf{r}_j + \mathbf{A}\mathbf{s}_j' - \mathbf{r}_i)\dot{\phi}_i^2 \\ -2\mathbf{u}_i'^T\mathbf{B}_i^T\mathbf{B}_j\mathbf{s}_j'\dot{\phi}_i\dot{\phi}_j + \mathbf{u}_i'^T\mathbf{A}_i^T\mathbf{A}_j\mathbf{s}_j'\dot{\phi}_j^2 \\ 0 \end{bmatrix}\end{aligned} \tag{6.1.22}$$

➧ **Example 6.1.5:** To illustrate use of Eq. 6.1.21, the right side of the acceleration equation for the slider–crank of Example 6.1.4 is (Prob. 6.1.9)

$$\begin{aligned}\boldsymbol{\gamma} = -(\mathbf{\Phi_q \dot{q}})_\mathbf{q}\mathbf{\dot{q}} &= -\begin{bmatrix} a\sin\phi_1\dot{\phi}_1 + \dot{x}_2 + \sin\phi_2\dot{\phi}_2 \\ -a\cos\phi_1\dot{\phi}_1 - \cos\phi_2\dot{\phi}_2 \end{bmatrix}_\mathbf{q}\mathbf{\dot{q}} \\ &= -\begin{bmatrix} a\cos\phi_1\dot{\phi}_1^2 + \cos\phi_2\dot{\phi}_2^2 \\ a\sin\phi_1\dot{\phi}_1^2 + \sin\phi_2\dot{\phi}_2^2 \end{bmatrix}\end{aligned}$$

Virtual Work and Generalized Force

The *virtual work* of applied forces and torques that act on a rigid body in the plane may be written in terms of its Cartesian virtual displacement and rotation $\delta\mathbf{r}$ and $\delta\phi$ as

$$\begin{aligned}\delta W^A &= \delta\mathbf{r}^T\mathbf{F}^A + \delta\phi n^A \\ &= [\delta\mathbf{r}^T, \delta\phi][\mathbf{F}^{A^T}, n^A]^T \\ &\equiv \delta\mathbf{q}^T\mathbf{Q}^A\end{aligned}$$

where $\mathbf{F}^A$ is the resultant applied force at the centroid of the body and n^A is the associated resultant applied torque. The *generalized applied force* $\mathbf{Q}^A$ is simply the coefficient of $\delta\mathbf{q}$; that is,

$$\mathbf{Q}^A = [\mathbf{F}^{A^T}, n^A]^T \tag{6.1.23}$$

A common *force element* that is encountered in mechanical system dynamics is a compliant connection between points P_i and P_j on bodies i and j, respectively, as shown in Fig. 6.1.10. The *compliant element* that connects these points may consist of a spring, a damper, or an actuator that exerts a force along the vector between points P_i and P_j. Such an element is called a *translational spring–damper–actuator*. Since this force element exerts equal and opposite collinear forces on bodies i and j, rather than calculating generalized forces as in the case of external applied forces, it is easier to make direct use of the virtual work definition of generalized force.

Consider first the vector $\mathbf{d}_{ij}$ from point P_i to point P_j,

$$\mathbf{d}_{ij} = \mathbf{r}_j + \mathbf{A}_j\mathbf{s}_j' - \mathbf{r}_i - \mathbf{A}_i\mathbf{s}_i' \tag{6.1.24}$$

The length l of this element is calculated from

$$l^2 = \mathbf{d}_{ij}^T\mathbf{d}_{ij} \tag{6.1.25}$$

In applications, the length of a force element is never zero, so l may be chosen

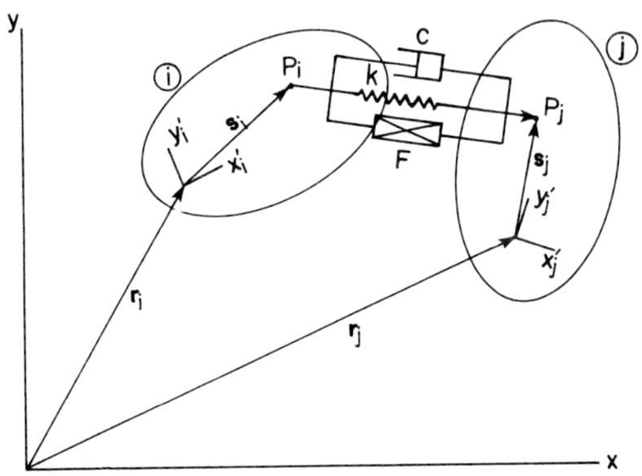

Figure 6.1.10 Planar translational spring–damper–actuator.

as positive. Differentiating Eq. 6.1.25 yields

$$2l\dot{l} = 2\mathbf{d}_{ij}^T\dot{\mathbf{d}}_{ij}$$

Since l is never equal to zero, both sides may be divided by l and Eq. 6.1.24 may be differentiated to obtain

$$\dot{l} = \left(\frac{\mathbf{d}_{ij}}{l}\right)^T (\dot{\mathbf{r}}_j + \mathbf{B}_j\mathbf{s}'_j\dot{\phi}_j - \dot{\mathbf{r}}_i - \mathbf{B}_i\mathbf{s}'_i\dot{\phi}_i) \tag{6.1.26}$$

Defining tension in the compliant element as positive (i.e., a force that tends to draw the bodies together is positive), the force that acts in the element is

$$f = k(l - l_0) + c\dot{l} + F(l, \dot{l}) \tag{6.1.27}$$

where k is the *spring coefficient*, l_0 the *free length of the spring*, c the *damping coefficient*, and F a general *actuator force* that may depend on l and $\dot{l}$.

Making direct use of the definition of virtual work, the virtual work of the force f is simply the product of the force and the variation in length of the element. Since a positive δl tends to separate the bodies and the sign convention on f is tension as positive, the virtual work is

$$\delta W^A = -f\,\delta l \tag{6.1.28}$$

Taking the differential of both sides of Eq. 6.1.25 and manipulating, as in derivation of the velocity relation of Eq. 6.1.26,

$$\delta l = \left(\frac{\mathbf{d}_{ij}}{l}\right)^T (\delta\mathbf{r}_j + \mathbf{B}_j\mathbf{s}'_j\,\delta\phi_j - \delta\mathbf{r}_i - \mathbf{B}_i\mathbf{s}'_i\,\delta\phi_i) \tag{6.1.29}$$

Substituting this result in Eq. 6.1.28, the virtual work of the spring–damper–actuator force is

$$\delta W^A = -\frac{f}{l}\mathbf{d}_{ij}^T(\delta\mathbf{r}_j + \mathbf{B}_j\mathbf{s}'_j\,\delta\phi_j - \delta\mathbf{r}_i - \mathbf{B}_i\mathbf{s}'_i\,\delta\phi_i) \tag{6.1.30}$$

By definition, the generalized forces that act on bodies i and j are simply the coefficients of generalized coordinate variations in Eq. 6.1.30; that is,

$$\begin{aligned}\mathbf{Q}_i^A &= \frac{f}{l}\begin{bmatrix}\mathbf{d}_{ij}\\ \mathbf{d}_{ij}^T\mathbf{B}_i\mathbf{s}'_i\end{bmatrix}\\ \mathbf{Q}_j^A &= -\frac{f}{l}\begin{bmatrix}\mathbf{d}_{ij}\\ \mathbf{d}_{ij}^T\mathbf{B}_j\mathbf{s}'_j\end{bmatrix}\end{aligned} \tag{6.1.31}$$

Example 6.1.6: A spring–damper is attached between bodies 1 and 2 of Example 6.1.4, as shown in Fig. 6.1.11. With $\mathbf{s}'_i = [a/2, 0]^T$ and $\mathbf{s}'_2 = [-(\frac{1}{2}), 0]^T$, the

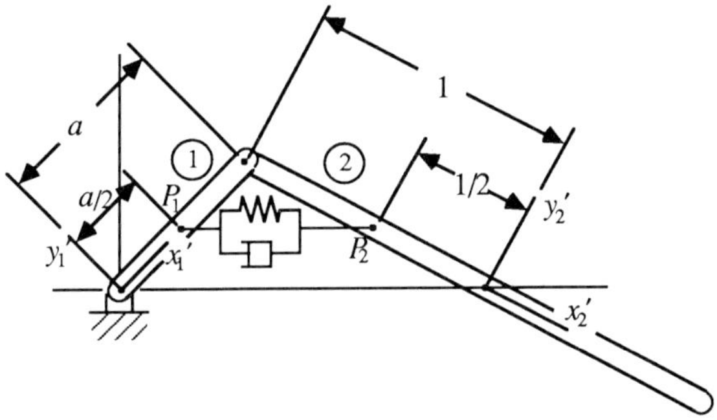

Figure 6.1.11 Elementary slider–crank with spring–damper.

generalized forces of Eq. 6.1.31 are

$$\mathbf{Q}_1^A = \frac{f}{l}\begin{bmatrix} x_2 - \frac{1}{2}\cos\phi_2 - x_1 - \frac{a}{2}\cos\phi_1 \\ y_2 - \frac{1}{2}\sin\phi_2 - y_1 - \frac{a}{2}\sin\phi_1 \\ \frac{a}{2}(-x_2 + x_1)\sin\phi_1 + (y_2 - y_1)\cos\phi_1 \\ + \frac{1}{2}\sin(\phi_1 - \phi_2) \end{bmatrix}$$

$$\mathbf{Q}_2^A = -\frac{f}{l}\begin{bmatrix} x_2 - \frac{1}{2}\cos\phi_2 - x_1 - \frac{a}{2}\cos\phi_1 \\ y_2 - \frac{1}{2}\sin\phi_2 - y_1 - \frac{a}{2}\sin\phi_1 \\ \frac{1}{2}(x_2 - x_1)\sin\phi_2 + (-y_2 + y_1)\cos\phi_2 \\ + \frac{a}{2}\sin(\phi_2 - \phi_1) \end{bmatrix}$$

where $f = k(l - l_0) + c\dot{l}$.

In Example 6.1.4, $x_1 = y_1 = y_2 = 0$, so $\delta x_1 = \delta y_1 = \delta y_2 = 0$. Thus, with $\mathbf{q} = [\phi_1, x_2, \phi_2]^T$,

$$l = \left[x_2^2 - x_2(\cos\phi_2 + a\cos\phi_1) + \left(\frac{1}{4} + \frac{a^2}{4}\right) + \frac{a}{2}\cos(\phi_2 - \phi_1)\right]^{1/2}$$

and the virtual work of the spring–damper force is

$$\begin{aligned}\delta W^A &= [0, 0, \delta\phi_1]\mathbf{Q}_1^A + [\delta x_2, 0, \delta\phi_2]\mathbf{Q}_2^A \\ &= [\delta\phi_1, \delta x_2, \delta\phi_2]\frac{f}{l}\begin{bmatrix} \frac{a}{2}\left(\frac{1}{2}\sin(\phi_1 - \phi_2) - x_2\sin\phi_1\right) \\ \frac{1}{2}\cos\phi_2 - x_2 + \frac{a}{2}\cos\phi_1 \\ \frac{1}{2}\left(-x_2\sin\phi_2 - \frac{a}{2}\sin(\phi_1 - \phi_2)\right) \end{bmatrix} \\ &\equiv \delta\mathbf{q}^T\mathbf{Q}^A\end{aligned}$$

Constraint Reaction Forces

Consider next the effect of *constraint reaction forces* between bodies at kinematic connections. If the kinematic connections or joints are friction free, forces of constraint are normal to surfaces of the joint that may move relative to one another, hence they do no work. Such joints are called *workless constraints*. Thus, even though substantial constraint forces act in these joints, their virtual work is zero, for all *kinematically admissible virtual displacements*.

Denote by $\mathbf{\Phi}^k = \mathbf{0}$ the equations of constraint of physical joint k that acts between a pair of bodies in a multibody system, $k = 1, \ldots, nhp$. Let $\mathbf{F}_l''^{C_k}$ and $T_l^{C_k}$ be the constraint force and torque due to this joint, acting on body l at the $x_l'' - y_l''$ joint reference frame, as shown in Fig. 6.1.12, with $l = i$ and j. The virtual work due to all constraint reaction forces associated with joint k is

$$\delta W^{C_k} = \sum_{l=1}^{nb} [\delta \mathbf{r}_l''^{PT} \mathbf{F}_l''^{C_k} + \delta\phi_l T_l^{C_k}] \tag{6.1.32}$$

where the only nonzero terms that arise are for $l = i$ and j.

A virtual displacement of the system is consistent with constraint k if

$$\sum_{l=1}^{nb} [\mathbf{\Phi}_{\mathbf{r}_l}^k \,\delta\mathbf{r}_l + \mathbf{\Phi}_{\phi_l}^k \,\delta\phi_l] = \mathbf{0} \tag{6.1.33}$$

To use Eqs. 6.1.32 and 6.1.33, it is desirable to write all virtual displacements in terms of $\delta\mathbf{r}_l''^P$ and $\delta\phi_l$. Recall that

$$\mathbf{r}_l^P = \mathbf{r}_l + \mathbf{A}_l \mathbf{s}_l' \tag{6.1.34}$$

whose differential yields

$$\delta\mathbf{r}_l^P = \delta\mathbf{r}_l + \mathbf{B}_l \mathbf{s}_l' \,\delta\phi_l \tag{6.1.35}$$

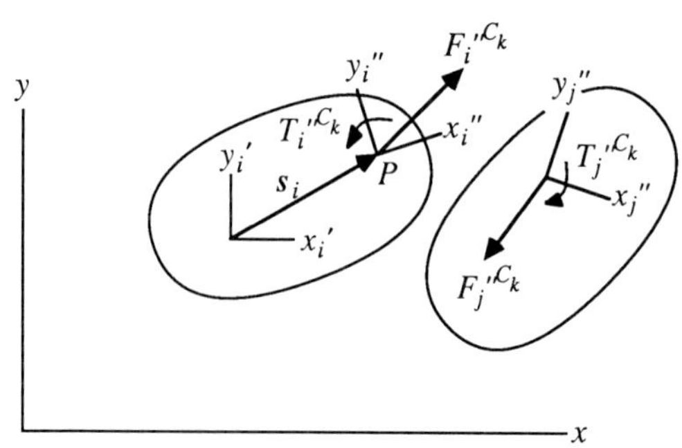

Figure 6.1.12 Constraint reaction force and torque.

Since virtual displacement is a vector quantity and the transformation matrix $\mathbf{C}_l$ transforms vectors from the x_l''-y_l'' frame to the x_l'-y_l' frame,

$$\delta\mathbf{r}_l'^P = \mathbf{A}_l\mathbf{C}_l\,\delta\mathbf{r}_l''^P \tag{6.1.36}$$

Substituting from Eq. 6.1.36 into Eq. 6.1.35 yields

$$\delta\mathbf{r}_l = \mathbf{A}_l\mathbf{C}_l\,\delta\mathbf{r}_l''^P - \mathbf{B}_l\mathbf{s}_l'\,\delta\phi_l \tag{6.1.37}$$

Substituting this result into Eq. 6.1.33 gives

$$\sum_{l=1}^{nb}[\mathbf{\Phi}_{\mathbf{r}_l}^k\mathbf{A}_l\mathbf{C}_l\,\delta\mathbf{r}_l''^P + (\mathbf{\Phi}_{\phi_l}^k - \mathbf{\Phi}_{\mathbf{r}_l}^k\mathbf{B}_l\mathbf{s}_l')\,\delta\phi_l] = \mathbf{0} \tag{6.1.38}$$

Since δW^{C_k} of Eq. 6.1.32 must be zero for all $\delta\mathbf{r}_l''^P$ and $\delta\phi_l$, $l = 1, \ldots, nb$, that satisfy Eq. 6.1.38, the Lagrange multiplier theorem guarantees existence of a multiplier $\boldsymbol{\lambda}^k$ such that

$$\sum_{l=1}^{nb}\{\delta\mathbf{r}_l''^{PT}[\mathbf{F}_l''^{C_k} + \mathbf{C}_l^T\mathbf{A}_l^T\mathbf{\Phi}_{\mathbf{r}_l}^{kT}\boldsymbol{\lambda}^k] + \delta\phi_l[T_l^{C_k} + \mathbf{\Phi}_{\phi_l}^{kT}\boldsymbol{\lambda}^k - \mathbf{s}_l'^T\mathbf{B}_l^T\mathbf{\Phi}_{\mathbf{r}_l}^{kT}\boldsymbol{\lambda}^k\} = 0 \tag{6.1.39}$$

for all $\delta\mathbf{r}_l''^P$ and $\delta\phi_l$. Thus,

$$\begin{aligned}\mathbf{F}_l''^{C_k} &= -\mathbf{C}_l^T\mathbf{A}_l^T\mathbf{\Phi}_{\mathbf{r}_l}^{kT}\boldsymbol{\lambda}^k \\ T_l^{C_k} &= -\mathbf{\Phi}_{\phi_l}^{kT}\boldsymbol{\lambda}^k + \mathbf{s}_l'^T\mathbf{B}_l^T\mathbf{\Phi}_{\mathbf{r}_l}^{kT}\boldsymbol{\lambda}^k\end{aligned} \tag{6.1.40}$$

Example 6.1.7: Constraint reaction force and torque, in the x_i''-y_i'' joint reference frame on body i of Fig. 6.1.12, for the revolute joint are given by Eqs. 6.1.11 and 6.1.40 as

$$\mathbf{F}_i''^{C_r} = \mathbf{C}_i^T\mathbf{A}_i^T\boldsymbol{\lambda}^r$$

$$T_i^{C_r} = \mathbf{s}_i'^T\mathbf{B}_i^T\boldsymbol{\lambda}^r - \mathbf{s}_i'^T\mathbf{B}_i^T\boldsymbol{\lambda}^r = 0$$

where $\boldsymbol{\lambda}^r$ is the Lagrange multiplier vector for the revolute joint. A zero value of torque about the center of the joint is, of course, expected. In the global x-y frame,

$$\mathbf{F}_i^{C_r} = \mathbf{A}_i\mathbf{C}_i\mathbf{F}_i''^r = \boldsymbol{\lambda}^r$$

Similar explicit relations may be derived for the translational joint and absolute constraints (Probs. 6.1.14 and 6.1.15).

To obtain the generalized constraint reaction forces in terms of Lagrange multipliers, both sides of Eq. 6.1.37 are multiplied on the left by $\mathbf{C}_l^T\mathbf{A}_l^T$ to obtain

$$\delta\mathbf{r}_l''^P = \mathbf{C}_l^T\mathbf{A}_l^T(\delta\mathbf{r}_l + \mathbf{B}_l\mathbf{s}_l'\delta\phi_l) \tag{6.1.41}$$

Substituting this result and Eq. 6.1.40 into Eq. 6.1.32 yields

$$\begin{aligned}
\delta W^{C_k} &= \sum_{l=1}^{nb} [-(\delta \mathbf{r}_l^T + \delta\phi_l \mathbf{s}_l'^T \mathbf{B}_l^T)\mathbf{A}_l \mathbf{C}_l \mathbf{C}_l^T \mathbf{A}_l^T \mathbf{\Phi}_{\mathbf{r}_l}^{k^T} \boldsymbol{\lambda}^k \\
&\quad - \delta\phi_l (\mathbf{\Phi}_{\phi_l}^{k^T}\boldsymbol{\lambda}^k - \mathbf{s}_l'^T \mathbf{B}_l^T \mathbf{\Phi}_{\mathbf{r}_l}^{k^T}\boldsymbol{\lambda}^k)] \\
&= -\sum_{l=1}^{nb} [\delta \mathbf{r}_l^T, \delta\phi_l][\mathbf{\Phi}_{\mathbf{r}_l}^k, \mathbf{\Phi}_{\phi_l}^k]^T \boldsymbol{\lambda}^k \qquad (6.1.42) \\
&= -\sum_{l=1}^{nb} \delta \mathbf{q}_l^T \mathbf{\Phi}_{\mathbf{q}_l}^{k^T} \boldsymbol{\lambda}^k \\
&\equiv -\sum_{l=1}^{nb} \delta \mathbf{q}_l^T \mathbf{Q}_l^{C_k}
\end{aligned}$$

Thus the generalized constraint force on body l due to joint k is

$$\mathbf{Q}_l^{C_k} = -\mathbf{\Phi}_{\mathbf{q}_l}^{k^T} \boldsymbol{\lambda}^k \qquad (6.1.43)$$

and the total generalized constraint force on body l is

$$\mathbf{Q}_l^{C} = -\sum_{k=1}^{nhp} \mathbf{\Phi}_{\mathbf{q}_l}^{k^T} \boldsymbol{\lambda}^k \qquad (6.1.44)$$

Equations of Motion of Constrained Planar Systems

For mechanical systems that are made up of a collection of rigid bodies in the plane, with kinematic constraints between bodies defined by Eq. 6.1.16, the variational equations of motion of Eq. 5.1.42, accounting for both applied and constraint forces, are

$$\delta \mathbf{q}_i^T [\mathbf{M}_i \ddot{\mathbf{q}}_i - \mathbf{Q}_i^A - \mathbf{Q}_i^C] = 0, \qquad i = 1, \ldots, nb \qquad (6.1.45)$$

or, using Eq. 6.1.44,

$$\delta \mathbf{q}_i^T [\mathbf{M}_i \ddot{\mathbf{q}}_i - \mathbf{Q}_i^A + \sum_{k=1}^{nh} \mathbf{\Phi}_{\mathbf{q}_i}^{k^T} \boldsymbol{\lambda}^k] = 0, \qquad i = 1, \ldots, nb \qquad (6.1.46)$$

for all $\delta \mathbf{q}_i$, where

$$\mathbf{M}_i = \mathrm{diag}(m_i, m_i, J_{0_i}) \qquad (6.1.47)$$

In order to treat the complete multibody system, it is convenient to define a composite vector of generalized coordinates, a composite mass matrix, a composite vector of applied generalized forces, and a composite vector of Lagrange multipliers as

$$\begin{aligned} \mathbf{q} &= [\mathbf{q}_1^T, \mathbf{q}_2^T, \ldots, \mathbf{q}_{nb}^T]^T \\ \mathbf{M} &= \text{diag}(\mathbf{M}_1, \mathbf{M}_2, \ldots, \mathbf{M}_{nb}) \\ \mathbf{Q}^A &= [\mathbf{Q}_1^{A^T}, \mathbf{Q}_2^{A^T}, \ldots, \mathbf{Q}_{nb}^{A^T}]^T \\ \boldsymbol{\lambda} &= [\boldsymbol{\lambda}^{1^T}, \boldsymbol{\lambda}^{2^T}, \ldots, \boldsymbol{\lambda}^{nh^T}]^T \end{aligned} \tag{6.1.48}$$

Using this notation, the variational equations of Eq. 6.1.46 may be summed and written in the more compact form

$$\delta\mathbf{q}^T[\mathbf{M}\ddot{\mathbf{q}} + \boldsymbol{\Phi}_{\mathbf{q}}^T\boldsymbol{\lambda} - \mathbf{Q}^A] = 0 \tag{6.1.49}$$

which must hold for all $\delta\mathbf{q}$. Thus the orthogonality theorem implies that

$$\mathbf{M}\ddot{\mathbf{q}} + \boldsymbol{\Phi}_{\mathbf{q}}^T\boldsymbol{\lambda} = \mathbf{Q}^A \tag{6.1.50}$$

In addition to these equations, the position, velocity, and acceleration equations of Eqs. 6.1.16, 6.1.20, and 6.1.21 must hold. Equations 6.1.50 and 6.1.21 may be written in matrix form as

$$\begin{bmatrix} \mathbf{M} & \boldsymbol{\Phi}_{\mathbf{q}}^T \\ \boldsymbol{\Phi}_{\mathbf{q}} & \mathbf{0} \end{bmatrix} \begin{bmatrix} \ddot{\mathbf{q}} \\ \boldsymbol{\lambda} \end{bmatrix} = \begin{bmatrix} \mathbf{Q}^A \\ \boldsymbol{\gamma} \end{bmatrix} \tag{6.1.51}$$

This is a system of *differential-algebraic equations*, since no derivatives of the Lagrange multiplier $\boldsymbol{\lambda}$ appear and the algebraic constraint equations of Eq. 6.1.16 must be satisfied.

Example 6.1.8: The equations of motion of the slider–crank mechanism of Fig. 6.1.9, with the spring–damper of Fig. 6.1.11, may be written in terms of the full 6-vector $[\mathbf{q}_1^T, \mathbf{q}_2^T]^T$ of Cartesian generalized coordinates and five constraints. As an alternative form of the equations of motion, x_1, y_1, and y_2 may be suppressed, as in Examples 6.1.4 and 6.1.6. In this reduced formulation, $\mathbf{q} = [\phi_1, x_2, \phi_2]^T$. For the force $\mathbf{F}$ in Fig. 6.1.9, $x_A = x_2 + \cos\phi_2$, so $\delta W^A = F\,\delta x_A = F(\delta x_2 - \delta\phi_2 \sin\phi_2)$. Since the centroids of bodies 1 and 2 do not move in the y direction, their generalized gravitational forces are zero. Substituting $\delta x_1 = \delta y_1 = \delta y_2 = 0$ in Eq. 6.1.49 and using results of Example 6.1.6,

$$\begin{aligned} &\delta\phi_1\left[J_{0_1}\ddot{\phi}_1 + \boldsymbol{\Phi}_{\phi_1}^T\boldsymbol{\lambda} - \frac{f}{l}\frac{a}{2}\left(\frac{1}{2}\sin(\phi_1 - \phi_2) - x_2\sin\phi_1\right)\right] \\ &\quad + \delta x_2\left[m_2\ddot{x}_2 + \boldsymbol{\Phi}_{x_2}^T\boldsymbol{\lambda} - \frac{f}{l}\left(\frac{1}{2}\cos\phi_2 - x_2 + \frac{a}{2}\cos\phi_1\right) - F\right] \\ &\quad + \delta\phi_2\left[J_{0_2}\ddot{\phi}_2 + \boldsymbol{\Phi}_{\phi_2}^T\boldsymbol{\lambda} + \frac{f}{l}\frac{1}{2}\left(x_2\sin\phi_2 + \frac{a}{2}\sin(\phi_1 - \phi_2)\right) + F\sin\phi_2\right] = 0 \end{aligned}$$

for all $\delta\mathbf{q} = [\delta\phi_1, \delta x_2, \delta\phi_2]^T$. With the Jacobian of Eq. 6.1.19 and γ from Example 6.1.5 (Prob. 6.1.12), this yields

$$\begin{bmatrix} J_{0_1} & 0 & 0 & a\sin\phi_1 & -a\cos\phi_1 \\ 0 & m_2 & 0 & 1 & 0 \\ 0 & 0 & J_{0_2} & \sin\phi_2 & -\cos\phi_2 \\ a\sin\phi_1 & 1 & \sin\phi_2 & 0 & 0 \\ -a\cos\phi_1 & 0 & -\cos\phi_2 & 0 & 0 \end{bmatrix} \begin{bmatrix} \ddot{\phi}_1 \\ \ddot{x}_2 \\ \ddot{\phi}_2 \\ \lambda_1 \\ \lambda_2 \end{bmatrix}$$

$$= \begin{bmatrix} \dfrac{f}{l}\dfrac{a}{2}\left(\dfrac{1}{2}\sin(\phi_1 - \phi_2) - x_2\sin\phi_1\right) \\ \dfrac{f}{l}\left(\dfrac{1}{2}\cos\phi_2 - x_2 + \dfrac{a}{2}\cos\phi_1\right) + F \\ \dfrac{f}{l}\dfrac{1}{2}\left(-x_2\sin\phi_2 - \dfrac{a}{2}\sin(\phi_1 - \phi_2)\right) - F\sin\phi_2 \\ -(a\dot{\phi}_1^2\cos\phi_1 + \dot{\phi}_2^2\cos\phi_2) \\ -(a\dot{\phi}_1^2\sin\phi_1 + \dot{\phi}_2^2\sin\phi_2) \end{bmatrix}$$

It is important to know when Eq. 6.1.51 has a unique solution for accelerations and Lagrange multipliers. A practical condition for the coefficient matrix in Eq. 6.1.51 to be nonsingular is that the constraints be independent (i.e., $\mathbf{\Phi_q}$ has full row rank) and that kinetic energy of the system be positive for any nonzero *virtual velocity* that is consistent with constraints; that is,

$$\delta\dot{\mathbf{q}}^T\mathbf{M}\,\delta\dot{\mathbf{q}} > 0 \tag{6.1.52}$$

for all nonzero *kinematically admissible virtual velocities*; that is, for all $\delta\dot{\mathbf{q}} \neq \mathbf{0}$ that satisfy

$$\mathbf{\Phi_q}\,\delta\dot{\mathbf{q}} = \mathbf{0} \tag{6.1.53}$$

This result is proved just as the constrained dynamic existence theorem of Section 3.4 (Prob. 6.1.13).

Initial Conditions

A set of additional conditions that, together with the kinematic constraint and velocity equations, uniquely determine initial position and velocity of the multibody system must be defined to initiate motion of a system.

Initial position conditions are defined in the form

$$\mathbf{\Phi}^I(\mathbf{q}(t_0)) = \mathbf{0} \tag{6.1.54}$$

where it is required that the number of equations in Eq. 6.1.54 be equal to the number of degrees of freedom. Furthermore, Eq. 6.1.16 at t_0 and Eq. 6.1.54 must be independent to uniquely determine the *initial position* $\mathbf{q}(t_0)$.

Initial velocity conditions may be given in the form

$$\mathbf{B}^I\dot{\mathbf{q}}(t_0) = \mathbf{v}^I \tag{6.1.55}$$

It is required that Eq. 6.1.20 at t_0 and Eq. 6.1.55 uniquely determine the *initial velocity* $\dot{\mathbf{q}}(t_0)$. Thus a complete set of initial conditions that are consistent with constraints is obtained.

Example 6.1.9: Since the slider–crank model of Examples 6.1.4 to 6.1.6 has one degree of freedom, one initial position condition and one initial velocity condition are required. If initial conditions on ϕ_1 are specified as $\Phi^I = \phi_1 - \phi_1^0 = 0$ and $\mathbf{B}^I\dot{\mathbf{q}}(t_0) = [1, 0, 0]\dot{\mathbf{q}}(t_0) = \dot{\phi}_1^0$, then both the combined position and velocity Jacobians are, from Eq. 6.1.19,

$$\begin{bmatrix} \mathbf{\Phi_q} \\ \Phi^I_{\mathbf{q}} \end{bmatrix} = \begin{bmatrix} \mathbf{\Phi_q} \\ \mathbf{B}^I \end{bmatrix} = \begin{bmatrix} a \sin \phi_1 & 1 & \sin \phi_2 \\ -a \cos \phi_1 & 0 & -\cos \phi_2 \\ 1 & 0 & 0 \end{bmatrix}$$

The determinant of this matrix is $-\cos \phi_2$. If $a < 1$, in which case the crank can turn a full revolution, then $\pi/2 < \phi_2 < \pi/2$ and $\cos \phi_2 \neq 0$. Thus a well-defined set of initial conditions is established.

As illustrated by the general form of the constrained equations of motion in Eq. 6.1.51 and with the slider–crank in Example 6.1.7, the Cartesian coordinate formulation of constrained system dynamics is systematic and generally applicable to large classes of mechanical systems. As developed in detail and illustrated in the computationally oriented text of Ref. 12, this method permits systematic coding of a general purpose mechanical system simulation software system that automatically generates and solves Eq. 6.1.51. The value of the Cartesian coordinate formulation is, therefore, principally associated with the systematic nature of the equations of motion it yields and the ease of direct numerical implementation.

In contrast to the attractiveness of the Cartesian coordinate formulation for computational implementation, it is of marginal value for analytical studies of realistic mechanical systems. It yields a large system of relatively simple equations of motion, using a maximal set of generalized coordinates and associated holonomic constraint equations. The resulting system of differential-algebraic equations of motion, Eqs. 6.1.16 and 6.1.51, is intractable for analytical solution in realistic applications. It is therefore of significant value only for systematic numerical implementation.

6.2 CARTESIAN COORDINATE FORMULATION FOR SPATIAL SYSTEMS

The same basic approach used in Section 6.1 to derive planar multibody equations of motion is extended in this section to spatial Cartesian coordinates and equations of constraint between bodies. The resulting formulation is well suited for computer implementation, but is impractical for analytical solution. Attention is therefore limited in this section to development of the formulation, which would subsequently need to be implemented for solution by a computer program [12].

Spatial Cartesian generalized coordinates for body i in a multibody system are selected as the vector $\mathbf{r}_i$ from the origin of the inertial reference frame to the origin of a centroidal body-fixed x_i'-y_i'-z_i' frame, as shown in Fig. 6.2.1, and the vector $\mathbf{p}_i$ of Euler parameters that define the orientation of this frame; that is,

$$\mathbf{q}_i = \begin{bmatrix} \mathbf{r}_i \\ \mathbf{p}_i \end{bmatrix}_{7\times 1}, \qquad i = 1, \ldots, nb \tag{6.2.1}$$

where nb is the number of bodies in the system. The composite vector of *Cartesian generalized coordinates* for the multibody system may thus be written as

$$\mathbf{q} = [\mathbf{q}_1^T, \mathbf{q}_2^T, \ldots, \mathbf{q}_{nb}^T]^T_{7nb\times 1} \tag{6.2.2}$$

Constraints on Vectors and Points

Kinematic constraints on relative position and orientation of pairs of points and vectors on two bodies are derived in this subsection. Body-fixed joint definition frames are defined and conditions for parallelism and orthogonality of pairs of vectors are defined, to serve as building blocks that characterize a variety of kinematic constraints. It must also be remembered that the Euler parameter vectors $\mathbf{p}_i$, $i = 1, \ldots, nb$, must satisfy the normalization constraint of Eq. 4.3.3,

$$\mathbf{p}_i^T\mathbf{p}_i - 1 = 0, \qquad i = 1, \ldots, nb$$

For the typical body, denoted body i in Fig. 6.2.1, the centroidal x_i'-y_i'-z_i' body-fixed reference frame is used to position and orient the body in space. A second x_i''-y_i''-z_i'' frame is attached to the body, with its origin at point P_i. To orient the x_i''-y_i''-z_i'' frame, unit vectors $\mathbf{f}_i$, $\mathbf{g}_i$, and $\mathbf{h}_i$ are defined along its coordinate axes.

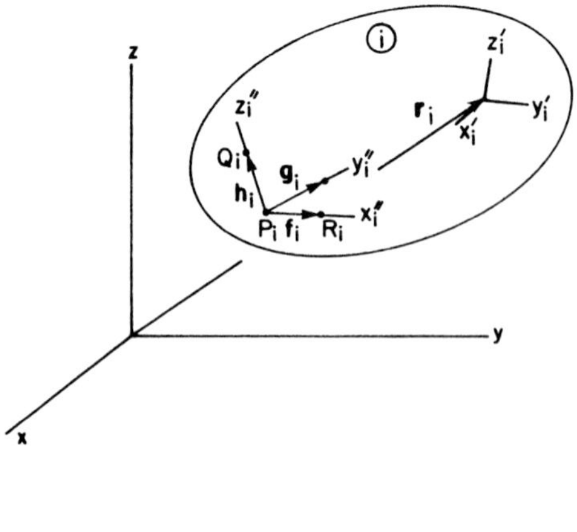

Figure 6.2.1 Body and joint definition frames.

The transformation matrix from the x_i''-y_i''-z_i'' frame to the x_i'-y_i'-z_i' frame, as in Eq. 2.5.13, is

$$\mathbf{C}_i^P = [\mathbf{f}_i', \mathbf{g}_i', \mathbf{h}_i'] \tag{6.2.3}$$

Consider the pair of rigid bodies, denoted bodies i and j, shown in Fig. 6.2.2. Reference points P_i and P_j and nonzero vectors $\mathbf{a}_i$ and $\mathbf{a}_j$ are fixed in bodies i and j, respectively. Kinematic constraints between pairs of bodies are often characterized by conditions that such pairs of vectors are orthogonal or parallel. The purpose here is to derive analytical conditions that require pairs of vectors to be orthogonal and parallel.

A necessary and sufficient condition that a pair of nonzero body-fixed vectors $\mathbf{a}_i$ and $\mathbf{a}_j$ on bodies i and j, respectively, are orthogonal is that their scalar product is zero; that is,

$$\Phi^{d1}(\mathbf{a}_i, \mathbf{a}_j) \equiv \mathbf{a}_i^T\mathbf{a}_j = 0 \tag{6.2.4}$$

Writing the vectors $\mathbf{a}_i$ and $\mathbf{a}_j$ in terms of their respective body orientation transformation matrices and body-fixed constant vectors, Eq. 6.2.4 may be written in the form

$$\Phi^{d1}(\mathbf{a}_i, \mathbf{a}_j) = \mathbf{a}_i'^T\mathbf{A}_i^T\mathbf{A}_j\mathbf{a}_j' = 0 \tag{6.2.5}$$

where the superscript $d1$ indicates the first form of scalar product, or dot product condition. This condition is called the *dot-1 constraint* between vectors $\mathbf{a}_i$ and $\mathbf{a}_j$. Note that it is written in terms of a pair of body-fixed constant vectors and the transformation matrices $\mathbf{A}_i$ and $\mathbf{A}_j$, which depend on the orientation generalized coordinates of bodies i and j, respectively. Thus the dot-1 constraint restricts the relative orientation of a pair of bodies.

The scalar product condition can also be used to prescribe orthogonality of a body-fixed vector $\mathbf{a}_i$ and the nonzero vector $\mathbf{d}_{ij}$ between bodies, shown in Fig.

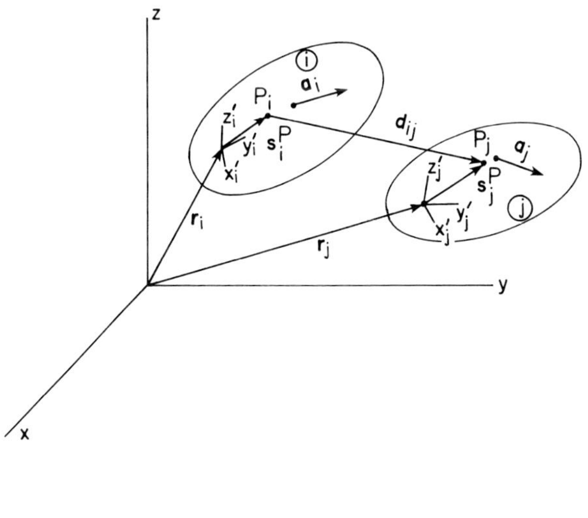

Figure 6.2.2 Vectors fixed in and between bodies.

6.2.2. This condition is defined, using the scalar product, as

$$\Phi^{d2}(\mathbf{a}_i, \mathbf{d}_{ij}) = \mathbf{a}_i^T\mathbf{d}_{ij} = 0 \tag{6.2.6}$$

where the superscript $d2$ indicates a second form of scalar product condition. This condition is called the *dot-2 constraint* between vectors $\mathbf{a}_i$ and $\mathbf{d}_{ij}$. Writing the vector $\mathbf{d}_{ij}$ as

$$\mathbf{d}_{ij} = \mathbf{r}_j + \mathbf{A}_j\mathbf{s}_j' - \mathbf{r}_i - \mathbf{A}_i\mathbf{s}_i' \tag{6.2.7}$$

Eq. 6.2.6 becomes

$$\Phi^{d2}(\mathbf{a}_i, \mathbf{d}_{ij}) = \mathbf{a}_i'^T\mathbf{A}_i^T(\mathbf{r}_j + \mathbf{A}_j\mathbf{s}_j' - \mathbf{r}_i) - \mathbf{a}_i'^T\mathbf{s}_i' = 0 \tag{6.2.8}$$

Note that the dot-2 constraint is not symmetric as regards bodies i and j. To require that the vector $\mathbf{a}_j$ on body j be orthogonal to $\mathbf{d}_{ij}$, the dot-2 constraint of Eq. 6.2.8 can be used, simply by interchanging indices i and j.

It is important to note that the dot-2 constraint fails to impose an orthogonality condition if $\mathbf{d}_{ij} = \mathbf{0}$. Thus if the dot-2 constraint is used to define constraint conditions for a joint, special care is required to assure that the analytical conditions selected indeed imply the geometry of the joint when $\mathbf{d}_{ij} = \mathbf{0}$.

It is often required that a pair of points P_i and P_j shown in Fig. 6.2.2 on two bodies coincide. Such a condition is defined by requiring that $\mathbf{d}_{ij} = \mathbf{0}$; that is,

$$\mathbf{\Phi}^s(P_i, P_j) = \mathbf{r}_j + \mathbf{A}_j\mathbf{s}_j' - \mathbf{r}_i - \mathbf{A}_i\mathbf{s}_i' = \mathbf{0} \tag{6.2.9}$$

where the superscript s anticipates use of this equation in defining a spherical, or ball-and-socket joint. It is called the *spherical constraint*. Note that this vector equation consists of three scalar equations.

The three constraint equations derived thus far form the basis for defining a library of kinematic constraints between bodies. In anticipation that differentials and derivatives of constraint functions with respect to generalized coordinates of bodies i and j need to be calculated, the vector and differential calculus methods of Chapters 2 and 4 are used to derive the needed differential and derivative expressions.

The differential of the dot-1 constraint function of Eq. 6.2.4 was derived in Section 4.2. From Eq. 4.2.29 it is

$$\delta\Phi^{d1}(\mathbf{a}_i, \mathbf{a}_j) = -(\mathbf{a}_j'^T\mathbf{A}_j^T\mathbf{A}_i\tilde{\mathbf{a}}_i'\,\delta\boldsymbol{\pi}_i' + \mathbf{a}_i'^T\mathbf{A}_i^T\mathbf{A}_j\tilde{\mathbf{a}}_j'\,\delta\boldsymbol{\pi}_j') \tag{6.2.10}$$

The differential of the dot-2 constraint function of Eq. 6.2.6 is

$$\delta\Phi^{d2}(\mathbf{a}_i, \mathbf{d}_{ij}) = \mathbf{a}_i^T\,\delta\mathbf{d}_{ij} + \mathbf{d}_{ij}^T\,\delta\mathbf{a}_i$$

Using Eqs. 6.2.7 and 4.2.20, this is (Prob. 6.2.1)

$$\begin{aligned}\delta\Phi^{d2}(\mathbf{a}_i, \mathbf{d}_{ij}) &= \mathbf{a}_i'^T\mathbf{A}_i^T(\delta\mathbf{r}_j - \mathbf{A}_j\tilde{\mathbf{s}}_j'\,\delta\boldsymbol{\pi}_j' - \delta\mathbf{r}_i + \mathbf{A}_i\tilde{\mathbf{s}}_i'\,\delta\boldsymbol{\pi}_i') - \mathbf{d}_{ij}^T\mathbf{A}_i\tilde{\mathbf{a}}_i'\,\delta\boldsymbol{\pi}_i' \\ &= \mathbf{a}_i'^T\mathbf{A}_i^T\,\delta\mathbf{r}_j - \mathbf{a}_i'^T\mathbf{A}_i^T\,\delta\mathbf{r}_i - \mathbf{a}_i'^T\mathbf{A}_i^T\mathbf{A}_j\tilde{\mathbf{s}}_j'\,\delta\boldsymbol{\pi}_j' + (\mathbf{a}_i'^T\tilde{\mathbf{s}}_i' - \mathbf{d}_{ij}^T\mathbf{A}_i\tilde{\mathbf{a}}_i')\,\delta\boldsymbol{\pi}_i'\end{aligned} \tag{6.2.11}$$

The differential of the spherical constraint of Eq. 6.2.9, using Eq. 4.2.20, is simply (Prob. 6.2.2)

$$\delta\mathbf{\Phi}^s(P_i, P_j) = \delta\mathbf{r}_j - \mathbf{A}_j\tilde{\mathbf{s}}_j'\,\delta\boldsymbol{\pi}_j' - \delta\mathbf{r}_i + \mathbf{A}_i\tilde{\mathbf{s}}_i'\,\delta\boldsymbol{\pi}_i' \tag{6.2.12}$$

The coefficients of $\delta\mathbf{r}_i$, $\delta\mathbf{r}_j$, $\delta\boldsymbol{\pi}_j'$, and $\delta\boldsymbol{\pi}_j'$ in the differentials of Eqs. 6.2.11 to 6.2.12 are tabulated in Table 6.2.1. The reader is cautioned that since $\delta\boldsymbol{\pi}'$ is not integrable, the Jacobians $\mathbf{\Phi}_{\boldsymbol{\pi}'}$ in Table 6.2.1 should be interpreted only as coefficients of $\delta\boldsymbol{\pi}'$ in Eqs. 6.2.10 to 6.2.12. Using the relationship $\delta\boldsymbol{\pi}' = 2\mathbf{G}\,\delta\mathbf{p}$ of Eq. 4.3.37, the differentials of Eqs. 6.2.10 to 6.2.12 can be written in terms of variations in Euler parameters. The coefficients of the differentials $\delta\mathbf{r}_i$, $\delta\mathbf{r}_j$, $\delta\mathbf{p}_i$, and $\delta\mathbf{p}_j$ are obtained from Table 6.2.1 by multiplying $\mathbf{\Phi}_{\boldsymbol{\pi}_i}$ and $\mathbf{\Phi}_{\boldsymbol{\pi}_j}$ on the right by $2\mathbf{G}_i$ and $2\mathbf{G}_j$, respectively.

Constraints between Pairs of Bodies

Some of the common kinematic constraints between bodies that are encountered in mechanical system dynamics are now derived. For a more complete treatment of general spatial multibody kinematics, see Ref. 12.

A *spherical joint*, or *ball-and-socket joint* is defined by the condition that the center of the ball at point P_i on body i coincide with the center of the socket at P_j on body j, as shown in Fig. 6.2.3. This condition is simply the spherical constraint of Eq. 6.2.9. These three scalar constraint equations restrict relative positions of the bodies.

The *universal joint* between bodies i and j shown in Fig. 6.2.4 is constructed with an intermediate body, or cross, that is pivoted in bodies i and j. The center P of the universal joint on bodies i and j must coincide, defined by points P_i and P_j on the respective bodies. Points Q_i and Q_j on the arms of the cross between bodies i and j, respectively, are specified to determine the z'' axes of the joint definition frames, as shown in Fig. 6.2.4.

The equations of constraint that characterize a universal joint are that points P_i and P_j coincide and that vectors $\mathbf{h}_i$ and $\mathbf{h}_j$ are orthogonal. These conditions are specified by the constraint equations

$$\begin{aligned} \mathbf{\Phi}^s(P_i, P_j) &= \mathbf{0} \\ \Phi^{d1}(\mathbf{h}_i, \mathbf{h}_j) &= 0 \end{aligned} \tag{6.2.13}$$

TABLE 6.2.1 Jacobians[a] of Basic Constraint Functions

Constraint function	$\mathbf{\Phi}_{\mathbf{r}_i}$	$\mathbf{\Phi}_{\mathbf{r}_j}$	$\mathbf{\Phi}_{\boldsymbol{\pi}_i'}$	$\mathbf{\Phi}_{\boldsymbol{\pi}_j'}$
$\Phi^{d1}(\mathbf{a}_i, \mathbf{a}_j)$	$\mathbf{0}$	$\mathbf{0}$	$-\mathbf{a}_j'^T\mathbf{A}_j^T\mathbf{A}_i\tilde{\mathbf{a}}_i'$	$-\mathbf{a}_i'^T\mathbf{A}_i^T\mathbf{A}_j\tilde{\mathbf{a}}_j'$
$\Phi^{d2}(\mathbf{a}_i, \mathbf{d}_{ij})$	$-\mathbf{a}_i'^T\mathbf{A}_i^T$	$\mathbf{a}_i'^T\mathbf{A}_i^T$	$\mathbf{a}_i'^T\tilde{\mathbf{s}}_i' - \mathbf{d}_{ij}^T\mathbf{A}_i\tilde{\mathbf{a}}_i'$	$-\mathbf{a}_i'^T\mathbf{A}_i^T\mathbf{A}_j\tilde{\mathbf{s}}_j'$
$\mathbf{\Phi}^s(P_i, P_j)$	$-\mathbf{I}_3$	$\mathbf{I}_3$	$\mathbf{A}_i\tilde{\mathbf{s}}_i'$	$-\mathbf{A}_j\tilde{\mathbf{s}}_j'$

[a] Technically, $\mathbf{\Phi}_{\boldsymbol{\pi}'}$ is not a partial derivative, but $\mathbf{\Phi}_{\mathbf{p}} = 2\mathbf{\Phi}_{\boldsymbol{\pi}'}\mathbf{G}$.

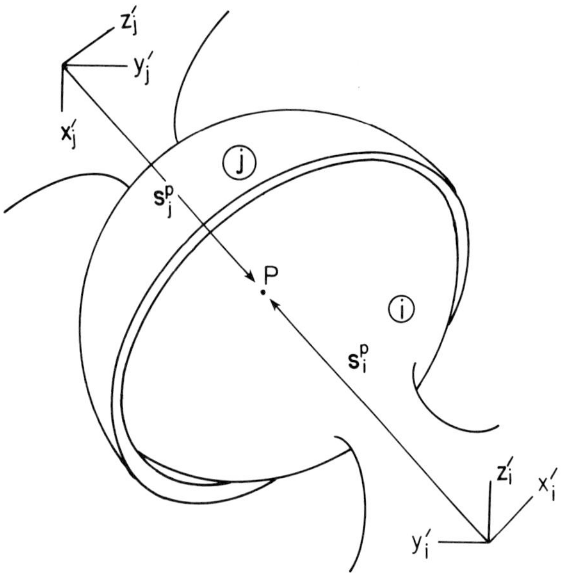

Figure 6.2.3 Spherical joint.

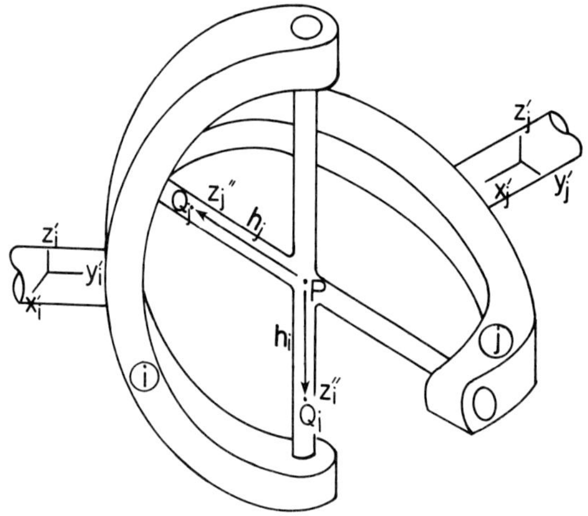

Figure 6.2.4 Universal joint.

These four constraint equations restrict the relative position of the bodies and rotation about the shafts (Prob. 6.2.3).

A *revolute joint* is constructed with bearings that allow relative rotation about a common axis in a pair of bodies, but preclude relative translation along this axis, as shown in Fig. 6.2.5. To define the revolute joint, the center of the joint is located on bodies i and j by points P_i and P_j, which must coincide. The axis of relative rotation is defined in bodies i and j by points Q_i and Q_j shown in Fig. 6.2.5, hence unit vectors $\mathbf{h}_i$ and $\mathbf{h}_j$ along the respective z'' axes.

The analytical formulation of the revolute joint is that points P_i and P_j co-

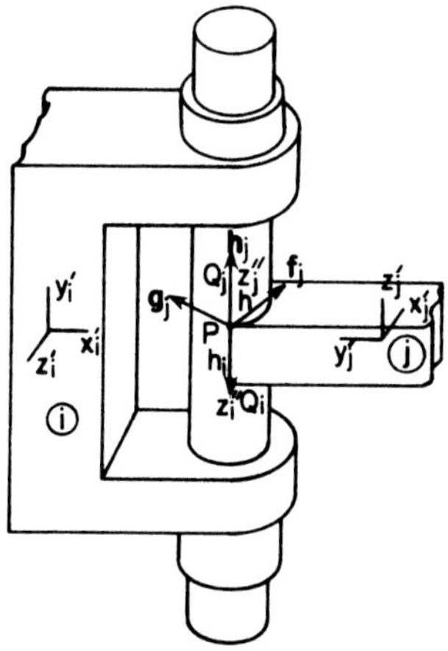

Figure 6.2.5 Revolute joint.

incide (Eq. 6.2.9) and that body-fixed vectors $\mathbf{h}_i$ and $\mathbf{h}_j$ are parallel. The latter geometric condition is equivalent to the analytical conditions that $\mathbf{h}_i$ is orthogonal to $\mathbf{f}_j$ and $\mathbf{g}_j$. Thus the five equations that define a revolute joint are

$$
\begin{aligned}
\boldsymbol{\Phi}^s(P_i, P_j) &= \mathbf{0} \\
\Phi^{d1}(\mathbf{h}_i, \mathbf{f}_j) &= 0 \\
\Phi^{d1}(\mathbf{h}_i, \mathbf{g}_j) &= 0
\end{aligned}
\tag{6.2.14}
$$

A *translational joint* shown in Fig. 6.2.6 allows relative translation along a common axis between a pair of bodies, but precludes relative rotation about this

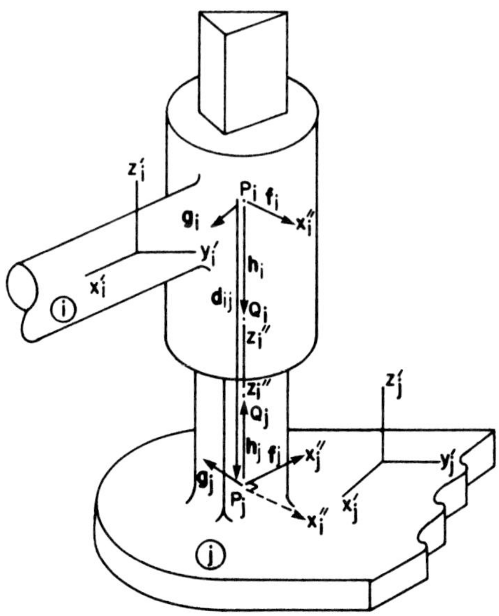

Figure 6.2.6 Translational joint.

axis. It is defined by joint definition points P_i, Q_i, P_j, and Q_j along the axis of translation. The x'' axes of the joint definition frames on bodies i and j are selected so that they are perpendicular, defined by vectors $\mathbf{f}_i$ and $\mathbf{f}_j$ shown in Fig. 6.2.6. The geometric definition of this joint is that $\mathbf{h}_i$ and $\mathbf{h}_j$ are collinear and $\mathbf{f}_i$ and $\mathbf{f}_j$ are orthogonal.

An analytical definition of a translational joint is that $\mathbf{h}_i$ and $\mathbf{d}_{ij}$ are orthogonal to $\mathbf{f}_j$ and $\mathbf{g}_j$ and that $\mathbf{f}_i$ and $\mathbf{f}_j$ are orthgonal; that is,

$$
\begin{aligned}
\Phi^{d1}(\mathbf{h}_i, \mathbf{f}_j) &= 0 \\
\Phi^{d1}(\mathbf{h}_i, \mathbf{g}_j) &= 0 \\
\Phi^{d2}(\mathbf{f}_j, \mathbf{d}_{ij}) &= 0 \\
\Phi^{d2}(\mathbf{g}_j, \mathbf{d}_{ij}) &= 0 \\
\Phi^{d1}(\mathbf{f}_i, \mathbf{f}_j) &= 0
\end{aligned}
\tag{6.2.15}
$$

Note that even if $\mathbf{d}_{ij} = \mathbf{0}$, the first two conditions of Eq. 6.2.15 imply that $\mathbf{h}_i$ and $\mathbf{h}_j$ are parallel. Thus when $\mathbf{d}_{ij} = \mathbf{0}$, these vectors have point P in common, so they are in fact collinear.

A *cylindrical joint* between a pair of bodies is shown in Fig. 6.2.7. It permits relative rotation between bodies about a common axis that is defined by points P_i, Q_i, P_j, and Q_j and relative translation along this axis. Constraint conditions that define the joint are just the first four equations of the translational joint in Eq. 6.2.15.

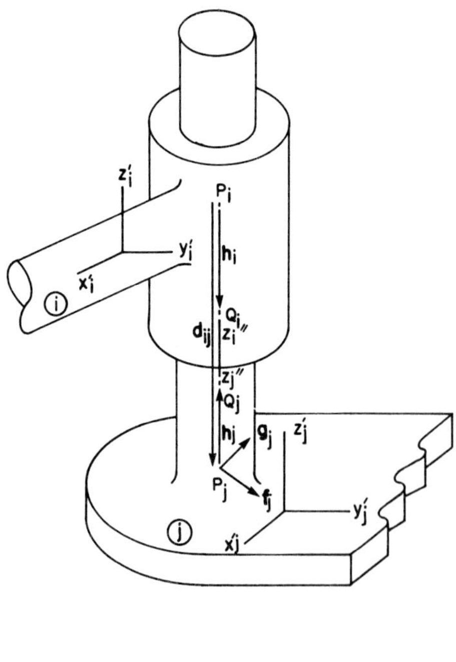

Figure 6.2.7 Cylindrical joint.

Example 6.2.1: A spatial slider–crank mechanism can be modeled by choosing combinations of kinematic joints to connect each body. Two models are presented here, to illustrate use of standard joints in modeling spatial systems. In the first model (Fig. 6.2.8), each body in the mechanism and ground is modeled as a body. Four types of kinematic joints are used to model joints between bodies, as follows:

BODIES	**NUMBER OF GENERALIZED COORDINATES**
4 bodies	28

CONSTRAINTS	**NUMBER OF CONSTRAINTS**
Revolute joint (crank, ground)	5
Spherical joint (crank, connecting rod)	3
Cylindrical joint (connecting rod, slider)	4
Translational joint (slider, ground)	5
Ground constraint (fixing ground)	6
Euler parameter normalization constraints	4
	27

$$\text{DOF} = 28 - 27 = 1$$

where DOF is an abbreviation for degrees of freedom.

A second model is shown in Fig. 6.2.9. A universal joint is used to replace the cylindrical joint between the slider and the connecting rod. Since a universal

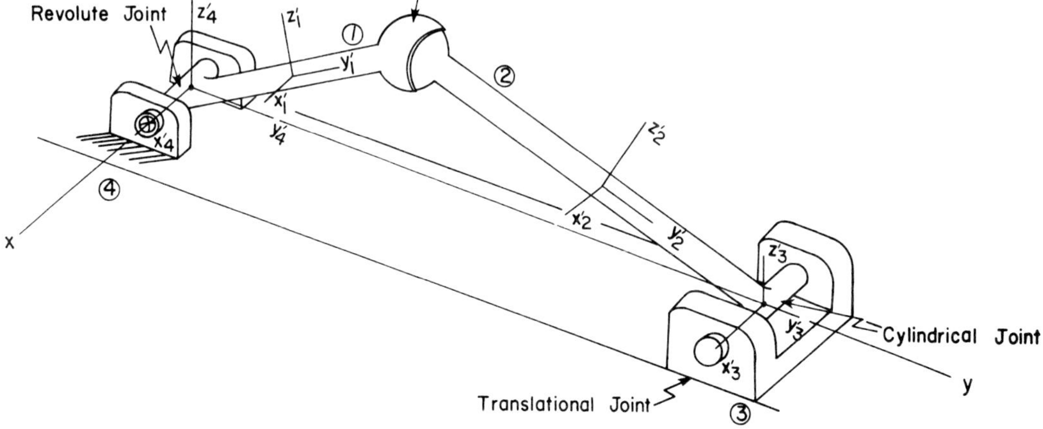

Figure 6.2.8 Spatial slider–crank, model 1.

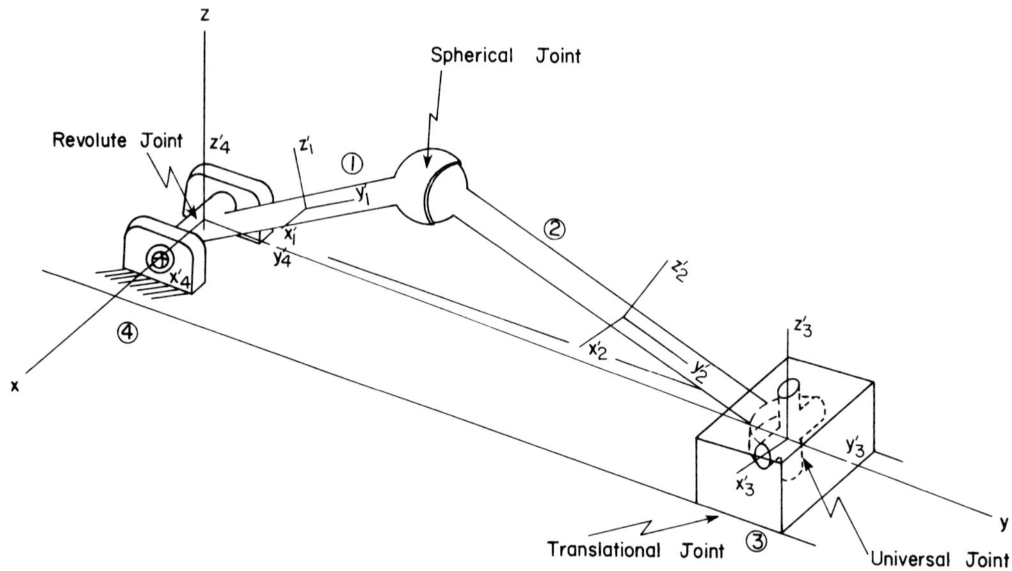

Figure 6.2.9 Spatial slider–crank, model 2.

joint also yields four constraint equations, the number of degrees of freedom is the same.

Velocity and Acceleration Equations

Velocity relations that are implied by constraint equations derived in this section are obtained by differentiating constraint equations $\mathbf{\Phi}(\mathbf{q}) = \mathbf{0}$ to obtain

$$\mathbf{\Phi}_{\mathbf{r}_i}\dot{\mathbf{r}}_i + \mathbf{\Phi}_{\mathbf{r}_j}\dot{\mathbf{r}}_j + \mathbf{\Phi}_{\pi_i}\boldsymbol{\omega}_i' + \mathbf{\Phi}_{\pi_j}\boldsymbol{\omega}_j' = \mathbf{0} \tag{6.2.16}$$

where the coefficient matrices of $\dot{\mathbf{r}}_i$, $\dot{\mathbf{r}}_j$, $\boldsymbol{\omega}_i'$, and $\boldsymbol{\omega}_j'$ are given in Table 6.2.1. Acceleration equations are obtained by differentiating the velocity equations of Eq. 6.2.16. Since terms arising in Eq. 6.2.16 involve the transformation matrix $\mathbf{A}$ for each body in the system, it is helpful to recall the first relation of Eq. 4.2.9,

$$\dot{\mathbf{A}} = \mathbf{A}\tilde{\boldsymbol{\omega}}' \tag{6.2.17}$$

Note also that no explicit time dependence appears in the constraint equations. Thus differentiation of Eq. 6.2.16 yields

$$\mathbf{\Phi}_{\mathbf{r}_i}\ddot{\mathbf{r}}_i + \mathbf{\Phi}_{\pi_i}\dot{\boldsymbol{\omega}}_i' + \mathbf{\Phi}_{\mathbf{r}_j}\ddot{\mathbf{r}}_j + \mathbf{\Phi}_{\pi_j}\dot{\boldsymbol{\omega}}_j' = -(\dot{\mathbf{\Phi}}_{\mathbf{r}_i}\dot{\mathbf{r}}_i + \dot{\mathbf{\Phi}}_{\mathbf{r}_j}\dot{\mathbf{r}}_j + \dot{\mathbf{\Phi}}_{\pi_i}\boldsymbol{\omega}_i' + \dot{\mathbf{\Phi}}_{\pi_j}\boldsymbol{\omega}_j') \equiv \boldsymbol{\gamma} \tag{6.2.18}$$

In order to evaluate terms in Eq. 6.2.18, the matrices presented in Table 6.2.1 may be differentiated to evaluate terms on the right of Eq. 6.2.18. For the dot-1 constraint, differentiation of terms arising in Table 6.2.1, using Eq. 6.2.17, yields (Prob. 6.2.4)

$$\begin{aligned}\gamma^{d1} &= \mathbf{a}_j'^T[\mathbf{A}_j^T\mathbf{A}_i\tilde{\boldsymbol{\omega}}_i' - \tilde{\boldsymbol{\omega}}_j'\mathbf{A}_j^T\mathbf{A}_i]\tilde{\mathbf{a}}_i'\boldsymbol{\omega}_i' + \mathbf{a}_i'^T[\mathbf{A}_i^T\mathbf{A}_j\tilde{\boldsymbol{\omega}}_j' - \tilde{\boldsymbol{\omega}}_i'\mathbf{A}_i^T\mathbf{A}_j]\tilde{\mathbf{a}}_j'\boldsymbol{\omega}_j' \\ &= -\mathbf{a}_j'^T[\mathbf{A}_j^T\mathbf{A}_i\tilde{\boldsymbol{\omega}}_i'\tilde{\boldsymbol{\omega}}_i' + \tilde{\boldsymbol{\omega}}_j'\tilde{\boldsymbol{\omega}}_j'\mathbf{A}_j^T\mathbf{A}_i]\mathbf{a}_i' + 2\boldsymbol{\omega}_j'^T\tilde{\mathbf{a}}_j'\mathbf{A}_j^T\mathbf{A}_i\tilde{\mathbf{a}}_i'\boldsymbol{\omega}_i'\end{aligned} \tag{6.2.19}$$

Similarly, for the dot-2 constraint (Prob. 6.2.5),

$$\begin{aligned}\gamma^{d2} &= -\mathbf{a}_i'^T\tilde{\boldsymbol{\omega}}_i\mathbf{A}_i^T(\dot{\mathbf{r}}_i - \dot{\mathbf{r}}_j) + [(\dot{\mathbf{r}}_j + \mathbf{A}_j\tilde{\boldsymbol{\omega}}_j'\mathbf{s}_j' - \dot{\mathbf{r}}_i - \mathbf{A}_i\tilde{\boldsymbol{\omega}}_i'\mathbf{s}_i')^T\mathbf{A}_i\tilde{\mathbf{a}}_i' \\ &\quad + \mathbf{d}_{ij}^T\mathbf{A}_i\tilde{\boldsymbol{\omega}}_i'\tilde{\mathbf{a}}_i']\boldsymbol{\omega}_i' + \mathbf{a}_i'^T[\mathbf{A}_i^T\mathbf{A}_j\tilde{\boldsymbol{\omega}}_j' - \tilde{\boldsymbol{\omega}}_i'\mathbf{A}_i^T\mathbf{A}_j]\tilde{\mathbf{s}}_j'\boldsymbol{\omega}_j' \\ &= 2\boldsymbol{\omega}_i'^T\tilde{\mathbf{a}}_i'\mathbf{A}_i^T(\dot{\mathbf{r}}_i - \dot{\mathbf{r}}_j) + 2\mathbf{s}_j'^T\tilde{\boldsymbol{\omega}}_j'\mathbf{A}_j^T\mathbf{A}_i\tilde{\boldsymbol{\omega}}_i'\mathbf{a}_i' \\ &\quad - \mathbf{s}_i'^T\tilde{\boldsymbol{\omega}}_i'\tilde{\boldsymbol{\omega}}_i'\mathbf{a}_i' - \mathbf{s}_j'^T\tilde{\boldsymbol{\omega}}_j'\tilde{\boldsymbol{\omega}}_j'\mathbf{A}_j^T\mathbf{A}_i\mathbf{a}_i' - \mathbf{d}_{ij}^T\mathbf{A}_i\tilde{\boldsymbol{\omega}}_i'\tilde{\boldsymbol{\omega}}_i'\mathbf{a}_i'\end{aligned} \tag{6.2.20}$$

Finally, for the spherical constraint (Prob. 6.2.6),

$$\begin{aligned}\gamma^s &= -\mathbf{A}_i\tilde{\boldsymbol{\omega}}_i'\tilde{\mathbf{s}}_i'\boldsymbol{\omega}_i' + \mathbf{A}_j\tilde{\boldsymbol{\omega}}_j'\tilde{\mathbf{s}}_j'\boldsymbol{\omega}_j' \\ &= \mathbf{A}_i\tilde{\boldsymbol{\omega}}_i'\tilde{\boldsymbol{\omega}}_i'\mathbf{s}_i' - \mathbf{A}_j\tilde{\boldsymbol{\omega}}_j'\tilde{\boldsymbol{\omega}}_j'\mathbf{s}_j'\end{aligned} \tag{6.2.21}$$

Equations of Motion of Constrained Spatial Systems

The use of Lagrange multipliers in deriving equations of motion in the literature differs somewhat from the approach used thus far in this book. Rather than using the condition that forces of constraint do no work to derive expressions for these forces in terms of Lagrange multipliers and then putting these expressions into the equations of motion, as has been done thus far in the book, a more conventional indirect approach is used here [9, 10, 12]. The variational equations of motion are first written only in terms of applied forces, but only for generalized coordinate variations that are consistent with constraints. This fact is used, with the Lagrange multiplier theorem, to introduce abstract Lagrange multipliers. A separate derivation is then required to show that the Lagrange multipliers yield expressions for constraint forces. Even though this approach is less direct than that employed thus far in the book, it is used here to illustrate the more conventional approach.

Consider nb bodies that form a constrained multibody system. The composite set of generalized coordinates for the system is

$$\begin{aligned}\mathbf{r} &= [\mathbf{r}_1^T, \mathbf{r}_2^T, \ldots, \mathbf{r}_{nb}^T]^T \\ \mathbf{p} &= [\mathbf{p}_1^T, \mathbf{p}_2^T, \ldots, \mathbf{p}_{nb}^T]^T\end{aligned} \tag{6.2.22}$$

The combined set of kinematic constraints that act on the system is written in the form

$$\boldsymbol{\Phi}(\mathbf{r}, \mathbf{p}) = \mathbf{0} \tag{6.2.23}$$

and the Euler parameter normalization constraints

$$\mathbf{p}_i^T\mathbf{p}_i - 1 = 0, \qquad i = 1, \ldots, nb \tag{6.2.24}$$

must be satisfied.

In order to implement the variational Newton–Euler equations, Eq. 5.1.26 is written for each body in the system and the resulting equations are added to obtain a Newton–Euler variational equation of motion for the system. To simplify notation, define

$$
\begin{aligned}
\delta \mathbf{r} &= [\delta \mathbf{r}_1^T, \delta \mathbf{r}_2^T, \ldots, \delta \mathbf{r}_{nb}^T]^T \\
\mathbf{M} &\equiv \operatorname{diag}(m_1 \mathbf{I}, m_2 \mathbf{I}, \ldots, m_{nb} \mathbf{I}) \\
\delta \boldsymbol{\pi}' &= [\delta \boldsymbol{\pi}_1'^T, \delta \boldsymbol{\pi}_2'^T, \ldots, \delta \boldsymbol{\pi}_{nb}'^T]^T \\
\mathbf{F}^A &\equiv [\mathbf{F}_1^{A^T}, \mathbf{F}_2^{A^T}, \ldots, \mathbf{F}_{nb}^{A^T}]^T \\
\mathbf{J}' &\equiv \operatorname{diag}(\mathbf{J}_1', \mathbf{J}_2', \ldots, \mathbf{J}_{nb}') \\
\boldsymbol{\omega}' &\equiv [\boldsymbol{\omega}_1'^T, \boldsymbol{\omega}_2'^T, \ldots, \boldsymbol{\omega}_{nb}'^T]^T \\
\mathbf{n}' &\equiv [\mathbf{n}_1'^T, \mathbf{n}_2'^T, \ldots, \mathbf{n}_{nb}'^T]^T \\
\tilde{\boldsymbol{\omega}}' &\equiv \operatorname{diag}(\tilde{\boldsymbol{\omega}}_1', \tilde{\boldsymbol{\omega}}_1', \ldots, \tilde{\boldsymbol{\omega}}_{nb}')
\end{aligned}
\tag{6.2.25}
$$

Using this notation, the sum of Eq. 5.1.26 over all bodies in the system may be written as (Prob. 6.2.7)

$$\delta \mathbf{r}^T[\mathbf{M}\ddot{\mathbf{r}} - \mathbf{F}^A] + \delta \boldsymbol{\pi}'^T[\mathbf{J}'\dot{\boldsymbol{\omega}}' + \tilde{\boldsymbol{\omega}}'\mathbf{J}'\boldsymbol{\omega}' - \mathbf{n}'^A] = 0 \tag{6.2.26}$$

which must hold for all virtual displacements $\delta \mathbf{r}$ and virtual rotations $\delta \boldsymbol{\pi}'$ that are consistent with constraints; that is, that satisfy

$$\boldsymbol{\Phi}_{\mathbf{r}}\, \delta \mathbf{r} + \boldsymbol{\phi}_{\boldsymbol{\pi}'}\, \delta \boldsymbol{\pi}' = \mathbf{0} \tag{6.2.27}$$

where $\boldsymbol{\Phi}_{\mathbf{r}}$ and $\boldsymbol{\Phi}_{\boldsymbol{\pi}'}$ are assembled using Table 6.2.1.

By the Lagrange multiplier theorem, there exists a Lagrange multiplier vector $\boldsymbol{\lambda}$ such that

$$\delta \mathbf{r}^T[\mathbf{M}\ddot{\mathbf{r}} - \mathbf{F}^A + \boldsymbol{\Phi}_{\mathbf{r}}^T\boldsymbol{\lambda}] + \delta \boldsymbol{\pi}'^T[\mathbf{J}'\dot{\boldsymbol{\omega}}' + \tilde{\boldsymbol{\omega}}'\mathbf{J}'\boldsymbol{\omega}' - \mathbf{n}'^A + \boldsymbol{\phi}_{\boldsymbol{\pi}'}^T\boldsymbol{\lambda}] = 0 \tag{6.2.28}$$

which must hold for arbitrary $\delta \mathbf{r}$ and $\delta \boldsymbol{\pi}'$. The coefficients of these arbitrary variations must be zero, yielding the *constrained Newton–Euler equations*,

$$
\begin{aligned}
\mathbf{M}\ddot{\mathbf{r}} + \boldsymbol{\Phi}_{\mathbf{r}}^T\boldsymbol{\lambda} &= \mathbf{F}^A \\
\mathbf{J}'\dot{\boldsymbol{\omega}}' + \boldsymbol{\Phi}_{\boldsymbol{\pi}'}^T\boldsymbol{\lambda} &= \mathbf{n}'^A - \tilde{\boldsymbol{\omega}}'\mathbf{J}'\boldsymbol{\omega}'
\end{aligned}
\tag{6.2.29}
$$

Note that at this point it is not clear how $\boldsymbol{\lambda}$ is related to constraint reaction forces.

To complete the equations of motion, acceleration equations associated with the kinematic constraints of Eq. 6.2.23 must be obtained. The *velocity equation* is obtained by taking the time derivative of Eq. 6.2.23,

$$\boldsymbol{\Phi}_{\mathbf{r}}\dot{\mathbf{r}} + \boldsymbol{\phi}_{\boldsymbol{\pi}'}\boldsymbol{\omega}' = \mathbf{0} \tag{6.2.30}$$

The time derivative of this equation yields the *acceleration equation*

$$\boldsymbol{\Phi}_{\mathbf{r}}\ddot{\mathbf{r}} + \boldsymbol{\Phi}_{\boldsymbol{\pi}'}\dot{\boldsymbol{\omega}}' = \boldsymbol{\gamma} \tag{6.2.31}$$

where components of the vector $\boldsymbol{\gamma}$ are defined in Eqs. 6.2.19 to 6.2.21 for each of the basic constraint equations.

Combining Eqs. 6.2.29 and 6.2.31, the *system acceleration equations* are

$$\begin{bmatrix} \mathbf{M} & \mathbf{0} & \boldsymbol{\Phi}_{\mathbf{r}}^T \\ \mathbf{0} & \mathbf{J}' & \boldsymbol{\Phi}_{\pi'}^T \\ \boldsymbol{\Phi}_{\mathbf{r}} & \boldsymbol{\Phi}_{\pi'} & \mathbf{0} \end{bmatrix} \begin{bmatrix} \ddot{\mathbf{r}} \\ \dot{\boldsymbol{\omega}}' \\ \boldsymbol{\lambda} \end{bmatrix} = \begin{bmatrix} \mathbf{F}^A \\ \mathbf{n}'^A - \tilde{\boldsymbol{\omega}}'\mathbf{J}'\boldsymbol{\omega}' \\ \boldsymbol{\gamma} \end{bmatrix} \tag{6.2.32}$$

These equations of motion, taken with the kinematic constraint equations of Eqs. 6.2.23 and 6.2.24 and the velocity equations of Eq. 6.2.30, yield the *differential-algebraic equations of motion* of the system.

Initial Conditions

Initial conditions on position and orientation and on velocity must be provided to define dynamics of a system. Since orientation of a body is specified by Euler parameters, the *initial position conditions* (conditions on position and orientation) must be specified in the form

$$\boldsymbol{\Phi}^I(\mathbf{r}(t_0), \mathbf{p}(t_0)) = \mathbf{0} \tag{6.2.33}$$

where $\mathbf{r}$ and $\mathbf{p}$ must satisfy Eqs. 6.2.23 and 6.2.24 at t_0. Equations 6.2.23, 6.2.24, and 6.2.33 must uniquely determine $\mathbf{r}(t_0)$ and $\mathbf{p}(t_0)$. *Initial velocity conditions* may be given in the form

$$\mathbf{B}^{\dot{\mathbf{r}}}\dot{\mathbf{r}} + \mathbf{B}^{\omega'}\boldsymbol{\omega}' = \boldsymbol{\nu}^I \tag{6.2.34}$$

at $t = t_0$. Equations 6.2.30 and 6.2.34 must uniquely determine $\dot{\mathbf{r}}(t_0)$ and $\boldsymbol{\omega}'(t_0)$.

As in Section 6.1 for planar systems, forces due to springs, dampers, and actuators between bodies can be accounted for using the principle of virtual work. The virtual work approach is even more attractive for spatial systems than for planar systems, since direct relationships for virtual rotations are available.

Spring, Damper, and Actuator Forces

Consider the *translational spring–damper–actuator* shown in Fig. 6.2.10 that connects points P_i and P_j on bodies i and j, respectively. The vector from P_i to P_j is

$$\mathbf{d}_{ij} = \mathbf{r}_j + \mathbf{A}_j\mathbf{s}_j' - \mathbf{r}_i - \mathbf{A}_i\mathbf{s}_i' \tag{6.2.35}$$

Thus the length l of the spring–damper–actuator is determined by

$$l^2 = \mathbf{d}_{ij}^T\mathbf{d}_{ij} \tag{6.2.36}$$

where it is presumed that in a sound design, $l > 0$. Differentiating both sides of Eq. 6.2.36 with respect to time yields

$$2l\dot{l} = 2\mathbf{d}_{ij}^T\dot{\mathbf{d}}_{ij} \tag{6.2.37}$$

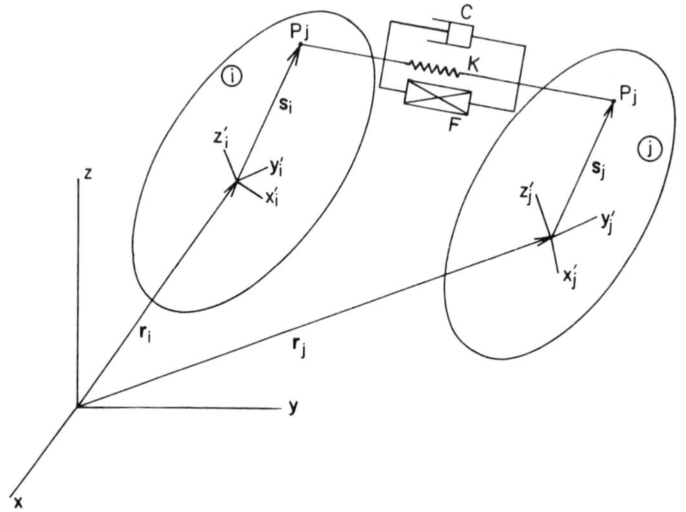

Figure 6.2.10 Spatial translational spring–damper–actuator.

which, since $l \neq 0$, may be solved for the time rate of change of length as

$$\begin{aligned} \dot{l} &= \left(\frac{\mathbf{d}_{ij}}{l}\right)^T (\dot{\mathbf{r}}_j + \dot{\mathbf{A}}_j\mathbf{s}_j' - \dot{\mathbf{r}}_i - \dot{\mathbf{A}}_i\mathbf{s}_i') \\ &= \left(\frac{\mathbf{d}_{ij}}{l}\right)^T (\dot{\mathbf{r}}_j - \mathbf{A}_j\tilde{\mathbf{s}}_j'\boldsymbol{\omega}_j' - \dot{\mathbf{r}}_i + \mathbf{A}_i\tilde{\mathbf{s}}_i'\boldsymbol{\omega}_i') \end{aligned} \tag{6.2.38}$$

where Eq. 6.2.17 has been used to write time derivatives of transformation matrices in terms of angular velocities.

The magnitude of the force in the spring–damper–actuator, with tension taken as positive, is

$$f = k(l - l_0) + c\dot{l} + F(l, \dot{l}) \tag{6.2.39}$$

where k is the *spring coefficient*, l_0 is the *spring free length*, c is the *damping coefficient*, and $F(l, \dot{l})$ is a general *actuator force*. The virtual work of this force is simply

$$\delta W^A = -f\,\delta l \tag{6.2.40}$$

where the variation δl in length is obtained by taking the differential of Eq. 6.2.36 and dividing by l, as in derivation of Eq. 6.2.38, to obtain

$$\delta l = \left(\frac{\mathbf{d}_{ij}}{l}\right)^T (\delta\mathbf{r}_j - \mathbf{A}_j\tilde{\mathbf{s}}_j'\,\delta\boldsymbol{\pi}_j' - \delta\mathbf{r}_i + \mathbf{A}_i\tilde{\mathbf{s}}_i'\,\delta\boldsymbol{\pi}_i') \tag{6.2.41}$$

where Eq. 4.2.19 has been used to expand the differential of the transformation matrix. Substituting this result into Eq. 6.2.40 yields

$$\delta W^A = -\frac{f}{l}\mathbf{d}_{ij}^T(\delta\mathbf{r}_j - \mathbf{A}_j\tilde{\mathbf{s}}_j'\,\delta\boldsymbol{\pi}_j' - \delta\mathbf{r}_i + \mathbf{A}_i\tilde{\mathbf{s}}_i'\,\delta\boldsymbol{\pi}_i') \tag{6.2.42}$$

The coefficients of virtual displacements and virtual rotations in Eq. 6.2.42 yield the generalized forces as

$$\begin{aligned}\mathbf{Q}_i^A &= \frac{f}{l}\begin{bmatrix}\mathbf{d}_{ij}\\ \tilde{\mathbf{s}}_i'\mathbf{A}_i^T\mathbf{d}_{ij}\end{bmatrix}\\ \mathbf{Q}_j^A &= -\frac{f}{l}\begin{bmatrix}\mathbf{d}_{ij}\\ \tilde{\mathbf{s}}_j'\mathbf{A}_j^T\mathbf{d}_{ij}\end{bmatrix}\end{aligned} \tag{6.2.43}$$

Translational spring–damper–actuator forces are accounted for in the equations of motion simply by inserting the generalized forces of Eq. 6.2.43 into the right side of Eq. 6.2.32.

Constraint Reaction Forces in Joints

Consider a typical constraint k, with constraint equations $\boldsymbol{\Phi}^k = \mathbf{0}$ and associated Lagrange multipliers $\boldsymbol{\lambda}^k$. In terms of the body-fixed x_i''-y_i''-z_i'' joint definition frame, the *constraint reaction forces* are $\mathbf{F}_i''^{C_k}$ and the *constraint reaction torques* are $\mathbf{T}_i''^{C_k}$. Equating the virtual work of these reaction forces and torques to the negative of the variational terms in the equations of motion associated with constraint k, as in Eqs. 6.1.38 and 6.1.39,

$$-(\delta\mathbf{r}_i^T\boldsymbol{\Phi}_{\mathbf{r}_i}^{kT}\boldsymbol{\lambda}^k + \delta\boldsymbol{\pi}_i'^T\boldsymbol{\Phi}_{\boldsymbol{\pi}_i}^{kT}\boldsymbol{\lambda}^k) = \delta\mathbf{r}_i''^{PT}\mathbf{F}_i''^{C_k} + \delta\boldsymbol{\pi}_i''^T\mathbf{T}_i''^{C_k} \tag{6.2.44}$$

which must hold for arbitrary virtual displacements.

In order to take advantage of Eq. 6.2.44, all virtual displacements must be represented in the x_i''-y_i''-z_i'' joint definition frame. From Eq. 4.2.20,

$$\delta\mathbf{r}_i^P = \delta\mathbf{r}_i - \mathbf{A}_i\tilde{\mathbf{s}}_i'\,\delta\boldsymbol{\pi}_i' \tag{6.2.45}$$

In terms of joint definition frame coordinates,

$$\begin{aligned}\delta\mathbf{r}_i^P &= \mathbf{A}_i\mathbf{C}_i\,\delta\mathbf{r}_i''^P\\ \delta\boldsymbol{\pi}_i' &= \mathbf{C}_i\,\delta\boldsymbol{\pi}_i''\end{aligned} \tag{6.2.46}$$

where $\mathbf{C}_i$ is the transformation matrix from the joint definition frame to the body-fixed centroidal frame. From Eqs. 6.2.45 and 6.2.46,

$$\delta\mathbf{r}_i = \mathbf{A}_i\mathbf{C}_i\,\delta\mathbf{r}_i''^P + \mathbf{A}_i\tilde{\mathbf{s}}_i'\mathbf{C}_i\,\delta\boldsymbol{\pi}_i'' \tag{6.2.47}$$

Substituting the virtual displacement from Eq. 6.2.47 into Eq. 6.2.44,

$$\delta\mathbf{r}_i''^{PT}(\mathbf{F}_i''^{C_k} + \mathbf{C}_i^T\mathbf{A}_i^T\boldsymbol{\Phi}_{\mathbf{r}_i}^{kT}\boldsymbol{\lambda}^k) + \delta\boldsymbol{\pi}_i''^T(\mathbf{T}_i''^{C_k} + \mathbf{C}_i^T\boldsymbol{\Phi}_{\boldsymbol{\pi}_i}^{kT}\boldsymbol{\lambda}^k - \mathbf{C}_i^T\tilde{\mathbf{s}}_i'\mathbf{A}_i^T\boldsymbol{\Phi}_{\mathbf{r}_i}^{kT}\boldsymbol{\lambda}^k) = 0 \tag{6.2.48}$$

which must hold for arbitrary $\delta\mathbf{r}_i''^P$ and $\delta\boldsymbol{\pi}_i''$. Therefore,

$$
\begin{aligned}
\mathbf{F}_i''^{C_k} &= -\mathbf{C}_i^T\mathbf{A}_i^T\boldsymbol{\Phi}_{\mathbf{r}_i}^{kT}\boldsymbol{\lambda}^k \\
\mathbf{T}_i''^{C_k} &= -\mathbf{C}_i^T(\boldsymbol{\Phi}_{\boldsymbol{\pi}_i}^{kT} - \tilde{\mathbf{s}}_i'\mathbf{A}_i^T\boldsymbol{\Phi}_{\mathbf{r}_i}^{kT})\boldsymbol{\lambda}^k
\end{aligned}
\qquad (6.2.49)
$$

which are the desired expressions for constraint reaction forces and torques on body i at joint k.

➧ **Example 6.2.2:** Constraint reaction force and torque that act on body i due to a spherical joint are, from Eq. 6.2.49 and Table 6.2.1,

$$\mathbf{F}_i''^{C_s} = \mathbf{C}_i^T\mathbf{A}_i^T\boldsymbol{\lambda}^s$$

$$\mathbf{T}_i''^{C_s} = -\mathbf{C}_i^T(-\tilde{\mathbf{s}}_i'\mathbf{A}_i^T + \tilde{\mathbf{s}}_i'\mathbf{A}_i^T)\boldsymbol{\lambda}^s = \mathbf{0}$$

As expected, the spherical joint creates no constraint reaction torque. In global coordinates, the constraint reaction force is

$$\mathbf{F}_i^{C_s} = \mathbf{A}_i\mathbf{C}_i\mathbf{F}_i''^{C_s} = \boldsymbol{\lambda}^s$$

Similar relations may be derived for other joints (Probs. 6.2.10 and 6.2.11).

Solving Spatial Differential-Algebraic Equations of Motion

Since the Newton–Euler form of the acceleration equations of Eq. 6.2.32 is solved for angular accelerations, some modification of the generalized coordinate integration method presented in Appendix D is required. One alternative is to solve Eq. 6.2.32 for $\ddot{\mathbf{r}}$ and $\dot{\boldsymbol{\omega}}'$ and to use Eq. 5.4.6 to obtain $\ddot{\mathbf{p}}$; that is,

$$\ddot{\mathbf{p}} = \tfrac{1}{2}\mathbf{G}^T\dot{\boldsymbol{\omega}}' - \tfrac{1}{4}\boldsymbol{\omega}'^T\boldsymbol{\omega}'\mathbf{p} \qquad (6.2.50)$$

and finally to integrate for Euler parameter generalized coordinates, including Eq. 6.2.24 in the algebraic equations of constraint.

Another possible use of Eq. 6.2.32 is to integrate angular acceleration $\dot{\boldsymbol{\omega}}'$ to obtain the angular velocity $\boldsymbol{\omega}'$ and then to use Eq. 4.3.31 (i.e., $\dot{\mathbf{p}} = \tfrac{1}{2}\mathbf{G}^T\boldsymbol{\omega}'$) to determine Euler parameter time derivatives. Equation 4.3.31 and the Euler parameter normalization constraint of Eq. 6.2.24 may then be integrated to obtain the Euler parameter vector $\mathbf{p}$. Although this approach appears to be attractive, intricate computations are involved in obtaining positive error control. For details regarding methods for integrating equations of constrained multibody dynamics, see Ref. 12.

Even more pronounced than in the case of the planar Cartesian coordinate formulation of Section 6.1, the spatial Cartesian equations of motion of Eq. 6.2.32 and the associated constraint equations are far too extensive and complicated to yield analytical solutions. This formulation is therefore practical for analysis of realistic mechanical systems only if numerical computations are used. For an extensive treatment of computational aspects of spatial Cartesian dynamics using this formulation, the reader is referred to Ref. 12.

6.3 RELATIVE COORDINATE FORMULATIONS FOR SIMPLE MACHINES

While the Cartesian coordinate formulation for multibody systems is systematic and generally applicable for both computer generation and solution of equations of motion [12], it leads to large systems of differential-algebraic equations of motion. Much as in Chapter 3 for systems of particles, it is possible to formulate the equations of motion of multibody systems in terms of a reduced set of generalized coordinates, using the variational form of the equations of motion. Simple machines are analyzed in this section to illustrate the method. More refined and extensive developments may be found in Refs. 10, 13, and 19.

♦ **Example 6.3.1:** A planar double pendulum is shown in Fig. 6.3.1. Each bar is 2 units long, with the centroid located at its center.

The transformation from the x_1'-y_1' frame to the x-y frame is

$$\mathbf{A}_1 = \begin{bmatrix} \sin q_1 & \cos q_1 \\ -\cos q_1 & \sin q_1 \end{bmatrix}$$

and the transformation from the x_2'-y_2' frame to the x_1'-y_1' frame is

$$\mathbf{A}_{12} = \begin{bmatrix} \cos q_2 & -\sin q_2 \\ \sin q_2 & \cos q_2 \end{bmatrix}$$

The position vectors to the centroids of bodies 1 and 2 are

$$\mathbf{r}_1 = \begin{bmatrix} \sin q_1 \\ -\cos q_1 \end{bmatrix}$$

$$\mathbf{r}_2 = \begin{bmatrix} 2\sin q_1 + \sin(q_1 + q_2) \\ -2\cos q_1 - \cos(q_1 + q_2) \end{bmatrix}$$

and, taking the differential, the associated virtual displacements are

$$\delta\mathbf{r}_1 = \begin{bmatrix} \cos q_1 \\ \sin q_1 \end{bmatrix} \delta q_1$$

$$\delta\mathbf{r}_2 = \begin{bmatrix} 2\cos q_1 + \cos(q_1 + q_2) \\ 2\sin q_1 + \sin(q_1 + q_2) \end{bmatrix} \delta q_1 + \begin{bmatrix} \cos(q_1 + q_2) \\ \sin(q_1 + q_2) \end{bmatrix} \delta q_2$$

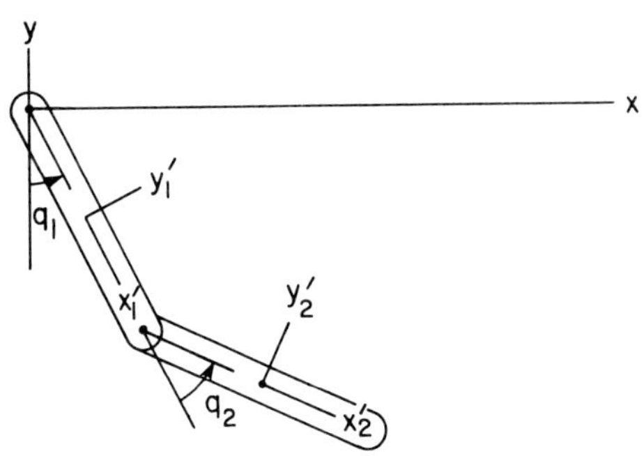

Figure 6.3.1 Planar double pendulum.

The accelerations of the centroids are

$$\ddot{\mathbf{r}}_1 = \begin{bmatrix} \cos q_1 \\ \sin q_1 \end{bmatrix} \ddot{\mathbf{q}}_1 + \begin{bmatrix} -\sin q_1 \\ \cos q_1 \end{bmatrix} \dot{q}_1^2$$

$$\ddot{\mathbf{r}}_2 = \begin{bmatrix} 2\cos q_1 + \cos(q_1 + q_2) \\ 2\sin q_1 + \sin(q_1 + q_2) \end{bmatrix} \ddot{q}_1 + \begin{bmatrix} \cos(q_1 + q_2) \\ \sin(q_1 + q_2) \end{bmatrix} \ddot{q}_2$$
$$+ \begin{bmatrix} -2\sin q_1 - \sin(q_1 + q_2) \\ 2\cos q_1 + \cos(q_1 + q_2) \end{bmatrix} \dot{q}_1^2 + \begin{bmatrix} -\sin(q_1 + q_2) \\ \cos(q_1 + q_2) \end{bmatrix} \dot{q}_2^2$$
$$+2 \begin{bmatrix} -\sin(q_1 + q_2) \\ \cos(q_1 + q_2) \end{bmatrix} \dot{q}_1 \dot{q}_2$$

Gravitational forces on the bars are $\mathbf{F}_i = -m_i g \mathbf{j}$, $i = 1, 2$. Since these forces act at the centers of mass, no external torques act (i.e., $n_1 = n_2 = 0$). Angular velocities, accelerations, and virtual rotations are

$$\dot{\phi}_1 = \dot{q}_1$$
$$\dot{\phi}_2 = \dot{q}_1 + \dot{q}_2$$
$$\ddot{\phi}_1 = \ddot{q}_1$$
$$\ddot{\phi}_2 = \ddot{q}_1 + \ddot{q}_2$$
$$\delta\phi_1 = \delta q_1$$
$$\delta\phi_2 = \delta q_1 + \delta q_2$$

Polar moments of inertia of the bars, from Table 5.2.1, are

$$J_1 = \frac{m_1}{3}$$

$$J_2 = \frac{m_2}{3}$$

where m_1 and m_2 are the masses of bodies 1 and 2, respectively.

Summing Eq. 5.1.42 over all bodies in the system, after substituting the foregoing relations, yields

$$\sum_{i=1}^{2} \{\delta\mathbf{r}_i^T[m_i\ddot{\mathbf{r}}_i - \mathbf{F}_i^A] + \delta\phi_i[J_i\ddot{\phi}_i - n_i^A]\}$$
$$= \delta q_1^T \left\{ \left[\frac{4m_1}{3} + m_2\left(\frac{16}{3} + 4\cos q_2\right)\right] \ddot{q}_1 \right.$$
$$+ m_2\left(\frac{4}{3} + 2\cos q_2\right)\ddot{q}_2$$
$$-2m_2 \sin q_2(\dot{q}_2^2 + 2\dot{q}_1\dot{q}_2)$$
$$\left. + m_2 g(2\sin q_1 + \sin(q_1 + q_2)) + m_1 g \sin q_1 \right\} \tag{6.3.1}$$

$$+\delta q_2 \left\{ m_2 \left(\frac{4}{3} + 2 \cos q_2 \right) \ddot{q}_1 + \frac{4}{3} m_2 \ddot{q}_2 \right.$$

$$\left. + 2m_2 \sin q_2 \dot{q}_1^2 + m_2 g \sin(q_1 + q_2) \right\}$$

$$= \delta \mathbf{q}^T \left\{ \begin{bmatrix} \frac{4m_1}{3} + m_2 \left(\frac{16}{3} + 4 \cos q_2 \right) & m_2 \left(\frac{4}{3} + 2 \cos q_2 \right) \\ m_2 \left(\frac{4}{3} + 2 \cos q_2 \right) & \frac{4}{3} m_2 \end{bmatrix} \ddot{\mathbf{q}} \right.$$

$$\left. + \begin{bmatrix} -2m_2 \sin q_2(\dot{q}_2^2 + 2\dot{q}_1\dot{q}_2) \\ \quad + m_2 g(2 \sin q_1 + \sin(q_1 + q_2)) + m_1 g \sin q_1 \\ 2m_2 \sin q_2 \dot{q}_1^2 + m_2 g \sin(q_1 + q_2) \end{bmatrix} \right\}$$

$$= 0$$

for arbitrary $\delta \mathbf{q}$, where $\mathbf{q} = [q_1, q_2]^T$. Thus the coefficient of $\delta \mathbf{q}^T$ must be zero, yielding the differential equations of motion as

$$\begin{bmatrix} \frac{4m_1}{3} + m_2 \left(\frac{16}{3} + 4 \cos q_2 \right) & m_2 \left(\frac{4}{3} + 2 \cos q_2 \right) \\ m_2 \left(\frac{4}{3} + 2 \cos q_2 \right) & \frac{4m_2}{3} \end{bmatrix} \ddot{\mathbf{q}}$$

$$= - \begin{bmatrix} -2m_2 \sin q_2(\dot{q}_2^2 + 2\dot{q}_1\dot{q}_2) + m_2 g(2 \sin q_1 + \sin(q_1 + q_2)) + m_1 g \sin q_1 \\ 2m_2 \sin q_2 \dot{q}_1^2 + m_2 g \sin(q_1 + q_2) \end{bmatrix}$$

Since the form of the left side of the equations of motion is the same as for the particle model of the double pendulum in Example 3.2.5, similar dynamic response is expected. As a numerical example that is integrated using the methods of Appendix D, let $m_1 = m_2 = 1$ kg and start the system from rest with $q_1(0) = q_2(0) = \pi/4$ rad. The dynamic response predicted by numerical integration is shown in Fig. 6.3.2. The differences in characteristics of response from Fig. 3.2.6 are due to different dimensions and rotary inertia effects.

➧ **Example 6.3.2:** The two body model of a slider–crank mechanism shown in Fig. 6.3.3 can be constructed from the double pendulum of Example 6.3.1, by requiring that the endpoint of body 2 slide along the x axis.

Equation 6.3.1 must hold for all $\delta \mathbf{q}$ that are consistent with the constraint equation

$$\Phi = -2 \cos q_1 - 2 \cos(q_1 + q_2) = 0$$

that is, that satisfy

$$\Phi_{\mathbf{q}} \, \delta \mathbf{q} = [2 \sin q_1 + 2 \sin(q_1 + q_2), 2 \sin(q_1 + q_2)] \, \delta \mathbf{q} = 0$$

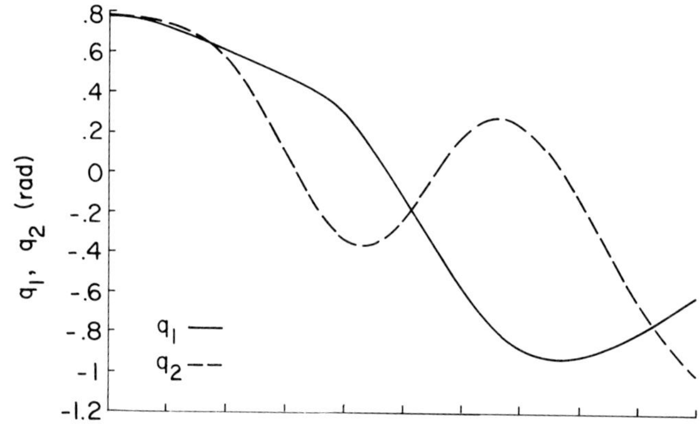

(a) q_1 and q_2 vs time

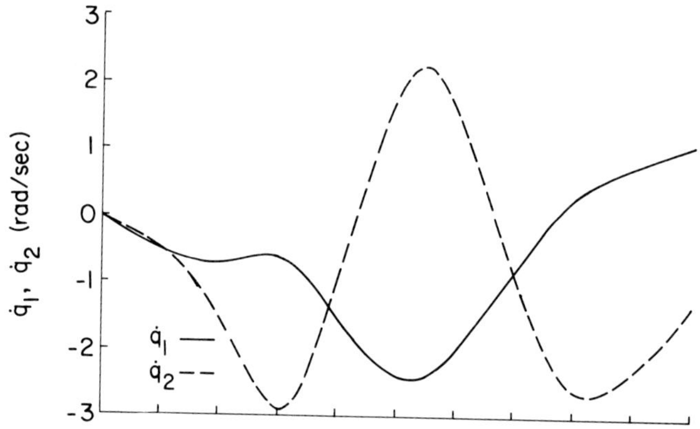

(b) $\dot{q}_1$ and $\dot{q}_2$ vs time

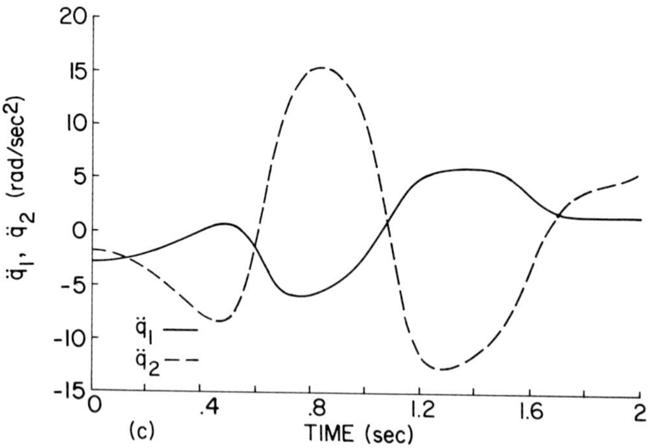

(c) $\ddot{q}_1$ and $\ddot{q}_2$ vs time

Figure 6.3.2 Dynamic response of double pendulum.

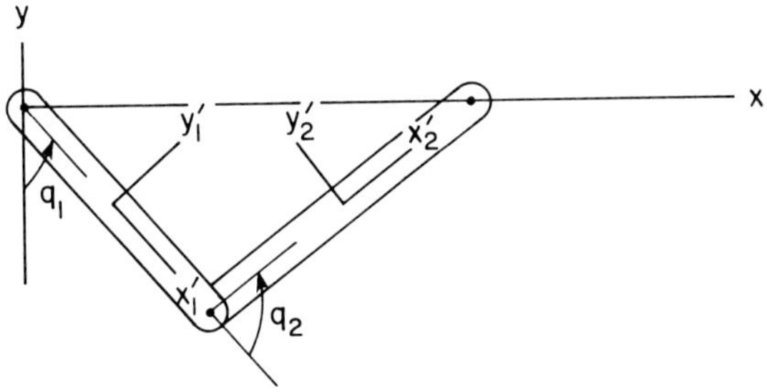

Figure 6.3.3 Two-body slider–crank.

By the Lagrange multiplier theorem, there exists a λ such that

$$\delta\mathbf{q}^T\left\{\begin{bmatrix}\frac{4m_1}{3}+m_2\left(\frac{16}{3}+4\cos q_2\right) & m_2\left(\frac{4}{3}+2\cos q_2\right)\\ m_2\left(\frac{4}{3}+2\cos q_2\right) & \frac{4m_2}{3}\end{bmatrix}\ddot{\mathbf{q}}\right.$$

$$+\begin{bmatrix}-2m_2\sin q_2(\dot{q}_2^2+2\dot{q}_1\dot{q}_2)+m_2g(2\sin q_1+\sin(q_1+q_2))+m_1g\sin q_1\\ 2m_2\sin q_2\dot{q}_1^2+m_2g\sin(q_1+q_2)\end{bmatrix}$$

$$\left.+\begin{bmatrix}2\sin q_1+2\sin(q_1+q_2)\\ 2\sin(q_1+q_2)\end{bmatrix}\lambda\right\}=0$$

for arbitrary $\delta\mathbf{q}$. Thus the coefficient of $\delta\mathbf{q}^T$ must be zero. To complete the equations of motion, the constraint acceleration equation must be obtained. Taking two time derivatives of the constraint equation yields the acceleration equation

$$\Phi_{\mathbf{q}}\ddot{\mathbf{q}}=-(2\cos q_1+2\cos(q_1+q_2))\dot{q}_1^2-2\cos(q_1+q_2)(2\dot{q}_1\dot{q}_2+\dot{q}_2^2)$$

Thus the differential-algebraic equations of motion of the system are

$$\begin{bmatrix}\frac{4m_1}{3}+m_2\left(\frac{16}{3}+4\cos q_2\right) & m_2\left(\frac{4}{3}+2\cos q_2\right) & 2\sin q_1+2\sin(q_1+q_2)\\ m_2\left(\frac{4}{3}+2\cos q_2\right) & \frac{4m_2}{3} & 2\sin(q_1+q_2)\\ 2\sin q_1+2\sin(q_1+q_2) & 2\sin(q_1+q_2) & 0\end{bmatrix}$$

$$\times\begin{bmatrix}\ddot{q}_1\\ \ddot{q}_2\\ \lambda\end{bmatrix}$$

$$=\begin{bmatrix}2m_2\sin q_2(\dot{q}_2^2+2\dot{q}_1\dot{q}_2)-m_2g(2\sin q_1+\sin(q_1+q_2))-m_1g\sin q_1\\ -2m_2\sin q_2\dot{q}_1^2-m_2g\sin(q_1+q_2)\\ -2(\cos q_1+\cos(q_1+q_2))\dot{q}_1^2-2\cos(q_1+q_2)(2\dot{q}_1\dot{q}_2+\dot{q}_2^2)\end{bmatrix}$$

As a numerical example, to be solved by the numerical method for solving differential-algebraic equations in Appendix D, let $m_1 = m_2 = 1$ kg. As in the particle model of Example 3.2.9, the initial conditions are selected as $q_1(0) = 3\pi/4$ rad and

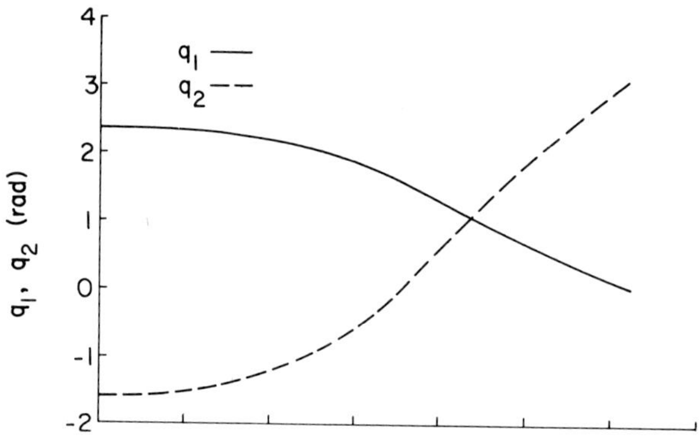

(a) q_1 and q_2 vs time

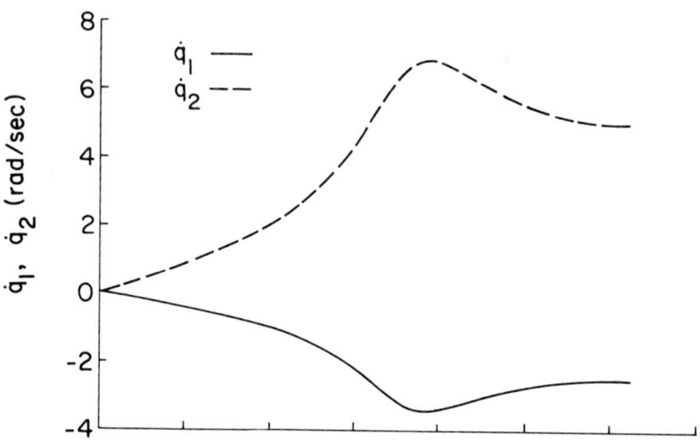

(b) $\dot{q}_1$ and $\dot{q}_2$ vs time

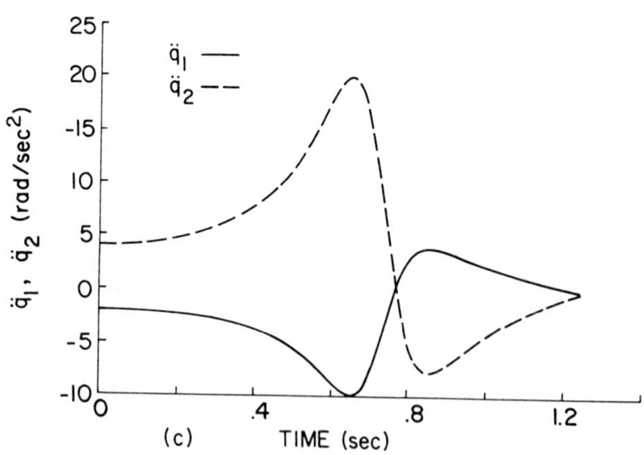

(c) $\ddot{q}_1$ and $\ddot{q}_2$ vs time

Figure 6.3.4 Dynamic response of slider–crank.

$\dot{q}_1(0) = 0$. Numerical results presented in Fig. 6.3.4 show the same qualitative behavior as in Fig. 3.2.13, with the singular configuration encountered at about $t = 1.23$ s.

♦ **Example 6.3.3:** If the mass of the slider plays an important role in dynamics of a slider–crank system, the three body model shown in Fig. 6.3.5 may be used.

The position of body 3 (the slider) is

$$\mathbf{r}_3 = \begin{bmatrix} q_3 \\ 0 \end{bmatrix}$$

The virtual displacement, acceleration, and gravitation force on body 3 are

$$\delta\mathbf{r}_3 = \begin{bmatrix} \delta q_3 \\ 0 \end{bmatrix}$$

$$\ddot{\mathbf{r}}_3 = \begin{bmatrix} \ddot{q}_3 \\ 0 \end{bmatrix}$$

$$\mathbf{F}_3^A = \begin{bmatrix} 0 \\ -m_3 g \end{bmatrix}$$

Since body 3 only slides along the x axis, there are no virtual rotation terms. Using the variational equations of bodies 1 and 2 in Example 6.3.1 and adding the variational equation of body 3, $\delta q_3\,(m_3, \ddot{q}_3) = 0$,

$$\delta\mathbf{q}\left\{\begin{bmatrix} \dfrac{4m_1}{3} + m_2\left(\dfrac{16}{3} + 4\cos q_2\right) & m_2\left(\dfrac{4}{3} + 2\cos q_2\right) & 0 \\ m_2\left(\dfrac{4}{3} + 2\cos q_2\right) & \dfrac{4m_2}{3} & 0 \\ 0 & 0 & m_3 \end{bmatrix}\ddot{\mathbf{q}}\right.$$

$$\left. + \begin{bmatrix} -2m_2 \sin q_2(\dot{q}_2 + 2\dot{q}_1\dot{q}_2) + m_2 g(2\sin q_1 + \sin(q_1 + q_2)) + m_1 g \sin q_1 \\ 2m_2 \sin q_2 \dot{q}_1^2 + m_2 g \sin(q_1 + q_2) \\ 0 \end{bmatrix}\right\}$$

$$= 0$$

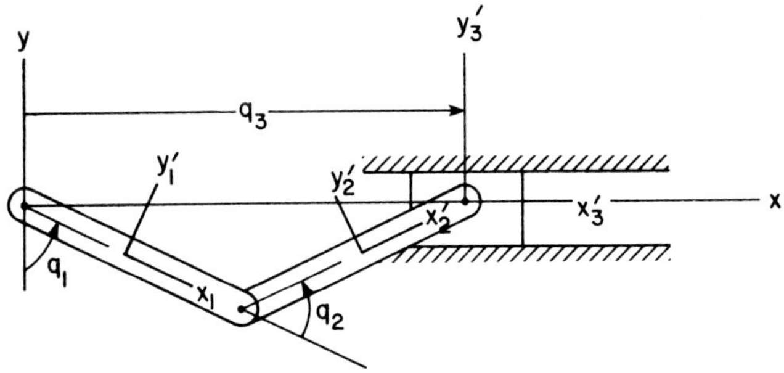

Figure 6.3.5 Three body slider–crank.

for all $\delta\mathbf{q}$ that are consistent with the revolute joint constraint between bodies 2 and 3; that is,

$$\mathbf{\Phi} = \begin{bmatrix} 2\sin q_1 + 2\sin(q_1+q_2) - q_3 \\ -2\cos q_1 - 2\cos(q_1+q_2) \end{bmatrix} = \mathbf{0}$$

Taking the variation of this equation yields

$$\mathbf{\Phi_q}\,\delta\mathbf{q} = \begin{bmatrix} 2\cos q_1 + 2\cos(q_1+q_2) & 2\cos(q_1+q_2) & -1 \\ 2\sin q_1 + 2\sin(q_1+q_2) & 2\sin(q_1+q_2) & 0 \end{bmatrix} \delta\mathbf{q} = \mathbf{0}$$

The acceleration equation corresponding to this constraint is

$$\mathbf{\Phi_q}\ddot{\mathbf{q}} = \begin{bmatrix} 2(\sin q_1 + \sin(q_1+q_2))\dot{q}_1^2 + 2\sin(q_1+q_2)(2\dot{q}_1\dot{q}_2 + \dot{q}_2^2) \\ -2(\cos q_1 + \cos(q_1+q_2))\dot{q}_1^2 - 2\cos(q_1+q_2)(2\dot{q}_1\dot{q}_2 + \dot{q}_2^2) \end{bmatrix}$$

Thus, by the Lagrange multiplier theorem, there exists a Lagrange multiplier vector $\boldsymbol{\lambda} = [\lambda_1, \lambda_2]^T$ such that the following differential-algebraic equation of motion holds:

$$\begin{bmatrix} \dfrac{4m_1}{3} + m_2\left(\dfrac{16}{3} + 4\cos q_2\right) & m_2\left(\dfrac{4}{3} + 2\cos q_2\right) & 0 & 2\cos q_1 + 2\cos(q_1+q_2) & 2\sin q_1 + 2\sin(q_1+q_2) \\ m_2\left(\dfrac{4}{3} + 2\cos q_2\right) & \dfrac{4m_2}{3} & 0 & 2\cos(q_1+q_2) & 2\sin(q_1+q_2) \\ 0 & 0 & m_3 & -1 & 0 \\ 2\cos q_1 + 2\cos(q_1+q_2) & 2\cos(q_1+q_2) & -1 & 0 & 0 \\ 2\sin q_1 + 2\sin(q_1+q_2) & 2\sin(q_1+q_2) & 0 & 0 & 0 \end{bmatrix} \begin{bmatrix} \ddot{\mathbf{q}} \\ \boldsymbol{\lambda} \end{bmatrix}$$

$$= \begin{bmatrix} 2m_2\sin q_2(\dot{q}_2^2 + 2\dot{q}_1\dot{q}_2) - m_2 g(2\sin q_1 + \sin(q_1+q_2)) - m_1 g\sin q_1 \\ -2m_2\sin q_2\dot{q}_1^2 - m_2 g\sin(q_1+q_2) \\ 0 \\ 2(\sin q_1 + \sin(q_1+q_2))\dot{q}_1^2 + 2\sin(q_1+q_2)(2\dot{q}_1\dot{q}_2 + \dot{q}_2^2) \\ -2(\cos q_1 + \cos(q_1+q_2))\dot{q}_1^2 - 2\cos(q_1+q_2)(2\dot{q}_1\dot{q}_2 + \dot{q}_2^2) \end{bmatrix}$$

While numerical solution of these differential-algebraic equations using the numerical method of Appendix D requires a little more computer time, no theoretical difficulties arise. As a numerical example, let $m_1 = m_2 = m_3 = 1$ kg and select initial conditions $q_1(0) = 3\pi/4$ rad and $\dot{q}_1(0) = 0$, as in Examples 3.2.9 and 6.3.2. In this case, constraint equations determine initial conditions on q_2 and q_3. Numerical results shown in Fig. 6.3.6 are similar to those of Fig. 6.3.4, except that the singular

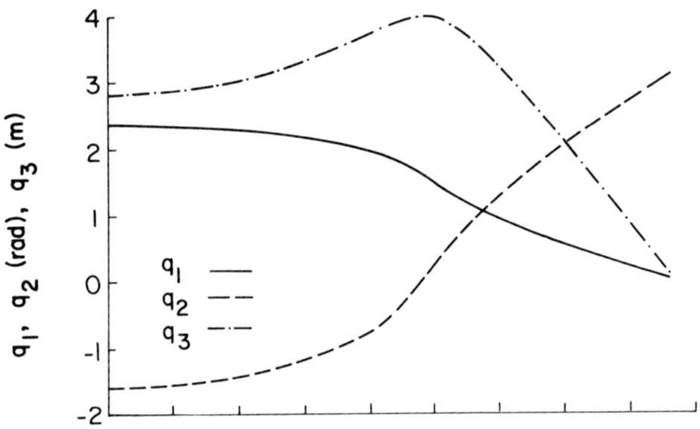

(a) q_1, q_2, and q_3 vs time

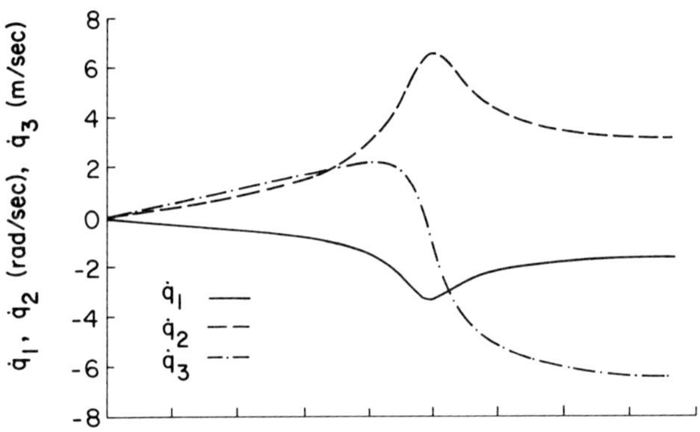

(b) $\dot{q}_1$, $\dot{q}_2$, and $\dot{q}_3$ vs time

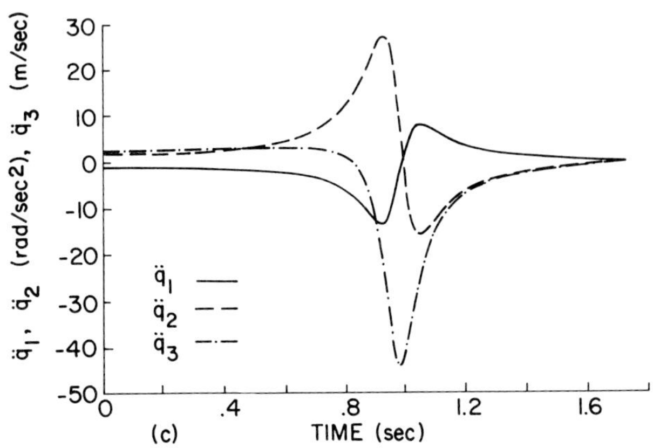

(c) $\ddot{q}_1$, $\ddot{q}_2$, and $\ddot{q}_3$ vs time

Figure 6.3.6 Dynamic response of three body slider–crank.

configuration is encountered later, at about $t = 1.73$ s. This is to be expected, since mass m_3 has been added to the model, which slows the dynamic response.

Example 6.3.4: Consider the two degree of freedom material handling device shown in Fig. 6.3.7. The kinematics of this mechanism are identical to Example 2.6.2. Each of the uniform bars in this mechanism is 2 units long, with its centroid located at the center.

The transformation matrix from the x_1'-y_1'-z_1' frame to the x-y-z frame is

$$\mathbf{A}_1 = \begin{bmatrix} \cos q_1 & -\sin q_1 & 0 \\ \sin q_1 & \cos q_1 & 0 \\ 0 & 0 & 1 \end{bmatrix}$$

Similarly, the transformation matrix from the x_2'-y_2'-z_2' frame to the x_1'-y_1'-z_1' is

$$\mathbf{A}_{12} = \begin{bmatrix} 1 & 0 & 0 \\ 0 & \sin q_2 & \cos q_2 \\ 0 & -\cos q_2 & \sin q_2 \end{bmatrix}$$

Using these transformation matrices, angular velocity information needed for the equations of motion of the system may be obtained. By direct observation, the angular velocity of body 1, in its centroidal body-fixed reference frame, is simply

$$\boldsymbol{\omega}_1' = [0, 0, \dot{q}_1]^T$$

The angular velocity of body 2, in the x_1'-y_1'-z_1' frame, is

$$\begin{aligned}(\boldsymbol{\omega}_2)_1' &= \boldsymbol{\omega}_1' + \boldsymbol{\omega}_{12}' \\ &= [0, 0, \dot{q}_1]^T + [\dot{q}_2, 0, 0]^T \\ &= [\dot{q}_2, 0, \dot{q}_1]^T\end{aligned}$$

Thus

$$\begin{aligned}\boldsymbol{\omega}_2' &= \mathbf{A}_{12}^T(\boldsymbol{\omega}_2)_1' \\ &= [\dot{q}_2, -\dot{q}_1 \cos q_2, \dot{q}_1 \sin q_2]^T\end{aligned}$$

Similarly, virtual rotations are

$$\delta\boldsymbol{\pi}_1' = [0, 0, \delta q_1]^T$$

$$\delta\boldsymbol{\pi}_2' = [\delta q_2, -\delta q_1 \cos q_2, \delta q_1 \sin q_2]^T$$

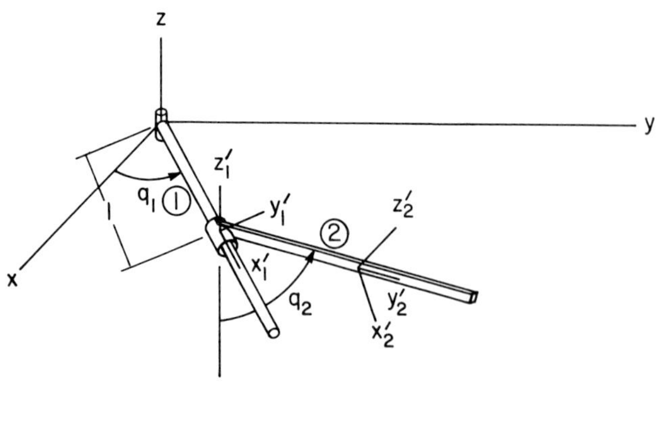

Figure 6.3.7 Spatial material handling device.

Angular accelerations of bodies 1 and 2 are obtained by direct differentiation as

$$\dot{\boldsymbol{\omega}}_1' = [0,\ 0,\ \ddot{q}_1]^T$$

$$\dot{\boldsymbol{\omega}}_2' = [\ddot{q}_2,\ -\ddot{q}_1 \cos q_2,\ \ddot{q}_1 \sin q_2]^T + [0,\ \dot{q}_1\dot{q}_2 \sin q_2,\ \dot{q}_1\dot{q}_2 \cos q_2]^T$$

Note that no quadratic terms arise in the angular acceleration expression for body 1, whereas they do arise in the expression for angular acceleration of body 2.

From Table 5.2.1, the inertia matrices of bodies 1 and 2 are

$$\mathbf{J}_1' = \text{diag}\left(0,\ \frac{m_1}{3},\ \frac{m_1}{3}\right)$$

$$\mathbf{J}_2' = \text{diag}\left(\frac{m_2}{3},\ 0,\ \frac{m_2}{3}\right)$$

Adding the variational equations of motion of Eq. 5.1.26 for bodies 1 and 2, noting that the gravitational forces $\mathbf{F}_i^A = -m_i g \mathbf{k}$ create no torque, $\mathbf{n}_1'^A = \mathbf{n}_2'^A = \mathbf{0}$,

$$\sum_{i=1}^{2} \{\delta \mathbf{r}_i^T [m_i \ddot{\mathbf{r}}_i - \mathbf{F}_i^A] + \delta \boldsymbol{\pi}_i'^T [\mathbf{J}_i' \dot{\boldsymbol{\omega}}_i' + \tilde{\boldsymbol{\omega}}_i' \mathbf{J}_i' \boldsymbol{\omega}_i']\} = 0$$

Since the centroids of bodies 1 and 2 are located in this example just as in Example 3.2.6, terms arising in the equations of motion for the particles in that derivation are identical to those that arise in the translational terms above. Rotational terms appearing above may be added to the variational equation of motion derived in Example 3.2.6, to obtain

$$\begin{aligned}
&\sum_{i=1}^{2} \{\delta \mathbf{r}_i^T [m_i \ddot{\mathbf{r}}_i - \mathbf{F}_i] + \delta \boldsymbol{\pi}_i'^T [\mathbf{J}_i' \dot{\boldsymbol{\omega}}_i' + \tilde{\boldsymbol{\omega}}_i' \mathbf{J}_i' \boldsymbol{\omega}_i']\} \\
&= \delta \mathbf{q}^T \left\{ \begin{bmatrix} m_1 + m_2(1 + \sin^2 q_2) & m_2 \cos q_2 \\ m_2 \cos q_2 & m_2 \end{bmatrix} \ddot{\mathbf{q}} \right. \\
&\quad \left. - \begin{bmatrix} m_2 \sin q_2(\dot{q}_2^2 - 2\dot{q}_1\dot{q}_2 \cos q_2) \\ -m_2 g \sin q_2 + m_2 \dot{q}_1^2 \cos q_2 \sin q_2 \end{bmatrix} \right\} + \delta q_1 \left[\frac{m_1}{3} \ddot{q}_1\right] \\
&\quad + [\delta q_2,\ -\delta q_1 \cos q_2,\ \delta q_1 \sin q_2] \left\{ \begin{bmatrix} \dfrac{m_2 \ddot{q}_2}{3} \\ 0 \\ \dfrac{m_2 \ddot{q}_1 \sin q_2}{3} \end{bmatrix} \right. \\
&\quad \left. + \begin{bmatrix} 0 \\ 0 \\ \dfrac{m_2 \dot{q}_1 \dot{q}_2 \cos q_2}{3} \end{bmatrix} + \begin{bmatrix} \dfrac{-(m_2 \dot{q}_1^2 \cos q_2 \sin q_2)}{3} \\ 0 \\ \dfrac{m_2 \dot{q}_1 \dot{q}_2 \cos q_2}{3} \end{bmatrix} \right\}
\end{aligned} \qquad (6.3.2)$$

$$= \delta \mathbf{q}^T \left\{ \begin{bmatrix} \dfrac{4m_1}{3} + m_2\left(1 + \dfrac{4}{3}\sin^2 q_2\right) & m_2 \cos q_2 \\ m_2 \cos q_2 & \dfrac{4m_2}{3} \end{bmatrix} \ddot{\mathbf{q}} - \begin{bmatrix} m_2 \sin q_2 \left(\dot{q}_2^2 - \dfrac{8}{3}\dot{q}_1\dot{q}_2 \cos q_2\right) \\ -m_2 g \sin q_2 + \dfrac{4}{3} m_2 \dot{q}_1^2 \cos q_2 \sin q_2 \end{bmatrix} \right\}$$

$$= 0$$

which holds for arbitrary $\delta \mathbf{q}$, since components of $\mathbf{q}$ are independent. Thus the coefficient of $\delta \mathbf{q}^T$ above must be zero, leading to the second order differential equations of motion

$$\begin{bmatrix} \dfrac{4m_2}{3} + m_2\left(1 + \dfrac{4\sin^2 q_2}{3}\right) & m_2 \cos q_2 \\ m_2 \cos q_2 & \dfrac{4m_2}{3} \end{bmatrix} \begin{bmatrix} \ddot{q}_1 \\ \ddot{q}_2 \end{bmatrix}$$

$$= \begin{bmatrix} m_2 \sin q_2 \left(\dot{q}_2^2 - \dfrac{8\dot{q}_1\dot{q}_2 \cos q_2}{3}\right) \\ -m_2 g \sin q_2 + \dfrac{4m_2\dot{q}_1^2 \cos q_2 \sin q_2}{3} \end{bmatrix}$$

Note that these differential equations are identical in form to those of Example 3.2.6. The only difference is the additional rotary inertia of the bars used in this model. On qualitative grounds, it is thus expected that slightly slower response would occur in the present example. To check this behavior, the same initial conditions are used as in Example 3.2.6 and the equations of motion are solved numerically using the method of Appendix D. Numerical results are shown in Fig. 6.3.8 for both this example and Example 3.2.6. Note that response in this example lags that of Example 3.2.6, as expected.

Example 6.3.5: Consider a spatial four-bar mechanism model, in which bodies 1 and 2 are the same as in Example 6.3.4 and a body with unit length and negligible mass connects the end of body 2 and point (0, 2, 0) on the y axis, as shown in Fig. 6.3.9.

Since the coupler between body 2 and ground has no inertia, Eq. 6.3.2 holds for all $\delta \mathbf{q}$ that are consistent with the constraint equation

$$\Phi = \mathbf{d}^T\mathbf{d} - 1 = 0$$

where

$$\mathbf{d} = \mathbf{r}_1 + \mathbf{A}_1 \begin{bmatrix} 0 \\ 2 \sin q_2 \\ -2 \cos q_2 \end{bmatrix} - \begin{bmatrix} 0 \\ 2 \\ 0 \end{bmatrix} = \begin{bmatrix} \cos q_1 - 2 \sin q_1 \sin q_2 \\ \sin q_1 + 2 \cos q_1 \sin q_2 - 2 \\ -2 \cos q_2 \end{bmatrix}$$

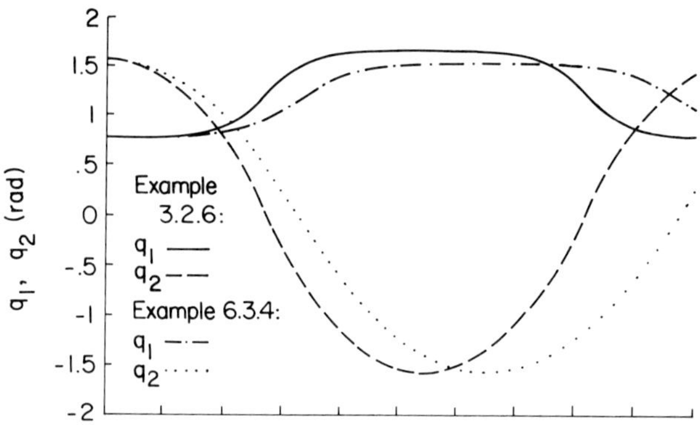

(a) q_1 and q_2 vs time, Examples 3.2.6 and 6.3.4

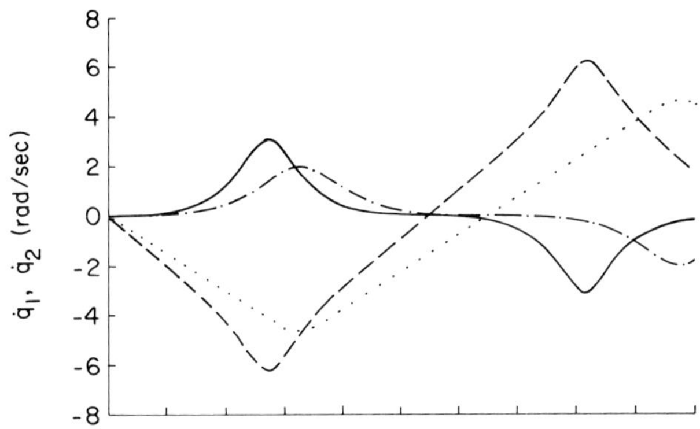

(b) $\dot{q}_1$ and $\dot{q}_2$ vs time, Examples 3.2.6 and 6.3.4

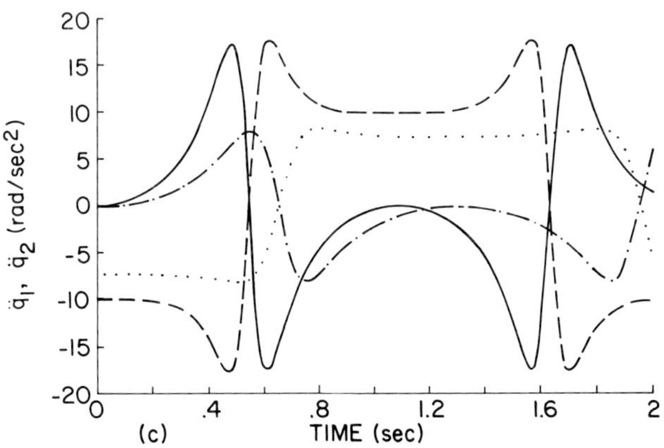

(c) $\ddot{q}_1$ and $\ddot{q}_2$ vs time, Examples 3.2.6 and 6.3.4

Figure 6.3.8 Dynamic response of two-degree-of-freedom material handler.

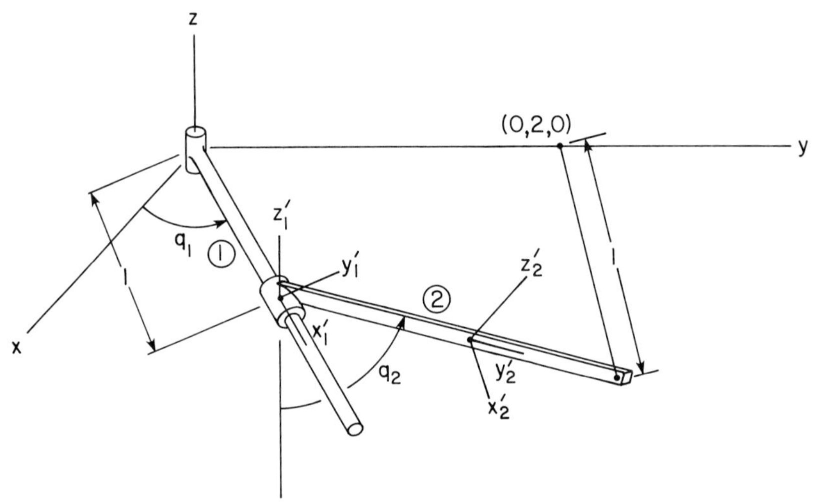

Figure 6.3.9 Spatial four-bar mechanism.

or

$$\Phi = 8 - 8\cos q_1 \sin q_2 - 4\sin q_1 = 0$$

Thus $\delta\mathbf{q}$ must satisfy

$$\Phi_{\mathbf{q}}\,\delta\mathbf{q} = [8\sin q_1 \sin q_2 - 4\cos q_1,\ -8\cos q_1 \cos q_2]\,\delta\mathbf{q} = 0$$

By the Lagrange multiplier theorem, there exists a Lagrange multiplier λ such that the augmented variational equation of motion holds for all $\delta\mathbf{q}$. The acceleration equation that completes the system equations of motion is

$$\Phi_{\mathbf{q}}\ddot{\mathbf{q}} = -[(8\cos q_1 \sin q_2 + 4\sin q_1)\dot{q}_1^2 + 16\sin q_1 \cos q_2 \dot{q}_1\dot{q}_2 + 8\cos q_1 \sin q_2 \dot{q}_2^2]$$

Thus the differential-algebraic equations of motion are

$$\begin{bmatrix} \dfrac{4m_1}{3} + m_2\left(1 + \dfrac{\sin^2 q_2}{3}\right) & m_2\cos q_2 & 8\sin q_1 \sin q_2 - 4\cos q_1 \\ m_2\cos q_2 & \dfrac{4m_2}{3} & -8\cos q_1 \cos q_2 \\ 8\sin q_1 \sin q_2 - 4\cos q_1 & -8\cos q_1 \cos q_1 & 0 \end{bmatrix} \begin{bmatrix} \ddot{\mathbf{q}} \\ \lambda \end{bmatrix}$$

$$= \begin{bmatrix} m_2 \sin q_2\left(\dot{q}_2^2 - \dfrac{8\dot{q}_1\dot{q}_2\cos q_2}{3}\right) \\ -m_2 g\sin q_2 + \dfrac{4m_2\dot{q}_1^2\cos q_2 \sin q_2}{3} \\ -(8\cos q_1 \sin q_2 + 4\sin q_1)\dot{q}_1^2 - 16\sin q_1 \cos q_2 \dot{q}_1\dot{q}_2 - 8\cos q_1 \sin q_2 \dot{q}_2^2 \end{bmatrix}$$

Example 6.3.6: Consider the three degree of freedom material handling mechanism of Examples 2.6.3 and 3.2.7, which is similar to the one treated in Example 6.3.4, but permitting body 2 to slide along the z_1' axis of body 1, as shown in Fig. 6.3.10. The additional generalized coordinate q_2 defines this translational degree of freedom.

Transformations from the x_1'-y_1'-z_1' frame to the x-y-z frame and from the x_2'-y_2'-z_2' frame to the x_1'-y_1'-z_1' frame are

$$\mathbf{A}_1 = \begin{bmatrix} -\sin q_1 & 0 & \cos q_1 \\ \cos q_1 & 0 & \sin q_1 \\ 0 & 1 & 0 \end{bmatrix}$$

$$\mathbf{A}_{12} = \begin{bmatrix} \cos q_3 & -\sin q_3 & 0 \\ \sin q_3 & \cos q_3 & 0 \\ 0 & 0 & 1 \end{bmatrix}$$

For body 1, note that

$$\mathbf{r}_1 = [\cos q_1, \sin q_1, 0]^T$$

$$\boldsymbol{\omega}_1' = [0, \dot{q}_1, 0]^T$$

$$\delta\mathbf{r}_1 = \delta q_1[-\sin q_1, \cos q_1, 0]^T$$

$$\delta\boldsymbol{\pi}_1' = [0, \delta q_1, 0]^T$$

$$\dot{\mathbf{r}}_1 = \dot{q}_1[-\sin q_1, \cos q_1, 0]^T$$

$$\ddot{\mathbf{r}}_1 = \ddot{q}_1[-\sin q_1, \cos q_1, 0]^T + \dot{q}_1^2[-\cos q_1, -\sin q_1, 0]^T$$

$$\dot{\boldsymbol{\omega}}_1' = [0, \ddot{q}_1, 0]^T$$

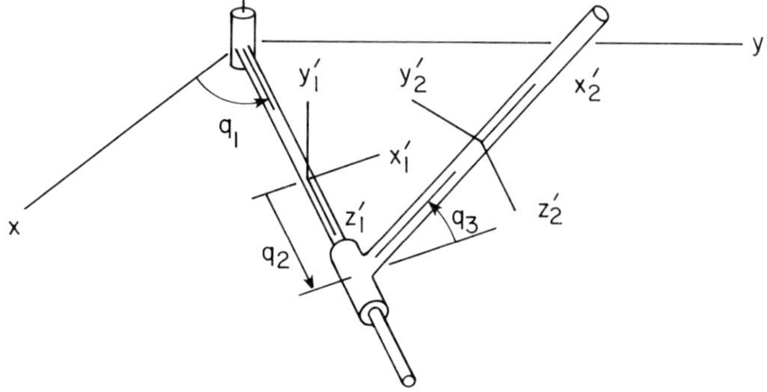

Figure 6.3.10 Three-degree-of-freedom two-body mechanism.

Similarly, for body 2 (Prob. 6.3.4),

$$\mathbf{r}_2 = \begin{bmatrix} (1 + q_2)\cos q_1 - \sin q_1 \cos q_3 \\ (1 + q_2)\sin q_1 + \cos q_1 \cos q_3 \\ \sin q_3 \end{bmatrix}$$

$$\boldsymbol{\omega}_2' = \begin{bmatrix} \sin q_3 \dot{q}_1 \\ \cos q_3 \dot{q}_1 \\ \dot{q}_3 \end{bmatrix}$$

$$\delta\mathbf{r}_2 = \begin{bmatrix} -(1 + q_2)\sin q_1 - \cos q_1 \cos q_3 & \cos q_1 & \sin q_1 \sin q_3 \\ (1 + q_2)\cos q_1 - \sin q_1 \cos q_3 & \sin q_1 & -\cos q_1 \sin q_3 \\ 0 & 0 & \cos q_3 \end{bmatrix} \delta\mathbf{q}$$

$$\delta\boldsymbol{\pi}_2' = \begin{bmatrix} \sin q_3\, \delta q_1 \\ \cos q_3\, \delta q_1 \\ \delta q_3 \end{bmatrix}$$

$$\dot{\mathbf{r}}_2 = \begin{bmatrix} -(1 + q_2)\sin q_1 - \cos q_1 \cos q_3 & \cos q_1 & \sin q_1 \sin q_3 \\ (1 + q_2)\cos q_1 - \sin q_1 \cos q_3 & \sin q_1 & -\cos q_1 \sin q_3 \\ 0 & 0 & \cos q_3 \end{bmatrix} \dot{\mathbf{q}}$$

$$\dot{\boldsymbol{\omega}}_2' = \begin{bmatrix} \sin q_3 \ddot{q}_1 + \cos q_3 \dot{q}_1 \dot{q}_3 \\ \cos q_3 \ddot{q}_1 - \sin q_3 \dot{q}_1 \dot{q}_3 \\ \ddot{q}_3 \end{bmatrix}$$

$$\begin{aligned} \ddot{\mathbf{r}}_2 = {} & \begin{bmatrix} -(1 + q_2)\sin q_1 - \cos q_1 \cos q_3 & \cos q_1 & \sin q_1 \sin q_3 \\ (1 + q_2)\cos q_1 - \sin q_1 \cos q_3 & \sin q_1 & -\cos q_1 \sin q_3 \\ 0 & 0 & \cos q_3 \end{bmatrix} \ddot{\mathbf{q}} \\ & + \dot{q}_1 \begin{bmatrix} -(1 + q_2)\cos q_1 + \sin q_1 \cos q_3 & -\sin q_1 & \cos q_1 \sin q_3 \\ -(1 + q_2)\sin q_1 - \cos q_1 \cos q_3 & \cos q_1 & \sin q_1 \sin q_3 \\ 0 & 0 & 0 \end{bmatrix} \dot{\mathbf{q}} \\ & + \dot{q}_2 \begin{bmatrix} -\sin q_1 & 0 & 0 \\ \cos q_1 & 0 & 0 \\ 0 & 0 & 0 \end{bmatrix} \dot{\mathbf{q}} \\ & + \dot{q}_3 \begin{bmatrix} \cos q_1 \sin q_3 & 0 & \sin q_1 \cos q_3 \\ \sin q_1 \sin q_3 & 0 & -\cos q_1 \cos q_3 \\ 0 & 0 & -\sin q_3 \end{bmatrix} \dot{\mathbf{q}} \end{aligned}$$

The inertia matrices with respect to the x_i'-y_i'-z_i' centroidal frames are

$$\mathbf{J}_1' = \operatorname{diag}\left(\frac{m_1}{3}, \frac{m_1}{3}, 0\right)$$

$$\mathbf{J}_2' = \operatorname{diag}\left(0, \frac{m_2}{3}, \frac{m_2}{3}\right)$$

Gravitational forces are $\mathbf{F}_i^A = [0, 0, -m_i g]^T$ and there is no externally applied torque (i.e., $\mathbf{n}_1'^A = \mathbf{n}_2'^A = \mathbf{0}$).

Substituting these equations into Eq. 6.2.26 yields the variational equation of motion

$$\sum_{i=1}^{2} \{\delta \mathbf{r}_i^T[m_i\ddot{\mathbf{r}}_i - \mathbf{F}_i^A] + \delta\boldsymbol{\pi}_i'^T[\mathbf{J}_i'\dot{\boldsymbol{\omega}}_i' + \tilde{\boldsymbol{\omega}}_i'\mathbf{J}_i'\boldsymbol{\omega}_i']\}$$

$$= \delta\mathbf{q}^T\left\{\begin{bmatrix} \frac{4}{3}(m_1 + m_2\cos^2 q_3) + m_2(1+q_2)^2 & -m_2\cos q_3 & -m_2(1+q_2)\sin q_3 \\ -m_2\cos q_3 & m_2 & 0 \\ -m_2(1+q_2)\sin q_3 & 0 & \frac{4}{3}m_2 \end{bmatrix}\ddot{\mathbf{q}}\right.$$

$$\left. + \begin{bmatrix} 2m_2(1+q_1)\dot{q}_1\dot{q}_2 - \frac{8}{3}m_2\sin q_3\cos q_3\dot{q}_1\dot{q}_3 - m_2(1+q_2)\cos q_3\dot{q}_3^2 \\ -m_2(1+q_2)\dot{q}_1^2 + 2m_2\sin q_3\dot{q}_1\dot{q}_3 \\ \frac{4}{3}m_2\sin q_3\cos q_3\dot{q}_1^2 - 2m_2\sin q_3\dot{q}_1\dot{q}_2 + m_2 g\cos q_3 \end{bmatrix}\right\} = 0$$

which must hold for arbitrary $\delta\mathbf{q}$. Thus the coefficient of $\delta\mathbf{q}^T$ must equal zero, yielding the differential equations of motion

$$\begin{bmatrix} \frac{4}{3}(m_1 + m_2\cos^2 q_3) + m_2(1+q_2)^2 & -m_2\cos q_3 & -m_2(1+q_2)\sin q_3 \\ -m_2\cos q_2 & m_2 & 0 \\ -m_2(1+q_2)\sin q_3 & 0 & \frac{4}{3}m_2 \end{bmatrix}\ddot{\mathbf{q}}$$

$$+ \begin{bmatrix} 2m_2(1+q_2)\dot{q}_1\dot{q}_2 - \frac{8}{3}m_2\sin q_3\cos q_3\dot{q}_1\dot{q}_3 - m_2(1+q_2)\cos q_3\dot{q}_3^2 \\ -m_2(1+q_2)\dot{q}_1^2 + 2m_2\sin q_3\dot{q}_1\dot{q}_3 \\ \frac{4}{3}m_2\sin q_3\cos q_3\dot{q}_1^2 - 2m_2\sin q_3\dot{q}_1\dot{q}_2 + m_2 g\cos q_3 \end{bmatrix} = \mathbf{0}$$

As a numerical example, these differential equations are solved numerically using the methods of Appendix D. Initial conditions are the same as those used in Example 3.2.7. Numerical results are shown in Fig. 6.3.11.

Since the geometry of the mechanism treated here is different from that of the particle model treated in Example 3.2.7, substantially different results might be expected. A comparison of Figs. 3.2.10 and 6.3.11 indeed shows considerably different quantitative results. The results, however, do show similar qualitative behavior.

The examples presented in this section illustrate a generally applicable method for deriving differential equations of motion for multibody systems, provided that independent generalized coordinates may be found. In all but Examples 6.3.2, 6.3.3, and 6.3.5, it was possible to define independent generalized coordinates that lead to nonlinear ordinary differential equations of motion. In Examples 6.3.2, 6.3.3, and 6.3.5, constraint equations among the generalized coordinates were retained and incorporated in the differential-algebraic equations of motion, using the Lagrange multiplier theorem.

Even in the relatively elementary examples in this section, for which independent generalized coordinates could be found, extremely nonlinear differential equations are obtained, precluding practical generation of analytical solutions. Numerical integration techniques (see Appendix D) are therefore needed to obtain solutions of the equations of motion of even elementary multibody mechanical system dynamics applications.

Finally, it is important to note that a unified, systematic approach has been used in all examples presented in this section to obtain equations of motion. First,

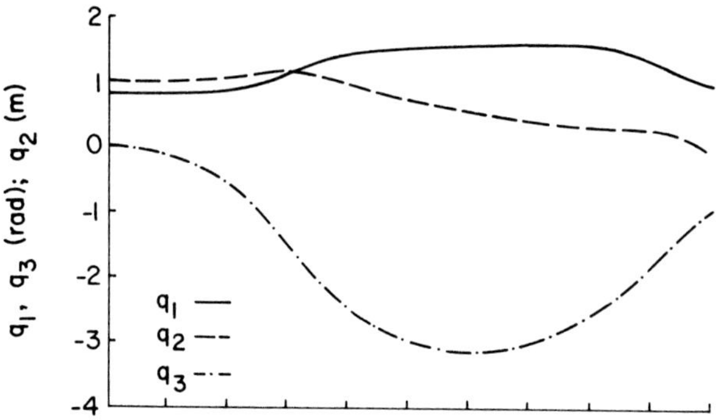

(a) q_1, q_2, and q_3 vs time

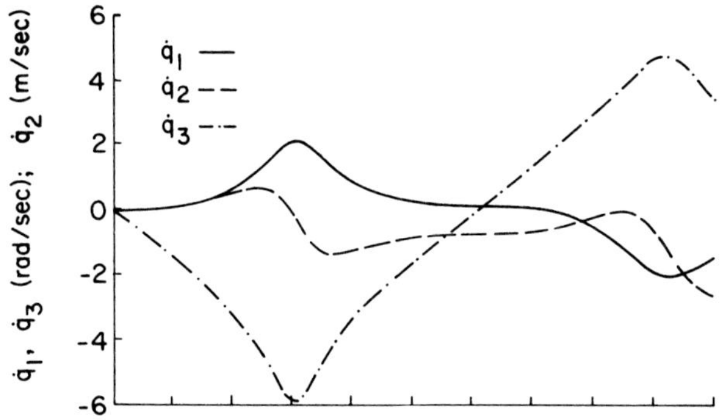

(b) $\dot{q}_1$, $\dot{q}_2$, and $\dot{q}_3$ vs time

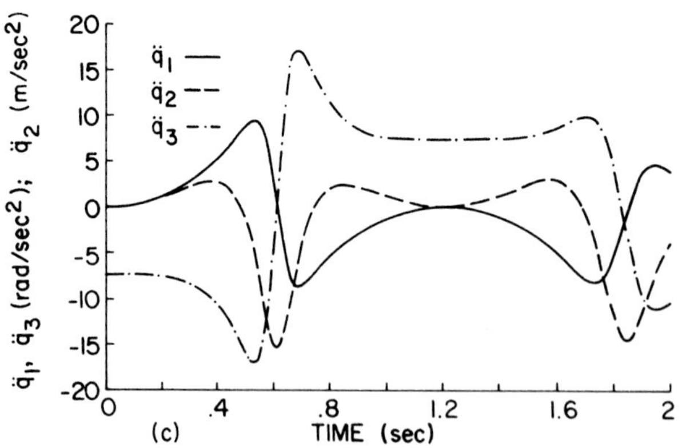

(c) $\ddot{q}_1$, $\ddot{q}_2$, and $\ddot{q}_3$ vs time

Figure 6.3.11 Dynamic response of three-degree-of-freedom material handler.

expressions for velocities, virtual displacements, and accelerations are obtained as functions of time derivatives and variations of generalized coordinates. These results are then substituted into the variational form of the equations of motion to derive differential equations if generalized coordinates are independent and differential-algebraic equations if the generalized coordinates are dependent. The prescription of a generally applicable algorithm follows very closely to the recipes presented for derivation for equations of motion of systems of particles in Section 3.7. The difference in the case of multibody systems is that angular velocities and virtual rotations must be dealt with, in addition to velocities and virtual displacements of the origins of body-fixed reference frames.

For extensive treatments of relative coordinate formulations of multibody dynamics, the reader is referred to Refs. 10, 13, and 19. This is a rapidly developing field of research and development at the time of writing this book.

6.4 RELATIVE COORDINATE VARIATIONAL FORMULATION FOR MOTION OF BODIES RELATIVE TO A BASE BODY

A two body subsystem whose kinematics is modeled with both Cartesian and relative coordinates is introduced here to illustrate a systematic method of formulating equations of motion that exploits properties of both vector and differential calculus methods and the variational Newton–Euler equations of motion. A pattern for construction of a system mass matrix and generalized force terms is developed and applied for derivation of the equations of motion of a vehicle model. More general relative coordinate approaches to multibody dynamics are presented in Sections 6.5 to 6.7 and in Refs. 10, 13, and 19.

Kinematics of a Pair of Bodies

A two body combination of machine components is shown in Fig. 6.4.1. The master body is designated as body c and may be thought of as the chassis of a vehicle or the central body of a spacecraft. It is located and oriented in space by the position vector $\mathbf{r}_c$ from the origin of the global reference frame to the origin of the body-fixed x'_c-y'_c-z'_c (or c') reference frame and by a set of generalized coordinates that define the orientation of this frame relative to the global frame (e.g., Euler angles or Euler parameters).

The attached body a shown in Fig. 6.4.1, which may be considered to be a vehicle suspension component or an antenna on a spacecraft, is permitted to translate and rotate relative to body c. To define the relationship between bodies a and c, body-fixed joint reference frames are defined on each of the bodies at joint connection points O''_c and O''_a, as shown in Fig. 6.4.1. The x''_c-y''_c-z''_c (or c'') frame is fixed in body c and the x''_a-y''_a-z''_a (or a'') frame is fixed in body a. A unit vector $\mathbf{b}_c$ is defined as fixed in the c'' frame to body c. Translation of the origin O''_a of the a'' frame is permitted to occur along the unit vector $\mathbf{b}_c$, with magnitude

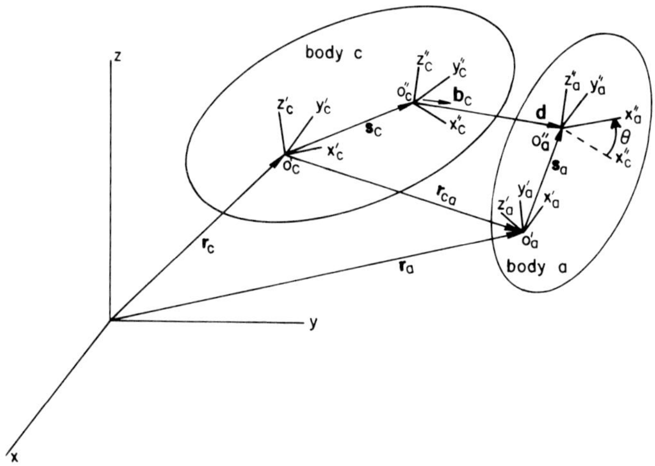

Figure 6.4.1 Two-degree-of-freedom attached body.

α (i.e., $\mathbf{d} = \alpha \mathbf{b}_c$ in Fig. 6.4.1). The kinematic connection between this pair of bodies is defined so that the z''_c and z''_a axes remain parallel and rotation of the a'' frame can occur relative to the c'' frame about the z''_c axis by an angle θ, as shown in Fig. 6.4.1.

A vector $\mathbf{u}'_c$ in the c' frame is transformed to its inertial x-y-z representation by

$$\mathbf{u}_c = \mathbf{A}_c \mathbf{u}'_c \tag{6.4.1}$$

where $\mathbf{A}_c$ depends on the orientation generalized coordinates of body c. A vector $\mathbf{u}''_c$ in the c'' frame is represented in the c' frame by the transformation

$$\mathbf{u}'_c = \mathbf{C}_c \mathbf{u}''_c \tag{6.4.2}$$

where $\mathbf{C}_c$ is the constant orthogonal transformation matrix from the c'' frame to the c' frame. Similarly, a vector $\mathbf{u}''_a$ in the a'' frame is transformed to the x'_a-y'_a-z'_a (or a') frame by

$$\mathbf{u}'_a = \mathbf{C}_a \mathbf{u}''_a \tag{6.4.3}$$

where $\mathbf{C}_a$ is the constant orthogonal transformation matrix from the a'' frame to the a' frame. Finally, vector $\mathbf{u}''_a$ in the a'' frame is transformed to the c'' frame by

$$\mathbf{u}''_c = \mathbf{A}''_{ca} \mathbf{u}''_a \tag{6.4.4}$$

where the orthogonal transformation matrix $\mathbf{A}''_{ca}$ from the a'' frame to the c'' frame depends only on the angle θ. As in Example 4.2.2 with $\omega t = \theta$,

$$\mathbf{A}''_{ca} = \begin{bmatrix} \cos\theta & -\sin\theta & 0 \\ \sin\theta & \cos\theta & 0 \\ 0 & 0 & 1 \end{bmatrix} \tag{6.4.5}$$

Since the direction of vector **d** from O''_c to O''_a is fixed in the c'' frame, it is represented in terms of the constant unit vector $\mathbf{b}''_c$ that defines the direction of translation and the distance α of translation; that is,

$$\mathbf{d}''_c = \alpha\mathbf{b}''_c \tag{6.4.6}$$

Virtual Displacement, Velocity, and Acceleration Relations

The vector $\mathbf{r}_a$ to the centroid of the body shown in Fig. 6.4.1 may be written as

$$\mathbf{r}_a = \mathbf{r}_c + \mathbf{A}_c\mathbf{r}'_{ca} \tag{6.4.7}$$

where $\mathbf{r}'_{ca}$ is the vector from O'_c to O'_a, represented in the c' frame. In terms of the body-fixed vectors $\mathbf{s}'_c$ and $\mathbf{s}'_a$ that define the location of joint attachment points O''_c and O''_a in bodies c and a, respectively, and the transformation matrices and vectors of Eqs. 6.4.1 to 6.4.6, the vector $\mathbf{r}'_{ca}$ may be written as

$$\mathbf{r}'_{ca} = \mathbf{s}'_c + \mathbf{C}_c\mathbf{b}''_c\alpha - \mathbf{C}_c\mathbf{A}''_{ca}\mathbf{C}_a^T\mathbf{s}'_a \tag{6.4.8}$$

A virtual displacement of the centroid of body a can be written, by taking the differential of Eq. 6.4.7, as

$$\delta\mathbf{r}_a = \delta\mathbf{r}_c + \delta\mathbf{A}_c\mathbf{r}'_{ca} + \mathbf{A}_c\,\delta\mathbf{r}'_{ca} \tag{6.4.9}$$

where, from Eq. 6.4.8,

$$\delta\mathbf{r}'_{ca} = \mathbf{C}_c\mathbf{b}''_c\,\delta\alpha - \mathbf{C}_c\,\delta\mathbf{A}''_{ca}\mathbf{C}_a^T\mathbf{s}'_a \tag{6.4.10}$$

Since rotation of the a'' frame with respect to the c'' frame is about the coincident z''_c and z''_a axes, the virtual relative rotation vector is simply

$$\delta\boldsymbol{\pi}''_{ca} = \mathbf{h}''_c\,\delta\theta \tag{6.4.11}$$

where $\mathbf{h}''_c = [0, 0, 1]^T$ is the unit vector along the z''_c axis, in the c'' frame. Substituting Eq. 6.4.11 into Eq. 4.2.19 yields

$$\delta\mathbf{A}''_{ca} = \mathbf{A}''_{ca}\,\delta\tilde{\boldsymbol{\pi}}''_{ca} = \mathbf{A}''_{ca}\tilde{\mathbf{h}}''_c\,\delta\theta \tag{6.4.12}$$

Equations 6.4.10 to 6.4.12 may now be substituted into Eq. 6.4.9, to obtain

$$\delta\mathbf{r}_a = \delta\mathbf{r}_c - \mathbf{A}_c\tilde{\mathbf{r}}'_{ca}\,\delta\boldsymbol{\pi}'_c + \mathbf{A}_c(\mathbf{C}_c\mathbf{b}''_c\,\delta\alpha - \mathbf{C}_c\mathbf{A}''_{ca}\tilde{\mathbf{h}}''_c\mathbf{C}_a^T\mathbf{s}'_a\,\delta\theta) \tag{6.4.13}$$

This relation defines the virtual displacement of the centroid of body a in terms of virtual displacements of body c and variations $\delta\alpha$ and $\delta\theta$ in the relative coordinates.

The derivation of the virtual displacement $\delta\mathbf{r}_a$ may be repeated, using the time derivative operator d/dt in place of the differential operator δ, to obtain the velocity of the centroid body a as (Prob. 6.4.1)

$$\dot{\mathbf{r}}_a = \dot{\mathbf{r}}_c - \mathbf{A}_c\tilde{\mathbf{r}}'_{ca}\boldsymbol{\omega}'_c + \mathbf{A}_c(\mathbf{C}_c\mathbf{b}''_c\dot{\alpha} - \mathbf{C}_c\mathbf{A}''_{ca}\tilde{\mathbf{h}}''_c\mathbf{C}_a^T\mathbf{s}'_a\dot{\theta}) \tag{6.4.14}$$

The acceleration of the centroid of body a may be obtained by differentiating both sides of Eq. 6.4.14,

$$\ddot{\mathbf{r}}_a = \ddot{\mathbf{r}}_c - \dot{\mathbf{A}}_c\tilde{\mathbf{r}}'_{ca}\boldsymbol{\omega}'_c - \mathbf{A}_c\dot{\tilde{\mathbf{r}}}'_{ca}\boldsymbol{\omega}'_c - \mathbf{A}_c\tilde{\mathbf{r}}'_{ca}\dot{\boldsymbol{\omega}}'_c + \dot{\mathbf{A}}_c(\mathbf{C}_c\mathbf{b}''_c\dot{\alpha} - \mathbf{C}_c\mathbf{A}''_{ca}\tilde{\mathbf{h}}''_c\mathbf{C}_a^T\mathbf{s}'_a\dot{\theta})$$
$$+ \mathbf{A}_c(\mathbf{C}_c\mathbf{b}''_c\ddot{\alpha} - \mathbf{C}_c\dot{\mathbf{A}}''_{ca}\tilde{\mathbf{h}}''_c\mathbf{C}_a^T\mathbf{s}'_a\dot{\theta} - \mathbf{C}_c\mathbf{A}''_{ca}\tilde{\mathbf{h}}''_c\mathbf{C}_a^T\mathbf{s}'_a\ddot{\theta})$$

Using Eqs. 6.4.5 and 6.4.8 and Eq. 4.2.8, this relation may be written as

$$\ddot{\mathbf{r}}_a = \ddot{\mathbf{r}}_c - \mathbf{A}_c\tilde{\mathbf{r}}'_{ca}\dot{\boldsymbol{\omega}}'_c + \mathbf{A}_c(\mathbf{C}_c\mathbf{b}''_c\ddot{\alpha} - \mathbf{C}_c\mathbf{A}''_{ca}\tilde{\mathbf{h}}''_c\mathbf{C}_a^T\mathbf{s}'_a\ddot{\theta}) + \mathbf{A}_c\mathbf{R}'_c \quad (6.4.15)$$

where

$$\mathbf{R}'_c = \tilde{\boldsymbol{\omega}}'_c\tilde{\boldsymbol{\omega}}'_c\mathbf{r}'_{ca} + 2\tilde{\boldsymbol{\omega}}'_c(\mathbf{C}_c\mathbf{b}''_c\dot{\alpha} - \mathbf{C}_c\mathbf{A}''_{ca}\tilde{\mathbf{h}}''_c\mathbf{C}_a^T\mathbf{s}'_a\dot{\theta}) - \mathbf{C}_c\mathbf{A}''_{ca}\tilde{\mathbf{h}}''_c\tilde{\mathbf{h}}''_c\mathbf{C}_a^T\mathbf{s}'_a\dot{\theta}^2 \quad (6.4.16)$$

Similar relationships for virtual rotation and angular acceleration are needed for use in the variational equations of motion. Since virtual rotation is a vector quantity, the virtual rotation of body a may be written directly, using Eq. 6.4.11, as

$$\delta\boldsymbol{\pi}_a = \delta\boldsymbol{\pi}_c + \delta\boldsymbol{\pi}_{ca} = \delta\boldsymbol{\pi}_c + \mathbf{A}_c\mathbf{C}_c\mathbf{A}''_{ca}\mathbf{h}''_c\,\delta\theta \quad (6.4.17)$$

The virtual rotation of body a, in the a' frame, is (Prob. 6.4.2)

$$\delta\boldsymbol{\pi}'_a = \mathbf{C}_a\mathbf{A}''^T_{ca}\mathbf{C}_c^T\mathbf{A}_c^T\,\delta\boldsymbol{\pi}_a$$

Substituting Eq. 6.4.17 into this result yields

$$\delta\boldsymbol{\pi}'_a = \mathbf{C}_a\mathbf{A}''^T_{ca}\mathbf{C}_c^T\,\delta\boldsymbol{\pi}'_c + \mathbf{C}_a\mathbf{h}''_c\,\delta\theta \quad (6.4.18)$$

This provides the desired relationship between virtual rotation of bodies a and c and the relative virtual rotation $\delta\theta$.

Using analogous calculations, the angular velocity of body a, in the a' frame, is (Prob. 6.4.3)

$$\boldsymbol{\omega}'_a = \mathbf{C}_a\mathbf{A}^T_{ca}\mathbf{C}_c^T\boldsymbol{\omega}'_c + \mathbf{C}_a\mathbf{h}''_c\dot{\theta} \quad (6.4.19)$$

which is written in terms of the angular velocity of body c and the relative angular velocity. The angular acceleration of body a is obtained by differentiating Eq. 6.4.19, to obtain

$$\dot{\boldsymbol{\omega}}'_a = \mathbf{C}_a\mathbf{A}'^T_{ca}\mathbf{C}_c^T\dot{\boldsymbol{\omega}}'_c + \mathbf{C}_a\mathbf{h}''_c\ddot{\theta} - \mathbf{C}_a\tilde{\mathbf{h}}''_c\mathbf{A}''^T_{ca}\mathbf{C}_c^T\boldsymbol{\omega}'_c\,\dot{\theta} \quad (6.4.20)$$

Note that the last term in Eq. 6.4.20 is quadratic in the system velocities, representing terms analogous to $\mathbf{R}'_c$ in Eqs. 6.4.15 and 6.4.16.

The virtual displacement relationships of Eqs. 6.4.13 and 6.4.18 may be written in matrix form as

$$\begin{bmatrix} \delta\mathbf{r}_a \\ \delta\boldsymbol{\pi}'_a \end{bmatrix} = \mathbf{B}\begin{bmatrix} \delta\mathbf{r}_c \\ \delta\boldsymbol{\pi}'_c \\ \delta\alpha \\ \delta\theta \end{bmatrix} \equiv [\mathbf{B}_1 \,\vdots\, \mathbf{B}_2]\begin{bmatrix} \delta\mathbf{r}_c \\ \delta\boldsymbol{\pi}'_c \\ \hline \delta\alpha \\ \delta\theta \end{bmatrix} \quad (6.4.21)$$

where the partitioning of $\mathbf{B}$ is given explicitly as

$$\mathbf{B} \equiv [\mathbf{B}_1 \,\vdots\, \mathbf{B}_2] = \left[\begin{array}{cc:cc} \mathbf{I} & -\mathbf{A}_c\tilde{\mathbf{r}}'_{ca} & \mathbf{A}_c\mathbf{C}_c\mathbf{b}''_c & -\mathbf{A}_c\mathbf{C}_c\mathbf{A}''_{ca}\tilde{\mathbf{h}}''_c\mathbf{C}_a^T\mathbf{s}'_a \\ \mathbf{0} & \mathbf{C}_a\mathbf{A}''^T_{ca}\mathbf{C}_c^T & \mathbf{0} & \mathbf{C}_a\mathbf{h}''_c \end{array}\right] \tag{6.4.22}$$

Similarly, the acceleration relations of Eqs. 6.4.15 and 6.4.20 may be written in matrix form as

$$\begin{bmatrix} \ddot{\mathbf{r}}_a \\ \dot{\boldsymbol{\omega}}'_a \end{bmatrix} = [\mathbf{B}_1 \quad \mathbf{B}_2] \begin{bmatrix} \ddot{\mathbf{r}}_c \\ \dot{\boldsymbol{\omega}}'_c \\ \hdashline \ddot{\alpha} \\ \ddot{\theta} \end{bmatrix} + \begin{bmatrix} \mathbf{A}_c\mathbf{R}'_c \\ -\mathbf{C}_a\tilde{\mathbf{h}}_c^T\mathbf{A}''^T_{ca}\mathbf{C}_c^T\boldsymbol{\omega}'_c\,\dot{\theta} \end{bmatrix} \tag{6.4.23}$$

It is important to note that the coefficient matrices in the transformations of Eqs. 6.4.21 and 6.4.23 are the same. This result yields essential simplification of the equations of motion of the combined system, when the variational form of the equations of motion is used.

Generalized Forces

Generalized forces that act on body a and between bodies a and c are associated with both externally applied forces and forces and torques that are generated by springs and dampers. Consider first a force $\mathbf{F}_a^A$ that is applied at a general point P on body a, as shown in Fig. 6.4.2. To incorporate force $\mathbf{F}_a^A$ into the equations of motion, the moment of this force in the body-fixed a' frame is calculated as

$$\mathbf{n}'_a = \tilde{\mathbf{v}}_a'^P\mathbf{F}_a'^A \tag{6.4.24}$$

Consider next a translational force element that exerts a force $\mathbf{F}_{ca}^A$ (tension as positive) between points P_c and P_a on bodies c and a, respectively, and a torsional element that exerts a torque T_{ca}^A (positive if it tends to increase θ) between the bodies, about their common z''_c and z''_a axes, as shown in Fig. 6.4.3. It is presumed that the magnitude of the force $\mathbf{F}_{ca}^A$ is known as a function of distance and distance rate between points P_c and P_a, and perhaps time. In the c' frame, the vector $\mathbf{u}_{ca}$ may be written as

$$\mathbf{u}'_{ca} \equiv (\mathbf{u}_{ca})'_c = \mathbf{C}_c\mathbf{b}''_c\alpha + \mathbf{C}_c\mathbf{A}''_{ca}\mathbf{C}_a^T(-\mathbf{s}'_a + \mathbf{v}'_a) + (\mathbf{s}'_c - \mathbf{v}'_c) \tag{6.4.25}$$

where Eq. 6.4.6 has been used.

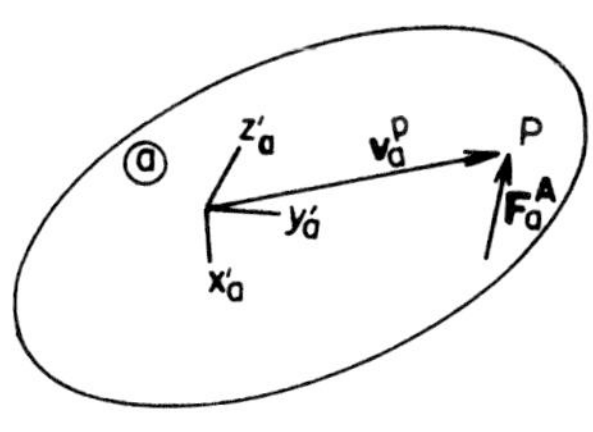

Figure 6.4.2 External force.

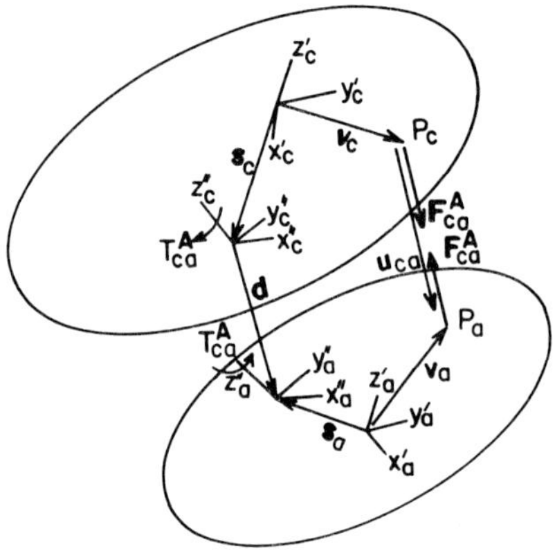

Figure 6.4.3 Internal force and torque.

The force vector $\mathbf{F}_{ca}^{A}$ may be written in the c' frame as

$$\mathbf{F}_{ca}^{\prime A} \equiv (\mathbf{F}_{ca}^{A})_c' = \frac{F_{ca}^{A}}{|\mathbf{u}_{ca}'|}\mathbf{u}_{ca}' \tag{6.4.26}$$

where $F_{ca}^{A} = |\mathbf{F}_{ca}^{A}|$. The virtual work of this force and the torque T_{ca}^{A} is

$$\delta W^{A} = -\delta\mathbf{u}_{ca}^{\prime T}\mathbf{F}_{ca}^{\prime A} + T_{ca}^{A}\,\delta\theta = -\frac{F_{ca}^{A}}{|\mathbf{u}_{ca}'|}\delta\mathbf{u}_{ca}^{\prime T}\mathbf{u}_{ca}' + \mathbf{T}_{ca}^{A}\,\delta\theta \tag{6.4.27}$$

Taking the differential of $\mathbf{u}_{ca}'$ in Eq. 6.4.25,

$$\delta\mathbf{u}_{ca}' = \mathbf{C}_c\mathbf{b}_c''\,\delta\alpha - \mathbf{C}_c\mathbf{A}_{ca}''\tilde{\mathbf{h}}_c''\mathbf{C}_a^{T}(\mathbf{s}_a' - \mathbf{v}_a')\,\delta\theta \tag{6.4.28}$$

Thus the virtual work of Eq. 6.4.27 may be written explicitly in terms of virtual displacement $\delta\alpha$ and virtual rotation $\delta\theta$ as

$$\begin{aligned}\delta W^{A} &= \delta\alpha\left\{-\frac{F_{ca}^{A}}{|\mathbf{u}_{ca}'|}\mathbf{b}_c^{\prime\prime T}\mathbf{C}_c^{T}\mathbf{u}_{ca}'\right\}\\ &\quad + \delta\theta\left\{-\frac{F_{ca}^{A}}{|\mathbf{u}_{ca}'|}(\mathbf{s}_a' - \mathbf{v}_a')^{T}\mathbf{C}_a\tilde{\mathbf{h}}_c''\mathbf{A}_{ca}^{\prime\prime T}\mathbf{C}_c^{T}\mathbf{u}_{ca}' + T_{ca}^{A}\right\}\end{aligned} \tag{6.4.29}$$

Variational Equation of Motion

The variational form of the equation of motion of Eq. 5.1.26 permits use of the acceleration and virtual displacement vector relations of Eqs. 6.4.21 and 6.4.23 and the virtual work of Eq. 6.4.29 to assemble terms in the equation of motion, in a systematic way. The variational equation of motion of the system, in matrix form, is

$$\sum_{i=a}^{c}\begin{bmatrix}\delta\mathbf{r}_i\\ \delta\boldsymbol{\pi}_i'\end{bmatrix}^{T}\left\{\mathbf{M}_i\begin{bmatrix}\ddot{\mathbf{r}}_i\\ \dot{\boldsymbol{\omega}}_i'\end{bmatrix} - \begin{bmatrix}\mathbf{F}_i^{A}\\ \mathbf{n}_i^{\prime A} - \tilde{\boldsymbol{\omega}}_i'\mathbf{J}_i\boldsymbol{\omega}_i'\end{bmatrix}\right\} - \delta W^{A} = 0 \tag{6.4.30}$$

where δW^A accounts for applied forces and forces due to compliant elements in the model and the mass matrix is

$$\mathbf{M}_i = \begin{bmatrix} m_i \mathbf{I} & \mathbf{0} \\ \mathbf{0} & \mathbf{J}'_i \end{bmatrix}, \qquad i = a, c \tag{6.4.31}$$

Equation 6.4.30 must hold for all kinematically admissible virtual displacements $\delta\mathbf{r}_i$ and virtual rotations $\delta\boldsymbol{\pi}'_c$.

Substituting from Eqs. 6.4.21, 6.4.23, and 6.4.29 into Eq. 6.4.30 and expanding matrix products yields

$$\begin{bmatrix} \delta\mathbf{r}_c \\ \delta\boldsymbol{\pi}'_c \\ \delta\alpha \\ \delta\theta \end{bmatrix}^T \left\{ \begin{bmatrix} \mathbf{M}_c + \mathbf{B}_1^T\mathbf{M}_a\mathbf{B}_1 & \mathbf{B}_1^T\mathbf{M}_a\mathbf{B}_2 \\ \mathbf{B}_2^T\mathbf{M}_a\mathbf{B}_1 & \mathbf{B}_2^T\mathbf{M}_a\mathbf{B}_2 \end{bmatrix} \begin{bmatrix} \ddot{\mathbf{r}}_c \\ \dot{\boldsymbol{\omega}}'_c \\ \ddot{\alpha} \\ \ddot{\theta} \end{bmatrix} \right.$$
$$\left. - \begin{bmatrix} \mathbf{B}_1^T \begin{bmatrix} \hat{\mathbf{F}}_a^A \\ \hat{\mathbf{n}}_a'^A - \tilde{\boldsymbol{\omega}}'_a\mathbf{J}'_a\boldsymbol{\omega}'_a \end{bmatrix} + \begin{bmatrix} \mathbf{F}_c^A \\ \mathbf{n}_c'^A - \tilde{\boldsymbol{\omega}}'_c\mathbf{J}'_c\boldsymbol{\omega}'_c \end{bmatrix} \\ \mathbf{B}_2^T \begin{bmatrix} \hat{\mathbf{F}}_a^A \\ \hat{\mathbf{n}}_a'^A - \tilde{\boldsymbol{\omega}}'_a\mathbf{J}'_a\boldsymbol{\omega}'_a \end{bmatrix} + \begin{bmatrix} 0 \\ T_{ca} \end{bmatrix} - \left(\dfrac{F_{ca}^A}{|\mathbf{u}'_{ca}|} \right) \begin{bmatrix} \mathbf{b}_c''^T\mathbf{C}_c^T\mathbf{u}'_{ca} \\ (\mathbf{s}'_a - \mathbf{v}'_a)^T\mathbf{C}_a\tilde{\mathbf{h}}''_c\boldsymbol{\Lambda}''^T_{ca}\mathbf{C}_c^T\mathbf{u}'_{ca} \end{bmatrix} \end{bmatrix} \right\} = 0 \tag{6.4.32}$$

where

$$\hat{\mathbf{F}}_a^A = \mathbf{F}_a^A - m_a\mathbf{A}_c\mathbf{R}'_c$$

$$\hat{\mathbf{n}}_a'^A = \mathbf{n}_a'^A + \mathbf{J}'_a\mathbf{C}_a\mathbf{n}''_c\mathbf{A}''^T_{ca}\mathbf{C}_c^T\boldsymbol{\omega}'_c\dot{\theta}$$

Equation 6.4.32 must hold for all kinematically admissible virtual displacements of body c and joint relative coordinate variations.

Note the systematic structure of the system mass matrix in Eq. 6.4.32. The upper left mass submatrix represents the inertia matrix of body c plus the transformed inertia matrix of body a. The lower right mass submatrix is a transformed mass matrix that is associated with body a and the relative generalized coordinates. The off-diagonal mass submatrices represent coupling effects between generalized coordinates of body c and the relative coordinates. Having derived Eq. 6.4.32 for a two-body set, if additional bodies are attached to body c, their associated mass submatrices are simply calculated and added appropriately in the overall system mass matrix. This organization of the system mass matrix parallels that of structural finite element mass and stiffness matrices, which are derived using variational methods of structural mechanics. This systematic property of the variational approach, both for dynamics and structural mechanics, follows from the additive property of energies in the variational formulation.

Even though the equations derived here are highly nonlinear, use of variational equations of motion in the development permits systematic organization of the governing equations, much as in finite element structural analysis. Similarly, generalized force terms in Eq. 6.4.32 contain external forces that act on body c, Coriolis and centrifugal forces, transformed external forces that act on body a, and internal forces and torques between bodies.

Example 6.4.1: To demonstrate extension of the variational equations of motion to multiple components, consider a vehicle with a solid front axle suspension and two trailing arms in the rear, as shown in Fig. 6.4.4. The solid front axle is modeled as an attached body that is kinematically constrained to the chassis so that

(a) Light truck

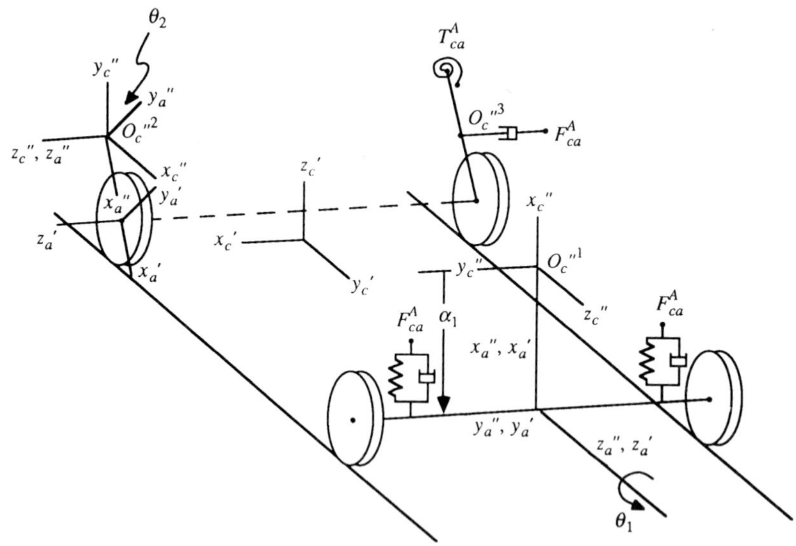

(b) Model

Figure 6.4.4 Model of a vehicle.

its motion relative to the chassis consists of relative roll about a fore–aft z_c'' axis that is fixed in the chassis and relative translation in the body-fixed x_c'' direction. Each trailing arm is kinematically constrained to have only one rotational degree of freedom relative to the chassis, about an axis that is parallel to the x_c' axis. Forces that act on the system include ground reaction force on each tire, gravity, two translational spring–damper elements that act on the solid axle, and torsional springs and translational dampers on the trailing arms.

Since there are no absolute constraints on body c, its virtual displacements and rotations are arbitrary. Furthermore, variations in joint relative coordinates are arbitrary. Thus the expression inside the parentheses in Eq. 6.4.32, expanded to include all suspension components, must be zero. This provides the coupled equations of motion for the chassis and the three attached suspension bodies. Note that there are no translational degrees of freedom for the trailing arms, so the matrix $\mathbf{B}_2$ for each is of dimension 6×1 rather than 6×2. The resulting equations of motion for this vehicle are (Prob. 6.4.4)

$$
\left[\begin{array}{c|c|c|c}
\mathbf{M}_c + \sum_{i=1}^{3} \mathbf{B}_1^{i^T}\mathbf{M}_a^i\mathbf{B}_1^i & \mathbf{B}_1^{1^T}\mathbf{M}_a^1\mathbf{B}_2^1 & \mathbf{B}_1^{2^T}\mathbf{M}_a^2\mathbf{B}_2^2 & \mathbf{B}_1^{3^T}\mathbf{M}_a^3\mathbf{B}_2^3 \\
\hline
\mathbf{B}_2^{1^T}\mathbf{M}_a^1\mathbf{B}_1^1 & \mathbf{B}_2^{1^T}\mathbf{M}_a^2\mathbf{B}_2 & \mathbf{0} & \mathbf{0} \\
\hline
\mathbf{B}_2^{2^T}\mathbf{M}_a^2\mathbf{B}_1^2 & \mathbf{0} & \mathbf{B}_2^{2^T}\mathbf{M}_a^2\mathbf{B}_2 & \mathbf{0} \\
\hline
\mathbf{B}_2^{3^T}\mathbf{M}_a^3\mathbf{B}_1^3 & \mathbf{0} & \mathbf{0} & \mathbf{B}_2^{3^T}\mathbf{M}_a^3\mathbf{B}_2^3
\end{array}\right]
\begin{bmatrix} \ddot{\mathbf{r}}_c \\ \dot{\boldsymbol{\omega}}_{c'} \\ \ddot{\alpha}_1 \\ \ddot{\theta}_1 \\ \ddot{\theta}_2 \\ \ddot{\theta}_3 \end{bmatrix}
$$

$$
= \begin{bmatrix}
\begin{bmatrix} \mathbf{F}_c^A \\ \mathbf{n}_c'^A - \tilde{\boldsymbol{\omega}}_c'\mathbf{J}_c'\boldsymbol{\omega}_c' \end{bmatrix} + \sum_{i=1}^{3}\left\{ \mathbf{B}_1^{i\,T} \begin{bmatrix} \hat{\mathbf{F}}_a^A \\ \hat{\mathbf{n}}_a'^A - \boldsymbol{\omega}_a'\mathbf{J}_a'\boldsymbol{\omega}_a \end{bmatrix}^i \right\} \\
\mathbf{B}_2^{1^T} \begin{bmatrix} \hat{\mathbf{F}}_a^A \\ \hat{\mathbf{n}}_a'^A - \boldsymbol{\omega}_a'\mathbf{J}_a'\boldsymbol{\omega}_a \end{bmatrix}^1 - \sum_{j=1}^{2} \left\{ \dfrac{F_{ca}^{jA}}{|\mathbf{u}_{ca}'^j|} \begin{bmatrix} \mathbf{b}_c''^T\mathbf{C}_c^T\mathbf{u}_{ca}' \\ (\mathbf{s}_a' - \mathbf{v}_a')^T\mathbf{C}_a\tilde{\mathbf{h}}_c''\boldsymbol{\Lambda}_{ca}''^T\mathbf{C}_c\mathbf{u}_{ca}' \end{bmatrix}^j \right. \\
\mathbf{B}_2^{2^T} \begin{bmatrix} \hat{\mathbf{F}}_a^A \\ \hat{\mathbf{n}}_a'^A - \boldsymbol{\omega}_a'\mathbf{J}_a'\boldsymbol{\omega}_a' \end{bmatrix}^2 - \left\{ \dfrac{F_{ca}^{3A}}{|\mathbf{u}'\mathrm{ca}^3|} [(\mathbf{s}_a' - \mathbf{v}_a')^T\mathbf{C}_a\tilde{\mathbf{h}}_c''\boldsymbol{\Lambda}_{ca}''^T\mathbf{C}_c\mathbf{u}_{ca}']^3 - k\theta_2 \right. \\
\mathbf{B}_2^{3^T} \begin{bmatrix} \hat{\mathbf{F}}_a^A \\ \hat{\mathbf{n}}_a'^A - \boldsymbol{\omega}_a'\mathbf{J}_a'\boldsymbol{\omega}_a' \end{bmatrix}^3 - \dfrac{F_{ca}^{4A}}{|\mathbf{u}_{ca}'^4|} [(\mathbf{s}_a' - \mathbf{v}_a')^T\mathbf{C}_a\mathbf{h}_c''\boldsymbol{\Lambda}_{ca}''^T\mathbf{C}_c\mathbf{u}_{ca}']^4 - k\theta_3
\end{bmatrix}
\tag{6.4.33}
$$

where torque in the rear arm torsion bars is $T_i = -k\theta_i$, $i = 2, 3$; $\hat{\mathbf{F}}_c^A$ contains gravitational, Coriolis, and centrifugal forces on the chassis; $\hat{\mathbf{F}}_a^{iA}$ contains gravitational, Coriolis, centrifugal, and tire forces on the ith suspension component, $i = 1, 2, 3$; and $\mathbf{F}_{ca}^{jA}$ are the suspension spring and damping forces, $i = 1, \ldots, 4$.

The remaining task for implementation of these equations is definition of the attachment point vectors $(\mathbf{s}_c')^i$, the translation direction vector $\mathbf{b}_c''$ for the front axle, the transformation matrices $\mathbf{C}_c$ and $(\mathbf{C}_a)^i$, and the suspension attachment vector $(\mathbf{s}'_a)^i$. For the front axle,

$$(\mathbf{s}_c)^1 = [x'_c, y'_c, z'_c]^T_{O_c''1}$$

$$\mathbf{b}''_c = [-1, 0, 0]^T$$

$$(\mathbf{C}_c)^1 = \begin{bmatrix} 0 & 1 & 0 \\ 0 & 0 & 1 \\ 1 & 0 & 0 \end{bmatrix}$$

$$(\mathbf{C}_a)^1 = \begin{bmatrix} 1 & 0 & 0 \\ 0 & 1 & 0 \\ 0 & 0 & 1 \end{bmatrix}$$

$$(\mathbf{s}_a)' = \mathbf{0}$$

For the right rear trailing arm,

$$(\mathbf{s}'_c)^2 = [x'_c, y'_c, z'_c]^T_{O_c''2}$$

$$\mathbf{b}''_c = \mathbf{0}$$

$$(\mathbf{C}_c)^2 = \begin{bmatrix} 0 & 0 & 1 \\ 1 & 0 & 0 \\ 0 & 1 & 0 \end{bmatrix}$$

$$(\mathbf{C}_a)^2 = \begin{bmatrix} 1 & 0 & 0 \\ 0 & 1 & 0 \\ 0 & 0 & 1 \end{bmatrix}$$

$$(\mathbf{s}'_a)^2 = [0, b, 0]^T$$

where b is the arm length. Similar expressions hold for the left rear trailing arm.

Clearly, Eq. 6.4.33 cannot be solved analytically. The systematic structure of terms on both the left and right sides of this equation, however, permits development of computer programs to both generate and solve these equations of motion. In particular, the sparse pattern of the mass matrix on the left side of Eq. 6.4.33 suggests a sequence of computations that can be used in solving for accelerations. These numerically computed accelerations may then be integrated numerically, using methods of Appendix D, to predict motion of the vehicle system.

The classical approach to analysis of dynamic systems is to derive equations of motion in all system generalized coordinates. As seen frequently in this book and as illustrated vividly by Eq. 6.4.33 for the vehicle model in Example 6.4.1, these equations are often so complicated that the only hope is to numerically construct solutions. Given this unavoidable situation, it is important to consider methods of formulating equations of motion that have attractive computational properties. To illustrate such an approach, which is developed in greater detail in Ref. 13, consider an application such as the vehicle of Example 6.4.1, in which l bodies are connected to a common base body c, such as the chassis of a vehicle or the body of a spacecraft. Equation 6.4.32 can be derived for each of the l bodies that are attached to body c. For attached body k, denote the submatrices of the transformation matrix $\mathbf{B}$ of Eq. 6.4.21 as $\mathbf{B}_{ik}$, $i = 1$ and 2, where k denotes the body. Similarly, denote the mass matrix for attached body k as $\mathbf{M}_{ak}$. Similarly, a subscript k is appended to each of the force and inertia related quantities associated with attached body k.

To write Eq. 6.4.32 in a more convenient form for manipulation, define

$$\mathbf{N}_k^{ij} = \mathbf{B}_{ik}^T \mathbf{M}_{a_k} \mathbf{B}_{jk} \tag{6.4.34}$$

$$\mathbf{\Gamma}_k = \mathbf{B}_{lk}^T \begin{bmatrix} \hat{\mathbf{F}}_{ak}^A \\ \hat{\mathbf{n}}_{ak}^{\prime A} - \tilde{\boldsymbol{\omega}}_{ak}' \mathbf{J}_{ak}' \boldsymbol{\omega}_{ak}' \end{bmatrix} \tag{6.4.35}$$

$$\mathbf{\Gamma}_c = \begin{bmatrix} \mathbf{F}_c^A \\ \mathbf{n}_c' - \tilde{\boldsymbol{\omega}}_c' \mathbf{J}_c' \boldsymbol{\omega}_c' \end{bmatrix} \tag{6.4.36}$$

$$\begin{aligned} \mathbf{\Pi}_k = \mathbf{B}_{2k}^T & \begin{bmatrix} \hat{\mathbf{F}}_{ak}^A \\ \mathbf{n}_{ak}^{\prime A} - \tilde{\boldsymbol{\omega}}_{ak}' \mathbf{J}_{ak}' \boldsymbol{\omega}_{ak}' \end{bmatrix} + \begin{bmatrix} \mathbf{0} \\ \mathbf{T}_{cak} \end{bmatrix} \\ & - \left(\frac{F_{cak}^A}{|\mathbf{u}_{cak}'|} \right) \begin{bmatrix} \mathbf{b}_{ck}^{\prime\prime T} \mathbf{C}_{ck}^T \mathbf{u}_{cak}' \\ (\mathbf{s}_{ak}' - \mathbf{v}_{ak}')^T \mathbf{C}_{ak} \tilde{\mathbf{h}}_{ck}'' \mathbf{A}_{cak}^{\prime\prime T} \mathbf{C}_{ck}^T \mathbf{u}_{cak}' \end{bmatrix} \end{aligned} \tag{6.4.37}$$

Using this notation, with $k = 1$, Eq. 6.4.32 is written in the more convenient form

$$\begin{aligned} \begin{bmatrix} \delta \mathbf{r}_c \\ \delta \boldsymbol{\pi}_c' \end{bmatrix}^T & \left\{ (\mathbf{M}_c + \mathbf{N}_1^{11}) \begin{bmatrix} \ddot{\mathbf{r}}_c \\ \dot{\boldsymbol{\omega}}_c' \end{bmatrix} + \mathbf{N}_1^{12} \ddot{\mathbf{q}}_1 - \mathbf{\Gamma}_1 - \mathbf{\Gamma}_c \right\} \\ & + \delta \mathbf{q}_1 \left\{ \mathbf{N}_1^{21} \begin{bmatrix} \ddot{\mathbf{r}}_c \\ \dot{\boldsymbol{\omega}}_c' \end{bmatrix} + \mathbf{N}_1^{22} \ddot{\mathbf{q}}_1 - \mathbf{\Pi}_1 \right\} = 0 \end{aligned} \tag{6.4.38}$$

Note that adding terms associated with attached bodies 2, . . . , l simply introduces additional terms of the form appearing in Eq. 6.4.32 or 6.4.38 associated with each of the attached bodies. Summing over all l of the attached bodies, the variational equation of motion for the system is obtained as (Prob. 6.4.5)

$$\begin{aligned} \begin{bmatrix} \delta \mathbf{r}_c^T \\ \delta \boldsymbol{\pi}_c' \end{bmatrix}^T & \left\{ (\mathbf{M}_c + \sum_{k=1}^{l} \mathbf{N}_k^{11}) \begin{bmatrix} \ddot{\mathbf{r}}_c \\ \dot{\boldsymbol{\omega}}_c' \end{bmatrix} + \sum_{k=1}^{l} \mathbf{N}_k^{12} \ddot{\mathbf{q}}_k - \sum_{k=1}^{l} \mathbf{\Gamma}_k - \mathbf{\Gamma}_c \right\} \\ & + \sum_{k=1}^{l} \delta \mathbf{q}_k^T \left\{ \mathbf{N}_k^{21} \begin{bmatrix} \ddot{\mathbf{r}}_c \\ \dot{\boldsymbol{\omega}}_c' \end{bmatrix} + \mathbf{N}_k^{22} \ddot{\mathbf{q}}_k - \mathbf{\Pi}_k \right\} = 0 \end{aligned} \tag{6.4.39}$$

for all kinematically admissible virtual displacements and rotations of body c and kinematically admissible variations $\delta \mathbf{q}_k$. If no additional kinematic constraints act on the attached bodies, $\delta \mathbf{q}_k$ are arbitrary, $k = 1, \ldots, l$. Therefore, their coefficients in Eq. 6.4.39 must be zero, yielding

$$\mathbf{N}_k^{22} \ddot{\mathbf{q}}_k = \mathbf{\Pi}_k - \mathbf{N}_k^{21} \begin{bmatrix} \ddot{\mathbf{r}}_c \\ \dot{\boldsymbol{\omega}}_c' \end{bmatrix}, \qquad k = 1, \ldots, l \tag{6.4.40}$$

Providing the joint formulation has been properly defined, $\mathbf{N}_k^{22}$ is nonsingular.

Multiplying both sides of Eq. 6.4.40 on the left by the inverse of this matrix yields

$$\ddot{\mathbf{q}}_k = \mathbf{N}_k^{22^{-1}}\left(\mathbf{\Pi}_k - \mathbf{N}_k^{21}\begin{bmatrix}\ddot{r}_c \\ \dot{\boldsymbol{\omega}}_c'\end{bmatrix}\right), \qquad k = 1, \ldots, l \tag{6.4.41}$$

Recalling that the sum of terms in Eq. 6.4.39 over all attached bodies is zero, Eq. 6.4.41 may be substituted into the first term of Eq. 6.4.39, to obtain

$$\begin{bmatrix}\delta\mathbf{r}_c \\ \delta\boldsymbol{\pi}_c'\end{bmatrix}^T \left\{\left[\mathbf{M}_c + \sum_{k=1}^{l}(\mathbf{N}_k^{11} - \mathbf{N}_k^{12}\mathbf{N}_k^{22^{-1}}\mathbf{N}_k^{21})\right]\begin{bmatrix}\ddot{\mathbf{r}}_c \\ \dot{\boldsymbol{\omega}}_c'\end{bmatrix} - \mathbf{\Gamma}_c - \sum_{l=1}^{l}(\mathbf{\Gamma}_k - \mathbf{N}_k^{12}\mathbf{N}_k^{22^{-1}}\mathbf{\Pi}_k)\right\} = 0 \tag{6.4.42}$$

which must hold for all kinematically admissible virtual displacements and virtual rotations of body c.

If, as in the case of the vehicle of Example 6.4.1, or a spacecraft that is free to move in space, the virtual displacement and rotation of body c are arbitrary, their coefficients in Eq. 6.4.42 must be zero, yielding

$$\left[\mathbf{M}_c + \sum_{l=l}^{l}(\mathbf{N}_k^{11} - \mathbf{N}_k^{12}\mathbf{N}_k^{22^{-1}}\mathbf{N}_k^{21})\right]\begin{bmatrix}\ddot{\mathbf{r}}_c \\ \dot{\boldsymbol{\omega}}_c'\end{bmatrix} = \mathbf{\Gamma}_c + \sum_{k=1}^{l}(\mathbf{\Gamma}_k - \mathbf{N}_k^{12}\mathbf{N}_k^{22^{-1}}\mathbf{\Pi}_k) \tag{6.4.43}$$

This reduced equation of motion for body c, comprising six equations in six unknowns, may now be solved for its acceleration and angular acceleration. This contrasts with the large system of equations of the form of Eq. 6.4.33 derived in Example 6.4.1. Thus substantial computational efficiency can be achieved.

After obtaining the acceleration and angular acceleration of body c, they may be substituted in Eq. 6.4.41 to recover relative coordinate accelerations in each of the joints with attached bodies. Finally, if desired, the Cartesian acceleration and angular acceleration of each of the attached bodies can be recovered from Eq. 6.4.23.

The sequence of computations defined by Eqs. 6.4.34 through 6.4.43 and the subsequent acceleration recovery for the attached bodies is called a *recursive algorithm with elimination* for dynamic analysis of multibody systems. The reader may find it interesting to apply this sequence of computations to the vehicle example of Example 6.4.1, to gain confidence in the procedure. The method is developed in greater generality and detail in Ref. 13.

While attention in this section has been focused on a specific example, the basic approach presented is generally applicable to large scale mechanical systems. Armed with insights gained in treating a specific application, more general and computationally efficient methods for forming and solving equations of motion of general mechanical systems may be developed [13, 19, 20, 21, 22].

6.5 RELATIVE COORDINATE KINEMATICS FOR MULTIBODY SYSTEMS

The relative coordinate approach used in Sections 6.3 and 6.4 can be generalized for analysis of broad classes of multibody systems, using graph theoretic methods. Concepts of graph theory are introduced in this section, to define connectivity between bodies in multibody systems. Graph theoretic ideas lead to definition of computational sequences of transformations between bodies, using the transformation approach derived in Section 6.4 for pairs of bodies.

System Graphs

Individual bodies in a multibody system are represented schematically in a diagram as *nodes*. Kinematic couplings between bodies are represented schematically as *edges* between nodes on a schematic diagram called a *graph*. Nodes are numbered as the bodies they represent. An edge that represents kinematic coupling between bodies i and j is denoted (i, j). If more than one coupling between bodies is allowed, a somewhat more intricate notation is required. A directed sense on edge (i, j) is defined as beginning from node (body) i and proceeding onto node (body) j, called a *directed edge*.

➧ **Example 6.5.1:** To illustrate these ideas, consider the double pendulum shown in Fig. 6.5.1(a). Ground is denoted as body 1 and moving links in the double pendulum are bodies 2 and 3. The node and edge representation of the double pendulum is shown in Fig. 6.5.1(b), in which individual bodies are nodes and directed edges that connect nodes represent kinematic couplings between bodies, revolute joints in this case.

The node and edge diagram associated with a system is called its *system graph*, or simply *graph*. In Example 6.5.1 it is possible to move from any body in the system across nodes and edges to reach any other body in the system. A system for which this is true is called a *connected graph*. There may be kinematically independent subsystems of a mechanical system, in which case it is not possible to move from each body across a path to reach any other body in the

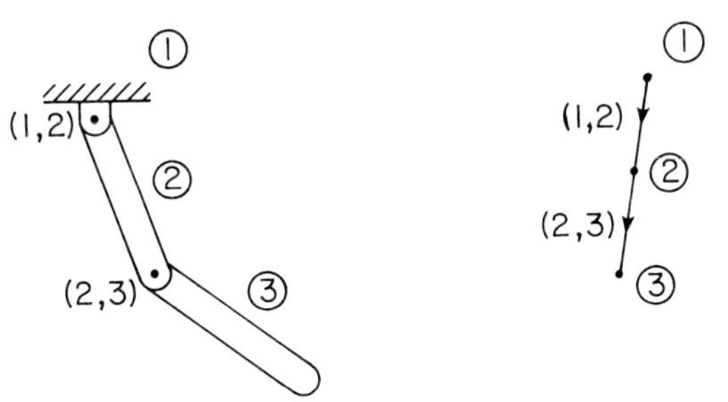

(a) Double pendulum (b) Graph

Figure 6.5.1 System graph of double pendulum.

system. In such cases the graph is called *disconnected*. In such a case, the system is made up of subsystems, each of which is represented by a connected subsystem graph.

In each connected graph of a system or subsystem, one body in the graph may be designated as the *base body*. In this book the base body is designated as body 1 and kinematic connections between bodies are defined as directed edges that form a *directed graph*, which provides an ordered path from the base body out to each body in the system. A node with no outgoing edges is called a *tree end node*, or a *tree end body*. Body 3 is the only tree end body for the double pendulum of Example 6.5.1.

Example 6.5.2: As a second example, consider the slider–crank mechanism of Fig. 6.5.2(a). Ground is designated as the base body for this system. Note that both bodies 2 and 4 are kinematically connected to ground in this system. The system graph is shown in Fig. 6.5.2(b).

Note that it is possible to begin from body 1 and traverse a path around the system that arrives back at body 1 in the graph of Example 6.5.2 [Fig. 6.5.2(b)]. This is a typical example of a *closed loop system*. The distinction between systems with closed loops, which permit paths that begin at a node and return to the same node, and the graph of Fig. 6.5.1(a), in which this is impossible, is important. A graph that does not contain a closed loop is called a *tree*. A system for which the graph is a tree is called a system with *tree structure*.

As will become apparent in kinematic analysis with the aid of relative coordinate transformations and system graphs, there are advantages in working with a tree structure. A technique of cutting joints to open loops permits systems with closed loops to be modified to form tree structures, called *spanning trees* [10, 13, 19]. The technique is illustrated, for the slider–crank mechanism of Fig. 6.5.2, in Fig. 6.5.3. Some kinematic joint in the system is imagined to be cut, called a *cut-joint*. In the example of Fig. 6.5.3, the translational joint between body 4 and ground, joint (4, 1), is cut. The kinematic constraint equations that define this joint are retained as constraints that act on the system and are called *cut-joint constraints*. Graph theoretic methods that are presented in detail in Refs. 10, 13, and 19 can be used to computationally identify closed loops in the system graph of the mechanism being studied and to cut a minimal number of joints to form a

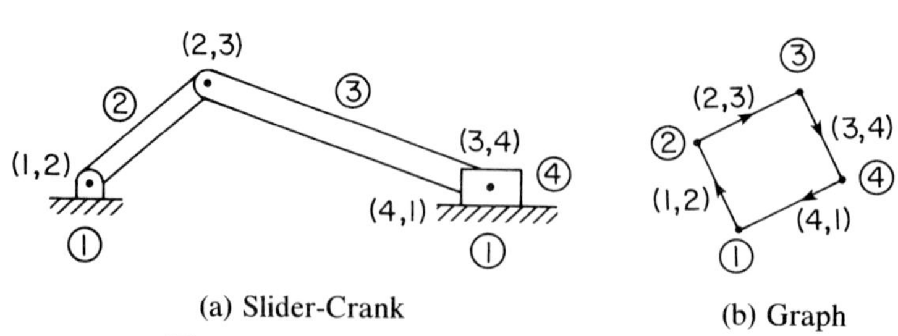

(a) Slider-Crank (b) Graph

Figure 6.5.2 System graph of slider–crank.

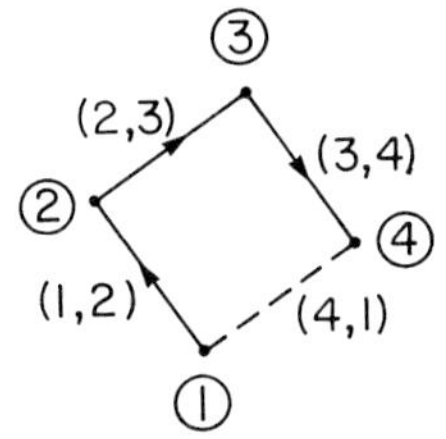

Figure 6.5.3 Cut-joint graph of slider–crank.

spanning tree. Constraint equations that represent cut-joints are then treated as kinematic constraint equations for the system.

➧ **Example 6.5.3:** An additional feature of system graphs that is important in dynamics is illustrated by the compound double pendulum shown in Fig. 6.5.4(a). Two outboard bodies are pivoted to body 2, which is in turn pivoted to body 1 (ground). The system graph for this mechanism is shown in Fig. 6.5.4(b). Even though the joints between body 2 and bodies 3 and 4 do not coincide, since body 2 is a single node in the graph, edges (2, 3) and (2, 4) are shown emanating from the common node that represents body 2.

A node that has two or more outgoing edges is called a *junction node*. For a system with tree structure, in which directed edges are along paths that begin at the base body and proceed to tree end bodies, only multiple edges can emanate from a single node. It is not possible for multiple edges to converge on a single node, since if this could occur each such convergent edge could be followed backward to arrive at the base body, leading to a closed loop. A *directed path* that begins at the base body or a junction node and proceeds to another junction node or a tree end body is called a *chain*. Chains are thus directed paths between pairs of nodes that are not junction nodes. The path 1-2-3 in Fig. 6.5.1 is a chain. In the graph of Fig. 6.5.4(b), paths 1-2, 2-3, and 2-4 are all chains.

➧ **Example 6.5.4:** Figure 6.5.5(a) shows a schematic representation of a system that consists of 10 bodies, connected by nine kinematic joints. The associated system graph is shown in Fig. 6.5.5(b). Note that it contains four distinct chains: 1-2-3, 3-6, 3-4-5, and 1-7-8-9-10.

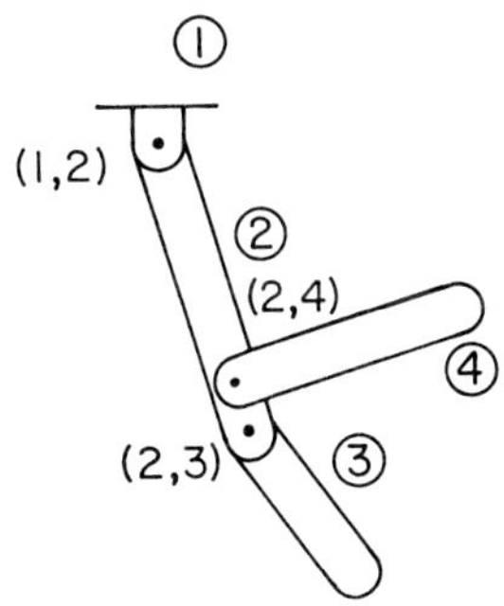

(a) Compound double pendulum

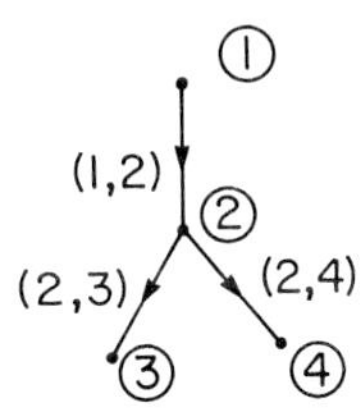

(b) Graph

Figure 6.5.4 System graph of compound double pendulum.

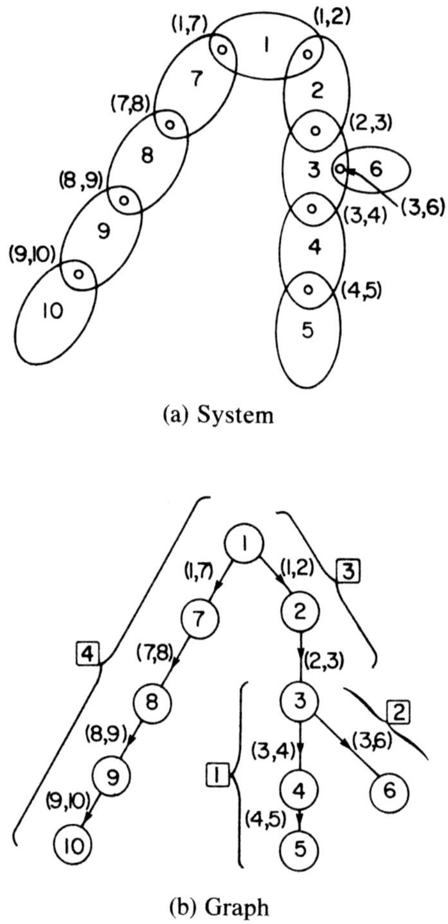

Figure 6.5.5 Tree structure mechanical system.

Joint Relative Coordinate Kinematics

Pairs of coupled bodies whose kinematics are defined in terms of Cartesian coordinates of one of the bodies and rotational and translational relative coordinates are presented in Section 6.4. Bodies a and c in Section 6.4 are denoted bodies i and j here, as shown in Fig. 6.5.6, for a more systematic notation in multibody systems. Relative translational and rotational generalized coordinates between bodies i and j are denoted as subscripted variables α_{ij} and θ_{ij}, respectively. Vectors that locate joint (i, j) attachment points in body i and j reference frames are denoted $\mathbf{s}'_{ij}$ and $\mathbf{s}'_{ji}$, respectively. Translational and rotational axes of the joint in the x''_i-y''_i-z''_i *joint definition frame* are denoted $\mathbf{b}''_{ij}$ and $\mathbf{h}''_{ij}$. Similarly, orthogonal matrices $\mathbf{C}_{ij}$, $\mathbf{C}_{ji}$, and $\mathbf{A}''_{ij}$ are transformation matrices from the $''$ frames to the centroidal $'$ frames on bodies i and j and from the $''$ frame on body j to the $''$ frame on body i, respectively.

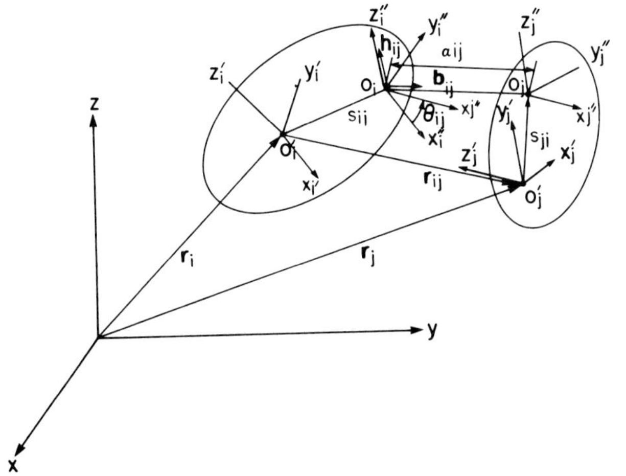

Figure 6.5.6 Two-degree-of-freedom relative coordinates.

Vectors in the global reference frame are used in this section to express velocity, virtual displacement, and acceleration relations, in order to avoid the appearance of many transformation matrices. In particular, from Fig. 6.5.6,

$$\mathbf{r}_j = \mathbf{r}_i + \mathbf{r}_{ij} \tag{6.5.1}$$

$$\mathbf{r}_{ij} = \mathbf{s}_{ij} + \mathbf{b}_{ij}\alpha_{ij} - \mathbf{s}_{ji} \tag{6.5.2}$$

where the translational unit vector $\mathbf{b}_{ij}$ and the vector $\mathbf{s}_{ji}$ in the global frame are obtained from their body-fixed constant values by the transformations

$$\mathbf{b}_{ij} = \mathbf{A}_i\mathbf{C}_{ij}\mathbf{b}''_{ij} \tag{6.5.3}$$

$$\mathbf{s}_{ji} = \mathbf{A}_i\mathbf{C}_{ij}\mathbf{A}''_{ij}\mathbf{C}^T_{ji}\mathbf{s}'_{ji} \tag{6.5.4}$$

where $\mathbf{s}'_{ji}$ is a constant vector in the O'_j frame. Note from Eq. 6.5.4 that the orientation transformation matrix for body j is

$$\mathbf{A}_j = \mathbf{A}_i\mathbf{C}_{ij}\mathbf{A}''_{ij}\mathbf{C}^T_{ji} \tag{6.5.5}$$

The angular velocity of body j is

$$\boldsymbol{\omega}_j = \boldsymbol{\omega}_i + \mathbf{h}_{ij}\dot{\theta}_{ij} \tag{6.5.6}$$

where $\mathbf{h}_{ij}$ is a unit vector along the rotational axis of the joint and can be written as

$$\mathbf{h}_{ij} = \mathbf{A}_i\mathbf{C}_{ij}\mathbf{h}''_{ij} \tag{6.5.7}$$

and $\mathbf{h}''_{ij} = [0, 0, 1]^T$, as shown in Fig. 6.5.6.

The time derivative of Eq. 6.5.1 is

$$\dot{\mathbf{r}}_j = \dot{\mathbf{r}}_i + \dot{\mathbf{s}}_{ij} + \dot{\mathbf{b}}_{ij}\alpha_{ij} + \mathbf{b}_{ij}\dot{\alpha}_{ij} - \dot{\mathbf{s}}_{ji} \tag{6.5.8}$$

Since $\mathbf{s}_{ij}$ is a vector that is fixed in body i, its time derivative is

$$\dot{\mathbf{s}}_{ij} = \tilde{\boldsymbol{\omega}}_i \mathbf{s}_{ij} \tag{6.5.9}$$

Similarly,

$$\dot{\mathbf{s}}_{ji} = \tilde{\boldsymbol{\omega}}_j \mathbf{s}_{ji} \tag{6.5.10}$$

Substituting for $\boldsymbol{\omega}_j$ from Eq. 6.5.6 into Eq. 6.5.10 gives

$$\dot{\mathbf{s}}_{ji} = \tilde{\boldsymbol{\omega}}_i \mathbf{s}_{ji} + \tilde{\mathbf{h}}_{ij} \mathbf{s}_{ji} \dot{\theta}_{ij} \tag{6.5.11}$$

The translational axis $\mathbf{b}_{ij}$ is fixed in body i, so its time derivative is

$$\dot{\mathbf{b}}_{ij} = \tilde{\boldsymbol{\omega}}_i \mathbf{b}_{ij} \tag{6.5.12}$$

Substituting Eqs. 6.5.9 to 6.5.12 into Eq. 6.5.8, the velocity of the centroid of body j is

$$\dot{\mathbf{r}}_j = \dot{\mathbf{r}}_i + \tilde{\boldsymbol{\omega}}_i \mathbf{r}_{ij} - \tilde{\mathbf{h}}_{ij} \mathbf{s}_{ji} \dot{\theta}_{ij} + \mathbf{b}_{ij} \dot{\alpha}_{ij} \tag{6.5.13}$$

The virtual rotation of body j can be written, as in Eq. 6.5.6, as

$$\delta\boldsymbol{\pi}_j = \delta\boldsymbol{\pi}_i + \mathbf{h}_{ij} \delta\theta_{ij} \tag{6.5.14}$$

A virtual displacement relationship for body j can be derived by replacing the time derivative operator by the variation operator in Eq. 6.5.13,

$$\delta\mathbf{r}_j = \delta\mathbf{r}_i - \tilde{\mathbf{r}}_{ij} \delta\boldsymbol{\pi}_i - \tilde{\mathbf{h}}_{ij} \mathbf{s}_{ji} \delta\theta_{ij} + \mathbf{b}_{ij}\, \delta\alpha_{ij} \tag{6.5.15}$$

The angular acceleration of body j is obtained by differentiating Eq. 6.5.6,

$$\dot{\boldsymbol{\omega}}_j = \dot{\boldsymbol{\omega}}_i + \mathbf{h}_{ij} \ddot{\theta}_{ij} + \dot{\mathbf{h}}_{ij} \dot{\theta}_{ij} \tag{6.5.16}$$

Since $\mathbf{h}_{ij}$ is fixed in body i,

$$\dot{\mathbf{h}}_{ij} = \tilde{\boldsymbol{\omega}}_i \mathbf{h}_{ij} \tag{6.5.17}$$

Therefore, the angular acceleration of body j is

$$\dot{\boldsymbol{\omega}}_j = \dot{\boldsymbol{\omega}}_i + \mathbf{h}_{ij} \ddot{\theta}_{ij} + \tilde{\boldsymbol{\omega}}_i \mathbf{h}_{ij} \dot{\theta}_{ij} \tag{6.5.18}$$

The acceleration of body j is obtained by taking the time derivative of Eq. 6.5.8,

$$\ddot{\mathbf{r}}_j = \ddot{\mathbf{r}}_i + \ddot{\mathbf{s}}_{ij} + \ddot{\mathbf{b}}_{ij} \alpha_{ij} + 2\dot{\mathbf{b}}_{ij} \dot{\alpha}_{ij} + \mathbf{b}_{ij} \ddot{\alpha}_{ij} - \ddot{\mathbf{s}}_{ji} \tag{6.5.19}$$

From Eqs. 6.5.9, 6.5.10, and 6.5.12,

$$\begin{aligned} \ddot{\mathbf{s}}_{ij} &= \dot{\tilde{\boldsymbol{\omega}}}_i \mathbf{s}_{ij} + \tilde{\boldsymbol{\omega}}_i \dot{\mathbf{s}}_{ij} \\ &= \dot{\tilde{\boldsymbol{\omega}}}_i \mathbf{s}_{ij} + \tilde{\boldsymbol{\omega}}_i \tilde{\boldsymbol{\omega}}_i \mathbf{s}_{ij} \end{aligned} \tag{6.5.20}$$

$$\ddot{\mathbf{s}}_{ji} = \dot{\tilde{\boldsymbol{\omega}}}_j \mathbf{s}_{ji} + \tilde{\boldsymbol{\omega}}_j \tilde{\boldsymbol{\omega}}_j \mathbf{s}_{ji} \tag{6.5.21}$$

$$\ddot{\mathbf{b}}_{ij} = \dot{\tilde{\boldsymbol{\omega}}}_i \mathbf{b}_{ij} + \tilde{\boldsymbol{\omega}}_i \tilde{\boldsymbol{\omega}}_i \mathbf{b}_{ij} \tag{6.5.22}$$

After substituting Eq. 6.5.16 into Eq. 6.5.21, Eqs. 6.5.20 to 6.5.22 are substituted into Eq. 6.5.19, to obtain

$$\ddot{\mathbf{r}}_j = \ddot{\mathbf{r}}_i + \dot{\tilde{\boldsymbol{\omega}}}_i\mathbf{r}_{ij} - \tilde{\mathbf{h}}_{ij}\mathbf{s}_{ji}\ddot{\theta}_{ij} + \mathbf{b}_{ij}\ddot{\alpha}_{ij} + \mathbf{R}_{ij} \tag{6.5.23}$$

where

$$\mathbf{R}_{ij} \equiv \tilde{\mathbf{s}}_{ij}\tilde{\boldsymbol{\omega}}_i\mathbf{h}_{ij}\dot{\theta}_{ij} - \tilde{\boldsymbol{\omega}}_j\tilde{\boldsymbol{\omega}}_j\mathbf{s}_{ji} + \tilde{\boldsymbol{\omega}}_i\tilde{\boldsymbol{\omega}}_i\mathbf{s}_{ij} + \tilde{\boldsymbol{\omega}}_i\tilde{\boldsymbol{\omega}}_i\mathbf{b}_{ij}\alpha_{ij} + 2\tilde{\boldsymbol{\omega}}_i\mathbf{b}_{ij}\dot{\alpha}_{ij} \tag{6.5.24}$$

The virtual displacement relationships of Eqs. 6.5.14 and 6.5.15 may be written in matrix form as

$$\begin{aligned} \begin{bmatrix} \delta\mathbf{r}_j \\ \delta\boldsymbol{\pi}_j \end{bmatrix} &= \begin{bmatrix} \mathbf{I} & -\tilde{\mathbf{r}}_{ij} \\ \mathbf{0} & \mathbf{I} \end{bmatrix} \begin{bmatrix} \delta\mathbf{r}_i \\ \delta\boldsymbol{\pi}_i \end{bmatrix} + \begin{bmatrix} -\tilde{\mathbf{h}}_{ij}\mathbf{s}_{ji} & \mathbf{b}_{ij} \\ \mathbf{h}_{ij} & \mathbf{0} \end{bmatrix} \begin{bmatrix} \delta\theta_{ij} \\ \delta\alpha_{ij} \end{bmatrix} \\ &\equiv \mathbf{B}_{ij1} \begin{bmatrix} \delta\mathbf{r}_i \\ \delta\boldsymbol{\pi}_i \end{bmatrix} + \mathbf{B}_{ij2} \begin{bmatrix} \delta\theta_{ij} \\ \delta\alpha_{ij} \end{bmatrix} \end{aligned} \tag{6.5.25}$$

where $\mathbf{B}_{ij1}$ and $\mathbf{B}_{ij2}$ are defined as the coefficient matrices in Eq. 6.5.25. The matrix form of the velocity relations of Eqs. 6.5.6 and 6.5.13 is

$$\begin{bmatrix} \dot{\mathbf{r}}_j \\ \boldsymbol{\omega}_j \end{bmatrix} = \mathbf{B}_{ij1} \begin{bmatrix} \dot{\mathbf{r}}_i \\ \boldsymbol{\omega}_i \end{bmatrix} + \mathbf{B}_{ij2} \begin{bmatrix} \dot{\theta}_{ij} \\ \dot{\alpha}_{ij} \end{bmatrix} \tag{6.5.26}$$

Similarly, the matrix form of acceleration relations of Eqs. 6.5.18 and 6.5.23 is

$$\begin{bmatrix} \ddot{\mathbf{r}}_j \\ \dot{\boldsymbol{\omega}}_j \end{bmatrix} = \mathbf{B}_{ij1} \begin{bmatrix} \ddot{\mathbf{r}}_i \\ \dot{\boldsymbol{\omega}}_i \end{bmatrix} + \mathbf{B}_{ij2} \begin{bmatrix} \ddot{\theta}_{ij} \\ \ddot{\alpha}_{ij} \end{bmatrix} + \mathbf{D}_{ij} \tag{6.5.27}$$

where

$$\mathbf{D}_{ij} \equiv \begin{bmatrix} \mathbf{R}_{ij} \\ \tilde{\boldsymbol{\omega}}_i\mathbf{h}_{ij}\dot{\theta}_{ij} \end{bmatrix} \tag{6.5.28}$$

For a revolute joint, $\alpha_{ij} = 0$. As a result, the relative rotational axes z_i'' in body i and z_j'' in body j coincide. For a translational joint, $\theta_{ij} = 0$. As a result, the transformation matrix $\mathbf{A}_{ij}''$ for a translational joint is the identity matrix. A cylindrical joint requires both relative rotation and translation coordinates, hence the full generality of the relative coordinate kinematics derived in this section. Although not all the joints defined in Section 6.2 can be represented by this formulation, it does represent connections that are most commonly encountered in machine dynamics. More general joint formulations are found in Ref. 13.

Cartesian coordinates and velocities of each body in a system must be found, in order to derive the system equations of motion. Relative coordinates and their time derivatives are obtained during a simulation by integrating relative coordinate velocities and accelerations. Cartesian coordinates of the base body and relative coordinates between pairs of bodies in a system with tree structure are adequate to position and orient each body in space, using Eqs. 6.5.1 to 6.5.5. Similarly, Eqs. 6.5.6 and 6.5.8 define the velocity and angular velocity of each body, in terms of base body velocities and relative coordinate time derivatives.

As an example, position and orientation of body 2 in Fig. 6.5.5 can be obtained in terms of base body Cartesian coordinates and joint (1, 2) relative coordinates. Similarly, once position and orientation of bodies 2 and 7 are known, position and orientation of bodies 3 and 8 can be determined. Therefore, the sequence for position and velocity determination must be based on paths from the base body outward to all bodies in the tree.

To simplify dynamic analysis, the following compact notations that describe translational and rotational motion are adopted:

$$\begin{aligned} \mathbf{Y}_i &= \begin{bmatrix} \dot{\mathbf{r}}_i \\ \boldsymbol{\omega}_i \end{bmatrix} \\ \dot{\mathbf{Y}}_i &= \begin{bmatrix} \ddot{\mathbf{r}}_i \\ \dot{\boldsymbol{\omega}}_i \end{bmatrix} \\ \delta\mathbf{Z}_i &= \begin{bmatrix} \delta\mathbf{r}_i \\ \delta\boldsymbol{\pi}_i \end{bmatrix} \\ \mathbf{q}_{ij} &= \begin{bmatrix} \theta_{ij} \\ \alpha_{ij} \end{bmatrix} \end{aligned} \tag{6.5.29}$$

Equations 6.5.25 to 6.5.27 can be rewritten, in the notation of Eq. 6.5.29, as

$$\delta\mathbf{Z}_j = \mathbf{B}_{ij1}\,\delta\mathbf{Z}_i + \mathbf{B}_{ij2}\,\delta\mathbf{q}_{ij} \tag{6.5.30}$$

$$\mathbf{Y}_j = \mathbf{B}_{ij1}\mathbf{Y}_i + \mathbf{B}_{ij2}\dot{\mathbf{q}}_{ij} \tag{6.5.31}$$

$$\dot{\mathbf{Y}}_j = \mathbf{B}_{ij1}\dot{\mathbf{Y}}_i + \mathbf{B}_{ij2}\ddot{\mathbf{q}}_{ij} + \mathbf{D}_{ij} \tag{6.5.32}$$

Example 6.5.5: Consider the planar double pendulum of Example 6.3.1 and let ground be body 1 and the pendula be bodies 2 and 3, as shown in Fig. 6.5.7. Since there are only planar rotational joints in this system, $\alpha_{12} = \alpha_{23} = 0$, $\theta_{12} = q_1$, and $\theta_{23} = q_2$. Thus $\boldsymbol{\omega}_j = \dot{q}_j\mathbf{k}$. Note also that

$$\mathbf{r}_{12} = [\sin q_1,\ -\cos q_1,\ 0]^T$$

$$\mathbf{r}_{23} = [\sin q_1 + \sin(q_1 + q_2),\ -\cos q_1 - \cos(q_1 + q_2),\ 0]^T$$

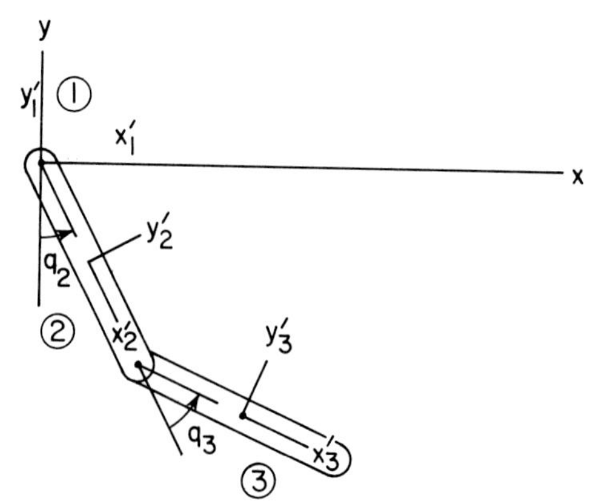

Figure 6.5.7 Planar double pendulum.

$$\mathbf{h}'_{12} = \mathbf{h}'_{23} = \mathbf{k}$$

$$\mathbf{s}_{21} = [-\sin q_1, \cos q_1, 0]^T$$

$$\mathbf{s}_{32} = [-\sin(q_1 + q_2), \cos(q_1 + q_2), 0]^T$$

The coefficient matrices of Eq. 6.5.25 are identified as

$$\mathbf{B}_{121} = \left[\begin{array}{c|ccc} & 0 & 0 & \cos q_1 \\ \mathbf{I} & 0 & 0 & \sin q_1 \\ & -\cos q_1 & -\sin q_1 & 0 \\ \hline \mathbf{0} & & \mathbf{I} & \end{array}\right]$$

$$\mathbf{B}_{122} = \left[\begin{array}{cc} \cos q_1 & 0 \\ \sin q_1 & 0 \\ 0 & 0 \\ \hline \mathbf{k} & \mathbf{0} \end{array}\right]$$

$$\mathbf{B}_{231} = \left[\begin{array}{c|ccc} & 0 & 0 & \cos q_1 + \cos(q_1 + q_2) \\ \mathbf{I} & 0 & 0 & \sin q_1 + \sin(q_1 + q_2) \\ & -\cos q_1 - \cos(q_1 + q_2) & -\sin q_1 - \sin(q_1 + q_2) & 0 \\ \hline \mathbf{0} & & \mathbf{I} & \end{array}\right]$$

$$\mathbf{B}_{232} = \left[\begin{array}{cc} \cos(q_1 + q_2) & 0 \\ \sin(q_1 + q_2) & 0 \\ 0 & 0 \\ \hline \mathbf{k} & \mathbf{0} \end{array}\right]$$

Thus the matrix form of velocity relations is

$$\begin{bmatrix} \dot{\mathbf{r}}_2 \\ \boldsymbol{\omega}_2 \end{bmatrix} = \mathbf{B}_{121} \begin{bmatrix} \mathbf{0} \\ \mathbf{0} \end{bmatrix} + \mathbf{B}_{122} \begin{bmatrix} \dot{q}_1 \\ 0 \end{bmatrix}$$

$$= \dot{q}_1 [\cos q_1, \sin q_1, 0, 0, 0, 1]^T$$

$$\begin{bmatrix} \dot{\mathbf{r}}_3 \\ \boldsymbol{\omega}_3 \end{bmatrix} = \mathbf{B}_{231} \begin{bmatrix} \dot{\mathbf{r}}_2 \\ \boldsymbol{\omega}_2 \end{bmatrix} + \mathbf{B}_{232} \begin{bmatrix} \dot{q}_2 \\ 0 \end{bmatrix}$$

$$= \begin{bmatrix} 2 \cos q_1 \dot{q}_1 + \cos(q_1 + q_2)(\dot{q}_1 + q_2) \\ 2 \sin q_1 \dot{q}_1 + \sin(q_1 + q_2)(\dot{q}_1 + \dot{q}_2) \\ 0 \\ 0 \\ 0 \\ \dot{q}_1 + \dot{q}_2 \end{bmatrix}$$

This is the result obtained by different means in Example 6.3.1.

◆ **Example 6.5.6:** Consider the spatial two body mechanism of Example 6.3.4. Ground is body 1 and the first and second bars are bodies 2 and 3, respectively, as shown in Fig. 6.5.8. Since this mechanism has only rotational joints, $\alpha_{12} = \alpha_{23} = 0$, $\theta_{12} = q_1$, and $\theta_{23} = q_2$. For use in Eq. 6.5.25, note that

$$\mathbf{r}_{12} = [\cos q_1, \sin q_1, 0]^T$$

$$\mathbf{r}_{23} = [-\sin q_1 \sin q_2, \cos q_1 \sin q_2, -\cos q_2]^T$$

$$\mathbf{s}_{21} = [-\cos q_1, -\sin q_1, 0]^T$$

$$\mathbf{s}_{32} = [\sin q_1 \sin q_3, -\cos q_1 \sin q_3, \cos q_3]^T$$

$$\mathbf{h}_{12} = \mathbf{k}$$

$$\mathbf{h}_{23} = [\cos q_1, \sin q_1, 0]^T$$

$$\mathbf{b}_{12} = \mathbf{b}_{23} = \mathbf{0}$$

The coefficient matrices in Eq. 6.5.25 are thus

$$\mathbf{B}_{121} = \left[\begin{array}{c|ccc} & 0 & 0 & -\sin q_1 \\ \mathbf{I} & 0 & 0 & \cos q_1 \\ & \sin q_1 & -\cos q_1 & 0 \\ \hline \mathbf{0} & & \mathbf{I} & \end{array}\right]$$

$$\mathbf{B}_{122} = \left[\begin{array}{cc} -\sin q_1 & 0 \\ \cos q_1 & 0 \\ 0 & 0 \\ \hline \mathbf{k} & \mathbf{0} \end{array}\right]$$

$$\mathbf{B}_{231} = \left[\begin{array}{c|ccc} & 0 & -\cos q_2 & -\cos q_1 \sin q_2 \\ \mathbf{I} & \cos q_2 & 0 & -\sin q_1 \sin q_2 \\ & \cos q_1 \sin q_2 & \sin q_1 \sin q_2 & 0 \\ \hline \mathbf{0} & & \mathbf{I} & \end{array}\right]$$

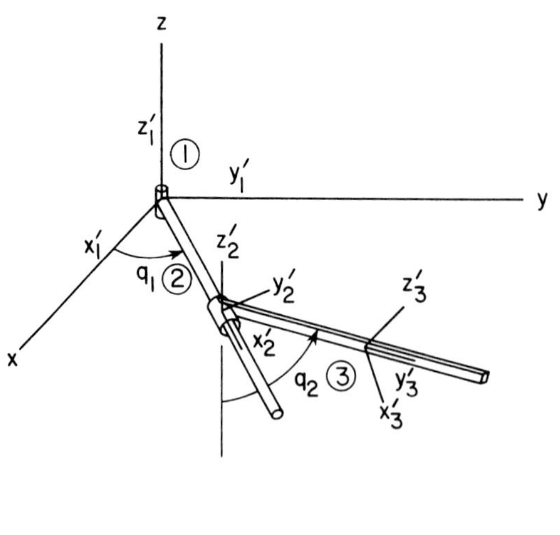

Figure 6.5.8 Spatial double pendulum.

$$\mathbf{B}_{232} = \begin{bmatrix} -\sin q_1 \cos q_2 & 0 \\ \cos q_1 \cos q_2 & 0 \\ \sin q_2 & 0 \\ \cos q_1 & 0 \\ \sin q_1 & 0 \\ 0 & 0 \end{bmatrix}$$

Substituting these expressions into Eq. 6.5.26, velocities are

$$\begin{bmatrix} \dot{\mathbf{r}}_2 \\ \boldsymbol{\omega}_2 \end{bmatrix} = \mathbf{B}_{121}\begin{bmatrix} \mathbf{0} \\ \mathbf{0} \end{bmatrix} + \mathbf{B}_{122}\begin{bmatrix} \dot{q}_1 \\ 0 \end{bmatrix}$$

$$= \dot{q}_1\,[-\sin q_1, \cos q_1, 0, 0, 0, 1]^T$$

$$\begin{bmatrix} \dot{\mathbf{r}}_3 \\ \boldsymbol{\omega}_3 \end{bmatrix} = \mathbf{B}_{231}\begin{bmatrix} \dot{\mathbf{r}}_2 \\ \boldsymbol{\omega}_2 \end{bmatrix} + \mathbf{B}_{232}\begin{bmatrix} \dot{q}_2 \\ 0 \end{bmatrix}$$

$$= \begin{bmatrix} (-\sin q_1 - \cos q_1 \sin q_2)\dot{q}_1 - \sin q_1 \cos q_2 \dot{q}_2 \\ (\cos q_1 - \sin q_1 \sin q_2)\dot{q}_1 + \cos q_1 \cos q_2 \dot{q}_2 \\ \sin q_2 \dot{q}_2 \\ \cos q_1 \dot{q}_2 \\ \sin q_1 \dot{q}_2 \\ \dot{q}_1 \end{bmatrix}$$

Note that these velocity expressions are the same as obtained in Example 4.2.4. The angular velocity here is in the global frame. It can be used to obtain the result in Example 6.3.4, through a transformation (Prob. 6.5.4).

➧ Example 6.5.7: Consider next the two body, three degree of freedom mechanism of Example 6.3.6, with bodies numbered as shown in Figure 6.5.9. Since the

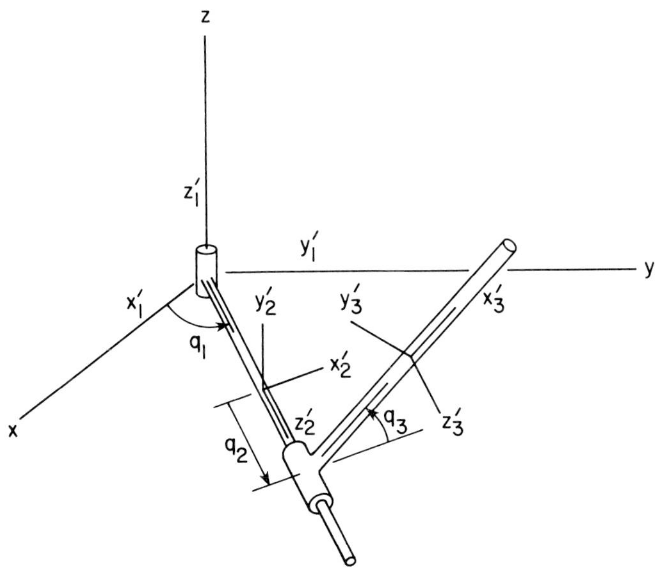

Figure 6.5.9 Three-degree-of-freedom two-body mechanism.

only translation that occurs is between bodies 2 and 3, $\alpha_{12} = 0$ and $\alpha_{23} = q_2$. Relative rotations are $\theta_{12} = q_1$ and $\theta_{23} = q_3$. For determining the coefficient matrices in Eq. 6.5.25, note that

$$\begin{aligned}
\mathbf{r}_{12} &= [\cos q_1,\ \sin q_1,\ 0]^T \\
\mathbf{r}_{23} &= [q_2 \cos q_1 - \sin q_1 \cos q_3,\ q_2 \sin q_1 + \cos q_1 \cos q_3,\ \sin q_3]^T \\
\mathbf{s}_{21} &= [-\cos q_1,\ -\sin q_1,\ 0]^T \\
\mathbf{s}_{32} &= [\sin q_1 \cos q_3,\ -\cos q_1 \cos q_3,\ -\sin q_3]^T \\
\mathbf{h}_{12} &= \mathbf{k} \\
\mathbf{h}_{23} &= [\cos q_1,\ \sin q_1,\ 0]^T \\
\mathbf{b}_{23} &= [\cos q_1,\ \sin q_1,\ 0]^T
\end{aligned}$$

The transformation matrices of Eq. 6.5.25 are

$$\mathbf{B}_{121} = \left[\begin{array}{c|ccc} & 0 & 0 & -\sin q_1 \\ \mathbf{I} & 0 & 0 & \cos q_1 \\ & \sin q_1 & -\cos q_1 & 0 \\ \hline \mathbf{0} & & \mathbf{I} & \end{array}\right]$$

$$\mathbf{B}_{122} = \left[\begin{array}{cc} -\sin q_1 & 0 \\ \cos q_1 & 0 \\ 0 & 0 \\ \hline \mathbf{k} & \mathbf{0} \end{array}\right]$$

$$\mathbf{B}_{231} = \left[\begin{array}{c|ccc} & 0 & \sin q_3 & -q_2 \sin q_1 - \cos q_1 \cos q_3 \\ \mathbf{I} & -\sin q_3 & 0 & q_2 \cos q_1 - \sin q_1 \cos q_3 \\ & q_2 \sin q_1 + \cos q_1 \cos q_3 & -q_2 \cos q_1 + \sin q_1 \cos q_3 & 0 \\ \hline \mathbf{0} & & \mathbf{I} & \end{array}\right]$$

$$\mathbf{B}_{232} = \begin{bmatrix} \sin q_1 \sin q_3 & \cos q_1 \\ -\cos q_1 \sin q_3 & \sin q_1 \\ \cos q_3 & 0 \\ \cos q_1 & 0 \\ \sin q_1 & 0 \\ 0 & 0 \end{bmatrix}$$

Substituting into Eq. 6.5.26, the matrix forms of the velocities of bodies 2 and 3 are

$$\begin{bmatrix} \dot{\mathbf{r}}_2 \\ \boldsymbol{\omega}_2 \end{bmatrix} = \mathbf{B}_{121} \begin{bmatrix} \mathbf{0} \\ \mathbf{0} \end{bmatrix} + \mathbf{B}_{122} \begin{bmatrix} \dot{q}_1 \\ \mathbf{0} \end{bmatrix} = \begin{bmatrix} -\sin q_1 \dot{q}_1 \\ \cos q_1 \dot{q}_1 \\ 0 \\ \dot{q}_1\, \mathbf{k} \end{bmatrix}$$

$$\begin{bmatrix} \dot{\mathbf{r}}_3 \\ \boldsymbol{\omega}_3 \end{bmatrix} = \mathbf{B}_{231} \begin{bmatrix} \dot{\mathbf{r}}_2 \\ \boldsymbol{\omega}_2 \end{bmatrix} + \mathbf{B}_{232} \begin{bmatrix} \dot{q}_n \\ \dot{q}_2 \end{bmatrix}$$

$$
= \begin{bmatrix} [-(1 + q_2)\sin q_1 - \cos q_1 \cos q_3]\dot{q}_1 + \cos q_1 \dot{q}_2 + \sin q_1 \sin q_3 \dot{q}_3 \\ [(1 + q_2)\cos q_1 - \sin q_1 \cos q_3]\dot{q}_1 + \sin q_1 \dot{q}_2 - \cos q_1 \sin q_3 \dot{q}_3 \\ \cos q_3 \dot{q}_3 \\ \cos q_1 \dot{q}_3 \\ \sin q_1 \dot{q}_3 \\ \dot{q}_1 \end{bmatrix}
$$

This velocity is the same as that obtained in Example 6.3.6.

Using the transformation matrix from the x_3'-y_3'-z_3' frame to the global frame, the same angular velocity vector in Example 6.3.6 can be obtained (Prob. 6.5.5).

System Relative Coordinate Kinematics

Joint relative coordinate kinematic relations derived thus far in this section relate position, velocity, acceleration, and virtual displacement of pairs of constrained bodies in a mechanical system. These relations, together with a system graph for a mechanical system with tree structure, permit systematic calculation of the Cartesian position, orientation, velocity, and acceleration of bodies in the system, in terms of base body Cartesian coordinates and relative coordinates between pairs of bodies.

Example 6.5.8: To illustrate use of these relations, consider the tree structure system of Fig. 6.5.5. With known base body position and orientation and relative coordinates, the position and orientation of bodies 2 and 7 may be calculated, using Eqs. 6.5.1 and 6.5.5. Equations 6.5.31 and 6.5.32 then yield the velocity and acceleration of bodies 2 and 7. The process may be continued to calculate position, velocity, and acceleration of bodies 3 and 8. Next, the same relations can be used to evaluate position, velocity, and acceleration of bodies 4, 6, and 9. Finally, position, velocity, and acceleration of bodies 5 and 10 may be calculated. Thus with the system graph and relative coordinate information, Eqs. 6.5.1, 6.5.5, 6.5.31, and 6.5.32 may be used in a systematic way to evaluate Cartesian position, orientation, velocity, and acceleration of each body in the system. While this process is conceptually simple and can easily be implemented in the form of a computer program, greater efficiency can be achieved by taking advantage of properties of the transformation matrices and Eqs. 6.5.31 and 6.5.32.

To develop a systematic matrix implementation of the relative coordinate velocity and acceleration relations outlined in Example 6.5.8, consider the typical chain shown in Fig. 6.5.10, which begins at body l and terminates at tree-end body n. To simplify notation, bodies are numbered consecutively in this chain.

Applying Eq. 6.5.31 with $j = l + 1$ and $i = l$ yields

$$
\mathbf{Y}_{l+1} = \mathbf{B}_{l(l+1)1}\mathbf{Y}_l + \mathbf{B}_{l(l+1)2}\dot{\mathbf{q}}_{l(l+1)} \tag{6.5.33}
$$

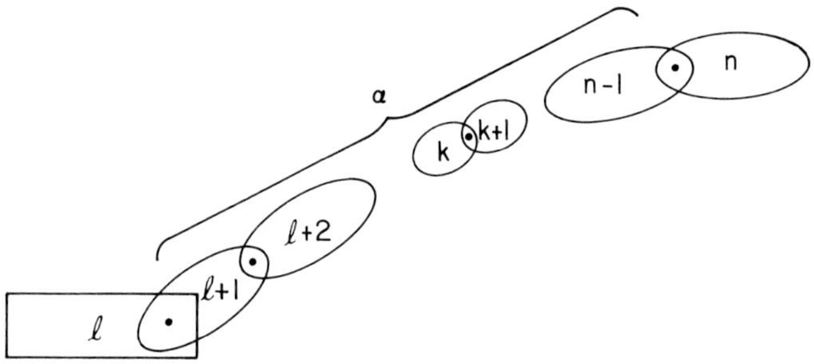

Figure 6.5.10 Single chain mechanism.

Similarly, writing Eq. 6.5.31 with $j = l + 2$ and $i = l + 1$ and substituting from Eq. 6.5.33 yields

$$
\begin{aligned}
\mathbf{Y}_{l+2} &= \mathbf{B}_{(l+1)(l+2)1}\mathbf{Y}_{l+1} + \mathbf{B}_{(l+1)(l+2)2}\dot{\mathbf{q}}_{(l+1)(l+2)} \\
&= \mathbf{B}_{(l+1)(l+2)1}[\mathbf{B}_{l(l+1)1}\mathbf{Y}_l + \mathbf{B}_{l(l+1)2}\dot{\mathbf{q}}_{l(l+1)}] \\
&\quad + \mathbf{B}_{(l+1)(l+2)2}\dot{\mathbf{q}}_{(l+1)(l+2)} \qquad (6.5.34) \\
&= \mathbf{B}_{(l+1)(l+2)1}\mathbf{B}_{l(l+1)1}\mathbf{Y}_l + \mathbf{B}_{(l+1)(l+2)1}\mathbf{B}_{l(l+1)2}\dot{\mathbf{q}}_{l(l+1)} \\
&\quad + \mathbf{B}_{(l+1)(l+2)2}\dot{\mathbf{q}}_{(l+1)(l+2)}
\end{aligned}
$$

This process can be continued by making successive substitutions of Eq. 6.5.31. For purposes of proof by induction, assume that the result of this calculation for body k is the *recursive velocity relation*

$$
\begin{aligned}
\mathbf{Y}_k = {}& \mathbf{B}_{(k-1)k1}\mathbf{B}_{(k-2)(k-1)1} \times \cdots \times \mathbf{B}_{l(l+1)1}\mathbf{Y}_l \\
&+ \mathbf{B}_{(k-1)k1}\mathbf{B}_{(k-2)(k-1)1} \times \cdots \times \mathbf{B}_{(l+1)(l+2)1}\mathbf{B}_{l(l+1)2}\dot{\mathbf{q}}_{l(l+1)} \qquad (6.5.35) \\
&+ \cdots + \mathbf{B}_{(k-1)k2}\dot{\mathbf{q}}_{(k-1)k}
\end{aligned}
$$

Writing Eq. 6.5.31 with $j = k + 1$ and $i = k$, and substituting the result of Eq. 6.5.35 yields

$$
\begin{aligned}
\mathbf{Y}_{k+1} &= \mathbf{B}_{k(k+1)1}\mathbf{Y}_k + \mathbf{B}_{k(k+1)2}\dot{\mathbf{q}}_{k(k+1)} \\
&= \mathbf{B}_{k(k+1)1}\mathbf{B}_{(k-1)k1} \times \cdots \times \mathbf{B}_{l(l+1)1}\mathbf{Y}_l \qquad (6.5.36) \\
&\quad + \mathbf{B}_{k(k+1)1}\mathbf{B}_{(k-1)k1} \times \cdots \times \mathbf{B}_{(l+1)(l+2)1}\mathbf{B}_{l(l+1)}\dot{\mathbf{q}}_{l(l+1)} \\
&\quad + \cdots + \mathbf{B}_{k(k+1)2}\dot{\mathbf{q}}_{k(k+1)}
\end{aligned}
$$

Note that this result is identical in form to Eq. 6.5.35, verifying the induction hypothesis and proving that the recursive formula of Eq. 6.5.35 is correct.

While the strings of products in Eq. 6.5.35 could be computed numerically, exploiting the form of matrices involved permits substantial simplification. More specifically, the first coefficient matrix in Eq. 6.5.35 may be expanded to obtain

$$\begin{aligned}
&\mathbf{B}_{(k-1)k1}\mathbf{B}_{(k-2)(k-1)1} \times \cdots \times \mathbf{B}_{(l+1)(l+2)1}\mathbf{B}_{l(l+1)1} \\
&= \begin{bmatrix} \mathbf{I} & \tilde{\mathbf{r}}_k - \tilde{\mathbf{r}}_{k-1} \\ \mathbf{0} & \mathbf{I} \end{bmatrix} \begin{bmatrix} \mathbf{I} & \tilde{\mathbf{r}}_{k-1} - \tilde{\mathbf{r}}_{k-2} \\ \mathbf{0} & \mathbf{I} \end{bmatrix} \times \cdots \times \begin{bmatrix} \mathbf{I} & \tilde{\mathbf{r}}_{l+1} - \tilde{\mathbf{r}}_l \\ \mathbf{0} & \mathbf{I} \end{bmatrix} \qquad (6.5.37) \\
&= \begin{bmatrix} \mathbf{I} & \tilde{\mathbf{r}}_k - \tilde{\mathbf{r}}_l \\ \mathbf{0} & \mathbf{I} \end{bmatrix}
\end{aligned}$$

Similarly, a typical coefficient of the relative coordinate time derivative terms in Eq. 6.5.35 may be expanded as

$$\begin{aligned}
&\mathbf{B}_{(k-1)k1}\mathbf{B}_{(k-2)(k-1)1} \times \cdots \times \mathbf{B}_{(p+1)(p+2)1}\mathbf{B}_{p(p+1)2} \\
&= \begin{bmatrix} \mathbf{I} & \tilde{\mathbf{r}}_k - \tilde{\mathbf{r}}_{k-1} \\ \mathbf{0} & \mathbf{I} \end{bmatrix} \begin{bmatrix} \mathbf{I} & \tilde{\mathbf{r}}_{k-1} - \tilde{\mathbf{r}}_{k-2} \\ \mathbf{0} & \mathbf{I} \end{bmatrix} \times \cdots \\
&\quad \times \begin{bmatrix} \mathbf{I} & \tilde{\mathbf{r}}_{p+2} - \tilde{\mathbf{r}}_{p+1} \\ \mathbf{0} & \mathbf{I} \end{bmatrix} \begin{bmatrix} -\tilde{\mathbf{h}}_{p(p+1)}\mathbf{s}_{(p+1)p} & \mathbf{b}_{p(p+1)} \\ \mathbf{h}_{p(p+1)} & \mathbf{0} \end{bmatrix} \qquad (6.5.38) \\
&= \begin{bmatrix} \mathbf{I} & \tilde{\mathbf{r}}_k - \tilde{\mathbf{r}}_{p+1} \\ \mathbf{0} & \mathbf{I} \end{bmatrix} \begin{bmatrix} -\tilde{\mathbf{h}}_{p(p+1)}\mathbf{s}_{(p+1)p} & \mathbf{b}_{p(p+1)} \\ \mathbf{h}_{p(p+1)} & \mathbf{0} \end{bmatrix} \\
&= \begin{bmatrix} (\tilde{\mathbf{r}}_k - \tilde{\mathbf{r}}_{p+1} + \tilde{\mathbf{s}}_{(p+1)p})\mathbf{h}_{p(p+1)} & \mathbf{b}_{p(p+1)} \\ \mathbf{h}_{p(p+1)} & \mathbf{0} \end{bmatrix}
\end{aligned}$$

Using Eqs. 6.5.37 and 6.5.38 to expand the sum of terms in Eq. 6.5.35 yields

$$\begin{aligned}
\mathbf{Y}_k &= \begin{bmatrix} \mathbf{I} & \tilde{\mathbf{r}}_k - \tilde{\mathbf{r}}_l \\ \mathbf{0} & \mathbf{I} \end{bmatrix} \mathbf{Y}_l + \sum_{p=l}^{k-1} \begin{bmatrix} \mathbf{b}_{p(p+1)} + (\tilde{\mathbf{r}}_k - \tilde{\mathbf{r}}_{p+1} + \tilde{\mathbf{s}}_{(p+1)p})\mathbf{h}_{p(p+1)} \\ \mathbf{h}_{p(p+1)} \end{bmatrix} \dot{\mathbf{q}}_{p(p+1)} \\
&\equiv \hat{\mathbf{E}}_{lk}\mathbf{Y}_l + \hat{\mathbf{B}}_{lk}\dot{\mathbf{q}}_{ln} \qquad (6.5.39)
\end{aligned}$$

where the composite vector of relative coordinates in the chain from bodies l to n is

$$\mathbf{q}_{ln} = \begin{bmatrix} \mathbf{q}_{l(l+1)} \\ \vdots \\ \mathbf{q}_{(n-1)n} \end{bmatrix} \qquad (6.5.40)$$

Differentiating both sides of Eq. 6.5.39 yields the acceleration relation

$$\begin{aligned}
\dot{\mathbf{Y}}_k &= \hat{\mathbf{E}}_{lk}\dot{\mathbf{Y}}_l + \hat{\mathbf{B}}_{lk}\ddot{\mathbf{q}}_{ln} + \dot{\hat{\mathbf{E}}}_{lk}\mathbf{Y}_l + \dot{\hat{\mathbf{B}}}_{lk}\dot{\mathbf{q}}_{ln} \\
&= \hat{\mathbf{E}}_{lk}\dot{\mathbf{Y}}_l + \hat{\mathbf{B}}_{lk}\ddot{\mathbf{q}}_{ln} + \begin{bmatrix} \mathbf{0} & \dot{\tilde{\mathbf{r}}}_k - \dot{\tilde{\mathbf{r}}}_l \\ \mathbf{0} & \mathbf{0} \end{bmatrix} \mathbf{Y}_l + \sum_{p=l}^{k-1} \\
&\quad \begin{bmatrix} \dot{\mathbf{b}}_{p(p+1)} + (\dot{\tilde{\mathbf{r}}}_k - \dot{\tilde{\mathbf{r}}}_{p+1} + \dot{\tilde{\mathbf{s}}}_{(p+1)p})\mathbf{h}_{p(p+1)} + (\tilde{\mathbf{r}}_k - \tilde{\mathbf{r}}_{p+1} + \tilde{\mathbf{s}}_{(p+1)p})\dot{\mathbf{h}}_{p(p+1)} \\ \dot{\mathbf{h}}_{p(p+1)} \end{bmatrix} \\
&\quad \times \dot{\mathbf{q}}_{p(p+1)}
\end{aligned} \qquad (6.5.41)$$

Using the relations $\dot{\mathbf{b}}_{p(p+1)} = \tilde{\boldsymbol{\omega}}_p \mathbf{b}_{p(p+1)}$ and $\dot{\mathbf{h}}_{p(p+1)} = \tilde{\boldsymbol{\omega}}_p \mathbf{h}_{p(p+1)}$, the quadratic velocity terms in Eq. 6.5.41 may be written as

$$\dot{\hat{\mathbf{E}}}_{lk}\mathbf{Y}_l + \dot{\hat{\mathbf{B}}}_{lk}\dot{\mathbf{q}}_{ln} = \begin{bmatrix} \mathbf{0} & \dot{\tilde{\mathbf{r}}}_k - \dot{\tilde{\mathbf{r}}}_l \\ \mathbf{0} & \mathbf{0} \end{bmatrix} \mathbf{Y}_l + \sum_{p=1}^{k-1} \begin{bmatrix} \tilde{\boldsymbol{\omega}}_p \mathbf{b}_{p(p+1)} + (\dot{\tilde{\mathbf{r}}}_k - \dot{\tilde{\mathbf{r}}}_{p+1} + \dot{\tilde{\mathbf{s}}}_{(p+1)p})\mathbf{h}_{p(p+1)} + (\tilde{\mathbf{r}}_k - \tilde{\mathbf{r}}_{p+1} + \tilde{\mathbf{s}}_{(p+1)p})\tilde{\boldsymbol{\omega}}_p \dot{\mathbf{h}}_{p(p+1)} \\ \tilde{\boldsymbol{\omega}}_p \mathbf{h}_{p(p+1)} \end{bmatrix} \times \dot{\mathbf{q}}_{p(p+1)} \tag{6.5.42}$$

The velocity relation of the Eq. 6.5.39 may be extended by moving outward on individual chains from the base body, passing junction nodes, to obtain matrix relationships for Cartesian components of velocity of each body in a system with tree structure, in terms of the base body Cartesian velocities and relative coordinate time derivatives. Using the vector notation

$$\mathbf{Y} = \begin{bmatrix} \mathbf{Y}_1 \\ \vdots \\ \mathbf{Y}_n \end{bmatrix} \tag{6.5.43}$$

for Cartesian components of velocity of all bodies in the system and

$$\dot{\mathbf{q}} = \begin{bmatrix} \mathbf{Y}_1 \\ \dot{\mathbf{q}}_{12} \\ \vdots \\ \dot{\mathbf{q}}_{ij} \end{bmatrix} \tag{6.5.44}$$

for the combination of base body velocity and relative coordinate time derivatives throughout the system, a composite matrix relationship between relative coordinate velocities and Cartesian velocities may be written in the form

$$\mathbf{Y} = \hat{\mathbf{B}}\dot{\mathbf{q}} \tag{6.5.45}$$

where the coefficient matrix $\hat{\mathbf{B}}$ is a composite matrix that is made up of submatrices defined in Eq. 6.5.39. It is called the *velocity transformation* matrix.

Differentiating Eq. 6.5.45 yields the acceleration relation

$$\dot{\mathbf{Y}} = \hat{\mathbf{B}}\ddot{\mathbf{q}} + \dot{\hat{\mathbf{B}}}\dot{\mathbf{q}} \tag{6.5.46}$$

where the vector of quadratic terms in velocity is assembled using the form of terms in Eqs. 6.5.41 and 6.5.42.

Forming the composite vector of system Cartesian virtual displacements, as defined in Eqs. 6.5.29, yields

$$\delta\mathbf{Z} = \begin{bmatrix} \delta\mathbf{Z}_1 \\ \vdots \\ \delta\mathbf{Z}_n \end{bmatrix} \tag{6.5.47}$$

a relationship between Cartesian virtual displacements of the system and variations of the system relative coordinates may be obtained. Recalling that the derivative and variation operators yield the same relationship between velocities and virtual displacements, the velocity relation of Eq. 6.5.39, for an individual body, can be rewritten in terms of virtual displacements. Combining this matrix result for all components of the system yields the same coefficient matrix as in Eq. 6.5.45; that is,

$$\delta \mathbf{Z} = \hat{\mathbf{B}}\, \delta \mathbf{q} \tag{6.5.48}$$

which relates base body Cartesian virtual displacements and relative coordinate variations to system Cartesian virtual displacements.

6.6 VELOCITY TRANSFORMATION AND RELATIVE COORDINATE FORMULATION FOR DYNAMICS OF MULTIBODY SYSTEMS

The relative coordinate kinematics formulation of Section 6.5 may be used to reduce the variational equations of motion for multibody systems to a relative coordinate form that leads to a reduction in the number of equations of motion, as compared to the Cartesian coordinate formulation of Sections 6.1 and 6.2. The formulation presented in this section is a matrix generalization of the ad hoc method used in analysis of specific systems in Section 6.4.

Relative Coordinate Variational Equations of Motion

The variational Newton–Euler equation of motion of Eq. 5.1.26, for each body in a system, can be written in terms of global Cartesian coordinates and summed over all bodies in the system to yield (Prob. 6.6.1)

$$\begin{aligned}
&\sum_{i=1}^{n} [\delta \mathbf{r}_i^T (m_i \ddot{\mathbf{r}}_i - \mathbf{F}_i^A) + \delta \boldsymbol{\pi}_i^T (\mathbf{J}_i \dot{\boldsymbol{\omega}}_i + \tilde{\boldsymbol{\omega}}_i \mathbf{J}_i \boldsymbol{\omega}_i - \mathbf{n}_i^A)] \\
&\quad = \sum_{i=1}^{n} \delta \mathbf{Z}_i^T (\mathbf{M}_i \dot{\mathbf{Y}}_i - \mathbf{Q}_i^A) \\
&\quad = \delta \mathbf{Z}^T (\mathbf{M}\dot{\mathbf{Y}} - \mathbf{Q}^A) = 0
\end{aligned} \tag{6.6.1}$$

where $\mathbf{Y}$ and $\delta\mathbf{Z}$ are defined in Eqs. 6.5.43 and 6.5.47 and

$$\begin{aligned}
\mathbf{M}_i &= \begin{bmatrix} m_i \mathbf{I} & \mathbf{0} \\ \mathbf{0} & \mathbf{J}_i \end{bmatrix} \\
\mathbf{J}_i &= \mathbf{A}_i \mathbf{J}_i' \mathbf{A}_i^T \\
\mathbf{Q}_i^A &= \begin{bmatrix} \mathbf{F}_i^A \\ \mathbf{n}_i^A - \tilde{\boldsymbol{\omega}}_i \mathbf{J}_i \boldsymbol{\omega}_i \end{bmatrix} \\
\mathbf{M} &= \mathbf{diag}\,(\mathbf{M}_1, \mathbf{M}_2, \cdots, \mathbf{M}_n) \\
\mathbf{Q}^A &= \begin{bmatrix} \mathbf{Q}_1^A \\ \vdots \\ \mathbf{Q}_n^A \end{bmatrix}
\end{aligned} \tag{6.6.2}$$

Equation 6.6.1 must hold for all Cartesian virtual displacements $\delta\mathbf{Z}$ that are consistent with constraints that act on the system.

Equations 6.5.46 and 6.5.48 may be used to transform Eq. 6.6.1 to relative coordinate form; that is,

$$\delta\mathbf{q}^T\hat{\mathbf{B}}^T(\mathbf{M}\hat{\mathbf{B}}\ddot{\mathbf{q}} + \mathbf{M}\dot{\hat{\mathbf{B}}}\dot{\mathbf{q}} - \mathbf{Q}^A) = 0 \tag{6.6.3}$$

which must hold for all base body and relative coordinate variations $\delta\mathbf{q}$ that are consistent with constraints that act on the base body and any cut-joint constraints that may have been introduced to create a tree structure. Defining

$$\begin{aligned}\hat{\mathbf{M}} &= \hat{\mathbf{B}}^T\mathbf{M}\hat{\mathbf{B}} \\ \hat{\mathbf{Q}}^A &= \hat{\mathbf{B}}^T\mathbf{Q}^A - \hat{\mathbf{B}}^T\mathbf{M}\dot{\hat{\mathbf{B}}}\dot{\mathbf{q}}\end{aligned} \tag{6.6.4}$$

Eq. 6.6.3 may be written in the simplified form

$$\delta\mathbf{q}^T(\hat{\mathbf{M}}\ddot{\mathbf{q}} - \hat{\mathbf{Q}}^A) = 0 \tag{6.6.5}$$

which must hold for all $\delta\mathbf{q}$ that are kinematically admissible for cut-joint constraints and external constraints that act on the base body. Use of this variational equation of motion thus depends on whether there are closed loops in the graph of the system under consideration.

Differential Equations of Motion for Systems with Tree Structure

Two subcases in formulating equations of motion for systems with tree structure may be considered, both leading to essentially the same form of the equations of motion. First, if the base body is free to move in space, then $\delta\mathbf{q} = [\delta\mathbf{Z}_1^T, \delta\mathbf{q}_{ij}^T]^T$ contains the Cartesian virtual displacements of the base body, which are arbitrary, and relative coordinates $\mathbf{q}_{ij}$ that may be varied arbitrarily without violating constraints in the system. Thus, in the case of a tree structure system with free base body, $\delta\mathbf{q}$ is arbitrary.

If the system graph contains an edge that is connected to ground, ground will be taken as the base body in the system model. In this case, $\delta\mathbf{Z}_1 = 0$, $\mathbf{Y}_1 = \dot{\mathbf{Y}}_1 = \mathbf{0}$, and the relative coordinate vector may be compressed to the form $\mathbf{q} = [\mathbf{q}_{ij}]$ (i.e., only relative coordinates between bodies in the system need to be considered). In this special case, the reduced mass matrix $\hat{\mathbf{M}}$ and generalized force vector $\hat{\mathbf{Q}}^A$ of Eq. 6.6.4 may be reduced in dimension to suppress contributions from the base body. Thus for a grounded base body, the reduced relative coordinate variation $\delta\mathbf{q}$ is arbitrary.

In both the case of a free base body and a grounded base body, using the convention noted above, $\delta\mathbf{q}$ in Eq. 6.6.5 is arbitrary, so its coefficient must be zero. This leads to the following differential equations of motion for systems with tree structure:

$$\hat{\mathbf{M}}\ddot{\mathbf{q}} = \hat{\mathbf{Q}}^A \tag{6.6.6}$$

The approach presented here for deriving equations of motion for systems with tree structure is a generalization of the methods presented in Sections 6.3 and 6.4. More specifically, the equations of motion for systems in Examples 6.3.1, 6.3.4, 6.3.6, and 6.4.1 can be derived directly. In fact, the details of Example 6.4.1 are identical to the method presented here.

Example 6.6.1: Consider the planar double pendulum of Example 6.5.5, using the notations of Eq. 6.5.29. From kinematic relations derived in Example 6.5.5,

$$\mathbf{Y}_2 = [\cos q_1, \sin q_1, 0, 0, 0, 1]^T \dot{q}_1$$

$$\mathbf{Y}_3 = \begin{bmatrix} 2\cos q_1 \dot{q}_1 + \cos(q_1 + q_2)(\dot{q}_1 + \dot{q}_2) \\ 2\sin q_1 \dot{q}_1 + \sin(q_1 + q_2)(\dot{q}_1 + \dot{q}_2) \\ 0 \\ 0 \\ 0 \\ \dot{q}^1 + \dot{q}^2 \end{bmatrix}$$

Differentiating yields

$$\dot{\mathbf{Y}}_2 = \begin{bmatrix} \cos q_1 \ddot{q}_1 - \sin q_1 \dot{q}_1^2 \\ \sin q_1 \ddot{q}_1 + \cos q_1 \dot{q}_1^2 \\ 0 \\ 0 \\ 0 \\ \ddot{q}_1 \end{bmatrix}$$

$$\dot{\mathbf{Y}}_3 = \begin{bmatrix} [2\cos q_1 + \cos(q_1 + q_2)]\ddot{q}_1 + \cos(q_1 + q_2)\ddot{q}_2 + [-2\sin q_1 - \sin(q_1 + q_2)]\dot{q}_1^2 \\ - \sin(q_1 + q_2)\dot{q}_2^2 - 2\sin(q_1 + q_2)\dot{q}_1\dot{q}_2 \\ [2\sin q_1 + \sin(q_1 + q_2)]\ddot{q}_1 + \sin(q_1 + q_2)\ddot{q}_2 + [2\cos q_1 + \cos(q_1 + q_2)]\dot{q}_1^2 \\ + \cos(q_1 + q_2)\dot{q}_2^2 + 2\cos(q_1 + q_2)\dot{q}_1\dot{q}_2 \\ 0 \\ 0 \\ 0 \\ \ddot{q}_1 + \ddot{q}_2 \end{bmatrix}$$

Virtual displacement vectors are

$$\delta\mathbf{Z}_1 = \mathbf{0}$$

$$\delta\mathbf{Z}_2 = \begin{bmatrix} \cos q_1 \, \delta q_1 \\ \sin q_1 \, \delta q_1 \\ 0 \\ 0 \\ 0 \\ \delta q_1 \end{bmatrix}$$

$$\delta\mathbf{Z}_3 = \begin{bmatrix} 2\cos q_1 \, \delta q_1 + \cos(q_1 + q_2)(\delta q_1 + \delta q_2) \\ 2\sin q_1 \, \delta q_1 + \sin(q_1 + q_2)(\delta q_1 + \delta q_2) \\ 0 \\ 0 \\ 0 \\ \delta q_1 + \delta q_2 \end{bmatrix}$$

and

$$\mathbf{M}_2 = \text{diag}\left[m_2,\ m_2,\ m_2,\ 0,\ 0,\ \frac{m_2}{3}\right]$$

$$\mathbf{M}_3 = \text{diag}\left[m_3,\ m_3,\ m_3,\ 0,\ 0,\ \frac{m_3}{3}\right]$$

$$\mathbf{Q}_2^A = [0,\ -m_2 g,\ 0,\ 0,\ 0,\ 0]^T$$

$$\mathbf{Q}_3^A = [0,\ -m_3 g,\ 0,\ 0,\ 0,\ 0]^T$$

Substituting, these results into Eq. 6.6.1, the variational equation of motion is

$$\sum_{i=1}^{3} \delta\mathbf{Z}_i^T(\mathbf{M}_i\dot{\mathbf{Y}}_i - \mathbf{Q}_i^A)$$

$$= \delta q_1\left\{\frac{4m_2}{3}\ddot{q}_1 + m_2 g\sin q_1 + m_3\left[\left(\frac{16}{3} + 4\cos q_2\right)\ddot{q}_1 + \left(\frac{4}{3} + 2\cos q_2\right)\ddot{q}_2\right.\right.$$

$$\left.\left. - 2\sin q_2(\dot{q}_1^2 + 2\dot{q}_1\dot{q}_2) + 2g\sin q_1 + g\sin(q_1 + q_2)\right]\right\}$$

$$+ \delta q_2\left\{m_3\left[\left(\frac{4}{3} + 2\cos q_2\right)\ddot{q}_1 + \frac{4}{3}\ddot{q}_2 + 2\sin q_2\dot{q}_1^2 + g\sin(q_1 + q_2)\right]\right\}$$

$$= \delta\mathbf{q}\left\{\begin{bmatrix} \frac{4m_2}{3} + m_3\left(\frac{16}{3} + 4\cos q_2\right) & m_3\left(\frac{4}{3} + 2\cos q_2\right) \\ m_3\left(\frac{4}{3} + 2\cos q_2\right) & \frac{4}{3}m_3 \end{bmatrix}\ddot{\mathbf{q}}\right.$$

$$\left. + \begin{bmatrix} -2m_3\sin q_2(\dot{q}_1^2 + 2\dot{q}_1\dot{q}_2) + m_2 g\sin q_1 + m_3 g[2\sin q_1 + \sin(q_1 + q_2)] \\ 2m_3\sin q_2\dot{q}_1^2 + m_3 g\sin(q_1 + q_2) \end{bmatrix}\right\} = 0$$

which holds for arbitrary $\delta\mathbf{q}$. Therefore, the differential equations of motion are

$$\begin{bmatrix} \frac{4m_2}{3} + m_3\left(\frac{16}{3} + 4\cos q_2\right) & m_3\left(\frac{4}{3} + 2\cos q_2\right) \\ m_3\left(\frac{4}{3} + 2\cos q_2\right) & \frac{4}{3}m_3 \end{bmatrix}\ddot{\mathbf{q}}$$

$$= -\begin{bmatrix} -2m_3\sin q_3(\dot{q}_1^2 + 2\dot{q}_1\dot{q}_2) + m_2 g\sin q_1 + m_3 g[2\sin q_1 + \sin(q_1 + q_2)] \\ 2m_3\sin q_2\dot{q}_1^2 + m_3 g\sin(q_1 + q_2) \end{bmatrix}$$

This is the same equation of motion as was obtained in Example 6.3.1. Note that since bodies are renumbered here, masses are renamed accordingly.

♦ **Example 6.6.2:** Consider next the spatial double pendulum of Example

6.5.6. Using the results of Example 6.5.6 gives

$$\mathbf{Y}_2 = \begin{bmatrix} \dot{\mathbf{r}}_2 \\ \boldsymbol{\omega}_2 \end{bmatrix} = \begin{bmatrix} -\sin q_1 \dot{q}_1 \\ \cos q_1 \dot{q}_1 \\ 0 \\ 0 \\ 0 \\ \dot{q}_1 \end{bmatrix}$$

$$\mathbf{Y}_3 = \begin{bmatrix} \dot{\mathbf{r}}_3 \\ \boldsymbol{\omega}_3 \end{bmatrix} = \begin{bmatrix} (-\sin q_1 - \cos q_1 \sin q_2)\dot{q}_1 - \sin q_1 \cos q_2 \dot{q}_2 \\ (\cos q_1 - \sin q_1 \sin q_2)\dot{q}_1 + \cos q_1 \cos q_2 \, \dot{q}_2 \\ \sin q_2 \, \dot{q}_2 \\ \cos q_1 \, \dot{q}_2 \\ \sin q_1 \, \dot{q}_2 \\ \dot{q}_1 \end{bmatrix}$$

The associated virtual displacements are

$$\delta \mathbf{Z}_1 = \mathbf{0}$$

$$\delta \mathbf{Z}_2 = \begin{bmatrix} -\sin q_1 \, \delta q_1 \\ \cos q_1 \, \delta q_1 \\ 0 \\ 0 \\ 0 \\ \delta q_1 \end{bmatrix}$$

$$\delta \mathbf{Z}_3 = \begin{bmatrix} -(\sin q_1 + \cos q_1 \sin q_2) \, \delta q_1 - \sin q_1 \cos q_2 \, \delta q_2 \\ (\cos q_1 - \sin q_1 \sin q_2) \, \delta q_1 + \cos q_1 \cos q_2 \, \delta q_2 \\ \sin q_2 \, \delta q_2 \\ \cos q_1 \, \delta q_2 \\ \sin q_1 \, \delta q_2 \\ \delta q_1 \end{bmatrix}$$

Differentiating $\mathbf{Y}_i$, accelerations of bodies 2 and 3 are

$$\dot{\mathbf{Y}}_2 = \begin{bmatrix} -\sin q_1 \ddot{q}_1 - \cos q_1 \dot{q}_1^2 \\ \cos q_1 \ddot{q}_1 - \sin q_1 \dot{q}_1^2 \\ 0 \\ 0 \\ 0 \\ \ddot{q}_1 \end{bmatrix}$$

$$\dot{\mathbf{Y}}_3 = \begin{bmatrix} -(\sin q_1 + \cos q_1 \sin q_2)\ddot{q}_1 - \sin q_1 \cos q_2 \ddot{q}_2 + \sin q_1 \sin q_2 \dot{q}_2^2 \\ (-\cos q_1 + \sin q_1 \sin q_2)\dot{q}_1^2 - 2 \cos q_1 \cos q_2 \dot{q}_1 \dot{q}_2 \\ (\cos q_1 - \sin q_1 \sin q_2)\ddot{q}_1 + \cos q_1 \cos q_2 \ddot{q}_2 - \cos q_1 \sin q_2 \dot{q}_2^2 \\ (-\sin q_1 - \cos q_1 \sin q_2)\dot{q}_1^2 - 2 \sin q_1 \cos q_2 \dot{q}_1 \dot{q}_2 \\ \sin q_2 \ddot{q}_2 + \cos q_2 \dot{q}_2^2 \\ \cos q_1 \ddot{q}_2 - \sin q_1 \dot{q}_1 \dot{q}_2 \\ \sin q_1 \ddot{q}_2 + \cos q_1 \dot{q}_1 \dot{q}_2 \\ \ddot{q}_1 \end{bmatrix}$$

From Example 6.3.4,

$$\mathbf{M}_2 = \left[\begin{array}{c|c} m_2\mathbf{I} & \mathbf{0} \\ \hline \mathbf{0} & \dfrac{m_2}{3}\begin{bmatrix} \sin^2 q_1 & 2\cos q_1 \sin q_1 & 0 \\ -\cos q_1 \sin q_1 & \cos^2 q_1 & 0 \\ 0 & 0 & 1 \end{bmatrix} \end{array}\right]$$

$$\mathbf{M}_3 = \left[\begin{array}{c|c} m_3\mathbf{I} & \mathbf{0} \\ \hline \mathbf{0} & \dfrac{m_3}{3}\begin{bmatrix} \cos^2 q_1 + \sin^2 q_1 \cos^2 q_2 & \sin q_1 \cos q_1 \sin^2 q_2 & -\sin q_1 \sin q_2 \cos q_2 \\ \sin q_1 \cos q_1 \sin^2 q_2 & \sin^2 q_1 + \cos^2 q_1 \cos^2 q_2 & \cos q_1 \sin q_2 \cos q_2 \\ -\sin q_1 \sin q_2 \cos q_2 & \cos q_1 \sin q_2 \cos q_2 & \sin^2 q_2 \end{bmatrix} \end{array}\right]$$

$$\mathbf{Q}_2^A = [0,\ 0,\ -m_2 g,\ 0,\ 0,\ 0]^T$$

$$\mathbf{Q}_3^A = \Big[0,\ 0,\ -m_3 g,\ \frac{m_3}{3}(\dot{q}_1^2 \cos q_1 \sin q_2 \cos q_2 + \dot{q}_1\dot{q}_2 \sin q_1 \cos^2 q_2),$$

$$\frac{m_3}{3}(\dot{q}_1^2 \sin q_1 \sin q_2 \cos q_2 - \dot{q}_1\dot{q}_2 \cos q_1 \cos^2 q_2),\ -\frac{m_3}{3}\dot{q}_1\dot{q}_2 \sin q_2 \cos q_2\Big]^T$$

Substituting these results into Eq. 6.6.1 yields

$$\begin{aligned}
&\sum_{i=1}^{3} \delta\mathbf{Z}_i^T[\mathbf{M}_i\dot{\mathbf{Y}}_i - \mathbf{Q}_i^A] \\
&\quad = \delta q_1\left[\frac{4}{3}m_2\ddot{q}_1\right] + \delta q_1\Big\{m_3[(1 + \sin^2 q_2)\ddot{q}_1 + \cos \ddot{q}_2\ddot{q}_2 - \sin q_2\dot{q}_2^2 \\
&\qquad + 2\cos q_2 \sin q_2\dot{q}_1\dot{q}_2] + \frac{m_3}{3}[\sin^2 q_2\ddot{q}_1 + 2\sin q_2 \cos q_2\dot{q}_1\dot{q}_2]\Big\} \\
&\qquad + \delta q_2\Big\{m_3[\cos q_2\ddot{q}_1 + \ddot{q}_2 - \sin q_2 \cos q_2\dot{q}_1^2] + \frac{m_3}{3}[\ddot{q}_2 - \sin q_2 \cos q_2\dot{q}_1^2] \\
&\qquad + m_3 g \sin q_2\Big\} \\
&\quad = \delta\mathbf{q}^T\left\{\begin{bmatrix} \frac{4}{3}m_2 + m_3(1 + \frac{4}{3}\sin^2 q_2) & m_3\cos q_2 \\ m_3\cos q_2 & \frac{4}{3}m_3 \end{bmatrix}\ddot{\mathbf{q}}\right. \\
&\qquad \left. + \begin{bmatrix} -m_3 \sin q_2(\dot{q}_2^2 + \frac{8}{3}\cos q_2\dot{q}_1\dot{q}_2) \\ m_3 g \sin q_2 - \frac{4}{3}\sin q_2 \cos q_2\dot{q}_1^2 \end{bmatrix}\right\} = 0
\end{aligned}$$

which holds for arbitrary $\delta\mathbf{q}$. Thus the coefficient of $\delta\mathbf{q}^T$ must equal zero, leading to the differential equations of motion

$$\begin{bmatrix} \frac{4}{3}m_2 + m_3(1 + \frac{4}{3}\sin^2 q_2) & m_3 \cos q_2 \\ m_3 \cos q_2 & \frac{4}{3}m_3 \end{bmatrix} \ddot{\mathbf{q}} = \begin{bmatrix} m_3 \sin q_2 (\dot{q}_2^2 - \frac{8}{3}\cos q_2 \dot{q}_1 \dot{q}_2) \\ -m_3 g \sin q_2 + \frac{4}{3}\sin q_2 \sin q_2 \dot{q}_1^2 \end{bmatrix}$$

Note that this equation of motion is the same as the result in Example 6.3.4, with masses renumbered.

Example 6.6.3: Consider the two body, three degree of freedom mechanism of Example 6.5.7. Using the results of Example 6.5.7, virtual displacements are

$$\delta\mathbf{Z}_1 = \mathbf{0}$$

$$\delta\mathbf{Z}_2 = \begin{bmatrix} -\sin q_1 \,\delta q_1 \\ \cos q_1 \,\delta q_1 \\ 0 \\ \delta q_1 \,\mathbf{k} \end{bmatrix}$$

$$\delta\mathbf{Z}_3 = \begin{bmatrix} [-(1 + q_2)\sin q_1 - \cos q_1 \cos q_3]\,\delta q_1 + \cos q_1 \delta q_2 + \sin q_1 \sin q_3 \delta q_3 \\ [(1 + q_2)\cos q_1 - \sin q_1 \cos q_3]\,\delta q_1 + \sin q_1 \,\delta q_2 - \cos q_1 \sin q_3 \,\delta q_3 \\ \cos q_3 \,\delta q_3 \\ \cos q_1 \,\delta q_3 \\ \sin q_1 \,\delta q_3 \\ \delta q_1 \end{bmatrix}$$

Differentiating $\mathbf{Y}_i$ yields

$$\dot{\mathbf{Y}}_2 = \begin{bmatrix} -\sin q_1 \ddot{q}_1 - \cos q_1 \dot{q}_1^2 \\ \cos q_1 \ddot{q}_1 - \sin q_1 \dot{q}_1^2 \\ 0 \\ \ddot{q}_1 \,\mathbf{k} \end{bmatrix}$$

$$\dot{\mathbf{Y}}_3 = \begin{bmatrix} [-(1 + q_2)\sin q_1 - \cos q_1 \cos q_3]\ddot{q}_1 + \cos q_1 \ddot{q}_2 + \sin q_1 \sin q_3 \ddot{q}_3 \\ \quad +[-(1 + q_2)\cos q_1 + \sin q_1 \cos q_3]\dot{q}_1^2 + \sin q_1 \cos q_3 \dot{q}_3^2 \\ \quad - 2\sin q_1 \dot{q}_1 \dot{q}_2 + 2\cos q_1 \sin q_3 \dot{q}_1 \dot{q}_3 \\ [(1 + q_2)\cos q_1 - \sin q_1 \cos q_3]\ddot{q}_1 + \sin q_1 \ddot{q}_2 - \cos q_1 \sin q_3 \ddot{q}_3 \\ \quad + [-(1 + q_2)\sin q_1 - \cos q_1 \cos q_3]\dot{q}_1^2 - \cos q_1 \cos q_3 \dot{q}_3^2 \\ \quad + 2\cos q_1 \dot{q}_1 \dot{q}_2 + 2\sin q_1 \sin q_3 \dot{q}_1 \dot{q}_3 \\ \cos q_3 \ddot{q}_3 - \sin q_3 \dot{q}_3^2 \\ \cos q_1 \ddot{q}_3 - \sin q_1 \dot{q}_1 \dot{q}_3 \\ \sin q_1 \ddot{q}_3 + \cos q_1 \dot{q}_1 \dot{q}_3 \\ \ddot{q}_1 \end{bmatrix}$$

Using the transformations and matrices $\mathbf{J}_i'$ and $\mathbf{F}_i$ from Example 6.3.6,

$$\mathbf{M}_2 = \left[\begin{array}{c|c} m_2\mathbf{I} & \mathbf{0} \\ \hline \mathbf{0} & \dfrac{m_2}{3}\begin{bmatrix} \sin^2 q_1 & -\sin q_1\cos q_1 & 0 \\ \sin q_1\cos q_2 & \cos^2 q_1 & 0 \\ 0 & 0 & 1 \end{bmatrix} \end{array}\right]$$

$$\mathbf{M}_3 = \left[\begin{array}{c|c} m_3\mathbf{I} & \mathbf{0} \\ \hline \mathbf{0} & \dfrac{m_3}{3}\begin{bmatrix} \sin q_1^2\sin q_3^2 + \cos^2 q_1 & \sin q_1\cos q_1\cos^2 q_3 & \sin q_1\sin q_3\cos q_3 \\ \sin q_1\cos q_1\cos^2 q_3 & \cos q_1^2\sin^2 q_3 + \sin^2 q_1 & -\cos q_1\sin q_3\cos q_3 \\ \sin q_1\sin q_3\cos q_3 & -\cos q_1\sin q_3\cos q_3 & \cos^2 q_3 \end{bmatrix} \end{array}\right]$$

$$\mathbf{Q}_2 = [0, 0, -m_2 g, 0, 0, 0]^T$$

$$\mathbf{Q}_3 = \Big[0, 0, -m_3 g, -\frac{m_3}{3}(\dot{q}_1^2\cos q_1\sin q_3\cos q_3 - \dot{q}_1\dot{q}_3\sin q_1\sin^2 q_3),$$
$$-\frac{m_3}{3}(\dot{q}_1^2\sin q_1\sin q_3\cos q_3 + \dot{q}_1\dot{q}_3\cos q_1\sin^2 q_3), \frac{m_3}{3}\dot{q}_1\dot{q}_3\sin q_3\cos q_3\Big]^T$$

Substituting these results into Eq. 6.6.1, the variational equation of motion is

$$\sum_{i=1}^{3}\delta\mathbf{Z}_i^T(\mathbf{M}_i\dot{\mathbf{Y}}_i - \mathbf{Q}_i^A)$$
$$= \delta q_1(\tfrac{4}{3}m_2\ddot{q}_1) + m_3\,\delta q_1\{[(1+q_2)^2 + \tfrac{4}{3}\cos^2 q_3]\ddot{q}_1 - \cos q_3\ddot{q}_2$$
$$- (1+q_2)\sin q_3\ddot{q}_3 - (1+q_2)\cos q_3\dot{q}_3^2 - \tfrac{8}{3}\sin q_3\cos q_3\dot{q}_1\dot{q}_3 + 2(1+q_2)\dot{q}_1\dot{q}_2\}$$
$$+ m_3\,\delta q_2\{-\cos q_3\ddot{q}_1 + \ddot{q}_2 - (1+q_2)\dot{q}_1^2 + 2\sin q_3\dot{q}_1\dot{q}_3\}$$
$$+ m_3\,\delta q_3\{-(1+q_2)\sin q_3\ddot{q}_1 + \tfrac{4}{3}\ddot{q}_3 + \tfrac{4}{3}\sin q_3\cos q_3\dot{q}_1^2 - 2\sin q_3\dot{q}_1\dot{q}_2$$
$$+ m_3 g\cos q_3\}$$

$$= \delta\mathbf{q}^T\left\{\begin{bmatrix} \tfrac{4}{3}(m_2 + m_3\cos^2 q_3) + m_3(1+q_2)^2 & -m_3\cos q_3 & -m_3(1+q_2)\sin q_3 \\ -m_3\cos q_3 & m_3 & 0 \\ -m_3(1+q_2)\sin q_3 & 0 & \tfrac{4}{3}m_3 \end{bmatrix}\ddot{\mathbf{q}}\right.$$
$$\left. + \begin{bmatrix} 2m_3(1+q_2)\dot{q}_1\dot{q}_2 - \tfrac{8}{3}m_3\sin q_3\cos q_3\dot{q}_1\dot{q}_2 - m_3(1+q_2)\cos q_3\dot{q}_3^2 \\ -m_3(1+q_2)\dot{q}_1^2 + 2m_3\sin q_3\dot{q}_1\dot{q}_3 \\ \tfrac{4}{3}m_3\sin q_3\cos q_3\dot{q}_1^2 - 2m_3\sin q_3\dot{q}_1\dot{q}_2 + m_3 g\cos q_3 \end{bmatrix}\right\} = 0$$

which must hold for arbitrary $\delta\mathbf{q}$. Thus the coefficient of $\delta\mathbf{q}^T$ must be zero, yielding the differential equations of motion

$$\begin{bmatrix} \frac{4}{3}(m_2 + m_3\cos^2 q_3) + m_3(1+q_2)^2 & -m_3\cos q_3 & -m_3(1+q_2)\sin q_3 \\ -m_3\cos q_3 & m_3 & 0 \\ -m_3(1+q_2)\sin q_3 & 0 & \frac{4}{3}m_3 \end{bmatrix}\ddot{\mathbf{q}}$$

$$= -\begin{bmatrix} 2m_3(1+q_2)\dot{q}_1\dot{q}_3 - \frac{8}{3}m_3\sin q_3\cos q_3\dot{q}_1\dot{q}_3 - m_3(1+q_2)\cos q_3\dot{q}_3^2 \\ -m_3(1+q_2)\dot{q}_1^2 + 2m_3\sin q_3\dot{q}_1\dot{q}_3 \\ \frac{4}{3}m_3\sin q_3\cos q_3\dot{q}_1^2 - 2m_3\sin q_3\dot{q}_1\dot{q}_2 + m_3 g\cos q_3 \end{bmatrix}$$

Note that this equation of motion is the same as the result in Example 6.3.6, with the masses renumbered.

Differential-Algebraic Equations of Motion for Systems with Closed Loops

As in Section 6.5, if the graph of a system has closed loops, a tree structure is formed by defining cut-joints. The relative coordinate kinematics formulation of Section 6.5 may then be developed for the resulting spanning tree, so it neglects the effect of kinematic constraints between bodies that were connected by cut-joints. While the variational equation of motion of Eq. 6.6.5 is still valid, it holds only for relative coordinate variations $\delta\mathbf{q}$ that are consistent with cut-joint constraints. It is, therefore, necessary to introduce Lagrange multipliers associated with cut-joint constraint equations in the equations of motion.

Consider first a cut-joint between bodies i and j. Kinematic constraint equations for this cut-joint are of the form

$$\mathbf{\Phi}^{(ij)}(\mathbf{r}_i, \mathbf{A}_i, \mathbf{r}_j, \mathbf{A}_j) = \mathbf{0} \tag{6.6.7}$$

which may be written in terms of generalized coordinates of bodies i and j only. Taking the variation of Eq. 6.6.7 yields

$$\mathbf{\Phi}^{(ij)}_{\mathbf{r}_i}\,\delta\mathbf{r}_i + \mathbf{\Phi}^{(ij)}_{\boldsymbol{\pi}_i}\,\delta\boldsymbol{\pi}_i + \mathbf{\Phi}^{(ij)}_{\mathbf{r}_j}\,\delta\mathbf{r}_j + \mathbf{\Phi}^{(ij)}_{\boldsymbol{\pi}_j}\,\delta\boldsymbol{\pi}_j = \mathbf{0} \tag{6.6.8}$$

which is written in terms of the global representation of virtual rotation. Using the relation $\delta\boldsymbol{\pi}_i' = \mathbf{A}_i^T\,\delta\boldsymbol{\pi}_i$, the variation of a constraint function, in terms of global representation of virtual rotation, may be written in the form

$$\mathbf{\Phi}^{(ij)}_{\boldsymbol{\pi}_i} = \mathbf{\Phi}^{(ij)}_{\boldsymbol{\pi}_i'}\mathbf{A}_i^T \tag{6.6.9}$$

where the variation of the constraint function with respect to virtual rotation in the body-fixed reference frame is defined for a variety of kinematic constraints in Section 6.2.

To take advantage of the relations developed in Section 6.5, the constraint Jacobian is written in terms of variation $\delta\mathbf{Z}_i$ as

$$\mathbf{\Phi}^{(ij)}_{\mathbf{Z}_i} \equiv [\mathbf{\Phi}^{(ij)}_{\mathbf{r}_i}, \mathbf{\Phi}^{(ij)}_{\boldsymbol{\pi}_i}] \tag{6.6.10}$$

and similarly for $\delta \mathbf{Z}_j$. Thus the constraint Jacobian with respect to system Cartesian coordinates is

$$\mathbf{\Phi}_{\mathbf{Z}}^{(ij)} = [\mathbf{0}, \ldots, \mathbf{\Phi}_{\mathbf{Z}_i}^{(ij)}, \ldots, \mathbf{\Phi}_{\mathbf{Z}_j}^{(ij)}, \ldots, \mathbf{0}] \tag{6.6.11}$$

which contains mostly zeros.

The collection of all *cut-joint constraint equations* may be written in vector form as

$$\mathbf{\Phi} = [\mathbf{\Phi}^{(ij)}] = \mathbf{0} \tag{6.6.12}$$

Combining the variation of individual constraints of Eq. 6.6.8, using the notation of Eqs. 6.6.11 and 6.6.12, and taking the variation of Eq. 6.6.12 yields

$$\mathbf{\Phi}_{\mathbf{Z}}\, \delta\mathbf{Z} = \mathbf{0} \tag{6.6.13}$$

Using the relation of Eq. 6.5.48 between Cartesian virtual displacement and relative coordinate variations, Eq. 6.6.13 may be written in the form

$$\mathbf{\Phi}_{\mathbf{Z}}\hat{\mathbf{B}}\, \delta\mathbf{q} \equiv \hat{\mathbf{\Phi}}_{\mathbf{q}}\, \delta\mathbf{q} = \mathbf{0} \tag{6.6.14}$$

which must be satisfied by kinematically admissible relative coordinate variations $\delta\mathbf{q}$.

Taking the time derivative of both sides of Eq. 6.6.12 yields the *cut-joint velocity equation*

$$\mathbf{\Phi}_{\mathbf{Z}}\mathbf{Y} = \mathbf{0} \tag{6.6.15}$$

where the notation of Eqs. 6.5.31 and 6.5.43 has been used. Using Eq. 6.5.45, Eq. 6.6.15 may be written in the form

$$\mathbf{\Phi}_{\mathbf{Z}}\hat{\mathbf{B}}\dot{\mathbf{q}} = \hat{\mathbf{\Phi}}_{\mathbf{q}}\dot{\mathbf{q}} = \mathbf{0} \tag{6.6.16}$$

Similarly, differentiating Eq. 6.6.15 with respect to time yields the *cut-joint acceleration equation*

$$\mathbf{\Phi}_{\mathbf{Z}}\dot{\mathbf{Y}} = -\{(\mathbf{\Phi}_{\mathbf{Z}}\mathbf{Y})_{\mathbf{Z}}\mathbf{Y}\} \equiv \boldsymbol{\gamma} \tag{6.6.17}$$

Using the relation of Eq. 6.5.46, Eq. 6.6.17 may be written in terms of relative coordinates as

$$\mathbf{\Phi}_{\mathbf{Z}}\hat{\mathbf{B}}\ddot{\mathbf{q}} = \hat{\mathbf{\Phi}}_{\mathbf{q}}\ddot{\mathbf{q}} = \boldsymbol{\gamma} - \mathbf{\Phi}_{\mathbf{Z}}\dot{\hat{\mathbf{B}}}\dot{\mathbf{q}} \equiv \hat{\boldsymbol{\gamma}} \tag{6.6.18}$$

which is the cut-joint acceleration equation in terms of relative coordinates.

Adopting the convention presented in the preceding subsection of suppressing generalized coordinates of a grounded base body, for either free or grounded base bodies, $\delta\mathbf{q}$ is required to satisfy only variational cut-joint constraint equations of Eq. 6.6.14. Thus Eq. 6.6.5 must hold for all $\delta\mathbf{q}$ that satisfy Eq. 6.6.14. By the Lagrange multiplier theorem, there exists a vector $\boldsymbol{\lambda}$ such that

$$\delta\mathbf{q}^T(\hat{\mathbf{M}}\ddot{\mathbf{q}} - \hat{\mathbf{Q}}^A + \hat{\mathbf{\Phi}}_{\mathbf{q}}^T\boldsymbol{\lambda})\, \delta\mathbf{q} = 0 \tag{6.6.19}$$

for arbitrary $\delta\mathbf{q}$. Thus the coefficient of $\delta\mathbf{q}^T$ in Eq. 6.6.19 must be zero, yielding the Lagrange multiplier form of the equations of motion for systems with closed loops,

$$\hat{\mathbf{M}}\ddot{\mathbf{q}} - \hat{\mathbf{Q}}^A + \hat{\mathbf{\Phi}}_{\mathbf{q}}^T\boldsymbol{\lambda} = \mathbf{0} \tag{6.6.20}$$

Combining Eqs. 6.6.18 and 6.6.20 yields the differential-algebraic equations of motion in matrix form,

$$\begin{bmatrix} \hat{\mathbf{M}} & \hat{\mathbf{\Phi}}_{\mathbf{q}}^T \\ \hat{\mathbf{\Phi}}_{\mathbf{q}} & \mathbf{0} \end{bmatrix} \begin{bmatrix} \ddot{\mathbf{q}} \\ \boldsymbol{\lambda} \end{bmatrix} = \begin{bmatrix} \hat{\mathbf{Q}}^A \\ \hat{\boldsymbol{\gamma}} \end{bmatrix} \tag{6.6.21}$$

which determines relative coordinate accelerations and Lagrange multipliers. The solution $\mathbf{q}$ and $\dot{\mathbf{q}}$ of this equation must, of course, satisfy the cut-joint constraint equations of Eq. 6.6.12 and the cut-joint velocity equations of Eq. 6.6.16.

While the form of the equations of motion of Eq. 6.6.21 in relative coordinates is similar to the constrained Cartesian equations of motion of Eq. 6.2.32, significant differences in the nature of these equations lead to very different methods of solution. First, the coefficient matrix of Eq. 6.2.32 in the Cartesian formulation is made up primarily of zeros. Hence it is highly sparse and can be treated by modern sparse matrix methods of matrix factorization [14]. The number of equations in the Cartesian formulation, however, is quite large and the dimension of the Lagrange multiplier in Eq. 6.2.32 is often very large, leading to the requirement for solution of a large system of nonlinear kinematic constraint equations. Numerical formulation of Eq. 6.2.32 is elementary, since the form of individual terms is quite simple.

In contrast to characteristics of the Cartesian equations of Eq. 6.2.32, the relative coordinate equations of Eq. 6.6.21 involve far fewer variables and fewer constraint equations. The coefficient matrix in Eq. 6.6.21 is, however, less sparse and is amenable primarily to full matrix solution methods. Of greater importance, evaluation of terms in the coefficient matrix of Eq. 6.6.21 is a demanding computational task. Prior to the advent of symbolic computation methods [15], evaluation of these terms for realistic systems was oppressive. Modern symbolic computation programs, however, carry out the algebraic manipulations required to form executable computer code to evaluate such terms. Emergence of symbolic computation tools makes evaluation of the equations of motion in relative coordinates practical and computationally attractive.

6.7 RECURSIVE FORMULATION FOR DYNAMICS OF MULTIBODY SYSTEMS WITH TREE STRUCTURE

A recently developed recursive algorithm [13, 20] for formulating and solving equations of motion is presented in this section, for systems with tree structure. For extension of this algorithm for systems with closed loops, see Refs. 13 and 21. Presuming Cartesian coordinates and velocity of the base body and relative coordinates and their time derivatives are known, as would be the case during numerical integration, the principal task addressed in this section is formulation and solution of the equations of motion for second derivatives of relative coordinates and Cartesian accelerations of the base body. Since this process is the most computationally intensive component of dynamic analysis, computational efficiency is the goal of the method presented.

The algorithm presented here makes heavy use of the graph structure of a multibody system. It begins with position and velocity analysis that starts at the base body and moves out the tree, along all paths to tree-end bodies, to obtain Cartesian position and orientation information for each body, using Eqs. 6.5.2, 6.5.5, 6.5.6, and 6.5.8. Following completion of this computation, the variational equation of motion for the system, in terms of Cartesian coordinates, is reduced to relative coordinate form, beginning with tree-end bodies and moving systematically toward the base body rather than in the reverse order used in the velocity transformation method of Section 6.6. As part of this algorithm, relative coordinate accelerations are determined as functions of inboard body Cartesian accelerations, beginning at tree-end bodies and proceeding inward to the base body. Upon completion of this task, the base body equations of motion are solved and relations between inboard body Cartesian accelerations and relative coordinate accelerations are evaluated in a sequence that begins at the base body and moves outward to all tree-end bodies. This three pass computational sequence efficiently evaluates base body Cartesian accelerations and relative coordinate accelerations throughout the system, for use in subsequent numerical integration.

Equations of Motion for a Single Chain

A tree structure mechanism consists of several chains. The structure of a typical chain β, beginning at junction node l and proceeding to a tree-end node n, is shown schematically in Fig. 6.7.1, with sequential body numbering. This chain is used to derive recursive formulas for reduced variational equations of motion. Equations of motion for the entire tree structure system are obtained using results from each chain computation.

The variational equation of motion for bodies in this chain is obtained, using

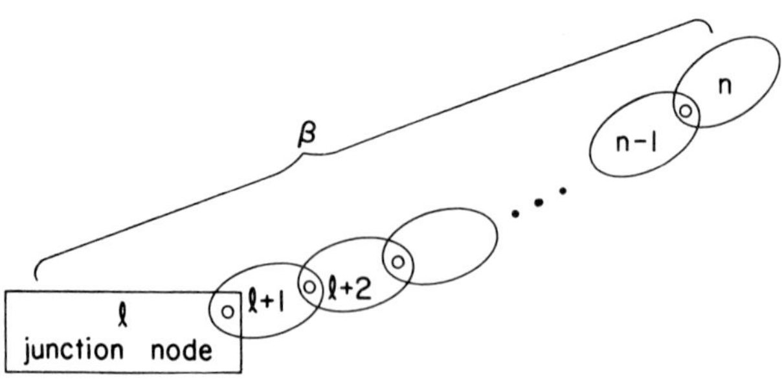

Figure 6.7.1 The βth chain.

notations of Eqs. 6.5.29 and 6.6.2. Adding variational equations for each body yields

$$\sum_{i=l}^{n} \delta \mathbf{Z}_i^T (\mathbf{M}_i \dot{\mathbf{Y}}_i - \hat{\mathbf{Q}}_i^A) = 0 \tag{6.7.1}$$

which must hold for all kinematically admissible virtual displacements $\delta \mathbf{Z}_i$.

Kinematically admissible virtual displacements $\delta \mathbf{Z}_i$ must satisfy constraints between bodies l to n in Fig. 6.7.1 and any other constraints that may act on junction body l. Even if not mentioned explicitly, these kinematic admissibility conditions on $\delta \mathbf{Z}_i$ must be satisfied throughout this section.

Subsituting Eqs. 6.5.30 and 6.5.32, with $i = n - 1$ and $j = n$, into Eq. 6.7.1 yields

$$\begin{aligned} &\sum_{i=l}^{n-1} \delta \mathbf{Z}_i^T (\mathbf{M}_i \dot{\mathbf{Y}}_i - \mathbf{Q}_i^A) + \delta \mathbf{Z}_{n-1}^T [\mathbf{B}_{(n-1)n1}^T (\mathbf{M}_n \mathbf{B}_{(n-1)n1} \dot{\mathbf{Y}}_{n-1} \\ &\quad + \mathbf{M}_n \mathbf{B}_{(n-1)n2} \ddot{\mathbf{q}}_{(n-1)n} + \mathbf{M}_n \mathbf{D}_{(n-1)n} - \hat{\mathbf{Q}}_n^A)] \\ &\quad + \delta \mathbf{q}_{(n-1)n}^T [\mathbf{B}_{(n-1)n2}^T (\mathbf{M}_n \mathbf{B}_{(n-1)n1} \dot{\mathbf{Y}}_{n-1} + \mathbf{M}_n \mathbf{B}_{(n-1)n2} \ddot{\mathbf{q}}_{(n-1)n} \\ &\quad + \mathbf{M}_n \mathbf{D}_{(n-1)n} - \hat{\mathbf{Q}}_n^A)] = 0 \end{aligned} \tag{6.7.2}$$

where $\delta \mathbf{Z}_i$ are kinematically admissible for joints $i = l, \ldots, n - 2$ and $\delta \mathbf{q}_{(n-1)n}$ is arbitrary. This is true since $\delta \mathbf{Z}_n$ of Eq. 6.5.30 is kinematically admissible for a kinematically admissible $\delta \mathbf{Z}_{n-1}$ and any relative coordinate variation $\delta \mathbf{q}_{(n-1)n}$. Thus the coefficient of $\delta \mathbf{q}_{(n-1)n}^T$ in Eq. 6.7.2 must be zero. As a result, the following expression for $\ddot{\mathbf{q}}_{(n-1)n}$ is obtained:

$$\begin{aligned} \ddot{\mathbf{q}}_{(n-1)n} = &-[\mathbf{B}_{(n-1)n2}^T \mathbf{M}_n \mathbf{B}_{(n-1)n2}]^{-1} [\mathbf{B}_{(n-1)n2}^T \mathbf{M}_n \mathbf{B}_{(n-1)n1} \dot{\mathbf{Y}}_{n-1} \\ &+ \mathbf{B}_{(n-1)n2}^T (\mathbf{M}_n \mathbf{D}_{(n-1)n} - \hat{\mathbf{Q}}_n^A)] \end{aligned} \tag{6.7.3}$$

where existence of $[\mathbf{B}_{(n-1)n2}^T \mathbf{M}_n \mathbf{B}_{(n-1)n2}]^{-1}$ is proved in Ref. 20.

Equation 6.7.3 can be substituted into Eq. 6.7.2 to obtain

$$\sum_{i=l}^{n-2} \delta \mathbf{Z}_i^T (\mathbf{M}_i \dot{\mathbf{Y}}_i - \hat{\mathbf{Q}}_i^A) + \delta \mathbf{Z}_{n-1}^T [(\mathbf{M}_{n-1} + \mathbf{K}_{n-1}) \dot{\mathbf{Y}}_{n-1} - (\hat{\mathbf{Q}}_{n-1}^A + \mathbf{L}_{n-1})] = 0 \tag{6.7.4}$$

where $\delta \mathbf{Z}_i$ are kinematically admissible for joints $l, \ldots, n - 2$ and

$$\begin{aligned} \mathbf{K}_{n-1} &= \mathbf{B}_{(n-1)n1}^T \mathbf{M}_n \mathbf{B}_{(n-1)n1} - \mathbf{B}_{(n-1)n1}^T \mathbf{M}_n \mathbf{B}_{(n-1)n2} \\ &\quad \times (\mathbf{B}_{(n-1)n2}^T \mathbf{M}_n \mathbf{B}_{(n-1)n2})^{-1} (\mathbf{B}_{(n-1)n2}^T \mathbf{M}_n \mathbf{B}_{(n-1)n1}) \\ \mathbf{L}_{n-1} &= \mathbf{B}_{(n-1)n1}^T (\hat{\mathbf{Q}}_n^A - \mathbf{M}_n \mathbf{D}_{(n-1)n}) + \mathbf{B}_{(n-1)n1}^T \mathbf{M}_n \mathbf{B}_{(n-1)n2} \\ &\quad \times (\mathbf{B}_{(n-1)n2}^T \mathbf{M}_n \mathbf{B}_{(n-1)n2})^{-1} \mathbf{B}_{(n-1)n2}^T (\mathbf{M}_n \mathbf{D}_{(n-1)n} - \hat{\mathbf{Q}}_n^A) \end{aligned} \tag{6.7.5}$$

It is proved in Ref. 20 that $\mathbf{K}_{n-1}$ is positive definite.

If this procedure is repeated down the chain to body $m + 1$, it can be assumed (for proof by induction) that the following variational form of the equations of motion is obtained:

$$\sum_{i=l}^{m} \delta \mathbf{Z}_i^T(\mathbf{M}_i\dot{\mathbf{Y}}_i - \hat{\mathbf{Q}}_i^A) + \delta \mathbf{Z}_{m+1}^T[(\mathbf{M}_{m+1} + \mathbf{K}_{m+1})\dot{\mathbf{Y}}_{m+1} - (\hat{\mathbf{Q}}_{m+1}^A + \mathbf{L}_{m+1})] = 0 \tag{6.7.6}$$

where virtual displacements $\delta \mathbf{Z}_i$ are kinematically admissible for joints between bodies l to $m + 1$, but are free from kinematic admissibility conditions imposed by constraints between bodies $m + 1$ and n in Fig. 6.7.1, because these kinematic constraints are satisfied by the relative coordinates.

Substituting Eqs. 6.5.30 and 6.5.32 into Eq. 6.7.6, with $i = m$ and $j = m + 1$,

$$\begin{aligned}
&\sum_{i=l}^{m-1} \delta \mathbf{Z}_i^T(\mathbf{M}_i\dot{\mathbf{Y}}_i - \hat{\mathbf{Q}}_i^A) + \delta \mathbf{Z}_m^T\{\mathbf{B}_{(m)(m+1)1}^T[\mathbf{M}_{m+1} + \mathbf{K}_{m+1})(\mathbf{B}_{m(m+1)1}\dot{\mathbf{Y}}_m \\
&\quad + \mathbf{B}_{m(m+1)2}\ddot{\mathbf{q}}_{m(m+1)} + \mathbf{D}_{m(m+1)}) - (\hat{\mathbf{Q}}_{m+1}^A + \mathbf{L}_{m+1})] + \mathbf{M}_m\dot{\mathbf{Y}}_m - \hat{\mathbf{Q}}_m^A\} \\
&\quad + \delta \mathbf{q}_{m(m+1)}^T\{\mathbf{B}_{m(m+1)2}^T[(\mathbf{M}_{m+1} + \mathbf{K}_{m+1})(\mathbf{B}_{m(m+1)1}\dot{\mathbf{Y}}_m \\
&\quad + \mathbf{B}_{m(m+1)2}\ddot{\mathbf{q}}_{m(m+1)} + \mathbf{D}_{m(m+1)}) - (\hat{\mathbf{Q}}_{m+1}^A + \mathbf{L}_{m+1})]\} = 0
\end{aligned} \tag{6.7.7}$$

where $\delta \mathbf{Z}_i$ are kinematically admissible for joints between bodies l to m and $\delta \mathbf{q}_{m(m+1)}$ is arbitrary. Therefore, the coefficient of $\delta \mathbf{q}_{m(m+1)}^T$ must be zero, yielding

$$\begin{aligned}
\ddot{\mathbf{q}}_{m(m+1)} = &-[\mathbf{B}_{m(m+1)2}^T(\mathbf{M}_{m+1} + \mathbf{K}_{m+1})\mathbf{B}_{m(m+1)2}]^{-1} \\
&\times \{\mathbf{B}_{m(m+1)2}^T(\mathbf{M}_{m+1} + \mathbf{K}_{m+1})\, \mathbf{B}_{m(m+1)1}\dot{\mathbf{Y}}_m \\
&+ \mathbf{B}_{m(m+1)2}^T[(\mathbf{M}_{m+1} + \mathbf{K}_{m+1})\mathbf{D}_{m(m+1)} - (\hat{\mathbf{Q}}_{m+1}^A + \mathbf{L}_{m+1})]\}
\end{aligned} \tag{6.7.8}$$

In Eq. 6.7.8, existence of $[\mathbf{B}_{m(m+1)2}^T(\mathbf{M}_{m+1} + \mathbf{K}_{m+1})\mathbf{B}_{m(m+1)2}]^{-1}$ can be proved, as in Ref. 20, where $\mathbf{K}_{m+1}$ is positive definite by the induction assumption.

If Eq. 6.7.8 is substituted into the coefficient of $\delta \mathbf{Z}_m^T$ in Eq. 6.7.6, to eliminate dependence on $\ddot{\mathbf{q}}_{m(m+1)}$, the following form of the variational equation of motion is obtained:

$$\sum_{i=l}^{m-1} \delta \mathbf{Z}_i^T(\mathbf{M}_i\dot{\mathbf{Y}}_i - \hat{\mathbf{Q}}_i^A) + \delta \mathbf{Z}_m^T[(\mathbf{K}_m + \mathbf{M}_m)\dot{\mathbf{Y}}_m - (\hat{\mathbf{Q}}_m^A + \mathbf{L}_m)] = 0 \tag{6.7.9}$$

where

$$\begin{aligned}
\mathbf{K}_m &= \mathbf{B}^T_{m(m+1)1}(\mathbf{M}_{m+1} + \mathbf{K}_{m+1})\mathbf{B}_{m(m+1)1} - \mathbf{B}^T_{m(m+1)1}(\mathbf{M}_{m+1} + \mathbf{K}_{m+1})\mathbf{B}_{m(m+1)2} \\
&\quad \times [\mathbf{B}^T_{m(m+1)2}(\mathbf{M}_{m+1} + \mathbf{K}_{m+1})\mathbf{B}_{m(m+1)2}]^{-1} \\
&\quad \times [\mathbf{B}^T_{m(m+1)2}(\mathbf{M}_{m+1} + \mathbf{K}_{m+1})\mathbf{B}_{m(m+1)1}] \\
\mathbf{L}_m &= \mathbf{B}^T_{m(m+1)1}\{-(\mathbf{M}_{m+1} + \mathbf{K}_{m+1})\mathbf{D}_{m(m+1)} + \hat{\mathbf{Q}}^A_{m+1} + \mathbf{L}_{m+1} \\
&\quad + (\mathbf{M}_{m+1} + \mathbf{K}_{m+1})\mathbf{B}_{m(m+1)2}[\mathbf{B}^T_{m(m+1)2}(\mathbf{M}_{m+1} + \mathbf{K}_{m+1})\mathbf{B}_{m(m+1)2}]^{-1} \\
&\quad \times \mathbf{B}^T_{m(m+1)2}[(\mathbf{M}_{m+1} + \mathbf{K}_{m+1})\mathbf{D}_{m(m+1)} - (\hat{\mathbf{Q}}^A_{m+1} + \mathbf{L}_{m+1})]\}
\end{aligned} \tag{6.7.10}$$

and $\delta\mathbf{Z}_i$ are kinematically admissible for joints between bodies l to m.

By induction, if the reduction is continued to junction body l, the equations of motion can be expressed as

$$\delta\mathbf{Z}_l^T[(\mathbf{M}_l + \mathbf{K}_l^\beta)\dot{\mathbf{Y}}_l - (\hat{\mathbf{Q}}_l^A + \mathbf{L}_l^\beta)] = 0 \tag{6.7.11}$$

where $\delta\mathbf{Z}_l$ is kinematically admissible for all constraints that act on body l, other than those associated with chain β from body l to body n in Fig. 6.7.1.

Equations of Motion for a System with Tree Structure

The reduced variational equation of motion for chain β has now been developed. Since a tree structure consists of several such chains, the variational equation of motion for the system can be obtained by combining variational equations of motion from all chains.

The variational equation of motion of Eq. 6.7.11 may be expanded to account for all chains that begin at junction node l and end at tree-end bodies, by adding variational terms for each such chain to form a reduced variational equation at node l. To illustrate this, the example in Fig. 6.5.5 is used. The variational equations from chains 1 and 2 of Fig. 6.5.5(b) are reduced to junction node 3, yielding

$$\delta\mathbf{Z}_3^T[(\mathbf{M}_3 + \mathbf{K}_3^1 + \mathbf{K}_3^2)\dot{\mathbf{Y}}_3 - (\hat{\mathbf{Q}}_3^A + \mathbf{L}_3^1 + \mathbf{L}_3^2)] = 0 \tag{6.7.12}$$

Equation 6.7.12 can be added to the variational equations of motion for bodies in chain 3 of Fig. 6.5.5(b), to obtain

$$\sum_{i=1}^{2} \delta\mathbf{Z}_i^T(\mathbf{M}_i\dot{\mathbf{Y}}_i - \hat{\mathbf{Q}}_i^A) + \delta\mathbf{Z}_3^T[(\mathbf{M}_3 + \mathbf{K}_3^1 + \mathbf{K}_3^2)\dot{\mathbf{Y}}_3 - (\hat{\mathbf{Q}}_3^A + \mathbf{L}_3^1 + \mathbf{L}_3^2)] = 0 \tag{6.7.13}$$

where $\delta\mathbf{Z}_i$, $i = 1, 2,$ and 3, are kinematically admissible for constraints between bodies 1 to 3 and all other constraints that act on body 1. Equation 6.7.13

can be reduced, using the chain recursive algorithm with $\mathbf{M}_3 + \mathbf{K}_3^1 + \mathbf{K}_3^2$ and $\mathbf{Q}_3^A + \mathbf{L}_3^1 + \mathbf{L}_3^2$ in place of $\mathbf{M}_n$ and $\hat{\mathbf{Q}}_n^A$, respectively, in Eq. 6.7.2, to obtain the reduced variational equation of body 1 as

$$\delta\mathbf{Z}_1^T[(\mathbf{M}_1 + \mathbf{K}_1^3)\dot{\mathbf{Y}}_1 - (\hat{\mathbf{Q}}_1^A + \mathbf{L}_1^3)] = 0 \tag{6.7.14}$$

where $\delta\mathbf{Z}_1$ is kinematically admissible for joints in chain 4 of Fig. 6.5.5(b) and any external constraints that act on body 1.

Following the preceding approach, reduced variational equations can be obtained by adding the variational equations for bodies along chain 4 of Fig. 6.5.5(b) (nodes 10-9-8-7) to Eq. 6.7.14, to obtain

$$\delta\mathbf{Z}_1^T[(\mathbf{M}_1 + \mathbf{K}_1^3 + \mathbf{K}_1^4)\dot{\mathbf{Y}}_1 - (\hat{\mathbf{Q}}_1^A + \mathbf{L}_1^3 + \mathbf{L}_1^4)] = 0 \tag{6.7.15}$$

which must hold for all $\delta\mathbf{Z}_1$ that are consistent with any external constraints that act on body 1 (i.e., other than those between body 1 and the attached chains).

Two types of external base body constraints can be considered. First, the base body may be free (i.e., it may move without external constraint), in which case $\delta\mathbf{Z}_1$ in Eq. 6.7.15 is arbitrary. As a result,

$$\mathbf{K}_1\dot{\mathbf{Y}}_1 = \mathbf{L}_1 \tag{6.7.16}$$

where

$$\begin{aligned} \mathbf{K}_1 &= \mathbf{M}_1 + \mathbf{K}_1^4 + \mathbf{K}_1^3 \\ \mathbf{L}_1 &= \hat{\mathbf{Q}}_1^A + \mathbf{L}_1^3 + \mathbf{L}_1^4 \end{aligned} \tag{6.7.17}$$

Equation 6.7.16 can now be solved to find base body accelerations.

Second, the base body may be ground, or motion of the base body may be fully determined by driving constraints, in which case $\mathbf{Y}_1$ is known.

The reduction procedure above, described for the example in Fig. 6.5.5, can be generalized. First, the reduced variational equations for first level junction nodes from tree-end bodies can be obtained by adding contributions from the variational equation for each chain that emanates from each junction node. The reduced variational equations of the next level of inboard junction nodes can be calculated by starting from first level junction nodes and, if any, tree-end chains that are connected to each junction node. This procedure is repeated for every chain in working back to the base body. The reduced variational equation for the base body is thus obtained as in Eq. 6.7.15.

Solution of the System Equations of Motion

Two stages of computation are employed to form and solve the reduced equations for accelerations of the system. First, reduced joint and base body equations are formed, starting from tree-end bodies and working back to the base body. Acceleration of the base body is then calculated for any of the types of external constraints considered. Second, after the base body acceleration is found, formulas for relative coordinate and Cartesian accelerations in Eqs. 6.7.8 and 6.5.27

are used to calculate relative coordinate accelerations, starting from the base body and proceeding toward tree-end bodies, along all chains that emanate from the base body. Base body and joint accelerations may now be integrated to find velocities and positions at the next time step.

As a computational consideration, parallel processors are beginning to appear that can be used to advantage with the algorithm presented here. In contrast to classical serial computers, parallel processors can carry out simultaneous independent computations. Independence in computing reduced inertias and right-side terms in different branches of a complicated mechanism lead naturally to use of parallel processors. To use a parallel processor in calculating reduced inertia and right-side terms, a reduction sequence must be defined to minimize waiting time for subsequent reductions. As an example, the inertia and right-side terms for bodies 5, 6, and 10 in Fig. 6.5.5 can simultaneously be reduced to equations for bodies 3, 4, and 9, due to their independence. Candidates for second level parallel reduction are bodies that are just inboard of the previous group (i.e., bodies 3, 4, and 9). However, variational terms associated with body 3 cannot be reduced further until terms from all chains emanating from body 3 are calculated. Thus the second group is bodies 4 and 9. The third group is bodies 3 and 8. The fourth group is bodies 2 and 7. Finally, the reduced inertia and right-side terms for body 1 can be calculated to find base body accelerations.

Parallel computations in reverse order can be applied to recover body and joint accelerations. For example, accelerations of bodies 2 and 7 and joint accelerations between these bodies and the base body in Fig. 6.5.5 can be computed in parallel, once base body accelerations are available. Therefore, bodies 2 and 7 become the first parallel group in the acceleration recovery sequence. After computing accelerations for bodies 2 and 7, joint accelerations between these bodies and bodies 3 and 8 can be computed in parallel. Similarly, the third group is bodies 4, 6, and 9 and the fourth group is bodies 5 and 10. For more details on parallel processor applications, see Ref. 22.

Example 6.7.1: The three body mechanism of Fig. 6.7.2 is presented to demonstrate use of the recursive formulation. Motion is planar, to simplify the derivation of formulas. The base body is designated as body 1, which has no externally applied constraints. A revolute joint connects bodies 1 and 2. Body 3 translates with respect to body 2. Therefore, this system has five degrees of freedom, including three base body Cartesian degrees of freedom.

Joint reference frames are parallel to centroidal reference frames, so matrices $\mathbf{C}_{ij}$ are identities. Joint definition points are unit distances from the centroidal reference frames for each body, to simplify vectors $\mathbf{s}_{ij}$ and $\mathbf{s}_{ji}$.

The position and orientation of body 2 can be expressed in terms of the position and orientation of body 1 and relative coordinate θ_{12} as

$$\begin{bmatrix} x_2 \\ y_2 \end{bmatrix} = \begin{bmatrix} x_1 \\ y_1 \end{bmatrix} + \begin{bmatrix} \cos\phi_1 & -\sin\phi_1 \\ \sin\phi_1 & \cos\phi_1 \end{bmatrix} \begin{bmatrix} 1 \\ 0 \end{bmatrix} - \begin{bmatrix} \cos\phi_2 & -\sin\phi_2 \\ \sin\phi_2 & \cos\phi_2 \end{bmatrix} \begin{bmatrix} -1 \\ 0 \end{bmatrix}$$

$$\phi_2 = \phi_1 + \theta_{12} \tag{6.7.18}$$

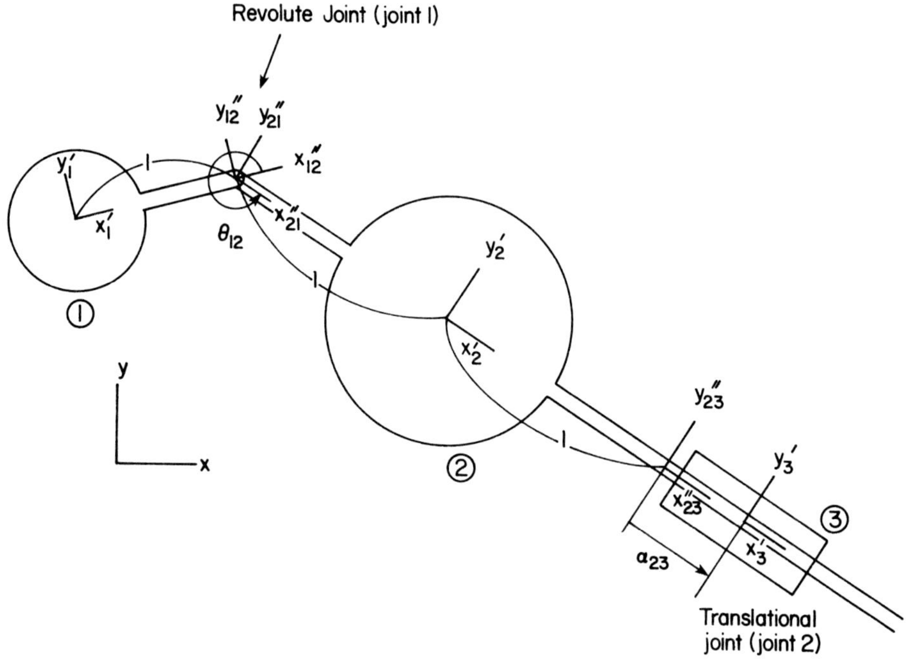

Figure 6.7.2 Three-body mechanism.

Similarly, the position and orientation of body 3 can be obtained as

$$\begin{bmatrix} x_3 \\ y_3 \end{bmatrix} = \begin{bmatrix} x_2 \\ y_2 \end{bmatrix} + \begin{bmatrix} \cos\phi_2 & -\sin\phi_2 \\ \sin\phi_2 & \cos\phi_2 \end{bmatrix} \begin{bmatrix} 1 + \alpha_{23} \\ 0 \end{bmatrix} \tag{6.7.19}$$
$$\phi_3 = \phi_2$$

Equations 6.7.18 and 6.7.19 can be differentiated to obtain the velocity relations between bodies as

$$\begin{bmatrix} \dot{x}_2 \\ \dot{y}_2 \end{bmatrix} = \begin{bmatrix} \dot{x}_1 \\ \dot{y}_1 \end{bmatrix} + \begin{bmatrix} -\sin\phi_1 \\ \cos\phi_1 \end{bmatrix} \dot{\phi}_1 + \begin{bmatrix} -\sin\phi_2 \\ \cos\phi_2 \end{bmatrix} (\dot{\phi}_1 + \dot{\theta}_{12})$$
$$\dot{\phi}_2 = \dot{\phi}_1 + \dot{\theta}_{12} \tag{6.7.20}$$

and

$$\begin{bmatrix} \dot{x}_3 \\ \dot{y}_3 \end{bmatrix} = \begin{bmatrix} \dot{x}_2 \\ \dot{y}_2 \end{bmatrix} + \begin{bmatrix} -\sin\phi_2 \\ \cos\phi_2 \end{bmatrix} (1 + \alpha_{23})\dot{\phi}_2 + \begin{bmatrix} \cos\phi_2 \\ \sin\phi_2 \end{bmatrix} \dot{\alpha}_{23}$$
$$\dot{\phi}_3 = \dot{\phi}_2 \tag{6.7.21}$$

Equations 6.7.20 and 6.7.21 can be expressed in matrix form as

$$\mathbf{Y}_2 = \begin{bmatrix} 1 & 0 & -(\sin\phi_1 + \sin\phi_2) \\ 0 & 1 & (\cos\phi_1 + \cos\phi_2) \\ 0 & 0 & 1 \end{bmatrix} \mathbf{Y}_1 + \begin{bmatrix} -\sin\phi_2 \\ \cos\phi_2 \\ 1 \end{bmatrix} \dot{\theta}_{12} \qquad (6.7.22)$$

$$\equiv \mathbf{B}_{121}\mathbf{Y}_1 + \mathbf{B}_{122}\dot{\theta}_{12}$$

and

$$\mathbf{Y}_3 = \begin{bmatrix} 1 & 0 & -\sin\phi_2(1 + \alpha_{23}) \\ 0 & 1 & \cos\phi_2(1 + \alpha_{23}) \\ 0 & 0 & 1 \end{bmatrix} \mathbf{Y}_2 + \begin{bmatrix} \cos\phi_2 \\ \sin\phi_2 \\ 0 \end{bmatrix} \dot{\alpha}_{23} \qquad (6.7.23)$$

$$\equiv \mathbf{B}_{231}\mathbf{Y}_2 + \mathbf{B}_{232}\dot{\alpha}_{23}$$

where $\mathbf{Y}_i = [x_i, y_i, \phi_i]^T$ and matrices $\mathbf{B}_{ij}$ are defined as in Eqs. 6.7.22 and 6.7.23.

Virtual displacement relations can be written, in analogy to Eqs. 6.7.22 and 6.7.23, in matrix form as

$$\begin{aligned} \delta\mathbf{Z}_2 &= \mathbf{B}_{121}\,\delta\mathbf{Z}_1 + \mathbf{B}_{122}\,\delta\theta_{12} \\ \delta\mathbf{Z}_3 &= \mathbf{B}_{231}\,\delta\mathbf{Z}_2 + \mathbf{B}_{232}\,\delta\alpha_{12} \end{aligned} \qquad (6.7.24)$$

where $\delta\mathbf{Z}_i = [\delta x_i, \delta y_i, \delta\phi_i]^T$.

Accelerations of bodies 2 and 3 can be obtained, by taking derivatives of Eqs. 6.7.22 and 6.7.23, as

$$\begin{aligned} \dot{\mathbf{Y}}_2 &= \mathbf{B}_{121}\dot{\mathbf{Y}}_1 + \mathbf{B}_{122}\ddot{\theta}_{12} + \dot{\mathbf{B}}_{121}\dot{\mathbf{Y}}_1 + \dot{\mathbf{B}}_{122}\dot{\alpha}_{12} \\ &\equiv \mathbf{B}_{121}\dot{\mathbf{Y}}_1 + \mathbf{B}_{122}\ddot{\theta}_{12} + \mathbf{D}_{12} \end{aligned} \qquad (6.7.25)$$

and

$$\begin{aligned} \dot{\mathbf{Y}}_3 &= \mathbf{B}_{231}\dot{\mathbf{Y}}_2 + \mathbf{B}_{232}\ddot{\alpha}_{23} + \dot{\mathbf{B}}_{121}\mathbf{Y}_2 + \dot{\mathbf{B}}_{232}\dot{\alpha}_{23} \\ &\equiv \mathbf{B}_{231}\dot{\mathbf{Y}}_2 + \mathbf{B}_{232}\ddot{\alpha}_{23} + \mathbf{D}_{23} \end{aligned} \qquad (6.7.26)$$

Time derivative $\dot{\mathbf{B}}_{ij}$ can be obtained by differentiating $\mathbf{B}_{ij}$ in Eqs. 6.7.22 and 6.7.23 as

$$\begin{aligned} \dot{\mathbf{B}}_{121} &= \begin{bmatrix} 0 & 0 & -(\cos\phi_1\dot{\phi}_1 + \cos\phi_2\dot{\phi}_2) \\ 0 & 0 & -(\sin\phi_1\dot{\phi}_1 + \sin\phi_2\dot{\phi}_2) \\ 0 & 0 & 0 \end{bmatrix} \\ \dot{\mathbf{B}}_{122} &= \begin{bmatrix} -\cos\phi_2\dot{\phi}_2 \\ -\sin\phi_2\dot{\phi}_2 \\ 0 \end{bmatrix} \\ \dot{\mathbf{B}}_{231} &= \begin{bmatrix} 0 & 0 & -\cos\phi_2\dot{\phi}_2(1 + \alpha_{23}) - \sin\phi_2\dot{\alpha}_{23} \\ 0 & 0 & -\sin\phi_2\dot{\phi}_2(1 + \alpha_{23}) + \cos\phi_2\dot{\alpha}_{23} \\ 0 & 0 & 0 \end{bmatrix} \\ \dot{\mathbf{B}}_{232} &= \begin{bmatrix} -\sin\phi_2\dot{\phi}_2 \\ \cos\phi_2\dot{\phi}_2 \\ 0 \end{bmatrix} \end{aligned} \qquad (6.7.27)$$

The variational equation of motion for this system is

$$\sum_{i=1}^{3} \delta \mathbf{Z}_i^T(\mathbf{M}_i\dot{\mathbf{Y}}_i - \mathbf{Q}_i^A) = 0 \tag{6.7.28}$$

where $\mathbf{M}_i = \text{diag}(m_i, m_i, J_i)$ and $\delta\mathbf{Z}_i$ must satisfy kinematic admissibility conditions corresponding to joints 1 and 2.

Substituting Eqs. 6.7.24 and 6.7.26 into Eq. 6.7.28 yields

$$\begin{aligned}\sum_{i=1}^{2} \delta \mathbf{Z}_i^T(\mathbf{M}_i\dot{\mathbf{Y}}_i - \mathbf{Q}_i^A) + \delta\mathbf{Z}_2^T(\mathbf{B}_{231}^T\mathbf{M}_3\mathbf{B}_{231}\dot{\mathbf{Y}}_2 + \mathbf{B}_{231}^T\mathbf{M}_3\mathbf{B}_{232}\ddot{\alpha}_{23} + \mathbf{B}_{231}^T\mathbf{M}_3\mathbf{D}_{23} - \mathbf{B}_{231}^T\mathbf{Q}_3^A)\\ + \delta\alpha_{23}(\mathbf{B}_{232}^T\mathbf{M}_3\mathbf{B}_{231}\dot{\mathbf{Y}}_2 + \mathbf{B}_{232}^T\mathbf{M}_3\mathbf{B}_{232}\ddot{\alpha}_{23} + \mathbf{B}_{232}^T\mathbf{M}_3\mathbf{D}_{23} - \mathbf{B}_{231}^T\mathbf{Q}_3^A) = 0\end{aligned} \tag{6.7.29}$$

where $\delta\mathbf{Z}_i$ must satisfy kinematic admissibility conditions corresponding to the revolute joint and $\delta\alpha_{23}$ is arbitrary. Therefore, the coefficient of $\delta\alpha_{23}$ must be zero. As a result, the relative coordinate acceleration $\ddot{\alpha}_{23}$ can be obtained as

$$\ddot{\alpha}_{23} = -(\mathbf{B}_{232}^T\mathbf{M}_3\mathbf{B}_{232})^{-1}(\mathbf{B}_{232}^T\mathbf{M}_3\mathbf{B}_{231}\dot{\mathbf{Y}}_2 + \mathbf{B}_{232}^T\mathbf{M}_3\mathbf{D}_{23} - \mathbf{B}_{232}^T\mathbf{Q}_3^A) \tag{6.7.30}$$

Substituting $\ddot{\alpha}_{23}$ into Eq. 6.7.29 yields

$$\sum_{i=1}^{2} \delta \mathbf{Z}_i^T(\mathbf{M}_i\dot{\mathbf{Y}}_i - \hat{\mathbf{Q}}_i^A) + \delta\mathbf{Z}_2^T(\mathbf{K}_2\dot{\mathbf{Y}}_2 - \mathbf{L}_2) = 0 \tag{6.7.31}$$

where $\delta\mathbf{Z}_2$ is kinematically admissible for joint 1 and

$$\begin{aligned}\mathbf{K}_2 &= \mathbf{B}_{231}^T\mathbf{M}_3\mathbf{B}_{231} - \mathbf{B}_{231}^T\mathbf{M}_3\mathbf{B}_{232}(\mathbf{B}_{232}^T\mathbf{M}_3\mathbf{B}_{232})^{-1}(\mathbf{B}_{232}\mathbf{M}_3\mathbf{B}_{231})\\ \mathbf{L}_2 &= \mathbf{B}_{231}^T(\mathbf{M}_3\mathbf{D}_{23} - \mathbf{Q}_3^A) - \mathbf{B}_{231}^T\mathbf{M}_3\mathbf{B}_{232}(\mathbf{B}_{232}^T\mathbf{M}_3\mathbf{B}_{232})^{-1}\\ &\quad\times(\mathbf{B}_{232}^T\mathbf{M}_3\mathbf{D}_{23} - \mathbf{B}_{232}^T\mathbf{Q}_3^A)\end{aligned} \tag{6.7.32}$$

Substituting Eqs. 6.7.24 and 6.7.25 into Eq. 6.7.31 yields

$$\begin{aligned}\delta\mathbf{Z}_1^T(\mathbf{M}_1\dot{\mathbf{Y}}_1 - \hat{\mathbf{Q}}_1^A) + \delta\mathbf{Z}_1^T\{\mathbf{B}_{121}^T(\mathbf{K}_2 + \mathbf{M}_2)(\mathbf{B}_{121}\dot{\mathbf{Y}}_1 + \mathbf{B}_{122}\ddot{\theta}_{12} + \mathbf{D}_{12}) - \mathbf{B}_{121}^T(\mathbf{L}_2 + \hat{\mathbf{Q}}_2^A)\}\\ + \delta\theta_{12}\{\mathbf{B}_{122}^T(\mathbf{K}_2 + \mathbf{M}_2)(\mathbf{B}_{121}\dot{\mathbf{Y}}_1 + \mathbf{B}_{122}\ddot{\theta}_{12} + \mathbf{D}_{12}) - \mathbf{B}_{122}^T(\mathbf{L}_2 + \hat{\mathbf{Q}}_2^A)\} = 0\end{aligned} \tag{6.7.33}$$

where $\delta\theta_{12}$ is arbitrary. Thus the coefficient of $\delta\theta_{12}$ must be zero, which yields

$$\begin{aligned}\ddot{\theta}_{12} = -(\mathbf{B}_{122}^T(\mathbf{K}_2 + \mathbf{M}_2)\mathbf{B}_{122})^{-1}\{\mathbf{B}_{122}^T(\mathbf{K}_2 + \mathbf{M}_2)\mathbf{B}_{121}\dot{\mathbf{Y}}_1\\ + \mathbf{B}_{122}^T(\mathbf{K}_2 + \mathbf{M}_2)\mathbf{D}_{12} - \mathbf{B}_{122}^T(\mathbf{L}_2 + \mathbf{Q}_2^A)\}\end{aligned} \tag{6.7.34}$$

Equation 6.7.34 can be substituted into Eq. 6.7.33 to obtain

$$\delta\mathbf{Z}_1^T\{(\mathbf{K}_1 + \mathbf{M}_1)\dot{\mathbf{Y}}_1 - (\mathbf{Q}_1^A + \mathbf{L}_1)\} = 0 \tag{6.7.35}$$

where

$$\begin{aligned}\mathbf{K}_1 &= \mathbf{B}_{121}^T(\mathbf{K}_2 + \mathbf{M}_2)\mathbf{B}_{121} - \mathbf{B}_{121}^T(\mathbf{K}_2 + \mathbf{M}_2)\mathbf{B}_{122}(\mathbf{B}_{122}^T(\mathbf{K}_2 + \mathbf{M}_2)\mathbf{B}_{122})^{-1}\\ &\quad\times[\mathbf{B}_{122}^T(\mathbf{K}_2 + \mathbf{M}_2)\mathbf{B}_{121}]\end{aligned}$$

$$\mathbf{L}_1 = \mathbf{B}_{121}^T(\mathbf{K}_2 + \mathbf{M}_2)\mathbf{D}_{12} - \mathbf{B}_{121}^T(\mathbf{L}_2 + \hat{\mathbf{Q}}_2^A) - \mathbf{B}_{121}^T(\mathbf{K}_2 + \mathbf{M}_2)\mathbf{B}_{122}$$
$$\times [\mathbf{B}_{122}^T(\mathbf{K}_2 + \mathbf{M}_2)\mathbf{B}_{122}]^{-1}[\mathbf{B}_{122}^T(\mathbf{K}_2 + \mathbf{M}_2)\mathbf{D}_{12} - \mathbf{B}_{122}^T(\mathbf{L}_2 + \hat{\mathbf{Q}}_2^A)] \quad (6.7.36)$$

and $\delta\mathbf{Z}_1$ is arbitrary, resulting in the base body differential equation of motion,

$$(\mathbf{K}_1 + \mathbf{M}_1)\dot{\mathbf{Y}}_1 = \hat{\mathbf{Q}}_1^A + \mathbf{L}_1 \quad (6.7.37)$$

Equation 6.7.37 can be solved for base body acceleration $\dot{\mathbf{Y}}_1$.

Once base body acceleration $\dot{\mathbf{Y}}_1$ is found, the acceleration of joint 1 can be obtained from Eq. 6.7.34. Base body and joint 1 accelerations can be substituted into Eq. 6.7.25 to obtain the acceleration $\dot{\mathbf{Y}}_2$ of body 2. The translational relative acceleration $\ddot{\alpha}_{23}$ can be calculated from Eq. 6.7.30, with known $\dot{\mathbf{Y}}_2$. Finally, acceleration $\dot{\mathbf{Y}}_3$ of body 3 can be obtained by substituting $\ddot{\alpha}_{23}$ and $\dot{\mathbf{Y}}_2$ into Eq. 6.7.26. Numerical values for $\dot{\mathbf{Y}}_1$ and relative coordinates accelerations can be integrated to obtain base body and joint velocities in the next time step.

Example 6.7.2: If the base body of Example 6.1.1 is fixed to ground, the system has only two degrees of freedom. The base body acceleration is thus zero and relative coordinate acceleration $\ddot{\theta}_{12}$, expressed in terms of only known quantities, can be obtained from Eq. 6.7.34. Cartesian acceleration $\dot{\mathbf{Y}}_2$ of body 2 can be calculated by substituting $\dot{\mathbf{Y}}_1 = \mathbf{0}$ and relative acceleration $\ddot{\theta}_{12}$ into Eq. 6.7.25. The acceleration of body 2 can now be substituted into Eq. 6.7.30 to obtain relative coordinate acceleration $\ddot{\alpha}_{23}$. The Cartesian acceleration of body 3 can be obtained by substituting $\ddot{\alpha}_{23}$ and $\dot{\mathbf{Y}}_2$ into Eq. 6.7.26.

A RECIPE FOR DERIVING MULTIBODY EQUATIONS OF MOTION

1. Define generalized coordinates $\mathbf{q}$ and determine if they are independent or dependent.
2. Write $\mathbf{r}_i = \mathbf{r}_i(\mathbf{q})$ and $\mathbf{A}_i = \mathbf{A}_i(\mathbf{q})$.
3. Write $\dot{\mathbf{r}}_i = \mathbf{r}_{i_\mathbf{q}}\dot{\mathbf{q}}$ and $\boldsymbol{\omega}_i' = \mathbf{B}_i(\mathbf{q})\dot{\mathbf{q}}$, using vector angular velocity relations, or $\tilde{\boldsymbol{\omega}}' = \mathbf{A}^T\dot{\mathbf{A}}$.
4. Write $\delta\mathbf{r}_i = \mathbf{r}_{i_\mathbf{q}}\,\delta\mathbf{q}$ and $\delta\boldsymbol{\pi}_i' = \mathbf{B}_i(\mathbf{q})\,\delta\mathbf{q}$.
5. Write $\ddot{\mathbf{r}}_i = \mathbf{r}_{i_\mathbf{q}}\ddot{\mathbf{q}} + \dot{\mathbf{r}}_{i_\mathbf{q}}\dot{\mathbf{q}}$ and $\dot{\boldsymbol{\omega}}_i' = \mathbf{B}_i(\mathbf{q})\ddot{\mathbf{q}} + \dot{\mathbf{B}}_i(\mathbf{q})\dot{\mathbf{q}}$.
6. Derive applied forces and torques $\mathbf{F}_i^A = \mathbf{F}_i^A(\mathbf{q})$ and $\mathbf{n}_i'^A = \mathbf{n}_i'^A(\mathbf{q})$.
7. Calculate masses m_i and inertia matrices $\mathbf{J}_i'$.
8. Substitute from steps 3 to 5 into

$$\sum_{i=1}^{nb} \{\delta\mathbf{r}_i^T[m_i\ddot{\mathbf{r}}_i - \mathbf{F}_i^A] + \delta\boldsymbol{\pi}_i'^T[\mathbf{J}_i'\dot{\boldsymbol{\omega}}_i' + \tilde{\boldsymbol{\omega}}_i'\mathbf{J}_i'\boldsymbol{\omega}_i - \mathbf{n}_i'^A]\} = 0$$

   and rewrite as $\delta\mathbf{q}^T[\mathbf{M}(\mathbf{q})\ddot{\mathbf{q}} - \mathbf{Q}^A(\mathbf{q}, \dot{\mathbf{q}}) - \mathbf{R}(\mathbf{q}, \dot{\mathbf{q}})] = 0$.
9. If $\mathbf{q}$ is independent, $\delta\mathbf{q}$ in step 8 is arbitrary and the equations of motion are $\mathbf{M}(\mathbf{q})\ddot{\mathbf{q}} = \mathbf{Q}^A(\mathbf{q}, \dot{\mathbf{q}}) + \mathbf{R}(\mathbf{q}, \dot{\mathbf{q}})$.

10. If $\mathbf{q}$ are dependent, derive constraint equations $\mathbf{\Phi}(\mathbf{q}) = \mathbf{0}$, the Jacobian $\mathbf{\Phi_q}$, and the constraint acceleration equation $\mathbf{\Phi_q}(\mathbf{q})\ddot{\mathbf{q}} = \boldsymbol{\gamma}(\mathbf{q}, \dot{\mathbf{q}}) \equiv -(\mathbf{\Phi_q}\dot{\mathbf{q}})_\mathbf{q}\dot{\mathbf{q}}$.
11. The constrained equations of motion are

$$\begin{bmatrix} \mathbf{M}(\mathbf{q}) & \mathbf{\Phi}_\mathbf{q}^T(\mathbf{q}) \\ \mathbf{\Phi_q}(\mathbf{q}) & 0 \end{bmatrix} \begin{bmatrix} \ddot{\mathbf{q}} \\ \boldsymbol{\lambda} \end{bmatrix} = \begin{bmatrix} \mathbf{Q}^A(\mathbf{q}, \dot{\mathbf{q}}) + \mathbf{R}(\mathbf{q}, \dot{\mathbf{q}}) \\ \boldsymbol{\gamma}(\mathbf{q}, \dot{\mathbf{q}}) \end{bmatrix}$$

PROBLEMS

Section 6.1

6.1.1 Use absolute constraints from among Eqs. 6.1.5 to 6.1.7 to write constraint equations on q_1 and their Jacobians for the body shown in Fig. P6.1.1, where points P_1 and P_2 are pins that slide in slots along x and y axes, respectively.

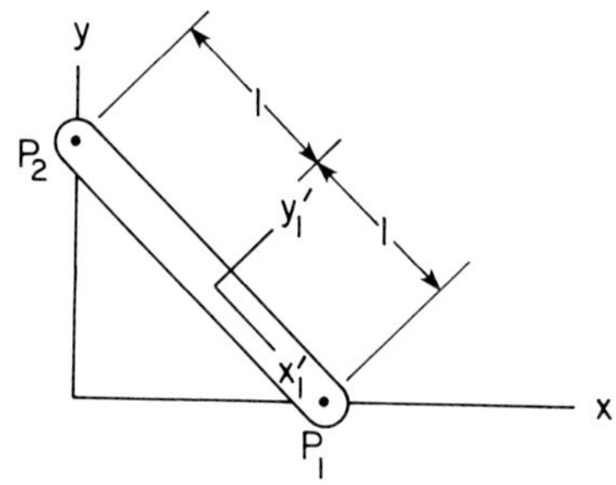

Figure P6.1.1

6.1.2 Verify that Eq. 6.1.11 is valid.

6.1.3 Verify that Eq. 6.1.15 is valid.

6.1.4 Write the explicit form of the constraint equations for the four-body slider-crank model of Example 6.1.3 (Fig. 6.1.7)

6.1.5 Derive constraint equations and the associated Jacobians for the following constraints:

(a) *Absolute distance constraint* [Fig. P6.1.5(a)]. $\Phi^{\mathrm{ad}(i j)}(\mathbf{q}_i) = 0$ between point P_i on body i and point (C_4, C_5) in the x-y frame (i.e., the distance is fixed as $C_6 > 0$). Show that the Jacobian fails to have full row rank if $C_6 = 0$.

(b) *Relative distance constraint* [Fig. P6.1.5(b)]. $\Phi^{\mathrm{rd}(i j)}(\mathbf{q}_i, \mathbf{q}_j) = 0$ between points P_i and P_j on bodies i and j (i.e., the distance is fixed as $C_7 > 0$). Show that the Jacobian fails to have full row rank if $C_7 = 0$.

6.1.6 Use the constraints of Section 6.1 and Problem 6.1.5 to formulate models of the slider–crank mechanism of Fig. 6.1.7 in Example 6.1.3, using the following combinations of bodies and constraints. (Draw a figure defining each.)

(a) Bodies 2 and 3, with absolute constraints for the connections between body 3 and ground, a revolute joint, and an absolute distance constraint to replace body 1

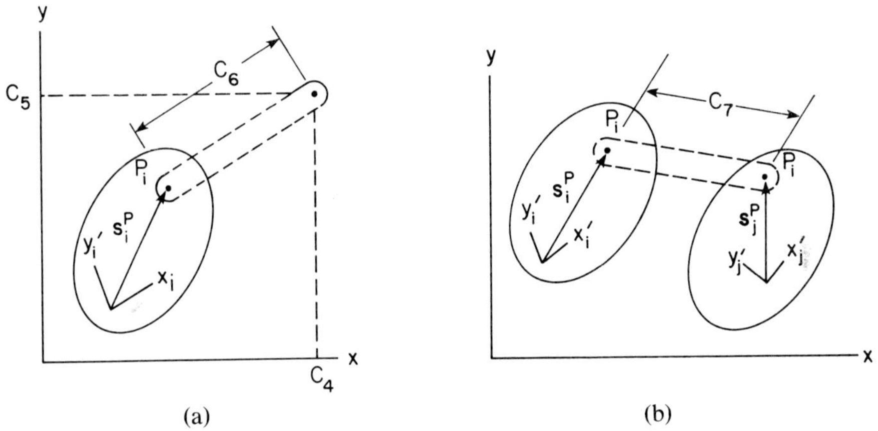

Figure P6.1.5

(b) Body 2 only, with an absolute constraint at the right end and an absolute distance constraint to replace body 1

(c) Bodies 1 and 3, with absolute constraints between the bodies and ground and a relative distance constraint to replace body 2

6.1.7 Write the complete set of constraint equations and their Jacobians for the models in Problem 6.1.6(b) and (c). Show that each Jacobian has full row rank.

6.1.8 Verify that Eq. 6.1.22 is correct.

6.1.9 Carry out calculations in Example 6.1.5 to verify the result given.

6.1.10 Derive the Lagrange multiplier form of the equations of motion (Eq. 6.1.50) for the slider–crank mechanism of Example 6.1.3, including the effect of force **F** in Fig. 6.1.9, the spring–damper in Fig. 6.1.11, and gravity acting in the negative y direction. In this formulation, $\mathbf{q} = [x_1, y_1, \phi_1, x_2, y_2, \phi_2]^T$ is subject to five constraints that may be written in the form $\mathbf{\Phi}(\mathbf{q}) = \mathbf{0}$. Generalized forces for the spring–damper evaluated in Example 6.1.5 may be used.

6.1.11 Use the virtual work definition of generalized force to derive the expression for the generalized force due to force **F** in Fig. 6.1.9, in terms of the reduced generalized coordinate vector $\mathbf{q} = [\phi_1, x_2, \phi_2]^T$.

6.1.12 Carry out the expansion required to obtain the equations of motion presented in Example 6.1.8.

6.1.13 Use the method of proof of the constrained dynamic existence theorem of Section 3.6 to show that full row rank of $\mathbf{\Phi_q}$ and Eqs. 6.1.52 and 6.1.53 imply that the coefficient matrix of Eq. 6.1.51 is nonsingular.

6.1.14 Derive explicit relations for reaction force and torque in a translational joint, in terms of its Lagrange multiplier $\boldsymbol{\lambda}^t$. Show that the component of reaction force along the translational joint is zero (i.e., $\mathbf{v}_i''^T\mathbf{F}_i''^t = 0$).

6.1.15 Derive explicit relations for reaction force and torque due to absolute x, y, and ϕ constraints, in terms of their Lagrange multipliers.

Section 6.2

6.2.1 Carry out the expansion to verify that Eq. 6.2.11 is valid.

6.2.2 Carry out the expansion to verify that Eq. 6.2.12 is valid.

6.2.3 Consider the universal joint shown in Fig. P6.2.3. Let the driving angle of the first body be θ_1 and the output angle of the second body be θ_2. Both bodies are constrained to rotate in the *x-y* plane as shown. Find a relation between variations in θ_1 and θ_2. Identify and analyze any singular configurations that may occur.

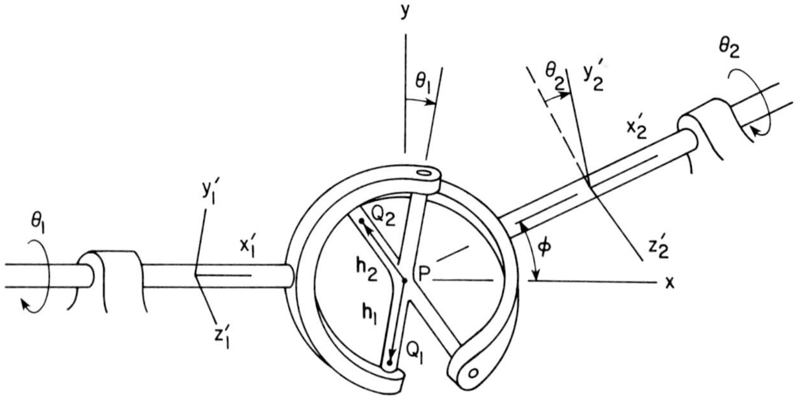

Figure P6.2.3

6.2.4 Verify that Eq. 6.2.19 is valid.

6.2.5 Verify that Eq. 6.2.20 is valid.

6.2.6 Verify that Eq. 6.2.21 is valid.

6.2.7 Expand Eq. 6.2.26, using the notation of Eq. 6.2.25, to show that it is the sum of Eqs. 5.1.26, over all bodies in the system.

6.2.8 Construct two alternate models of the slider–crank mechanism of Example 6.2.1, using kinematic joints of Section 6.2. Confirm that 27 constraint equations have been defined and present a physical argument that they define kinematics of the system.

6.2.9 Construct two alternative models of the four bar mechanism shown in Fig. P6.2.9, using kinematic joints of Section 6.2.

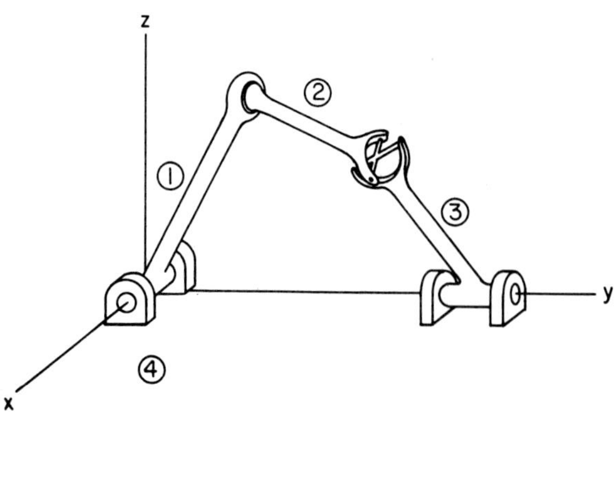

Figure P6.2.9

6.2.10 Derive explicit relations for reaction force and torque in a revolute joint, in terms of its Lagrange multipliers. Show that the component of reaction torque about the revolute joint axis is zero (i.e., $\mathbf{h}_i''^T\mathbf{T}_i''^r = 0$).

6.2.11 Derive explicit relations for reaction force and torque in a translational joint, in terms of its Lagrange multipliers. Show that the component of reaction force along the joint is zero (i.e., $\mathbf{h}_i''^T\mathbf{F}_i''^t = 0$).

Section 6.3

6.3.1 Derive differential equations of motion of the system shown in Fig. P6.3.1. A block of mass m_2 slides on another block of mass m_1, which slides on a horizontal surface. There is no friction between surfaces. Use q_1 and q_2 as generalized coordinates.

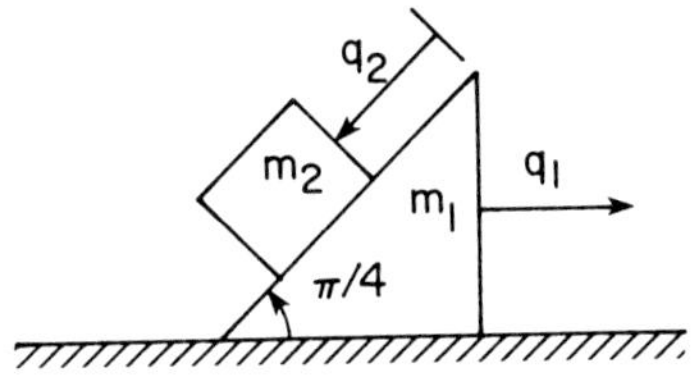

Figure P6.3.1

6.3.2 Consider a system that consists of a simple pendulum of length l and mass m_2 that is pivoted at point 0 on a block of mass m_1 that slides on a horizontal surface (Fig. P6.3.2). There is no fricton between surfaces. Derive differential equations of motion in terms of q_1 and q_2.

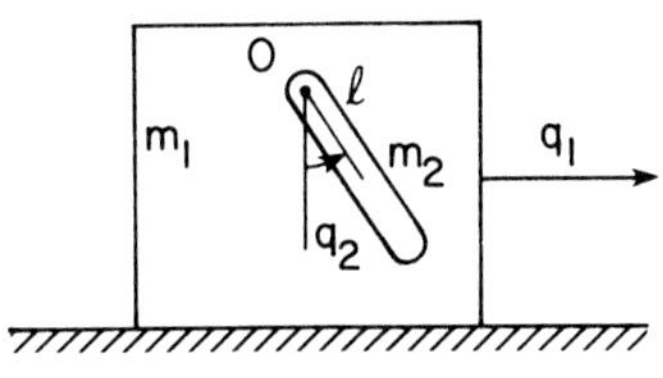

Figure P6.3.2

6.3.3 A small solid ring of mass m_2 slides along a rod that is pivoted at point O about the global z axis (Fig. P6.3.3). A linear spring is attached between the pivot and the ring. Derive differential equations of motion in terms of q_1 and q_2.

6.3.4 Verify expressions for $\mathbf{r}_2$, $\boldsymbol{\omega}_2'$, $\delta\mathbf{r}_2$, $\delta\boldsymbol{\pi}_2'$, $\dot{\mathbf{r}}_2$, $\dot{\boldsymbol{\omega}}_2'$, and $\ddot{\mathbf{r}}_2$ in Example 6.3.6.

6.3.5 Derive the equations of motion of the triple pendulum shown in Fig. P6.3.5. Use q_1, q_2, and q_3 as generalized coordinates. Let m_1, m_2, and m_3 be masses of the uniform bars and the length of each bar be 2 units. Note that the equations of motion are very messy.

6.3.6 A tube of mass m_2 slides along a massless guide that is hinged to a solid disk of mass m_1, as shown in Fig. P6.3.6. A linear spring is attached between the tube and the hinge. Derive differential equations of motion in terms of q_1, q_2, and q_3.

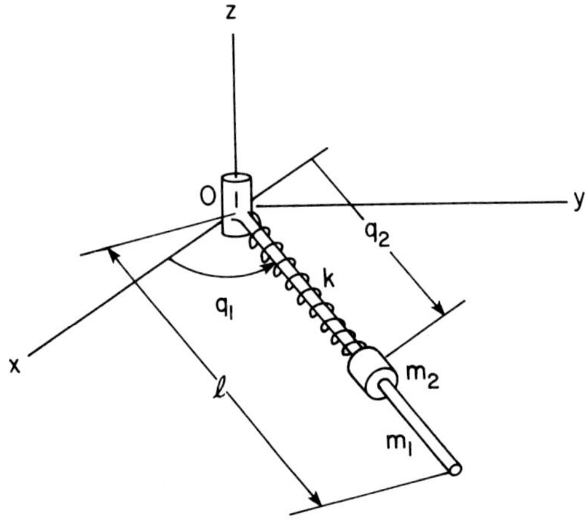

Figure P6.3.3

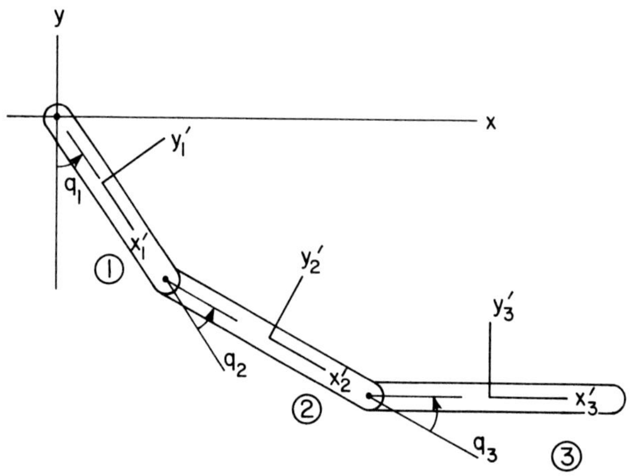

Figure P6.3.5

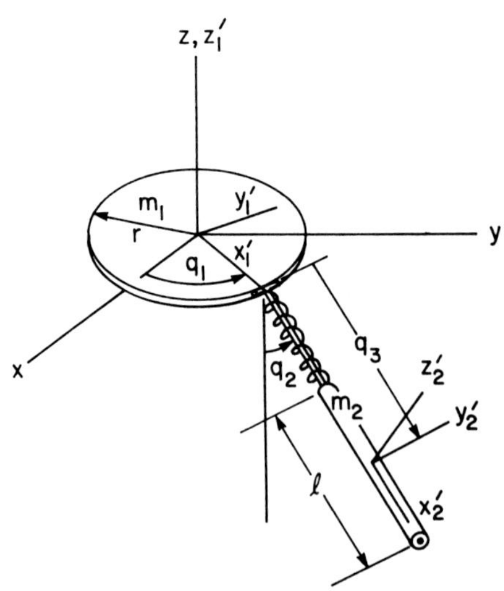

Figure P6.3.6

Section 6.4

6.4.1 Carry out a derivation analogous to that presented for Eq. 6.4.13 to obtain Eq. 6.4.14.

6.4.2 Verify that $\mathbf{A}_a = \mathbf{A}_c\mathbf{C}_c\mathbf{A}''_{ca}\mathbf{C}_a^T$ for the bodies in Fig. 6.4.1.

6.4.3 Carry out a derivation analogous to that presented for Eq. 6.4.18 to obtain Eq. 6.4.19.

6.4.4 Carry out the expansion of Eq. 6.4.32 for the vehicle of Example 6.4.1, accounting for the three attached suspension components, to obtain Eq. 6.4.33.

6.4.5 Verify that Eq. 6.4.39 is correct.

Section 6.5

6.5.1 Draw a graph that represents nodes and kinematic couplings between nodes of the mechanism of Problem 6.3.2.

6.5.2 Repeat Problem 6.5.1 for the mechanism of Problem 6.3.3.

6.5.3 Repeat Problem 6.5.1 for the quick return mechanism shown in Fig. P6.5.3.

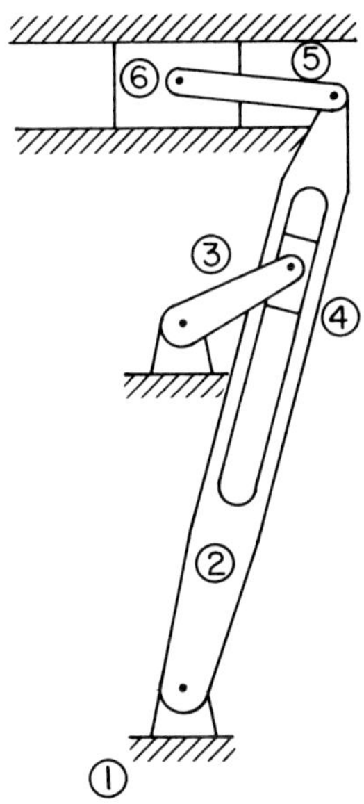

Figure P6.5.3

6.5.4 Find the transformation matrix from the x_2'-y_2'-z_2' frame to the x-y-z frame in Example 6.5.6 and verify that the angular velocity vector $\boldsymbol{\omega}_2'$ in Example 6.3.4 yields the same results as in Example 6.5.6.

6.5.5 Verify that the angular velocity obtained in Example 6.3.6 agrees with the result of Example 6.5.7, using the proper transformation matrix.

6.5.6 Verify that Eq. 6.5.37 is valid.

6.5.7 Verify that Eqs. 6.5.38 and 6.5.39 are valid.

6.5.8 Verify that Eqs. 6.5.41 and 6.5.42 are valid.

Section 6.6

6.6.1 Verify that Eq. 6.6.1 is valid.

6.6.2 Repeat Prob. 6.3.1, using the velocity transformation technique.

6.6.3 Repeat Prob. 6.3.2, using the velocity transformation technique.

6.6.4 Repeat Prob. 6.3.4, using the velocity transformation technique.

6.6.5 Repeat Example 6.3.2, using Eq. 6.6.21, and compare the result with that obtained in Example 6.3.2.

6.6.6 Repeat Example 6.3.5, using Eq. 6.6.21, and compare the result with that obtained in Example 6.3.5.

Appendix A
Vector and Matrix Algebra

The study of dynamics of particles, rigid bodies, systems of rigid bodies, and more general mechanical systems is based on systematic application of Newton's fundamental law of motion,

$$m\vec{a} = \vec{f}$$

This simple vector equation permits determination of the vector $\vec{a}$ of acceleration for a particle of mass $m \neq 0$, once the resultant $\vec{f}$ of all forces acting on the particle is known. If only motion of unconnected particles were of interest, this form of the equations of dynamics would be adequate. However, virtually all practical dynamic systems exhibit greater structure than unconnected particles.

The practical study of dynamics of mechanical systems requires a richer mathematical foundation than elementary vector analysis. To support a systematic development of governing equations for dynamics of mechanical systems, this appendix presents basic concepts from vector analysis and matrix algebra.

Vector and matrix algebra form the mathematical foundation for dynamics. Geometry of motion is at the heart of both kinematics and dynamics of mechanical systems. Vector analysis is the time honored tool for describing geometry [1]. In its geometric form, however, vector algebra is not well suited to implementation with the calculus of multivariable systems that is needed in formulation and computer solution of the equations of dynamics. In this appendix, a systematic matrix formulation of vector algebra, which is referred to as an *algebraic vector representation*, is summarized for use throughout the book. In contrast to the more traditional geometric form, this form of vector representation is easier to use for both analysis and computer implementation. Elementary properties of vector and matrix algebra are used in this development, without proofs. For a review of the theory of matrix algebra, the reader may refer to standard texts such as Ref. 2.

A.1 GEOMETRIC VECTORS

The concept of a vector may be introduced in a very general geometric setting, with no requirement for identification of a reference frame. In this setting a *geometric vector*, or simply a *vector*, is defined as the directed line segment from one point in space to another point in space. Its *direction* is therefore established and its *magnitude* is defined to be the distance between the points connected by the vector. This geometric setting for vector analysis, together with algebraic operations of addition, multiplication by a scalar, scalar product, and vector product, form the classical foundation for vector analysis. Coupling of vector analysis and calculus, however, requires that a *reference frame* be defined and that geometric vectors be characterized in terms of a set of variables called *generalized coordinates* that form the foundation for the study of motion of particles and systems. The geometric foundations of vector analysis are reviewed in this section, in preparation for subsequent reduction to a more suitable form for application of methods of calculus and digital computing.

The vector $\vec{a}$ in Fig. A.1.1 that begins at point A and ends at point B is denoted by the $(\vec{\ })$ notation in its geometrical sense. The *magnitude of a vector* $\vec{a}$ is the *length of the vector* (the distance between A and B) and is denoted by $|\vec{a}|$, or simply a. Note that the length of a vector is positive if points A and B do not coincide and it is zero only when they coincide. A vector with zero length is denoted $\vec{0}$ and is called the *zero vector*.

Multiplication of a vector by a scalar, for example, a vector $\vec{a}$ and a scalar $\alpha \geqslant 0$, is defined as a vector in the same direction as $\vec{a}$, but having magnitude αa. A *unit vector*, having a length of 1 unit, in the direction $\vec{a} \neq \vec{0}$ is $(1/a)\vec{a}$. Multiplication of a vector $\vec{a}$ by a negative scalar $\beta < 0$ is defined as the vector with magnitude $|\beta| a$ and direction opposite to that of $\vec{a}$. The *negative of a vector* is obtained by multiplying the vector by -1 (i.e., it is the vector with the same magnitude but opposite direction).

Example A.1.1: Let points A and B in Fig. A.1.1 be located in an orthogonal x-y-z reference frame, as shown in Fig. A.1.2. The distance between points A and

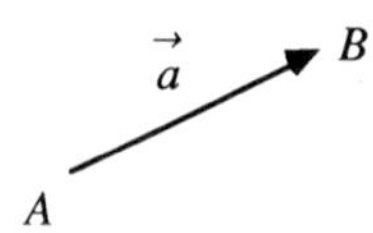

Figure A.1.1 Vector from point A to point B

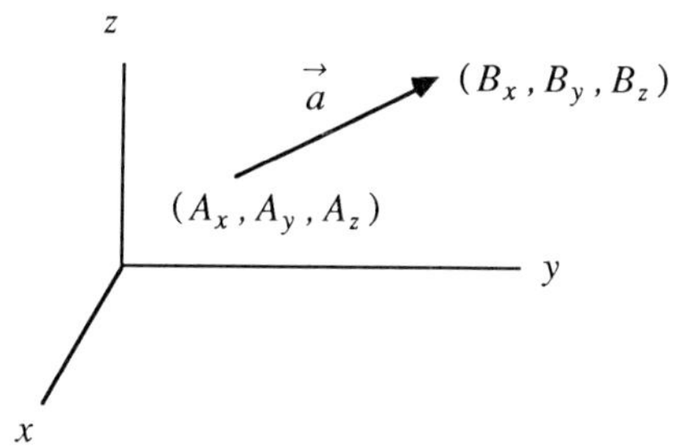

Figure A.1.2 Vector located in orthogonal reference frame.

B that have coordinates (A_x, A_y, A_z) and (B_x, B_y, B_z), respectively, is the length of $\vec{a}$; that is,

$$|\vec{a}| = [(B_x - A_x)^2 + (B_y - A_y)^2 + (B_z - A_z)^2]^{1/2}$$

Two vectors $\vec{a}$ and $\vec{b}$ are added according to the *parallelogram rule*, as shown in Fig. A.1.3. The parallelogram in this construction is formed in the plane that contains the intersecting vectors $\vec{a}$ and $\vec{b}$. The *vector sum* of $\vec{a}$ and $\vec{b}$ is written as

$$\vec{c} = \vec{a} + \vec{b} \tag{A.1.1}$$

Addition of vectors and multiplication of vectors by scalars obey the following rules [1]:

$$\begin{aligned} \vec{a} + \vec{b} &= \vec{b} + \vec{a} \\ (\alpha + \beta)\vec{a} &= \alpha\vec{a} + \beta\vec{a} \\ \alpha(\vec{a} + \vec{b}) &= \alpha\vec{a} + \alpha\vec{b} \end{aligned} \tag{A.1.2}$$

where $\vec{a}$ and $\vec{b}$ are vectors and α and β are scalars.

◆ **Example A.1.2:** Using the definition of vector addition, the plane formed by intersecting vectors $\vec{a}$ and $\vec{b}$ in Fig. A.1.3 may be called the x'-y' plane, with the x' axis along vector $\vec{a}$, as shown in Fig. A.1.4. Plane trigonometry can now be used to calculate the length of vector $\vec{a} + \vec{b}$ and the angle $\theta(\vec{a}, \vec{a} + \vec{b})$ between $\vec{a}$ and $\vec{a} + \vec{b}$,

$$|\vec{a} + \vec{b}| = [(a + b\cos\theta(\vec{a}, \vec{b}))^2 + (b\sin\theta(\vec{a}, \vec{b}))^2]^{1/2}$$

$$\theta(\vec{a}, \vec{a} + \vec{b}) = \arctan\frac{b\sin\theta(\vec{a}, \vec{b})}{a + b\cos\theta(\vec{a}, \vec{b})}$$

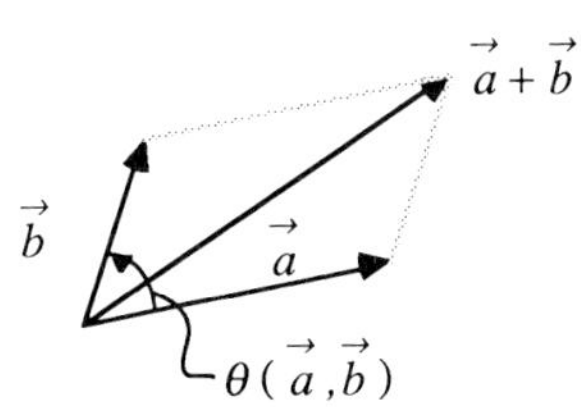

Figure A.1.3 Addition of vectors.

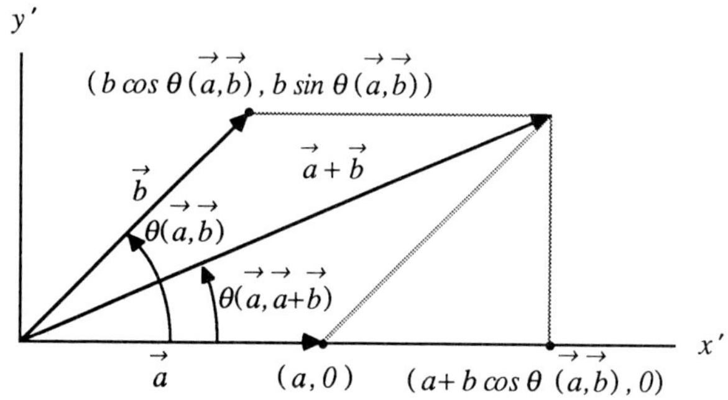

Figure A.1.4 Sum of vectors in their plane of intersection.

These calculations characterize the vector $\vec{a} + \vec{b}$, but they are not convenient and, more important, they are not well suited for computation.

Orthogonal reference frames are used extensively in representing vectors. Use in this book is limited to *right-handed x-y-z* orthogonal reference frames (i.e., with mutually orthogonal x, y, and z axes that are ordered by the finger structure of the right hand, as shown in Fig. A.1.5). Such a frame is called a *Cartesian reference frame*. Note that rotating the finger structure of the right hand from the x to the y axis means that the thumb is pointing in the positive z direction.

A vector $\vec{a}$ can be *resolved* into components a_x, a_y, and a_z along the x, y, and z axes of a Cartesian reference frame, as shown in Fig. A.1.6. These components are called the *Cartesian components* of the vector, or simply the *components* of the vector. The *unit coordinate vectors* $\vec{i}$, $\vec{j}$, and $\vec{k}$ are directed along the x, y, and z axes, respectively. In vector notation,

$$\vec{a} = a_x\vec{i} + a_y\vec{j} + a_z\vec{k} \tag{A.1.3}$$

Denote the angle from vector $\vec{a}$ to vector $\vec{b}$ in the plane that contains them by $\theta(\vec{a}, \vec{b})$, with counterclockwise as positive about the normal to the plane of

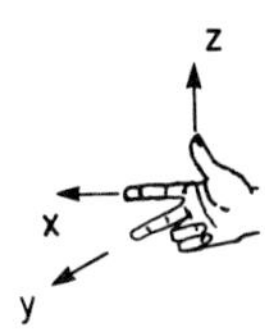

Figure A.1.5 Right-hand orthogonal reference frame.

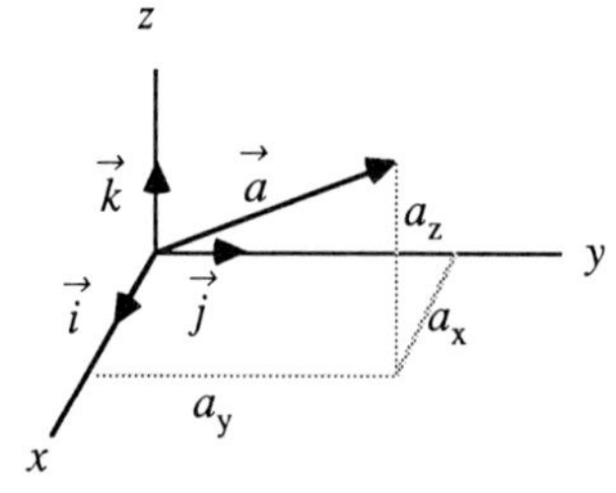

Figure A.1.6 Cartesian components of vector.

the vectors that points toward the viewer, as shown in Fig. A.1.3. In terms of the angles between the vector $\vec{a}$ and the positive x, y, and z axes, the components of vector $\vec{a}$ are

$$\begin{aligned} a_x &= a \cos \theta(\vec{i}, \vec{a}) \\ a_y &= a \cos \theta(\vec{j}, \vec{a}) \\ a_z &= a \cos \theta(\vec{k}, \vec{a}) \end{aligned} \tag{A.1.4}$$

The quantities $\cos \theta(\vec{i}, \vec{a})$, $\cos \theta(\vec{j}, \vec{a})$, and $\cos \theta(\vec{k}, \vec{a})$ are called the *direction cosines* of vector $\vec{a}$. Note that if the viewer had been on the back side of the plane of Fig. A.1.3, the counterclockwise angle from $\vec{a}$ to $\vec{b}$ would be $2\pi - \theta(\vec{a}, \vec{b})$. However, $\cos(2\pi - \theta) = \cos \theta$, so the viewpoint does not influence the value of the direction cosines.

The vector sum of $\vec{a}$ and $\vec{b}$ may be expressed in terms of their components, using Eq. A.1.2, as [1]

$$\begin{aligned} \vec{c} = \vec{a} + \vec{b} &= (a_x + b_x)\vec{i} + (a_y + b_y)\vec{j} + (a_z + b_z)\vec{k} \\ &\equiv c_x\vec{i} + c_y\vec{j} + c_z\vec{k} \end{aligned} \tag{A.1.5}$$

where c_x, c_y, and c_z are the components of vector $\vec{c}$. Thus addition of vectors occurs component by component. Using this idea, three vectors $\vec{a}$, $\vec{b}$, and $\vec{c}$ may be added, to show (Prob. A.1.1) that

$$(\vec{a} + \vec{b}) + \vec{c} = \vec{a} + (\vec{b} + \vec{c}) \tag{A.1.6}$$

The *scalar product*, or *dot product*, of two vectors $\vec{a}$ and $\vec{b}$ is defined as the product of their magnitudes and the cosine of the angle between them; that is,

$$\vec{a} \cdot \vec{b} = ab \cos \theta(\vec{a}, \vec{b}) \tag{A.1.7}$$

This definition is given in purely geometric terms, so it is independent of the reference frame in which the vectors are represented.

Note that if two vectors $\vec{a}$ and $\vec{b}$ are nonzero (i.e., $a \neq 0$ and $b \neq 0$), their scalar product is zero if and only if $\cos \theta(\vec{a}, \vec{b}) = 0$. Two nonzero vectors are said to be *orthogonal vectors* if their scalar product is zero. Since $\theta(\vec{b}, \vec{a}) = 2\pi - \theta(\vec{a}, \vec{b})$ and $\cos(2\pi - \theta) = \cos \theta$, the order of terms appearing on the right side of Eq. A.1.7 is immaterial. Thus

$$\vec{a} \cdot \vec{b} = \vec{b} \cdot \vec{a} \tag{A.1.8}$$

The scalar product of two vectors may be interpreted as the product of the magnitude of one of the vectors times the magnitude of the projection of the other vector onto that vector. To see this, refer to Fig. A.1.4, where the projection of vector $\vec{b}$ onto $\vec{a}$ has length $b \cos(\vec{a}, \vec{b})$.

Based on the definition of scalar product, the following relations hold for the unit coordinate vectors $\vec{i}$, $\vec{j}$, and $\vec{k}$:

$$\begin{aligned} \vec{i} \cdot \vec{j} = \vec{j} \cdot \vec{k} = \vec{k} \cdot \vec{i} = 0 \\ \vec{i} \cdot \vec{j} = \vec{j} \cdot \vec{j} = \vec{k} \cdot \vec{k} = 1 \end{aligned} \tag{A.1.9}$$

For any vector $\vec{a}$,

$$\vec{a} \cdot \vec{a} = aa \cos \theta = a^2$$

While not obvious on geometrical grounds, the scalar product satisfies the relation [1]

$$(\vec{a} + \vec{b}) \cdot \vec{c} = \vec{a} \cdot \vec{c} + \vec{b} \cdot \vec{c} \tag{A.1.10}$$

Using Eq. A.1.10 and the relations of Eq. A.1.9, a direct calculation (Prob. A.1.2) yields

$$\vec{a} \cdot \vec{b} = a_x b_x + a_y b_y + a_z b_z \tag{A.1.11}$$

The *vector product*, or *cross product*, of two vectors $\vec{a}$ and $\vec{b}$ is defined as the vector

$$\vec{a} \times \vec{b} = ab \sin \theta(\vec{a}, \vec{b})\vec{v} \tag{A.1.12}$$

where $\vec{v}$ is a unit vector that is orthogonal (perpendicular) to the plane of intersection of vectors $\vec{a}$ and $\vec{b}$, taken in the positive right-hand coordinate direction, as shown in Fig. A.1.7.

If the viewer were behind the plane of Fig. A.1.7, the unit normal to the plane would be $-\vec{v}$ and the counterclockwise angle from $\vec{a}$ to $\vec{b}$ would be $2\pi - \theta(\vec{a}, \vec{b})$. Then the vector product would be

$$\begin{aligned}\vec{a} \times \vec{b} &= ab \sin(2\pi - \theta(\vec{a}, \vec{b}))(-\vec{v}) \\ &= ab \sin \theta(\vec{a}, \vec{b})\vec{v}\end{aligned}$$

since $\sin(2\pi - \theta) = -\sin \theta$. This is the same result as in Eq. A.1.12, so the viewpoint does not influence evaluation of the vector product. Since the definition of vector product is purely geometric, the result is independent of the reference frame in which the vectors are represented.

Since reversing the order of vectors $\vec{a}$ and $\vec{b}$ in Eq. A.1.12 yields the opposite direction for the unit vector $\vec{v}$,

$$\vec{b} \times \vec{a} = -\vec{a} \times \vec{b} \tag{A.1.13}$$

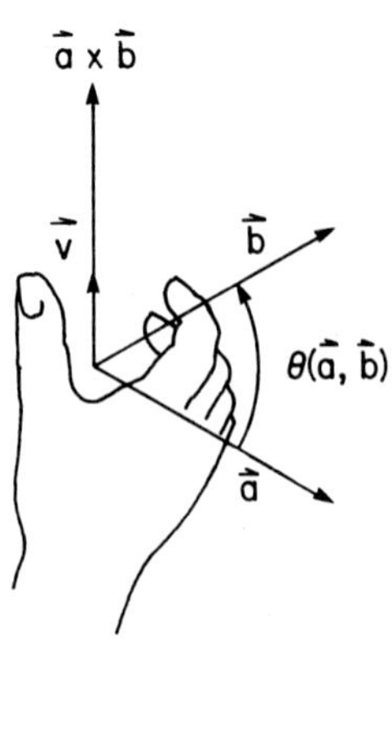

Figure A.1.7 Vector product.

Analogous to Eq. A.1.10, the vector product satisfies [1]

$$(\vec{a} + \vec{b}) \times \vec{c} = \vec{a} \times \vec{c} + \vec{b} \times \vec{c} \tag{A.1.14}$$

Since $\theta(\vec{a}, \vec{a}) = 0$, for any vector $\vec{a}$,

$$\vec{a} \times \vec{a} = a^2 \sin \theta \, \vec{u} = \vec{0}$$

From the definition of the unit coordinate vectors and the vector product, the following relations are valid:

$$\begin{aligned} \vec{i} \times \vec{i} &= \vec{j} \times \vec{j} = \vec{k} \times \vec{k} = \vec{0} \\ \vec{i} \times \vec{j} &= -\vec{j} \times \vec{i} = \vec{k} \\ \vec{j} \times \vec{k} &= -\vec{k} \times \vec{j} = \vec{i} \\ \vec{k} \times \vec{i} &= -\vec{i} \times \vec{k} = \vec{j} \end{aligned} \tag{A.1.15}$$

Using Eq. A.1.15 and the property of vector product of Eq. A.1.14, the vector product of two vectors may be expanded and written in terms of their components as (Prob. A.1.3)

$$\begin{aligned} \vec{c} = \vec{a} \times \vec{b} &= (a_y b_z - a_z b_y)\vec{i} + (a_z b_x - a_x b_z)\vec{j} + (a_x b_y - a_y b_x)\vec{k} \\ &\equiv c_x \vec{i} + c_y \vec{j} + c_z \vec{k} \end{aligned} \tag{A.1.16}$$

A.2 MATRIX ALGEBRA

Matrix notation permits the engineer to focus on the form of a system of equations and what it means, rather than on the details of its construction. Matrix manipulation also allows organized development and simplification of systems of equations and is ideal for organizing calculations for digital computer implementation. A review of basic matrix definitions and results needed in this book is presented in this section. For more detail, the reader is referred to Ref. 2 or any standard text on matrix theory. A *matrix* is defined as a rectangular array of numbers, limited here to real numbers, organized in rows and columns. A matrix with m rows and n columns is said to be of *dimension* $m \times n$, and is written in the form

$$\mathbf{A} \equiv [a_{ij}] \equiv \begin{bmatrix} a_{11} & a_{12} & \cdots & a_{1n} \\ a_{21} & a_{22} & \cdots & a_{2n} \\ \vdots & \vdots & & \vdots \\ a_{m1} & a_{m2} & \cdots & a_{mn} \end{bmatrix}_{m \times n} \tag{A.2.1}$$

where element a_{ij} is located in row i and column j. A matrix with more than one row and column (i.e., $m > 1$ and $n > 1$) is denoted by a boldface capital letter (e.g., $\mathbf{A}$ in Eq. A.2.1).

The *transpose of a matrix* is formed by interchanging rows and columns and is designated by the superscript T. Thus if a_{ij} is the i-j element of an $m \times n$ matrix $\mathbf{A}$, then a_{ji} is the i-j element of its transpose $\mathbf{A}^T$, whose dimension is $n \times m$.

Example A.2.1: The transpose of the 2×3 matrix

$$\mathbf{A} = \begin{bmatrix} 1 & 3 & 0 \\ 2 & 1 & 1 \end{bmatrix}$$

is the 3×2 matrix

$$\mathbf{A}^T = \begin{bmatrix} 1 & 2 \\ 3 & 1 \\ 0 & 1 \end{bmatrix}$$

A matrix with only one column is called a *column matrix* and is denoted by a boldface lowercase letter (e.g., $\mathbf{a} = [a_1, \ldots, a_k]^T$). A matrix with only one row is called a *row matrix* and is denoted as $\mathbf{a}^T = [a_1, \ldots, a_k]$ (i.e., it is the transpose of a column matrix). An $m \times n$ matrix can be considered as being constructed of n columns $\mathbf{a}_j$, $j = 1, \ldots, n$, or m rows $\mathbf{b}_i^T$, $i = 1, \ldots, m$; that is,

$$\mathbf{A} = [\mathbf{a}_1, \ldots, \mathbf{a}_n] = \begin{bmatrix} \mathbf{b}_1^T \\ \vdots \\ \mathbf{b}_m^T \end{bmatrix}$$

Example A.2.2: The matrix $\mathbf{A}$ of Example A.2.1 may be considered as made up of three 2×1 columns,

$$\mathbf{a}_1 = \begin{bmatrix} 1 \\ 2 \end{bmatrix}, \quad \mathbf{a}_2 = \begin{bmatrix} 3 \\ 1 \end{bmatrix}, \quad \mathbf{a}_3 = \begin{bmatrix} 0 \\ 1 \end{bmatrix}$$

that is,

$$\mathbf{A} = [\mathbf{a}_1, \mathbf{a}_2, \mathbf{a}_3]$$

The same matrix may be considered as made up of two 1×3 rows

$$\mathbf{b}_1^T = [1, 3, 0], \qquad \mathbf{b}_2^T = [2, 1, 1]$$

that is,

$$\mathbf{A} = \begin{bmatrix} \mathbf{b}_1^T \\ \mathbf{b}_2^T \end{bmatrix}$$

The transpose of $\mathbf{A}$ may be written in either of the forms

$$\mathbf{A}^T = \begin{bmatrix} \mathbf{a}_1^T \\ \mathbf{a}_2^T \\ \mathbf{a}_3^T \end{bmatrix} = [\mathbf{b}_1, \mathbf{b}_2]$$

A *square matrix* has an equal number of rows and columns. A *diagonal matrix* is a square matrix with $a_{ij} = 0$ for $i \neq j$. An $n \times n$ diagonal matrix $\mathbf{A}$ can be represented as

$$\mathbf{A} \equiv \text{diag}(a_{11}, a_{22}, \ldots, a_{nn})$$

If square *submatrices* $\mathbf{B}_i$, $i = 1, \ldots, k$, are arranged along the diagonal of a square matrix $\mathbf{D}$,

$$\mathbf{D} \equiv \begin{bmatrix} \mathbf{B}_1 & & & \mathbf{0} \\ & \mathbf{B}_2 & & \\ & & \ddots & \\ & \mathbf{0} & & \mathbf{B}_k \end{bmatrix}$$

then the matrix is called a *block-diagonal matrix* and is represented as

$$\mathbf{D} \equiv \text{diag}(\mathbf{B}_1, \mathbf{B}_2, \ldots, \mathbf{B}_k)$$

even though $\mathbf{D}$ is not a diagonal matrix.

Example A.2.3: Let

$$\mathbf{B}_1 = \begin{bmatrix} 1 & 0 \\ 2 & 1 \end{bmatrix}, \quad \mathbf{B}_2 = \begin{bmatrix} 1 & 0 & 1 \\ 0 & 1 & 0 \\ 0 & 2 & 1 \end{bmatrix}$$

A block-diagonal matrix of dimension 5×5 may be defined as

$$\mathbf{D} = \text{diag}(\mathbf{B}_1, \mathbf{B}_2) = \begin{bmatrix} \mathbf{B}_1 & \mathbf{0} \\ \mathbf{0} & \mathbf{B}_2 \end{bmatrix}$$

$$= \begin{bmatrix} 1 & 0 & 0 & 0 & 0 \\ 2 & 1 & 0 & 0 & 0 \\ 0 & 0 & 1 & 0 & 1 \\ 0 & 0 & 0 & 1 & 0 \\ 0 & 0 & 0 & 2 & 1 \end{bmatrix}$$

A direct calculation (Prob. A.2.1) verifies that

$$\mathbf{D}^T = \text{diag}(\mathbf{B}_1^T, \mathbf{B}_2^T) = \begin{bmatrix} \mathbf{B}_1^T & \mathbf{0} \\ \mathbf{0} & \mathbf{B}_2^T \end{bmatrix}$$

An $n \times n$ *identity matrix*, denoted as $\mathbf{I}$, or $\mathbf{I}_n$ to emphasize its dimension, is a diagonal matrix with $a_{ii} = 1$, $i = 1, \ldots, n$. A *null matrix*, or *zero matrix*, is designated as $\mathbf{0}$ and has $a_{ij} = 0$ for all i and j.

If two matrices $\mathbf{A}$ and $\mathbf{B}$ are of the same dimension, they are defined to be *equal matrices* if $a_{ij} = b_{ij}$, for all i and j. The *sum of two matrices* $\mathbf{A}$ and $\mathbf{B}$ that have the same dimension is a matrix with the same dimension, defined as

$$\mathbf{C} = \mathbf{A} + \mathbf{B} \tag{A.2.2}$$

where $c_{ij} = a_{ij} + b_{ij}$, for all i and j. The *difference between two matrices* $\mathbf{A}$ and $\mathbf{B}$ of the same dimension is defined as the matrix

$$\mathbf{D} = \mathbf{A} - \mathbf{B} \tag{A.2.3}$$

where $d_{ij} = a_{ij} - b_{ij}$, for all i and j. If three matrices have the same dimension, then (Prob. A.2.2)

$$(\mathbf{A} + \mathbf{B}) + \mathbf{C} = \mathbf{A} + (\mathbf{B} + \mathbf{C}) \equiv \mathbf{A} + \mathbf{B} + \mathbf{C} \tag{A.2.4}$$

Similarly, for matrices **A** and **B** with the same dimension (Prob. A.2.3),

$$\mathbf{A} + \mathbf{B} = \mathbf{B} + \mathbf{A} \tag{A.2.5}$$

Example A.2.4: The sum of the 2×3 matrices

$$\mathbf{A} = \begin{bmatrix} 1 & 2 & 1 \\ 0 & 3 & 1 \end{bmatrix}, \qquad \mathbf{B} = \begin{bmatrix} 0 & 1 & -2 \\ 2 & 2 & 1 \end{bmatrix}$$

is the 2×3 matrix

$$\mathbf{A} + \mathbf{B} = \begin{bmatrix} 1 & 3 & -1 \\ 2 & 5 & 2 \end{bmatrix}$$

The difference of **A** and **B** is

$$\mathbf{A} - \mathbf{B} = \begin{bmatrix} 1 & 1 & 3 \\ -2 & 1 & 0 \end{bmatrix}$$

Let **A** be an $m \times p$ matrix and **B** be a $p \times n$ matrix, written in the form

$$\mathbf{A} \equiv [a_{ij}] = \begin{bmatrix} \mathbf{a}_1^T \\ \mathbf{a}_2^T \\ \vdots \\ \mathbf{a}_m^T \end{bmatrix}, \qquad \mathbf{B} \equiv [b_{ij}] = [\mathbf{b}_1, \mathbf{b}_2, \ldots, \mathbf{b}_n] \tag{A.2.6}$$

where $\mathbf{a}_i^T$, $i = 1, \ldots, m$, are row matrices with p elements and $\mathbf{b}_j$, $j = 1, \ldots, n$, are column matrices with p elements. The *matrix product* of **A** and **B** is defined as the $m \times n$ matrix

$$\mathbf{C} = \mathbf{AB} \tag{A.2.7}$$

where

$$c_{ij} = \sum_{k=1}^{p} a_{ik} b_{kj}, \qquad \text{for all } i \text{ and } j \tag{A.2.8}$$

or, in terms of the rows of **A** and columns of **B**,

$$c_{ij} = \mathbf{a}_i^T \mathbf{b}_j, \qquad \text{for all } i \text{ and } j \tag{A.2.9}$$

Example A.2.5: The product of matrices

$$\mathbf{A} = \begin{bmatrix} 1 & 2 & 1 \\ 0 & 3 & 1 \end{bmatrix}, \qquad \mathbf{B} = \begin{bmatrix} 1 & 2 \\ 2 & 1 \\ 0 & 1 \end{bmatrix}$$

is the 2×2 matrix

$$\mathbf{AB} = \begin{bmatrix} 5 & 5 \\ 6 & 4 \end{bmatrix}$$

Example A.2.6: The product of matrices

$$\mathbf{A} = \begin{bmatrix} 1 & 1 \\ 2 & 2 \end{bmatrix}, \qquad \mathbf{B} = \begin{bmatrix} 2 & 1 \\ -2 & -1 \end{bmatrix}$$

is the 2 × 2 matrix

$$\mathbf{AB} = \begin{bmatrix} 0 & 0 \\ 0 & 0 \end{bmatrix}$$

Example A.2.6 shows that the product of two nonzero matrices can be a zero matrix. This is in sharp contrast to properties of multiplication of real numbers.

It is important to note that the product of two matrices is defined only if the number of columns in the first matrix equals the number of rows in the second matrix. From the definition of matrix multiplication, in general,

$$\mathbf{AB} \neq \mathbf{BA}$$

In fact, the products **AB** and **BA** are defined and have the same dimension only if both **A** and **B** are square and of equal dimension.

Example A.2.7: The products of the 2 × 2 matrices

$$\mathbf{A} = \begin{bmatrix} 1 & 2 \\ 2 & 1 \end{bmatrix}, \qquad \mathbf{B} = \begin{bmatrix} 1 & 0 \\ 1 & 2 \end{bmatrix}$$

in different orders are

$$\mathbf{AB} = \begin{bmatrix} 3 & 4 \\ 3 & 2 \end{bmatrix}$$

$$\mathbf{BA} = \begin{bmatrix} 1 & 2 \\ 5 & 4 \end{bmatrix}$$

Clearly, $\mathbf{AB} \neq \mathbf{BA}$. This is another contrast between properties of real and matrix multiplication.

If **A** and **B** are $m \times p$ matrices and **C** is a $p \times n$ matrix (Prob. A.2.4),

$$(\mathbf{A} + \mathbf{B})\mathbf{C} = \mathbf{AC} + \mathbf{BC} \tag{A.2.10}$$

Similarly, if **A** is an $m \times p$ matrix, **B** is a $p \times q$ matrix, and **C** is a $q \times n$ matrix (Prob. A.2.5),

$$(\mathbf{AB})\mathbf{C} = \mathbf{A}(\mathbf{BC}) \equiv \mathbf{ABC} \tag{A.2.11}$$

Multiplication of a matrix **A** by a scalar α is defined as

$$\mathbf{C} = \alpha\mathbf{A} \tag{A.2.12}$$

where

$$c_{ij} = \alpha a_{ij}, \qquad \text{for all } i \text{ and } j \tag{A.2.13}$$

If $a_{ij} = a_{ji}$ for all i and j, the square matrix $\mathbf{A}$ is called *symmetric* (i.e., $\mathbf{A} = \mathbf{A}^T$). If $a_{ij} = -a_{ji}$, for all i and j, the square matrix $\mathbf{A}$ is called *skew-symmetric* (i.e., $\mathbf{A} = -\mathbf{A}^T$). Note that in this case, $a_{ii} = 0$ for all i.

Example A.2.8: An example of a 2×2 symmetric matrix is

$$\mathbf{A} = \begin{bmatrix} -1 & 2 \\ 2 & 3 \end{bmatrix}$$

An example of a 2×2 skew-symmetric matrix is

$$\mathbf{B} = \begin{bmatrix} 0 & 1 \\ -1 & 0 \end{bmatrix}$$

The transpose of the sum of two matrices is the sum of their transposes (Prob. A.2.6); that is,

$$(\mathbf{A} + \mathbf{B})^T = \mathbf{A}^T + \mathbf{B}^T \tag{A.2.14}$$

Also, if $\mathbf{A}$ is an $m \times p$ matrix and $\mathbf{B}$ is a $p \times n$ matrix, then (Prob. A.2.8)

$$(\mathbf{AB})^T = \mathbf{B}^T\mathbf{A}^T \tag{A.2.15}$$

Any square matrix $\mathbf{A}$ can be uniquely expressed as the sum of symmetric and skew-symmetric matrices $\mathbf{B}$ and $\mathbf{C}$, respectively [2]; that is

$$\mathbf{A} = \mathbf{B} + \mathbf{C} \tag{A.2.16}$$

where

$$\mathbf{B} = \tfrac{1}{2}(\mathbf{A} + \mathbf{A}^T)$$
$$\mathbf{C} = \tfrac{1}{2}(\mathbf{A} - \mathbf{A}^T)$$

For any skew-symmetric matrix $\mathbf{C}$, $\mathbf{a}^T\mathbf{C}\mathbf{a} = 0$ (Prob. A.2.10).

A set of column matrices $\mathbf{a}_j$, $j = 1, \ldots, m$, of the same dimension is called *linearly dependent* if there are constants α_j, $j = 1, \ldots, m$, that are not all zero such that

$$\sum_{j=1}^{m} \alpha_j \mathbf{a}_j = \mathbf{0} \tag{A.2.17}$$

If a set of column matrices is linearly dependent, at least one of them can be written in terms of the others; for example, if $\alpha_k \neq 0$ is one of the nonzero constants in Eq. A.2.17, then

$$\mathbf{a}_k = -\frac{1}{\alpha_k} \sum_{j=1, j\neq k}^{m} \alpha_j \mathbf{a}_j$$

If a set of column matrices of the same dimension is not linearly dependent, it is called *linearly independent*. Equivalently, column matrices $\mathbf{a}_j$, $j = 1, \ldots, m$, of the same dimension are linearly independent if and only if Eq. A.2.17 implies that $\alpha_j = 0$, $j = 1, \ldots, m$.

If $\mathbf{a}_j$, $j = 1, \ldots, m$, are linearly dependent with $\alpha_k \neq 0$ in Eq. A.2.17, then $\mathbf{a}_k$ can be represented in terms of the remaining column matrices, so it is removed from the set. If the remaining column matrices $\mathbf{a}_j$, $j = 1, \ldots, k-1, k+1, \ldots, m$, are linearly dependent, then

$$\sum_{j=1, j\neq k}^{m} \beta_j \mathbf{a}_j = \mathbf{0}$$

with $\beta_l \neq 0$ for some l. Then

$$\mathbf{a}_l = -\frac{1}{\beta_l} \sum_{j=1, l\neq j\neq k}^{m} \beta_j \mathbf{a}_j$$

and $\mathbf{a}_l$ may be deleted. This process may be continued until a linearly independent set of column matrices remains. All the deleted column matrices can be written as sums of constants times the remaining linearly independent column matrices, so no information is lost.

Example A.2.9: The column matrices

$$\mathbf{a}_1 = \begin{bmatrix} 1 \\ 2 \\ -1 \end{bmatrix}, \quad \mathbf{a}_2 = \begin{bmatrix} 2 \\ 4 \\ -2 \end{bmatrix}$$

are linearly dependent, since

$$2\mathbf{a}_1 - \mathbf{a}_2 = \mathbf{0}$$

The column matrices

$$\mathbf{b}_1 = \begin{bmatrix} 1 \\ 2 \\ -1 \end{bmatrix}, \quad \mathbf{b}_2 = \begin{bmatrix} 1 \\ 2 \\ 1 \end{bmatrix}$$

are linearly independent, since

$$\alpha_1 \mathbf{b}_1 + \alpha_2 \mathbf{b}_2 = \mathbf{0}$$

is equivalent to the three scalar equations

$$\alpha_1 + \alpha_2 = 0$$

$$2\alpha_1 + 2\alpha_2 = 0$$

$$-\alpha_1 + \alpha_2 = 0$$

From the last equation, $\alpha_2 = \alpha_1$. Substituting this into the first equation yields $2\alpha_1 = 0$. Thus $\alpha_1 = \alpha_2 = 0$.

A general method of determining whether column matrices $\mathbf{a}_j$, $j = 1, \ldots,$

m, of the same dimension are linearly dependent is to form the matrix equation

$$\mathbf{0} = \sum_{j=1}^{m} \alpha_j \mathbf{a}_j = [\mathbf{a}_1, \ldots, \mathbf{a}_m] \begin{bmatrix} \alpha_1 \\ \vdots \\ \alpha_m \end{bmatrix} \equiv \mathbf{A}\boldsymbol{\alpha}$$

and to solve it for α_j, $j = 1, \ldots, m$. If the only solution is $\alpha_j = 0$, $j = 1, \ldots, m$, then the column matrices are linearly independent. If there is any solution for which some $\alpha_j \neq 0$, the column matrices are linearly dependent and the columns of the matrix $\mathbf{A}$ are said to be linearly dependent. Otherwise, they are linearly independent. Numerical methods for carrying out this process are presented in Appendix B.

Note that if a collection of column matrices $\mathbf{a}_j$, $j = 1, \ldots, m$, of the same dimension contains a zero column matrix, they are linearly dependent. To see this, let $\mathbf{a}_k = \mathbf{0}$. Then with $\alpha_j = 0$, $j \neq k$, and $\alpha_k = 1$,

$$\sum_{j=1}^{m} \alpha_j \mathbf{a}_j = \mathbf{a}_k = \mathbf{0}$$

so the column matrices are linearly dependent.

A set of row matrices $\mathbf{b}_i^T$, $i = 1, \ldots, m$, of the same dimension is linearly dependent if there are constants β_i, $i = 1, \ldots, m$, that are not all zero such that

$$\mathbf{0} = \sum_{i=1}^{m} \beta_i \mathbf{b}_i^T = [\beta_1, \ldots, \beta_m] \begin{bmatrix} \mathbf{b}_1^T \\ \vdots \\ \mathbf{b}_m^T \end{bmatrix} \equiv \boldsymbol{\beta}^T \mathbf{B}$$

or equivalently,

$$\mathbf{B}^T \boldsymbol{\beta} = \mathbf{0}$$

If $\mathbf{b}_1^T, \ldots, \mathbf{b}_m^T$ are linearly dependent, the rows of $\mathbf{B}$ are said to be linearly dependent. Otherwise, they are linearly independent.

The *row rank* (*column rank*) of a matrix $\mathbf{A}$ is defined as the largest number of linearly independent rows (columns) in the matrix. The row (column) rank may be determined by eliminating dependent rows (columns), using the procedure described for Eq. A.2.17. Although nontrival, it may be shown that the row and column ranks of any matrix are equal [2], hence defining the *rank* of the matrix. The rank of a matrix is also equal to the dimension of the largest square submatrix with nonzero determinant that can be obtained by deleting rows and columns of the original matrix [2]. A square matrix with linearly independent rows (columns) is said to have *full rank*. By the above, it must have a nonzero determinant. A numerical method for determining the rank of a matrix is presented in Appendix B.

When a square matrix does not have full rank, it is called *singular*. A square matrix $\mathbf{A}$ of full rank is called *nonsingular*. For such a matrix, there is an *inverse*

[2], denoted by $\mathbf{A}^{-1}$, such that

$$\mathbf{A}\mathbf{A}^{-1} = \mathbf{A}^{-1}\mathbf{A} = \mathbf{I} \tag{A.2.18}$$

where **I** is the identity matrix. Using the definition of Eq. A.2.18, it can be shown (Prob. A.2.12) that

$$(\mathbf{A}^{-1})^T = (\mathbf{A}^T)^{-1} \tag{A.2.19}$$

and if **A** and **B** are nonsingular (Prob. A.2.13) that

$$(\mathbf{AB})^{-1} = \mathbf{B}^{-1}\mathbf{A}^{-1} \tag{A.2.20}$$

A special nonsingular matrix that arises in describing the orientation of Cartesian reference frames is called an *orthogonal matrix*, with the property that

$$\mathbf{A}^{-1} = \mathbf{A}^T \tag{A.2.21}$$

Since constructing the inverse of a nonsingular matrix is generally time consuming, it is important to know whether a matrix is orthogonal. In this special case, the inverse is trivially constructed using Eq. A.2.21.

Example A.2.10: The matrix

$$\mathbf{A} = \begin{bmatrix} \cos\theta & -\sin\theta & 0 \\ \sin\theta & \cos\theta & 0 \\ 0 & 0 & 1 \end{bmatrix}$$

satisfies

$$\begin{aligned} \mathbf{A}\mathbf{A}^T &= \begin{bmatrix} \cos^2\theta + \sin^2\theta & 0 & 0 \\ 0 & \cos^2\theta + \sin^2\theta & 0 \\ 0 & 0 & 1 \end{bmatrix} \\ &= \mathbf{I} \end{aligned}$$

so it is orthogonal, for any value of θ. Thus,

$$\mathbf{A}^{-1} = \mathbf{A}^T = \begin{bmatrix} \cos\theta & \sin\theta & 0 \\ -\sin\theta & \cos\theta & 0 \\ 0 & 0 & 1 \end{bmatrix}$$

A symmetric $n \times n$ matrix **A** is called *positive definite* if $\mathbf{a}^T\mathbf{A}\mathbf{a} \geqslant 0$ for all $m \times 1$ column matrices **a** and if $\mathbf{a}^T\mathbf{A}\mathbf{a} = 0$ implies that $\mathbf{a} = \mathbf{0}$. A positive definite matrix is nonsingular [2].

A.3 ALGEBRAIC VECTORS

Recall from Eq. A.1.3 that a geometric vector $\vec{a}$ can be written in component form in a Cartesian *x-y-z* frame as

$$\vec{a} = a_x\vec{i} + a_y\vec{j} + a_z\vec{k}$$

The geometric vector $\vec{a}$ is thus uniquely defined by its Cartesian components, which may be written in matrix notation as

$$\mathbf{a} = [a_x, a_y, a_z]^T$$

This is the *algebraic representation of a vector.*

Note that the algebraic representation of vectors is dependent on the Cartesian reference frame selected (i.e., on vectors $\vec{i}$, $\vec{j}$, and $\vec{k}$). Some of the purely geometric properties of vectors are thus lost and properties of the reference frame that is used in defining components of vectors come into play. This apparent problem becomes a valuable tool in analysis of rigid body kinematics and dynamics.

An *algebraic vector* is defined as a column matrix. When an algebraic vector represents a geometric vector in three dimensional space, it has three components. Algebraic vectors with more than three components will also be employed in dynamics of multibody systems. In case $\mathbf{a} = [a_1, \ldots, a_n]^T$, the algebraic vector $\mathbf{a}$ is called an *n-vector* and is said to belong to *n-dimensional real space,* denoted R^n.

If two geometric vectors $\vec{a}$ and $\vec{b}$ are represented in algebraic form as

$$\mathbf{a} = [a_x, a_y, a_z]^T \tag{A.3.1}$$

and

$$\mathbf{b} = [b_x, b_y, b_z]^T \tag{A.3.2}$$

then their vector sum $\vec{c} = \vec{a} + \vec{b}$ of Eq. A.1.5 is represented in algebraic form by (Prob. A.3.1)

$$\mathbf{c} = \mathbf{a} + \mathbf{b} \tag{A.3.3}$$

Example A.3.1: The algebraic representation of the geometric vectors

$$\vec{a} = \vec{i} + 2\vec{j} + 3\vec{k}$$
$$\vec{b} = -\vec{i} + \vec{j} - \vec{k}$$

is

$$\mathbf{a} = [1, 2, 3]^T$$
$$\mathbf{b} = [-1, 1, -1]^T$$

The algebraic representation of the sum $\vec{c} = \vec{a} + \vec{b}$ *is*

$$\mathbf{c} = \mathbf{a} + \mathbf{b} = [0, 3, 2]^T$$

Two geometric vectors are equal, $\vec{a} = \vec{b}$, if and only if the Cartesian components of the vectors are equal (i.e., $\mathbf{a} = \mathbf{b}$). The last of Eqs. A.1.2 shows that multiplication of a vector $\vec{a}$ by a scalar α occurs component by component, so the geometric vector $\alpha\vec{a}$ is represented by the algebraic vector $\alpha\mathbf{a}$.

Since there is a one-to-one correspondence between geometric vectors and 3×1 algebraic vectors that are formed from their components in a specified

Cartesian reference frame, no distinction other than notation will be made between them in this book.

The scalar product of two geometric vectors may be expressed in algebraic form, using the result of Eq. A.1.11, as

$$\vec{a} \cdot \vec{b} = a_x b_x + a_y b_y + a_z b_z = \mathbf{a}^T \mathbf{b} \tag{A.3.4}$$

Example A.3.2: The scalar product of vectors $\vec{a}$ and $\vec{b}$ (or **a** and **b**) in Example A.3.1 is

$$\vec{a} \cdot \vec{b} = \mathbf{a}^T \mathbf{b} = [1, 2, 3] \begin{bmatrix} -1 \\ 1 \\ -1 \end{bmatrix} = -2$$

From the definition of scalar product,

$$\mathbf{a}^T \mathbf{b} = ab \cos \theta(\mathbf{a}, \mathbf{b})$$

and with $a = (\mathbf{a}^T\mathbf{a})^{1/2} = \sqrt{14}$ and $b = (\mathbf{b}^T\mathbf{b})^{1/2} = \sqrt{3}$, $\cos \theta(\mathbf{a}, \mathbf{b}) = -2/\sqrt{42}$. Thus $\theta(\mathbf{a}, \mathbf{b}) = 1.88$ or 4.40 rad.

A skew-symmetric matrix $\tilde{\mathbf{a}}$ associated with an algebraic 3-vector **a** is defined as

$$\tilde{\mathbf{a}} \equiv \begin{bmatrix} 0 & -a_z & a_y \\ a_z & 0 & -a_x \\ -a_y & a_x & 0 \end{bmatrix} \tag{A.3.5}$$

Note that an overhead (˜) (pronounced tilde) indicates that the components of the algebraic 3-vector **a** are used to generate a skew-symmetric 3×3 matrix $\tilde{\mathbf{a}}$. Conversely, any 3×3 skew-symmetric matrix **B** is of the form

$$\mathbf{B} = \begin{bmatrix} 0 & b_{12} & b_{13} \\ -b_{12} & 0 & b_{23} \\ -b_{13} & -b_{23} & 0 \end{bmatrix}$$

so it can be written as $\mathbf{B} = \tilde{\mathbf{b}}$, where $\mathbf{b} = [b_1, b_2, b_3]^T$, with $b_1 = -b_{23}$, $b_2 = b_{13}$, and $b_3 = -b_{12}$; that is,

$$\mathbf{B} = \tilde{\mathbf{b}} = \begin{bmatrix} -b_{23} \\ b_{13} \\ -b_{12} \end{bmatrix}$$

The vector product $\vec{c} = \vec{a} \times \vec{b}$, which is expanded in component form in Eq. A.1.16, can be written in algebraic vector form as

$$\mathbf{c} = \tilde{\mathbf{a}}\mathbf{b} = \begin{bmatrix} a_y b_z - a_z b_y \\ a_z b_x - a_x b_z \\ a_x b_y - a_y b_x \end{bmatrix} \tag{A.3.6}$$

This is an important result, from the point of view of both analytical manipulation and digital computation. It is used heavily throughout this book.

Example A.3.3: The algebraic representation of the vector product $\vec{c} = \vec{a} \times \vec{b}$ of $\vec{a}$ and $\vec{b}$ in Example A.3.1 is

$$\mathbf{c} = \tilde{\mathbf{a}}\mathbf{b} = \begin{bmatrix} 0 & -3 & 2 \\ 3 & 0 & -1 \\ -2 & 1 & 0 \end{bmatrix} \begin{bmatrix} -1 \\ 1 \\ -1 \end{bmatrix} = \begin{bmatrix} -5 \\ -2 \\ 3 \end{bmatrix}$$

For later use, it is helpful to develop some standard properties of the $(\tilde{\ })$ operation. First, note that

$$\tilde{\mathbf{a}}^T = \begin{bmatrix} 0 & a_z & -a_y \\ -a_z & 0 & a_x \\ a_y & -a_x & 0 \end{bmatrix} = -\tilde{\mathbf{a}} \tag{A.3.7}$$

Also, for a scalar α,

$$\alpha\tilde{\mathbf{a}} = \begin{bmatrix} 0 & -\alpha a_z & \alpha a_y \\ \alpha a_z & 0 & -\alpha a_x \\ -\alpha a_y & \alpha a_x & 0 \end{bmatrix} = (\widetilde{\alpha\mathbf{a}}) \tag{A.3.8}$$

For any vectors **a** and **b**, a direct calculation (Prob. A.3.2) shows that

$$\tilde{\mathbf{a}}\mathbf{b} = -\tilde{\mathbf{b}}\mathbf{a} \tag{A.3.9}$$

which agrees with the vector product relation $\vec{a} \times \vec{b} = -\vec{b} \times \vec{a}$ of Eq. A.1.13.

Direct calculation shows that

$$\tilde{\mathbf{a}}\mathbf{a} = \mathbf{0} \tag{A.3.10}$$

Hence, by Eq. A.3.7,

$$(\tilde{\mathbf{a}}\mathbf{a})^T = \mathbf{a}^T\tilde{\mathbf{a}}^T = -\mathbf{a}^T\tilde{\mathbf{a}} = \mathbf{0} \tag{A.3.11}$$

It can also be verified by direct calculation (Prob. A.3.3) that

$$\tilde{\mathbf{a}}\tilde{\mathbf{b}} = \mathbf{b}\mathbf{a}^T - \mathbf{a}^T\mathbf{b}\mathbf{I} \tag{A.3.12}$$

and that (Prob. A.3.4)

$$(\widetilde{\tilde{\mathbf{a}}\mathbf{b}}) = \mathbf{b}\mathbf{a}^T - \mathbf{a}\mathbf{b}^T \tag{A.3.13}$$

$$= \tilde{\mathbf{a}}\tilde{\mathbf{b}} - \tilde{\mathbf{b}}\tilde{\mathbf{a}} \tag{A.3.14}$$

From Eqs. A.3.13 and A.3.14,

$$\tilde{\mathbf{a}}\tilde{\mathbf{b}} + \mathbf{a}\mathbf{b}^T = \tilde{\mathbf{b}}\tilde{\mathbf{a}} + \mathbf{b}\mathbf{a}^T \tag{A.3.15}$$

Finally, it can be verified by direct calculation that

$$(\widetilde{\mathbf{a} + \mathbf{b}}) = \tilde{\mathbf{a}} + \tilde{\mathbf{b}} \tag{A.3.16}$$

Matrix implementation of vector operations permits systematic organization of calculations, which is especially valuable for computer calculations. A technique that takes advantage of matrix algebra is representation of vectors in convenient Cartesian reference frames that are fixed to bodies that move in space.

➧ **Example A.3.4:** To define a Cartesian x'-y'-z' reference frame with its origin at point P, with desired orientation in the x-y-z reference frame, it is essential to locate point P with vector $\mathbf{r}^P$, as shown in Fig. A.3.1. If a second point Q is located on the x' axis by vector $\mathbf{r}^Q$, the unit vector $\mathbf{f}$ along the x' axis is

$$\mathbf{f} = \frac{1}{|\mathbf{r}^Q - \mathbf{r}^P|}(\mathbf{r}^Q - \mathbf{r}^P)$$

If a third point R that is not on the x' axis is located in the x'-y' plane, by vector $\mathbf{r}^R$, the vector product of $\mathbf{f}$ and $\mathbf{r}^R - \mathbf{r}^P$ is a vector in the direction of the z' axis; that is,

$$\mathbf{h} = \frac{1}{|\tilde{\mathbf{f}}(\mathbf{r}^R - \mathbf{r}^P)|}\tilde{\mathbf{f}}(\mathbf{r}^R - \mathbf{r}^P)$$

Finally, the unit vector $\mathbf{g}$ along the y' axis is

$$\mathbf{g} = \tilde{\mathbf{h}}\mathbf{f}$$

Thus three noncollinear points P, Q, and R in the x'-y' plane define unit vectors along the x', y', and z' axes, hence defining the x'-y'-z' Cartesian reference frame.

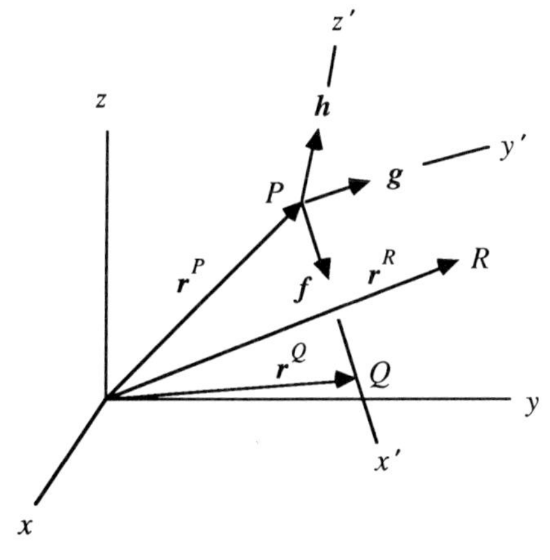

Figure A.3.1 Points that define Cartesian reference frame.

PROBLEMS

Section A.1

A.1.1 Use the Cartesian component forms of vectors $\vec{a}$, $\vec{b}$, and $\vec{c}$ to verify that Eq. A.1.6 is valid.

A.1.2 Carry out the expansion indicated to verify that Eq. A.1.11 is valid.

A.1.3 Carry out the expansion indicated to verify that Eq. A.1.16 is valid.

Section A.2

A.2.1 Use the definition of matrix transpose to verify that for
$\mathbf{D} = \text{diag}(\mathbf{B}_1, \mathbf{B}_2)$, $\quad \mathbf{D}^T = \text{diag}(\mathbf{B}_1^T, \mathbf{B}_2^T)$.

A.2.2 Verify that Eq. A.2.4 is valid.

A.2.3 Verify that Eq. A.2.5 is valid.

A.2.4 Verify that Eq. A.2.10 is valid.

A.2.5 Verify that Eq. A.2.11 is valid.

A.2.6 Verify that Eq. A.2.14 is valid.

A.2.7 Use Eqs. A.2.4 and A.2.14 to show that $(\mathbf{A} + \mathbf{B} + \mathbf{C})^T = \mathbf{A}^T + \mathbf{B}^T + \mathbf{C}^T$.

A.2.8 Verify that Eq. A.2.15 is valid.

A.2.9 Use Eqs. A.2.11 and A.2.15 to show that $(\mathbf{ABC})^T = \mathbf{C}^T\mathbf{B}^T\mathbf{A}^T$.

A.2.10 For any $n \times n$ skew-symmetric matrix $\mathbf{C}$ and $n \times 1$ matrix $\mathbf{a}$, show that $\mathbf{a}^T\mathbf{Ca} = 0$.

A.2.11 Determine whether the following column matrices are linearly dependent or independent:

(a)
$$\mathbf{a}_1 = \begin{bmatrix} 1 \\ 0 \\ 1 \end{bmatrix}, \quad \mathbf{a}_2 = \begin{bmatrix} 0 \\ 1 \\ 0 \end{bmatrix}, \quad \mathbf{a}_3 = \begin{bmatrix} 1 \\ 1 \\ 1 \end{bmatrix}$$

(b)
$$\mathbf{b}_1 = \begin{bmatrix} 1 \\ 0 \\ 1 \end{bmatrix}, \quad \mathbf{b}_2 = \begin{bmatrix} 0 \\ 1 \\ 0 \end{bmatrix}, \quad \mathbf{b}_3 = \begin{bmatrix} 1 \\ 0 \\ 0 \end{bmatrix}$$

A.2.12 Note that $\mathbf{I}^T = \mathbf{I}$, take the transpose of Eq. A.2.18, and use Eq. A.2.15 to verify that Eq. A.2.19 is valid.

A.2.13 Show that $(\mathbf{AB})(\mathbf{B}^{-1}\mathbf{A}^{-1}) = \mathbf{I}$, hence verifying that Eq. A.2.20 is valid.

Section A.3

A.3.1 Use the component representation of physical vectors and the associated properties of vector addition (Eq. A.1.5) to verify that Eq. A.3.3 is valid.

A.3.2 Expand both sides of Eq. A.3.9 and verify that it is valid.

A.3.3 Verify Eq. A.3.12 by expanding both sides and show that they are equal.

A.3.4 Verify Eqs. A.3.13 and A.3.14 by expanding both sides and show that they are equal.

A.3.5 Given points P with $\mathbf{r}^P = [1, 1, 1]^T$, Q with $\mathbf{r}^Q = [2, 1, 0]^T$, and R with $\mathbf{r}^R = [2, 2, 0]^T$, construct unit vectors $\mathbf{f}$, $\mathbf{g}$, and $\mathbf{h}$ that define an x'-y'-z' Cartesian reference frame with origin at point P, using the method of Example A.3.4.

SUMMARY OF KEY FORMULAS

Algebraic Vectors

$$\tilde{\mathbf{a}} = \begin{bmatrix} 0 & -a_z & a_y \\ a_z & 0 & -a_x \\ -a_y & a_x & 0 \end{bmatrix} \tag{A.3.5}$$

$$\tilde{\mathbf{a}}^T = -\tilde{\mathbf{a}}, \quad \tilde{\mathbf{a}}\mathbf{b} = -\tilde{\mathbf{b}}\mathbf{a}, \quad \tilde{\mathbf{a}}\mathbf{a} = \mathbf{0} \qquad \text{(A.3.7, 9, 10)}$$

$$\tilde{\mathbf{a}}\tilde{\mathbf{b}} = \mathbf{b}\mathbf{a}^T - \mathbf{a}^T\mathbf{b}\mathbf{I}, \quad \widetilde{(\tilde{\mathbf{a}}\mathbf{b})} = \mathbf{b}\mathbf{a}^T - \mathbf{a}\mathbf{b}^T = \tilde{\mathbf{a}}\tilde{\mathbf{b}} - \tilde{\mathbf{b}}\tilde{\mathbf{a}} \qquad \text{(A.3.12, 13, 14)}$$

$$\tilde{\mathbf{a}}\tilde{\mathbf{b}} + \mathbf{a}\mathbf{b}^T = \tilde{\mathbf{b}}\tilde{\mathbf{a}} + \mathbf{b}\mathbf{a}^T, \quad \widetilde{(\mathbf{a} + \mathbf{b})} = \tilde{\mathbf{a}} + \tilde{\mathbf{b}} \qquad \text{(A.3.15, 16)}$$

Appendix B
Solution of Linear and Nonlinear Equations

Matrix equations arise in both the theory and numerical solution of the equations of dynamics. The rank of Jacobian matrices of generalized coordinate mappings and constraint equations provide information on the number of degrees of freedom of a system and must usually be determined numerically in realistic applications. The purpose of this appendix is to provide an introduction to analysis and solution of matrix and nonlinear equations. The reader who is interested in more detail is referred to the literature on matrix theory and numerical methods (e.g., Ref. 2).

B.1 GAUSSIAN METHODS FOR LINEAR EQUATIONS

Consider a system of n linear algebraic equations in n unknowns, with real constant coefficients,

$$\begin{aligned} a_{11}x_1 + a_{12}x_2 + \cdots + a_{1n}x_n &= b_1 \\ a_{21}x_1 + a_{22}x_2 + \cdots + a_{2n}x_n &= b_2 \\ &\vdots \\ a_{n1}x_1 + a_{n2}x_2 + \cdots + a_{nn}x_n &= b_n \end{aligned} \tag{B.1.1}$$

which can be written in matrix form as

$$\mathbf{A}\mathbf{x} = \mathbf{b} \tag{B.1.2}$$

where

$$\mathbf{x} = [x_1, x_2, \ldots, x_n]^T$$

$$\mathbf{b} = [b_1, b_2, \ldots, b_n]^T$$

$$\mathbf{A} = \begin{bmatrix} a_{11} & a_{12} & \cdots & a_{2n} \\ a_{21} & a_{22} & \cdots & a_{2n} \\ & & \vdots & \\ a_{n1} & a_{n2} & \cdots & a_{nn} \end{bmatrix}$$

There are many methods of solving Eq. B.1.1. One of the easiest methods is *Gaussian elimination*, which is based on the elementary idea of eliminating variables one at a time. The Gaussian elimination method consists of two major steps, forward elimination and back substitution.

Forward Elimination

First, make the coefficient of x_1 in the first equation unity, by dividing that equation by a_{11} (presume for now that $a_{11} \neq 0$). Then eliminate the variable x_1 from the second to the nth equation, by multiplying the first modified equation by $-a_{j1}$ and adding the resulting equation to the jth equation, to obtain

$$\begin{bmatrix} 1 & a_{12}^{(1)} & a_{13}^{(1)} & \cdots & a_{1n}^{(1)} \\ 0 & a_{22}^{(1)} & a_{23}^{(1)} & \cdots & a_{2n}^{(1)} \\ \vdots & & & \vdots & \\ 0 & a_{n2}^{(1)} & a_{n3}^{(1)} & \cdots & a_{nn}^{(1)} \end{bmatrix} \begin{bmatrix} x_1 \\ x_2 \\ \vdots \\ x_n \end{bmatrix} = \begin{bmatrix} b_1^{(1)} \\ b_2^{(1)} \\ \vdots \\ b_n^{(1)} \end{bmatrix} \tag{B.1.3}$$

where the superscript (1) indicates values obtained after the first step of forward elimination.

Next, make the coefficient of x_2 in the second equation of Eq. B.1.3 equal to unity by dividing by $a_{22}^{(1)}$ (presume for now that $a_{22}^{(1)} \neq 0$). Then eliminate x_2 from the third to the nth equation by adding $-a_{j2}$ times the second modified equation to the jth equation, to obtain

$$\begin{bmatrix} 1 & a_{12}^{(1)} & a_{13}^{(1)} & \cdots & a_{1n}^{(1)} \\ 0 & 1 & a_{23}^{(2)} & \cdots & a_{2n}^{(2)} \\ 0 & 0 & a_{33}^{(2)} & \cdots & a_{3n}^{(2)} \\ \vdots & & & & \\ 0 & 0 & a_{n3}^{(2)} & \cdots & a_{nn}^{(2)} \end{bmatrix} \begin{bmatrix} x_1 \\ x_2 \\ \vdots \\ x_n \end{bmatrix} = \begin{bmatrix} b_1^{(1)} \\ b_2^{(2)} \\ \vdots \\ b_n^{(n)} \end{bmatrix} \tag{B.1.4}$$

After $n - 1$ steps of this algorithm, divide the nth row by $a_{nn}^{(n-1)}$ (presume for now it is not zero), to obtain the final result of forward elimination,

$$\begin{bmatrix} 1 & a_{12}^{(1)} & a_{13}^{(1)} & \cdots & a_{1n}^{(1)} \\ 0 & 1 & a_{23}^{(2)} & \cdots & a_{2n}^{(2)} \\ 0 & 0 & 1 & \cdots & a_{3n}^{(3)} \\ \vdots & & & & \\ 0 & 0 & 0 & \cdots & 1 \end{bmatrix} \begin{bmatrix} x_1 \\ x_2 \\ \vdots \\ x_n \end{bmatrix} = \begin{bmatrix} b_1^{(1)} \\ b_2^{(2)} \\ \vdots \\ b_n^{(n)} \end{bmatrix} \tag{B.1.5}$$

Since diagonal elements are unity and all elements below the diagonal of the matrix are zero, the determinant of the coefficient matrix in Eq. B.1.5 is unity, so it is nonsingular. Recall that to complete calculations leading to Eq. B.1.5, it was presumed that the modified diagonal elements $a_{jj}^{(j-1)}$ obtained after $j - 1$ steps were all nonzero. If this were not the case, the form of Eq. B.1.5 might be different. A refinement of the algorithm above is presented later in this section to overcome this limitation.

Back Substitution

Back substitution consists of $n - 1$ elementary solution steps, using Eq. B.1.5. From the nth equation of Eq. B.1.5, $x_n = b_n^{(n)}$. This value of x_n is then substituted into equation $n - 1$, which then gives the value of x_{n-1}. This process is repeated in the preceding equations, to solve for the remaining variables.

Example B.1.1: Solve the equations

$$\begin{bmatrix} 3 & 1 & -1 \\ -1 & 2 & 1 \\ 2 & -3 & 1 \end{bmatrix} \begin{bmatrix} x_1 \\ x_2 \\ x_3 \end{bmatrix} = \begin{bmatrix} 2 \\ 6 \\ -1 \end{bmatrix}$$

Forward elimination:

(1)
$$\begin{bmatrix} 1 & \frac{1}{3} & -\frac{1}{3} \\ -1 & 2 & 1 \\ 2 & -3 & 1 \end{bmatrix} \begin{bmatrix} x_1 \\ x_2 \\ x_3 \end{bmatrix} = \begin{bmatrix} \frac{2}{3} \\ 6 \\ -1 \end{bmatrix}$$

$$\begin{bmatrix} 1 & \frac{1}{3} & -\frac{1}{3} \\ 0 & \frac{7}{3} & \frac{2}{3} \\ 0 & -\frac{11}{3} & \frac{5}{3} \end{bmatrix} \begin{bmatrix} x_1 \\ x_2 \\ x_3 \end{bmatrix} = \begin{bmatrix} \frac{2}{3} \\ \frac{20}{3} \\ -\frac{7}{3} \end{bmatrix}$$

(2)
$$\begin{bmatrix} 1 & \frac{1}{3} & -\frac{1}{3} \\ 0 & 1 & \frac{2}{7} \\ 0 & -\frac{11}{3} & \frac{5}{3} \end{bmatrix} \begin{bmatrix} x_1 \\ x_2 \\ x_3 \end{bmatrix} = \begin{bmatrix} \frac{2}{3} \\ \frac{20}{7} \\ -\frac{7}{3} \end{bmatrix}$$

$$\begin{bmatrix} 1 & \frac{1}{3} & -\frac{1}{3} \\ 0 & 1 & \frac{2}{7} \\ 0 & 0 & \frac{19}{7} \end{bmatrix} \begin{bmatrix} x_1 \\ x_2 \\ x_3 \end{bmatrix} = \begin{bmatrix} \frac{2}{3} \\ \frac{20}{7} \\ \frac{57}{7} \end{bmatrix}$$

(3)
$$\begin{bmatrix} 1 & \frac{1}{3} & -\frac{1}{3} \\ 0 & 1 & \frac{2}{7} \\ 0 & 0 & 1 \end{bmatrix} \begin{bmatrix} x_1 \\ x_2 \\ x_3 \end{bmatrix} = \begin{bmatrix} \frac{2}{3} \\ \frac{20}{7} \\ 3 \end{bmatrix}$$

Back substitution:

(1) $x_3 = 3$

(2) $x_2 + \frac{2}{7}(3) = \frac{20}{7}$, so $x_2 = 2$

(3) $x_1 + \frac{1}{3}(2) - \frac{1}{3}(3) = \frac{2}{3}$, so $x_1 = 1$

It should be observed that the forward-elimination step requires division by $a_{jj}^{(j-1)}$ in steps 1 through n. The preceding operation is valid only if $a_{jj}^{(j-1)} \neq 0$. This element is called a *pivot*. Additional steps are required to handle a zero pivot.

In forward elimination, it is clear that the algorithm fails at the jth step if $a_{jj}^{(j-1)}$ is zero. Also, when the pivot element $a_{jj}^{(j-1)}$ becomes too small, *round-off error* in computation may lead to erroneous results. Therefore, the order in which the equations and variables are treated during forward elimination may significantly affect the accuracy of the algorithm. To circumvent this difficulty, the ordering of equations and variables may be determined by the algorithm (i.e., the ordering of equations and variables used need not necessarily be the original ordering 1, 2, . . . , n). The algorithm picks the equations and variables in an order that is determined by the actual system being solved.

Row Pivoting

In the jth forward-elimination step of the Gaussian algorithm, the equation with the largest coefficient (in absolute value) of x_j on or below the diagonal in the jth column is chosen for row pivoting in the jth step. Before the elimination step, this row of the matrix and the right-side vector are interchanged with the jth row and right side. This amounts simply to interchanging the order of equations, which does not affect the solution. This procedure is called *row pivoting*, or *partial pivoting*. The following example illustrates this procedure.

Example B.1.2: Perform Gaussian elimination with partial pivoting to solve the equations:

$$\begin{bmatrix} 4 & -3 & 5 & 2 \\ -3 & 1 & 1 & -6 \\ 5 & -5 & 10 & 0 \\ 2 & -3 & 9 & -7 \end{bmatrix} \begin{bmatrix} x_1 \\ x_2 \\ x_3 \\ x_4 \end{bmatrix} = \begin{bmatrix} -1.5 \\ 9 \\ 2.5 \\ 13.5 \end{bmatrix}$$

(1) The largest coefficient in column 1 is 5. Therefore, interchange the first and third equations to obtain

$$\begin{bmatrix} 5 & -5 & 10 & 0 \\ -3 & 1 & 1 & -6 \\ 4 & -3 & 5 & 2 \\ 2 & -3 & 9 & -7 \end{bmatrix} \begin{bmatrix} x_1 \\ x_2 \\ x_3 \\ x_4 \end{bmatrix} = \begin{bmatrix} 2.5 \\ 9 \\ -1.5 \\ 13.5 \end{bmatrix}$$

Forward elimination yields

$$\begin{bmatrix} 1 & -1 & 2 & 0 \\ 0 & -2 & 7 & -6 \\ 0 & 1 & -3 & 2 \\ 0 & -1 & 5 & -7 \end{bmatrix}\begin{bmatrix} x_1 \\ x_2 \\ x_3 \\ x_4 \end{bmatrix} = \begin{bmatrix} 0.5 \\ 10.5 \\ -3.5 \\ 12.5 \end{bmatrix}$$

(2) The largest coefficient in column 2 on or below the diagonal is -2, which is on the diagonal, so no interchange is necessary. Forward elimination yields

$$\begin{bmatrix} 1 & -1 & 2 & 0 \\ 0 & 1 & -3.5 & 3 \\ 0 & 0 & 0.5 & -1 \\ 0 & 0 & 1.5 & -4 \end{bmatrix}\begin{bmatrix} x_1 \\ x_2 \\ x_3 \\ x_4 \end{bmatrix} = \begin{bmatrix} 0.5 \\ -5.25 \\ 1.75 \\ 7.25 \end{bmatrix}$$

(3) The largest coefficient in column 3 on or below the diagonal is in the fourth row. Interchange third and fourth equations to obtain

$$\begin{bmatrix} 1 & -1 & 2 & 0 \\ 0 & 1 & -3.5 & 3 \\ 0 & 0 & 1.5 & -4 \\ 0 & 0 & 0.5 & -1 \end{bmatrix}\begin{bmatrix} x_1 \\ x_2 \\ x_3 \\ x_4 \end{bmatrix} = \begin{bmatrix} 0.5 \\ -5.25 \\ 7.25 \\ 1.75 \end{bmatrix}$$

Forward elimination yields

$$\begin{bmatrix} 1 & -1 & 2 & 0 \\ 0 & 1 & -3.5 & 3 \\ 0 & 0 & 1 & -2.66 \\ 0 & 0 & 0 & 0.33 \end{bmatrix}\begin{bmatrix} x_1 \\ x_2 \\ x_3 \\ x_4 \end{bmatrix} = \begin{bmatrix} 0.5 \\ -5.25 \\ 4.83 \\ -0.66 \end{bmatrix}$$

(4) Finally,

$$\begin{bmatrix} 1 & -1 & 2 & 0 \\ 0 & 1 & 3.5 & 3 \\ 0 & 0 & 1 & -2.66 \\ 0 & 0 & 0 & 1 \end{bmatrix}\begin{bmatrix} x_1 \\ x_2 \\ x_3 \\ x_4 \end{bmatrix} = \begin{bmatrix} 0.5 \\ -5.25 \\ 4.83 \\ -2 \end{bmatrix}$$

Back substitution yields $\mathbf{x} = [0.5, -1, -0.5, -2]^T$.

Full Pivoting

Full or complete pivoting is the selection of the largest element (in absolute value) on, below, or to the right of the diagonal element as the pivot for the next stage of forward elimination. In full pivoting, both row and column interchanges are carried out to bring the largest element to the diagonal. When two columns of the coefficient matrix are interchanged, their corresponding variables in vector **x** are interchanged. This process is called *full pivoting.*

Example B.1.3: Apply Gaussian elimination with full pivoting to solve the following set of equations:

$$\begin{bmatrix} 2 & -1 & 1 \\ -1 & 0 & 2 \\ 1 & 4 & -2 \end{bmatrix}\begin{bmatrix} x_1 \\ x_2 \\ x_3 \end{bmatrix} = \begin{bmatrix} 0 \\ 5 \\ -5 \end{bmatrix}$$

(1) The largest coefficient in the matrix is 4. Interchange columns 2 and 1 and the associated variable to bring the largest element into the first column,

$$\begin{bmatrix} -1 & 2 & 1 \\ 0 & -1 & 2 \\ 4 & 1 & -2 \end{bmatrix} \begin{bmatrix} x_2 \\ x_1 \\ x_3 \end{bmatrix} = \begin{bmatrix} 0 \\ 5 \\ -5 \end{bmatrix}$$

Interchange equations 3 and 1 to bring the largest element to the pivot position,

$$\begin{bmatrix} 4 & 1 & -2 \\ 0 & -1 & 2 \\ -1 & 2 & 1 \end{bmatrix} \begin{bmatrix} x_2 \\ x_1 \\ x_3 \end{bmatrix} = \begin{bmatrix} -5 \\ 5 \\ 0 \end{bmatrix}$$

Forward elimination yields

$$\begin{bmatrix} 1 & 0.25 & -0.5 \\ 0 & -1 & 2 \\ 0 & 2.25 & 0.5 \end{bmatrix} \begin{bmatrix} x_2 \\ x_1 \\ x_3 \end{bmatrix} = \begin{bmatrix} -1.25 \\ 5 \\ -1.25 \end{bmatrix}$$

(2) The largest eligible coefficient for the second pivot element is 2.25. Since it is in the second column, interchange rows 2 and 3 to obtain

$$\begin{bmatrix} 1 & 0.25 & -0.5 \\ 0 & 2.25 & 0.5 \\ 0 & -1 & 2 \end{bmatrix} \begin{bmatrix} x_2 \\ x_1 \\ x_3 \end{bmatrix} = \begin{bmatrix} -1.25 \\ -1.25 \\ 5 \end{bmatrix}$$

Forward elimination yields

$$\begin{bmatrix} 1 & 0.25 & -0.5 \\ 0 & 1 & 0.22 \\ 0 & 0 & 2.22 \end{bmatrix} \begin{bmatrix} x_2 \\ x_1 \\ x_3 \end{bmatrix} = \begin{bmatrix} -1.25 \\ -0.55 \\ 4.44 \end{bmatrix}$$

(3) Finally,

$$\begin{bmatrix} 1 & 0.25 & -0.5 \\ 0 & 1 & 0.22 \\ 0 & 0 & 1 \end{bmatrix} \begin{bmatrix} x_2 \\ x_1 \\ x_3 \end{bmatrix} = \begin{bmatrix} -1.25 \\ -0.55 \\ 2 \end{bmatrix}$$

Back substitution yields $\mathbf{x} = [-1, 0, 2]^T$.

In most computer programs, partial or full pivoting is carried out by interchanging two indices of the corresponding rows or columns, instead of interchanging coefficients in the matrix and variables. Two integer arrays hold the indices for column and row numbers of the matrix [2].

Forward Elimination with Nonsquare Matrices

The forward-elimination step of Gaussian method, with or without pivoting, can be applied to nonsquare matrices. Consider the case of m equations in n unknowns, with $m < n$; that is,

$$\begin{aligned} a_{11}x_1 + \cdots + a_{mn}x_n &= b_1 \\ &\vdots \\ a_{m1}x_1 + \cdots a_{mn}x_n &= b_m \end{aligned} \tag{B.1.6}$$

If no zero pivots are encountered in forward elimination, after m steps the system may be reduced to the form

$$\begin{bmatrix} 1 & a_{12}^{(m)} & \cdots & a_{1m}^{(m)} & a_{1m+1}^{(m)} & \cdots & a_{1n}^{(m)} \\ 0 & 1 & & & & & \\ \vdots & & & & & & \\ 0 & 0 & \cdots & 1 & a_{mm+1}^{(m)} & \cdots & a_{mn}^{(m)} \end{bmatrix} \begin{bmatrix} u_1 \\ \vdots \\ u_m \\ v_1 \\ \vdots \\ v_{n-m} \end{bmatrix} = \begin{bmatrix} b_1^{(m)} \\ \vdots \\ b_m^{(m)} \end{bmatrix} \tag{B.1.7}$$

where $\mathbf{u} = [u_1, \ldots, u_m]^T$ and $\mathbf{v} = [v_1, \ldots, v_{n-m}]^T$ are elements of $\mathbf{x}$ that are reordered when column pivoting is used. Equation B.1.7 may be written in partitioned form as

$$\mathbf{Uu} + \mathbf{Rv} = \hat{\mathbf{b}} \tag{B.1.8}$$

where

$$\begin{aligned} \mathbf{U} &= \begin{bmatrix} 1 & a_{12}^{(m)} & \cdots & a_{mm}^{(m)} \\ 0 & 1 & & \\ \vdots & & & \\ 0 & 0 & \cdots & 1 \end{bmatrix} \\ \mathbf{R} &= \begin{bmatrix} a_{1m+1}^{(m)} & \cdots & a_{mn}^{(m)} \\ \vdots & & \\ a_{mm+1}^{(m)} & & a_{mn}^{(m)} \end{bmatrix} \\ \hat{\mathbf{b}} &= \begin{bmatrix} b_1^{(m)} \\ \vdots \\ b_m^{(m)} \end{bmatrix} \end{aligned} \tag{B.1.9}$$

Note that if $\mathbf{v}$ in Eq. B.1.8 is fixed, $\mathbf{u}$ may be determined through back substitution, with

$$\mathbf{Uu} = \hat{\mathbf{b}} - \mathbf{Rv}$$

Thus $\mathbf{v}$ may be treated as a vector of *independent variables* and $\mathbf{u}$ as a vector of *dependent variables*. Note that this is just a computational implementation of the coordinate partitioning theorem of Section 2.3.

The factorization of Eq. B.1.7 is possible, using full pivoting, if and only if the coefficient matrix $\mathbf{A}$ of Eq. B.1.6 has full row rank [2]. Thus a method of determining the row rank of **A** is to carry out Gaussian forward elimination, using full pivoting, to obtain

$$\begin{bmatrix} 1 & a_{12}^{(r)} & \cdots & a_{(1r)}^{(r)} & a_{1r+1}^{(r)} & \cdots & a_{1,n}^{(r)} \\ 0 & 1 & & a_{2r}^{(r)} & a_{2r+1}^{(r)} & & \\ \vdots & & & & & & \\ 0 & 0 & & 1 & a_{rr+1}^{(r)} & & a_{rn}^{(r)} \\ 0 & 0 & & 0 & 0 & & 0 \\ \vdots & & & & & & \\ 0 & 0 & & 0 & 0 & & 0 \end{bmatrix} \begin{bmatrix} u_1 \\ \vdots \\ \\ u_r \\ v_1 \\ \vdots \\ v_{n-r} \end{bmatrix} = \begin{bmatrix} b_1^{(r)} \\ \vdots \\ \\ b_r^{(r)} \\ b_{r+1}^{(r)} \\ \vdots \\ b_m^{(r)} \end{bmatrix} \tag{B.1.10}$$

In partitioned matrix form, this is

$$\begin{aligned} \mathbf{U}_r\mathbf{u} + \mathbf{R}_r\mathbf{v} &= \hat{\mathbf{b}}_r \\ \mathbf{0} &= \hat{\mathbf{b}}_{m-r} \end{aligned} \tag{B.1.11}$$

where $\mathbf{U}_r$, $\mathbf{R}_r$, and $\hat{\mathbf{b}}_r$ are defined as in Eq. B.1.9, with m replaced by r, and

$$\hat{\mathbf{b}}_{m-r} = \begin{bmatrix} b_{r+1}^{(r)} \\ \vdots \\ b_m^{(r)} \end{bmatrix} \tag{B.1.12}$$

Thus the row rank of matrix **A** is r, and if $r < m$, the second of Eqs. B.1.11 shows that Eq. B.1.10, hence Eq. B.1.6, has a solution if and only if $\hat{\mathbf{b}}_{m-r} = \mathbf{0}$. If $\hat{\mathbf{b}}_{m-r} = \mathbf{0}$, the equations are said to be *consistent equations* and have many solutions. Otherwise, they are *inconsistent equations* and have no solution. If $\hat{\mathbf{b}}_{m-r} = \mathbf{0}$, the equations of Eq. B.1.6 that were interchanged into the last $m - r$ equations (rows) of Eq. B.1.7 are dependent and may be eliminated, without changing the solution of Eq. B.1.6.

Example B.1.4: Use Gaussian forward elimination to determine the row rank and dependent and independent variable partitioning for the equations

$$\mathbf{Ax} \equiv \begin{bmatrix} 1 & 0 & 2 & 1 & 4 \\ 1 & 1 & 2 & 0 & 3 \\ 3 & 1 & 6 & 2 & 11 \end{bmatrix} \begin{bmatrix} x_1 \\ x_2 \\ x_3 \\ x_4 \\ x_5 \end{bmatrix} = \begin{bmatrix} 3 \\ 2 \\ 8 \end{bmatrix} \equiv \mathbf{b}$$

The solution will be carried out without pivoting until a zero diagonal entry appears that requires pivoting.

(1) Forward elimination in the first column yields

$$\begin{bmatrix} 1 & 0 & 2 & 1 & 4 \\ 0 & 1 & 0 & -1 & -1 \\ 0 & 1 & 0 & -1 & -1 \end{bmatrix} \begin{bmatrix} x_1 \\ x_2 \\ x_3 \\ x_4 \\ x_5 \end{bmatrix} = \begin{bmatrix} 3 \\ -1 \\ -1 \end{bmatrix}$$

(2) Forward elimination in the second column yields

$$\begin{bmatrix} 1 & 0 & 2 & 1 & 4 \\ 0 & 1 & 0 & -1 & -1 \\ 0 & 0 & 0 & 0 & 0 \end{bmatrix} \begin{bmatrix} x_1 \\ x_2 \\ x_3 \\ x_4 \\ x_5 \end{bmatrix} = \begin{bmatrix} 3 \\ -1 \\ 0 \end{bmatrix}$$

Thus the rank of the coefficient matrix is just 2, and since $\hat{b}_3 = 0$, the equations are consistent. The partitioning defined by this process is $\mathbf{u} = [x_1, x_2]^T$ and $\mathbf{v} = [x_3, x_4, x_5]^T$. If a different right-side vector had been chosen, the rank of the coefficient matrix would still be 2, but the equations could be inconsistent (i.e., there may exist no solution). If full pivoting had been employed in the forward elimination process, the rank would still be determined to be 2, but a different partitioning of variables might result.

From a numerical point of view, the bottom right block of zeros in the coefficient matrix of Eq. B.1.10 will not contain precisely zero entries. The entries will be zero only to within the round-off error that has accumulated during forward elimination. Similarly, $\hat{\mathbf{b}}_{m-r}$ can only be zero to within the same numerical precision. A perplexing and numerically challenging situation arises when the bottom right factors in the coefficient matrix of Eq. B.1.10 are nearly zero, but not to within accumulated round-off error. If the entries in $\hat{\mathbf{b}}_{m-r}$ are a few orders of magnitude larger, numerical difficulties arise. In this case the coefficient matrix is called *ill-conditioned* and numerical difficulties will arise.

The column rank of a matrix $\mathbf{A}$ (e.g., the generalized coordinate mapping Jacobian $\mathbf{r_q}$, in Chapter 2) can be determined by applying Gaussian elimination to its transpose $\mathbf{A}^T$. The resulting row rank of $\mathbf{A}^T$ is the column rank of $\mathbf{A}$.

B.2 NEWTON–RAPHSON METHOD FOR NONLINEAR EQUATIONS

A frequently occurring problem in constrained system dynamics is to find solutions of n nonlinear algebraic equations of the form

$$\mathbf{\Phi}(\mathbf{q}) = \mathbf{0} \tag{B.2.1}$$

in variables $\mathbf{q}$. When $\mathbf{\Phi}(\mathbf{q})$ and its Jacobian $\mathbf{\Phi_q}$ can be evaluated, the iterative Newton–Raphson method may be used to find a solution of Eq. B.2.1.

Consider first a single equation that is nonlinear in the scalar variable q. Let $q = q^*$ be a solution of Eq. B.2.1 and let $q^{(i)}$ be an approximation of q^*. The Taylor series expansion of $\Phi(q)$ about $q^{(i)}$ is

$$\Phi(q) = \Phi(q^{(i)}) + \Phi_q(q^{(i)})(q - q^{(i)}) + \text{higher order terms} \tag{B.2.2}$$

Let $q^{(i+1)}$ be an improved approximation that is to be determined from the condition

$$\Phi(q^{(i+1)}) \approx \Phi(q^{i)}) + \Phi_q(q^{(i)})(q^{(i+1)} - q^{(i)}) = 0 \tag{B.2.3}$$

where if $q^{(i+1)} - q^{(i)}$ is small, higher-order terms are neglected. If $\Phi_q(q^{(i)}) \neq 0$, Eq. B.2.3 can be solved to obtain

$$q^{(i+1)} = q^{(i)} - \frac{\Phi(q^{(i)})}{\Phi_q(x^{(i)})} \tag{B.2.4}$$

Equation B.2.4 defines the *Newton–Raphson algorithm* that can be employed iteratively to produce a sequence of approximate solutions, beginning with an initial estimate $q^{(0)}$ of the solution of Eq. B.2.1:

1. Make an estimate $q^{(0)}$ of the solution of Eq. B.2.1.
2. In iteration i, $i = 0, 1, 2, \ldots$, evaluate $\Phi(q^{(i)})$ and $\Phi_q(q^{(i)})$. If $|\Phi(q^{(i)})| < \epsilon_e$, where ϵ_e is the equation error tolerance, and $|q^{(i)} - q^{(i-1)}| < \epsilon_s$, where ϵ_s is the solution error tolerance, terminate. If $\Phi_q(q^{(i)}) = 0$ and $\Phi(q^{(i)}) \neq 0$, return to step 1 with a new estimate. Otherwise, go to step 3.
3. Calculate $q^{(i+1)}$ from Eq. B.2.4 and return to step 2, with $i + 1$ in place of i.

The sequence $q^{(i)}$, in many problems, will approach a solution of $\Phi(q) = 0$. However, if $q^{(0)}$ is not adequately close to q^*, the sequence may diverge. If the Newton–Raphson algorithm converges and the Jacobian is nonsingular, it is said to be *quadratically convergent* [16]; that is, there is a constant c such that

$$|q^{(k+1)} - q^*| < c\,|q^{(k)} - q^*|^2 \tag{B.2.5}$$

As might be expected, if $|q^{(0)} - q^*|$ is large, the error may grow quadratically.

The geometry of Newton–Raphson iteration is illustrated in Fig. B.2.1 for a case in which it converges. Figure B.2.2 shows that the method may diverge

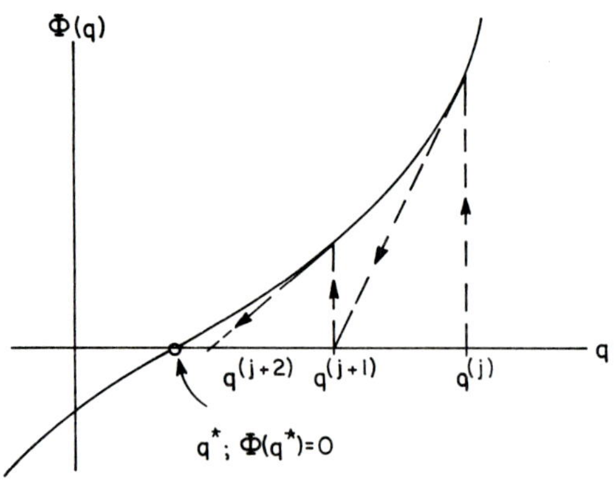

Figure B.2.1 Graphic representation of Newton–Raphson method.

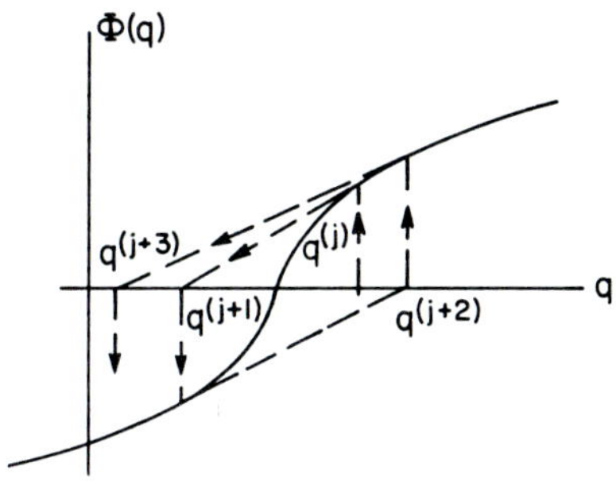

Figure B.2.2 Root at inflection point.

when the solution is an inflection point. If there are multiple solutions, the solution obtained is a function of the initial estimate, as shown in Fig. B.2.3. If the initial estimate is near a local minimum or maximum of Φ, as shown in Fig. B.2.4, the algorithm may fail to converge. It is essential that a reasonable estimate of the solution be used to initiate the iterative process.

Consider next Eq. B.2.1 as n nonlinear algebraic equations in n unknowns, where a solution vector is denoted as $\mathbf{q}^* \equiv [q_1^*, q_2^*, \ldots, q_n^*]^T$. A first order Taylor expansion about an estimate $\mathbf{q}^{(i)}$ of the solution can be applied to Eq. B.2.1. Neglecting higher order terms, the first order Taylor approximation for an improved approximation $\mathbf{q}^{(i+1)}$ is

$$\mathbf{\Phi}(\mathbf{q}^{(i)}) + \mathbf{\Phi_q}(\mathbf{q}^{(i)})[\mathbf{q}^{(i+1)} - \mathbf{q}^{(i)}] = \mathbf{0}$$

If the Jacobian matrix $\mathbf{\Phi_q}(\mathbf{q}^{(i)})$ is nonsingular, this equation may be solved for $\mathbf{q}^{(i+1)}$ by solving

$$\mathbf{\Phi_q}(\mathbf{q}^{(i)})\,\Delta\mathbf{q}^i = -\mathbf{\Phi}(\mathbf{q}^{(i)}) \tag{B.2.6}$$

and setting $\mathbf{q}^{(i+1)} = \mathbf{q}^{(i)} + \Delta\mathbf{q}_i$.

The *Newton–Raphson algorithm* for systems of equations is as follows:

1. Make an initial estimate $\mathbf{q}^{(0)}$ of the solution of Eq. B.2.1.
2. In iteration $i = 0, 1, 2, \ldots$, evaluate $\mathbf{\Phi}(\mathbf{q}^{(i)})$ and $\mathbf{\Phi_q}(\mathbf{q}^{(i)})$. If the magnitudes

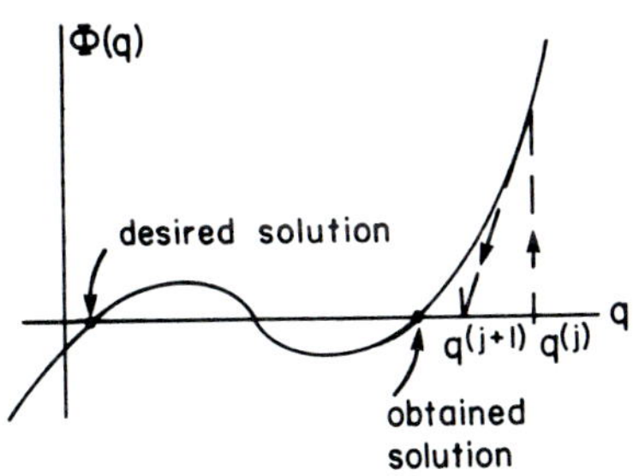

Figure B.2.3 Multiple roots.

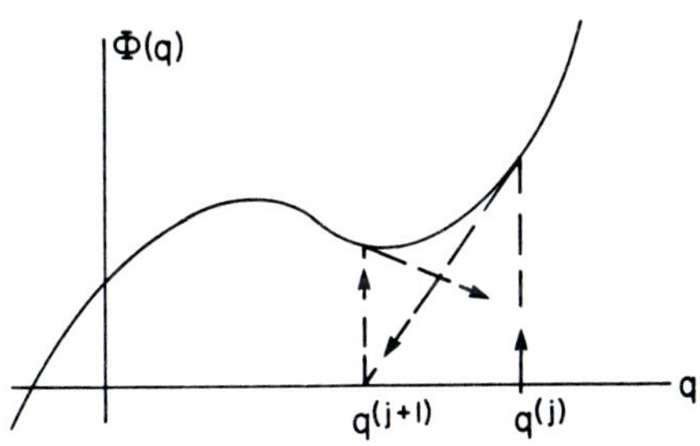

Figure B.2.4 Divergence near local minimum.

of all errors and solution changes satisfy

$$|\Phi_k(\mathbf{q}^{(i)})| \leq \epsilon_e, \qquad k = 1, \ldots, n$$

$$(\mathbf{q}^{(i)} - \mathbf{q}^{(i-1)})^T(\mathbf{q}^{(i)} - \mathbf{q}^{(i-1)}) \leq \epsilon_s^2$$

terminate. If $\mathbf{\Phi_q}(\mathbf{q}^{(i)})$ is singular and $\mathbf{\Phi}(\mathbf{q}^{(i)}) \neq \mathbf{0}$, return to step 1 with a new estimate. Otherwise, go to step 3. Here ϵ_e is equation error tolerance and ϵ_s is solution error tolerance.

3. Solve Eq. B.2.6 and set $\mathbf{q}^{(i+1)} = \mathbf{q}^{(i)} + \Delta\mathbf{q}^i$. Return to step 2, with $i + 1$ in place of i.

Convergence properties of this algorithm for systems of equations are essentially the same as for scalar equations [16].

Appendix C

Theory of Ordinary Differential and Differential-Algebraic Equations

Most of the attention in the book is devoted to derivation of differential equations of motion and differential-algebraic equations of motion, in the case of constrained dynamic systems. A wealth of theory exists on existence, uniqueness, and properties of solutions of ordinary differential equations, which is applicable to the study of dynamics [5]. Much less theory exists on differential-algebraic equations [17,23], even though such equations have been used by engineers for nearly two centuries. The purpose of this appendix is to summarize key results from the theory of initial-value problems and to adapt these theoretical results to the theory of second order differential equations and differential-algebraic equations of dynamics.

C.1 FIRST ORDER DIFFERENTIAL EQUATIONS

Consider a vector variable **x** that depends on time and a continuously differentiable vector function f that depends on **x** and time:

$$\begin{aligned} \mathbf{x}(t) &\equiv [x_1(t), \ldots, x_n(t)]^T \\ \mathbf{f}(\mathbf{x}, t) &\equiv [f_1(\mathbf{x}, t), \ldots, f_n(\mathbf{x}, t)]^T \end{aligned} \tag{C.1.1}$$

A system of *first order ordinary differential equations* may be written in the general form

$$\dot{\mathbf{x}} = \mathbf{f}(\mathbf{x}, t) \tag{C.1.2}$$

As is known [5], many solutions of differential equations of this form exist. To obtain a unique solution, *initial conditions* may be specified at some time t_0 in the form

$$\mathbf{x}(t_0) = \mathbf{x}^0 \tag{C.1.3}$$

where $\mathbf{x}^0$ is a specified vector of initial values of $\mathbf{x}$. The combination of the differential equation of Eq. C.1.2 and initial conditions of Eq. C.1.3 is an *initial-value problem* of ordinary differential equations. Existence and uniqueness theory for such initial-value problems, under rather weak hypotheses, is presented in the literature. An adequate result, for the purposes of the present study of dynamics, is the following theorem [5].

Initial-Value Existence Theorem: Let $\mathbf{f}(\mathbf{x}, t)$ be continuously differentiable in its arguments for $|x_i| \leq k$, $i = 1, \ldots, n$ and $t_0 \leq t \leq t_1$ and let $|x_i^0| < k$, $i = 1, \ldots, n$. Then there exists a unique solution $\mathbf{x}(t)$ of Eqs. C.1.2 and C.1.3 in an interval $t_0 \leq t \leq t^*$, with $t_0 < t^*$, such that $|x_i(t)| \leq k$, $i = 1, \ldots, n$.

It is important to note that this theorem guarantees a *local solution* of the initial-value problem of Eqs. C.1.2 and C.1.3, under rather mild hypotheses. The local nature of the solution is dictated by the fact that instabilities in solutions of nonlinear differential equations may arise. Globally valid existence theorems for general nonlinear ordinary differential equations are not generally available. Nevertheless, this theorem yields valuable information and permits initiation of a numerical integration process, using methods outlined in Appendix D. Providing the solution process continues to the desired terminal time, confidence in validity of the solution is obtained. Only if the integration process diverges must deeper theoretical analysis of properties of the solution be investigated.

C.2 REDUCTION OF SECOND ORDER EQUATIONS TO FIRST ORDER FORM

As noted throughout the book, second order differential equations of motion are obtained. To utilize the theoretical results available for first order systems, it is possible to reduce systems of second order equations to larger systems of first order equations. Let the independent generalized coordinates of the system under investigation be denoted as $\mathbf{v}$ and its velocity as $\dot{\mathbf{v}}$ and let the right side of the differential equations be $\mathbf{g}$, which depend on $\mathbf{v}$ and $\dot{\mathbf{v}}$; that is,

$$\begin{aligned} \mathbf{v} &= [v_1, \ldots, v_k]^T \\ \mathbf{g}(\mathbf{v}, \dot{\mathbf{v}}, t) &= [g_1(\mathbf{v}, \dot{\mathbf{v}}, t), \ldots, g_k(\mathbf{v}, \dot{\mathbf{v}}, t)]^T \end{aligned} \tag{C.2.1}$$

With this notation, a system of *second order differential equations* may be written in the form

$$\ddot{\mathbf{v}} = \mathbf{g}(\mathbf{v}, \dot{\mathbf{v}}, t) \tag{C.2.2}$$

As noted from elementary applications, to obtain a unique solution of the equations of dynamics, *initial conditions* must be given on both the independent generalized coordinates and their time derivatives; that is,

$$\begin{aligned} \mathbf{v}(t_0) &= \mathbf{v}^0 \\ \dot{\mathbf{v}}(t_0) &= \dot{\mathbf{v}}^0 \end{aligned} \tag{C.2.3}$$

The second order differential equations of Eq. C.2.2 and the initial conditions of Eq. C.2.3 constitute a second order *initial-value problem*.

In order to treat the second order initial-value problem of Eqs. C.2.2 and C.2.3, define the *state variable* of the system as a composit vector of position and velocity,

$$\mathbf{z} \equiv \begin{bmatrix} \mathbf{v} \\ \dot{\mathbf{v}} \end{bmatrix} \tag{C.2.4}$$

which is a vector of dimension $2k$. Defining the vector function

$$\mathbf{f}(\mathbf{z}, t) \equiv \begin{bmatrix} z_{k+1} \\ \vdots \\ z_{2k} \\ \mathbf{g}(z_1, \ldots, z_k, z_{k+1}, \ldots, z_{2k}, t) \end{bmatrix} \tag{C.2.5}$$

of dimension $2k$, the second order differential equation of Eq. C.2.2 and the initial conditions of Eq. C.2.3 may be written in first order form as

$$\begin{aligned} \dot{\mathbf{z}} &= \mathbf{f}(\mathbf{z}, t) \\ \mathbf{z}(t_0) &= \begin{bmatrix} \mathbf{v}^0 \\ \dot{\mathbf{v}}^0 \end{bmatrix} \end{aligned} \tag{C.2.6}$$

Note that the first k equations in the differential equation of Eq. C.2.6 are simply the definition of velocities as state variables and the last k equations are equivalent to Eq. C.2.2.

The initial-value existence theorem may now be applied to Eq. C.2.6 to assure existence of a unique solution, provided that the function **g** is continuously differentiable in its arguments. The result may then be translated as an existence and uniqueness result for the second order initial-value problem of Eqs. C.2.2 and C.2.3. Thus a rather complete theory of existence and uniqueness for second order differential equations of dynamics is obtained.

C.3 MIXED DIFFERENTIAL-ALGEBRAIC EQUATIONS OF MOTION

It is important to realize that the initial-value existence theorem yields information only on differential equations of motion and does not yield direct information on differential-algebraic equations. The coordinate partitioning theorem is applied here to provide a theoretical reduction of differential-algebraic equations to differential equation form, to obtain valid existence and uniqueness results for differential-algebraic equations of motion for constrained systems.

Constrained equations of motion derived in Chapters 3, 5, and 6 are summarized in matrix form as

$$\begin{bmatrix} \mathbf{M} & \mathbf{\Phi}_{\mathbf{q}}^T \\ \mathbf{\Phi}_{\mathbf{q}} & \mathbf{0} \end{bmatrix} \begin{bmatrix} \ddot{\mathbf{q}} \\ \boldsymbol{\lambda} \end{bmatrix} = \begin{bmatrix} \mathbf{Q} \\ \boldsymbol{\gamma} \end{bmatrix} \tag{C.3.1}$$

where the mass matrix $\mathbf{M}$ may depend on generalized coordinates, nonlinear terms in velocities and generalized coordinates are included in the right side, and constraints are of the form

$$\mathbf{\Phi}(\mathbf{q}, t) = \mathbf{0} \tag{C.3.2}$$

where $\mathbf{\Phi}(\mathbf{q}, t)$ is presumed to have two continuous derivatives with respect to its arguments. In addition, velocity equations must also hold,

$$\mathbf{\Phi}_{\mathbf{q}}\dot{\mathbf{q}} = \boldsymbol{\nu} \tag{C.3.3}$$

The differential and algebraic equations of Eqs. C.3.1 to C.3.3 comprise the *differential-algebraic equations of motion.*

In addition to the position and velocity equations of Eqs. C.3.2 and C.3.3, which must hold at the initial time t_0, additional *initial conditions* on position and velocity must be introduced,

$$\mathbf{\Phi}^I(\mathbf{q}(t_0), t_0) = \mathbf{0} \tag{C.3.4}$$

$$\mathbf{B}^I\dot{\mathbf{q}}(t_0) = \boldsymbol{\nu}^I \tag{C.3.5}$$

It is essential that the initial position condition of Eq. C.3.4, taken with the kinematic constraints of Eq. C.3.2 at t_0, uniquely determine the initial position $\mathbf{q}(t_0)$. Similarly, it is essential that the initial velocity conditions of Eq. C.3.5, taken with the velocity equation of C.3.3 at t_0, uniquely determine the initial velocity $\dot{\mathbf{q}}(t_0)$.

As shown by the constrained dynamic existence theorem in Section 3.4, the coefficient matrix in Eq. C.3.1 is nonsingular for meaningful physical systems. Therefore, it may theoretically be inverted to obtain an explicit solution of the form

$$\begin{aligned} \ddot{\mathbf{q}} &= \mathbf{f}(\mathbf{q}, \dot{\mathbf{q}}, t) \\ \boldsymbol{\lambda} &= \mathbf{g}(\mathbf{q}, \dot{\mathbf{q}}, t) \end{aligned} \tag{C.3.6}$$

where, by the implicit function theorem, the vector functions on the right are continuously differentiable with respect to their arguments. This result is theoretical, since explicit inversion of a large coefficient matrix with variable terms is impractical.

Direct numerical integration for $\mathbf{q}$, from Eq. C.3.6 and initial conditions, is both theoretically and computationally nontrivial. Since $\mathbf{q}$ and $\dot{\mathbf{q}}$ must satisfy Eqs. C.3.2 and C.3.3, all components of $\mathbf{q}$ are not independent. One algorithm for integration that is used in the literature is to ignore this dependence and simply integrate accelerations that are obtained by numerically solving Eq. C.3.1. There is no guarantee, however, that the function $\mathbf{q}(t)$ obtained will satisfy the constraints of Eqs. C.3.2 and C.3.3. This difficulty arises because the equations of motion are in fact differential-algebraic equations [23].

To obtain insight into the basic differential-algebraic equations of motion, consider a virtual displacement $\delta\mathbf{q}$ that satisfies the constraint equations to first order (i.e., $\mathbf{\Phi_q}\,\delta\mathbf{q} = \mathbf{0}$, where $\mathbf{\Phi_q}$ has full row rank). At a value of $\mathbf{q}$ that satisfies Eq. C.3.2, the Gaussian reduction technique of Appendix A may be employed to transform this equation to the form of Eq. B.1.8; that is,

$$\mathbf{U}\,\delta\mathbf{u} + \mathbf{R}\,\delta\mathbf{v} = \mathbf{0} \tag{C.3.7}$$

Since the coefficient matrix $\mathbf{U}$ is triangular with unit values on the diagonal, it is nonsingular and Eq. C.3.7 uniquely determines $\delta\mathbf{u}$, once values are selected for $\delta\mathbf{v}$. Therefore, $\mathbf{v}$ is interpreted as a vector of *independent generalized coordinates* and $\mathbf{u}$ is a vector of *dependent generalized coordinates*.

Since $\mathbf{U}$ in Eq. C.3.7 is a matrix that is obtained by elementary row operations [2] acting on the matrix $\mathbf{\Phi_u}$, $\mathbf{\Phi_u}$ is nonsingular. Therefore, by the implicit function theorem, Eq. C.3.2 can theoretically be solved for $\mathbf{u}$ as a function of $\mathbf{v}$ and t; that is,

$$\mathbf{u} = \mathbf{h}(\mathbf{v}, t) \tag{C.3.8}$$

where the vector function $\mathbf{h}(\mathbf{v}, t)$ is twice continuously differentiable in its arguments. Since Eq. C.3.2 is highly nonlinear, explicit construction of the functional relationship in Eq. C.3.8 is generally impractical. The result, however, is of great theoretical importance in analyzing the nature of solutions of differential-algebraic equations of motion.

The equations of motion of Eq. C.3.1 may be reordered and partitioned, according to the decomposition of $\mathbf{q}$ into $\mathbf{u}$ and $\mathbf{v}$, as

$$\begin{aligned}
\mathbf{M}^{\mathbf{uu}}\ddot{\mathbf{u}} + \mathbf{M}^{\mathbf{uv}}\ddot{\mathbf{v}} + \mathbf{\Phi}_{\mathbf{u}}^{T}\boldsymbol{\lambda} &= \mathbf{Q}^{\mathbf{u}} \\
\mathbf{M}^{\mathbf{vu}}\ddot{\mathbf{u}} + \mathbf{M}^{\mathbf{vv}}\ddot{\mathbf{v}} + \mathbf{\Phi}_{\mathbf{v}}^{T}\boldsymbol{\lambda} &= \mathbf{Q}^{\mathbf{v}} \\
\mathbf{\Phi_u}\ddot{\mathbf{u}} + \mathbf{\Phi_v}\ddot{\mathbf{v}} &= \boldsymbol{\gamma}
\end{aligned} \tag{C.3.9}$$

where the mass matrices in the first two equations are submatrices of $\mathbf{M}$ and the vector functions on the right are a partitioning of the vector $\mathbf{Q}$. Similarly, the velocity equation of Eq. C.3.3 may be written in the form

$$\mathbf{\Phi_u}\dot{\mathbf{u}} + \mathbf{\Phi_v}\dot{\mathbf{v}} = \boldsymbol{\nu} \tag{C.3.10}$$

For purposes of theoretical analysis, Eq. C.3.8 may be used to write all expressions that involve the dependent generalized coordinates $\mathbf{u}$ as functions of the independent coordinates $\mathbf{v}$ and time. Therefore, all terms in Eqs. C.3.9 and C.3.10 that depend on $\mathbf{u}$ may be interpreted as functions of $\mathbf{v}$. Since the coefficient matrix $\mathbf{\Phi_u}$ of $\dot{\mathbf{u}}$ in Eq. C.3.10 is nonsingular, the dependent velocity $\dot{\mathbf{u}}$ may theoretically be written as

$$\dot{\mathbf{u}} = \mathbf{\Phi_u}^{-1}[\boldsymbol{\nu} - \mathbf{\Phi_v}\dot{\mathbf{v}}] \tag{C.3.11}$$

which is a function only of $\mathbf{v}$ and $\dot{\mathbf{v}}$. Similarly, the third of Eqs. C.3.9 may theoretically be solved for $\ddot{\mathbf{u}}$, to obtain

$$\ddot{\mathbf{u}} = \mathbf{\Phi_u}^{-1}[\boldsymbol{\gamma} - \mathbf{\Phi_v}\ddot{\mathbf{v}}] \tag{C.3.12}$$

which, after using Eq. C.3.11, may be written in terms of only $\mathbf{v}$, $\dot{\mathbf{v}}$, and $\ddot{\mathbf{v}}$.

Since the coefficient matrix $\mathbf{\Phi_u}^T$ of $\boldsymbol{\lambda}$ in the first of Eqs. C.3.9 is nonsingular, this equation may theoretically be solved for $\boldsymbol{\lambda}$ to obtain

$$\boldsymbol{\lambda} = (\mathbf{\Phi_u}^{-1})^T[\mathbf{Q^u}(\mathbf{v}, \dot{\mathbf{v}}, t) - \mathbf{M^{uv}}\ddot{\mathbf{v}} - \mathbf{M^{uu}}\ddot{\mathbf{u}}] \tag{C.3.13}$$

Substituting this result and Eq. C.3.12 into the second of Eqs. C.3.9 yields

$$\hat{\mathbf{M}}^{\mathbf{v}}(\mathbf{v}, \dot{\mathbf{v}}, t)\ddot{\mathbf{v}} = \hat{\mathbf{Q}}^{\mathbf{v}}(\mathbf{v}, \dot{\mathbf{v}}, t) \tag{C.3.14}$$

where

$$\hat{\mathbf{M}}^{\mathbf{v}} = \mathbf{M^{vv}} - \mathbf{M^{vu}}\mathbf{\Phi_u}^{-1}\mathbf{\Phi_v} - \mathbf{\Phi_v}^T(\mathbf{\Phi_u}^{-1})^T[\mathbf{M^{uv}} - \mathbf{M^{uu}}\mathbf{\Phi_u}^{-1}\mathbf{\Phi_v}] \tag{C.3.15}$$

$$\hat{\mathbf{Q}}^{\mathbf{v}} = \mathbf{Q^v} - \mathbf{M^{vu}}\mathbf{\Phi_u}^{-1}\boldsymbol{\gamma} - \mathbf{\Phi_v}^T(\mathbf{\Phi_u}^{-1})^T[\mathbf{Q^u} - \mathbf{M^{uu}}\mathbf{\Phi_u}^{-1}\boldsymbol{\gamma}] \tag{C.3.16}$$

The result is a set of differential equations in only the independent generalized coordinates $\mathbf{v}$ that are consistent with position and velocity constraints that act on the system.

Example C.3.1: To illustrate the foregoing theoretical reduction process, consider the simple pendulum shown in Fig. C.3.1, with $\mathbf{q} = [x, y, \phi]^T$.

For this elementary system, the constraint equations are

$$\mathbf{\Phi}(\mathbf{q}) = \begin{bmatrix} x - \cos\phi \\ y - \sin\phi \end{bmatrix}$$

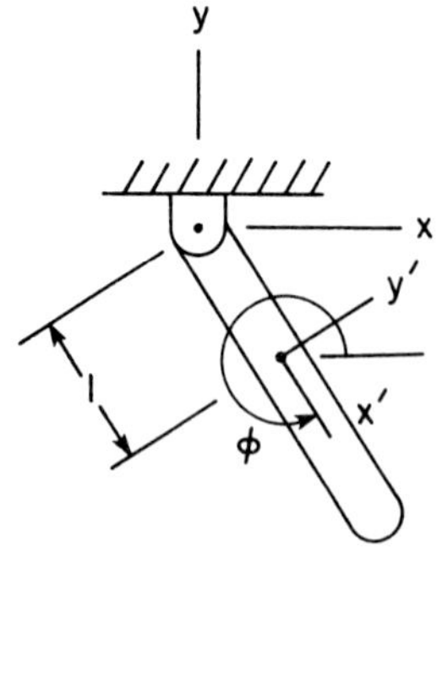

Figure C.3.1 Simple pendulum.

The constraint Jacobian is simply

$$\mathbf{\Phi_q(q)} = \begin{bmatrix} 1 & 0 & \sin\phi \\ 0 & 1 & -\cos\phi \end{bmatrix}$$

and virtual displacements $\delta\mathbf{q}$ satisfy

$$\mathbf{I}\begin{bmatrix} \delta x \\ \delta y \end{bmatrix} + \begin{bmatrix} \sin\phi \\ -\cos\phi \end{bmatrix}\delta\phi = \mathbf{0}$$

which is of the form of Eq. C.3.7, with $\mathbf{u} = [x, y]^T$ and $v = \phi$. Thus

$$\mathbf{\Phi_u} = \mathbf{I}$$

$$\mathbf{\Phi}_v = \begin{bmatrix} \sin\phi \\ -\cos\phi \end{bmatrix}$$

In this elementary example, the constraint equations are trivially solved for $\mathbf{u}$ as

$$\mathbf{u} = \begin{bmatrix} \cos\phi \\ \sin\phi \end{bmatrix} \equiv \mathbf{h}(v)$$

The right sides of the velocity and acceleration equations are simply $\boldsymbol{\nu} = \mathbf{0}$ and

$$\boldsymbol{\gamma} = \begin{bmatrix} -\cos\phi \\ -\sin\phi \end{bmatrix}\dot{\phi}^2$$

Thus Eqs. C.3.11 and C.3.12 are

$$\dot{\mathbf{u}} = \begin{bmatrix} -\sin\phi \\ \cos\phi \end{bmatrix}\dot{v}$$

$$\ddot{\mathbf{u}} = \begin{bmatrix} -\cos\phi \\ -\sin\phi \end{bmatrix}\dot{v}^2 - \begin{bmatrix} \sin\phi \\ -\cos\phi \end{bmatrix}\ddot{v}$$

The partitioned form of the equations of motion of Eq. C.3.9 is

$$\begin{bmatrix} m & 0 \\ 0 & m \end{bmatrix}\ddot{\mathbf{u}} + \boldsymbol{\lambda} = \begin{bmatrix} 0 \\ -mg \end{bmatrix}$$

$$J'\ddot{v} + [\sin\phi, \ -\cos\phi]\boldsymbol{\lambda} = 0$$

$$\ddot{\mathbf{u}} + \begin{bmatrix} \sin\phi \\ -\cos\phi \end{bmatrix}\ddot{v} = \begin{bmatrix} -\cos\phi \\ -\sin\phi \end{bmatrix}\dot{v}^2$$

Carrying out the reduction that leads to Eqs. C.3.14, C.3.15, and C.3.16, or simply evaluating terms in Eqs. C.3.15 and C.3.16 yields

$$\hat{M}^v = J' + m$$

$$\hat{Q}^v = -mg\cos v$$

Thus the reduced equation of motion of Eq. C.3.14 is simply

$$(J' + m)\ddot{v} = -mg\cos v$$

The reader is cautioned that while the theoretical reduction presented in this section can be explicitly carried out for this simple pendulum, it is not practical for more realistic systems. Even in the case of the two body model of a slider–crank mechanism of Example 6.3.2, this reduction procedure is impractical.

To assure existence of solutions of equations of Eq. C.3.14, it is important to show that the coefficient matrix on the left is nonsingular. To see that this is true, first observe that *kinematically admissible virtual velocities* must satisfy

$$\mathbf{\Phi_u}\,\delta\dot{\mathbf{u}} + \mathbf{\Phi_v}\,\delta\dot{\mathbf{v}} = \mathbf{0} \tag{C.3.17}$$

which is essentially Eq. C.3.10 with the right side equal to zero. Using the definition of the reduced mass matrix $\hat{\mathbf{M}}^{\mathbf{v}}$ of Eq. C.3.15 and the identity of Eq. C.3.17, the quadratic form associated with kinetic energy and independent virtual velocities may be expanded to obtain

$$\delta\dot{\mathbf{v}}^T\hat{\mathbf{M}}^{\mathbf{v}}\,\delta\dot{\mathbf{v}} = \delta\dot{\mathbf{q}}^T\mathbf{M}\,\delta\dot{\mathbf{q}} > 0 \tag{C.3.18}$$

which must hold for all $\delta\dot{\mathbf{v}} \neq \mathbf{0}$, since if $\delta\dot{\mathbf{v}} \neq \mathbf{0}$, $\delta\dot{\mathbf{q}} \neq \mathbf{0}$. Thus the matrix $\hat{\mathbf{M}}^{\mathbf{v}}$ is positive definite, hence nonsingular. From the initial-value existence theorem, the differential equation of Eq. C.3.14, with initial conditions on $\mathbf{v}$ defined by Eqs. C.3.2 to C.3.5, has a unique solution $\mathbf{v}(t)$. This $\mathbf{v}(t)$ and the associated $\mathbf{u}(t)$ that is defined by Eq. C.3.8 satisfy the differential-algebraic equations of motion (i.e., Eqs. C.3.1 to C.3.3).

The preceding argument provides a proof of existence of a unique solution of the equations of motion, under the modest hypotheses of the constrained dynamic existence theorem of Section 3.4 on kinematic and kinetic structure of the system. Computational implementation of the foregoing reduction to an independent set of equations, however, is not practical. Rather, implicit algorithms that in effect carry out this reduction are outlined in Appendix D.

Appendix D

Numerical Solution of Differential and Differential-Algebraic Equations

Basic concepts of polynomial interpolation and their use in deriving numerical algorithms for integrating initial-value problems of ordinary differential equations are developed. The resulting methods are applied, using the idea of generalized coordinate partitioning from Appendix C to numerically solve differential-algebraic equations.

D.1 POLYNOMIAL INTERPOLATION

Numerous algorithms for solving first order differential equations are presented in the literature [16]. The purpose of this section is to summarize basic ideas and numerical algorithms that may be used for this purpose. Consider the *initial-value problem*

$$\begin{aligned} \dot{\mathbf{x}} &= \mathbf{f}(\mathbf{x}, t) \\ \mathbf{x}(t_0) &= \mathbf{x}^0 \end{aligned} \tag{D.1.1}$$

where $\mathbf{x} = [x_1, x_2, \ldots, x_m]^T$ is the vector of variables to be integrated and the function $\mathbf{f}(\mathbf{x}, t)$ on the right of Eq. D.1.1 is continuously differentiable. The vector $\mathbf{x}^0$ defines initial values of $\mathbf{x}$.

Equation D.1.1 is an initial-value problem that is to be integrated numerically. Methods of polynomial interpolation, upon which most numerical integration algorithms are based, are summarized in this appendix and used to derive and demonstrate properties of numerical integration algorithms. Error estimates are presented to give an indication of methods that can be used for error control during the integration process. For a more detailed survey of numerical integration methods in dynamics, the reader may consult the survey article by Enright [18].

The fundamental idea in numerical integration is approximation of a function $f(t)$ by a polynomial in t (i.e., a polynomial that agrees with f and perhaps some of its derivatives at one or more points t_i). There are numerous methods by which approximating polynomials can be derived, each with its inherent advantages and disadvantages.

Taylor Expansion

The Taylor expansion method is based on approximating a function $f(t)$ near a point t_0 by the polynomial

$$T_k(t) \approx f(t_0) + f^{(1)}(t_0)(t - t_0) + \frac{f^{(2)}(t_0)(t - t_0)^2}{2!} + \cdots + \frac{f^{(k-1)}(t_0)(t - t_0)^{k-1}}{(k - 1)!} \tag{D.1.2}$$

where $f^{(i)}(t_0)$ denotes the ith derivative of $f(t)$, evaluated at t_0. This polynomial agrees with f and its first $k - 1$ derivatives at t_0; that is,

$$T_k^{(j)}(t_0) = f^{(j)}(t_0), \qquad j = 0, 1, \ldots, k - 1 \tag{D.1.3}$$

Example D.1.1: Consider the use of Eq. D.1.2 to approximate e^t, by expanding $f(t) = e^t$ about $t_0 = 0$. Since $f^{(j)} = e^t$, Eq. D.1.2 yields

$$\begin{aligned} T_k(t) &= e^0 + e^0(t - 0) + \frac{e^0(t - 0)^2}{2!} + \cdots + \frac{e^0(t - 0)^{k-1}}{(k - 1)!} \\ &= 1 + t + \frac{t^2}{2!} + \frac{t^3}{3!} + \cdots + \frac{t^{(k-1)}}{(k - 1)!} \end{aligned}$$

If $t = 1$, then $f(1) = e$ is approximated by

$$e \approx T_k(1) = 1 + 1 + \frac{1^2}{2!} + \frac{1^3}{3!} + \cdots + \frac{1^{(k-1)}}{(k - 1)!}$$

The value of e to 10-place accuracy is $e = 2.718281828$. Table D.1.1 lists $T_k(1)$ and the error $E_k = T_k(1) - e$ for various values of k

From this example, it is clear that a Taylor expansion may be used to approximate a function at nearby points, to any desired accuracy, provided that the function and $k - 1$ of its derivatives are known.

In general, values of a function $f(t)$ at several values of t are easier to obtain than high order derivatives of f at a single value of t, so the Taylor expansion is not well suited for general purpose integration algorithms. It is, however, useful

TABLE D.1.1 TAYLOR APPROXIMATE VALUES OF *e*

k	$T_k(1)$	E_k
1	1.000000000	−1.718281828
2	2.000000000	−0.718281828
3	2.500000000	−0.218281828
4	2.666666667	−0.0516151617
5	2.708333333	−0.0099484951
6	2.716666667	−0.0016151617
7	2.718055556	−0.0002262729
8	2.718253968	−0.0000278602
9	2.718278770	−0.0000030586
10	2.718281526	−0.0000003028
11	2.718281801	−0.0000000273
12	2.718281826	−0.0000000022
13	2.718281828	−0.0000000001

in understanding how other interpolating polynomials are developed into effective integration algorithms and how error estimates can be made.

Interpolating Polynomials

A polynomial $P_k(t)$ of degree $k - 1$ can be constructed that interpolates $f(t)$ at k distinct points $t_{n-k+1}, \ldots, t_{n-1}, t_n$; that is,

$$P_k(t_i) = f(t_i), \qquad i = n - k + 1, \ldots, n - 1, n$$

There is one and only one *interpolation polynomial* of degree $k - 1$ that satisfies these conditions [16]. For equally spaced points, $t_i - t_{i-1} = h$, this polynomial is constructed in the form of a *Newton backward difference polynomial* [16] as

$$P_k(t) = f_n + \sum_{i=1}^{k-1} \frac{\nabla^i f_n}{(i)!\, h^i} \prod_{j=0}^{i-1} (t - t_{n-j}) \tag{D.1.4}$$

where the *backward differences* $\nabla^i f_n$ are defined by

$$f_i = f(t_i), \qquad i = n - k + 1, \ldots, n$$

$$\nabla^1 f_n = f_n - f_{n-1}$$

$$\nabla^2 f_n = \nabla^1 f_n - \nabla^1 f_{n-1}$$

$$\vdots$$

$$\nabla^{k-1} f_n = \nabla^{k-2} f_n - \nabla^{k-2} f_{n-1}$$

and

$$\prod_{j=0}^{i-1} (t - t_{n-j}) = (t - t_n)(t - t_{n-1}) \cdots (t - t_{n-j+1})$$

➧ **Example D.1.2:** Consider the use of Eq. D.1.4 to approximate $e^{1.1}$, using Newton backward difference polynomials. Let $t_n = 1$ and $h = 0.1$. From Eq. D.1.4, $P_k(1.1)$ is evaluated for $k = 1, 2, \ldots, 10$ and is tabulated in Table D.1.2. The third column contains values of the approximation error $E_k = P_k(1.1) - e^{1.1}$. As expected, the higher the order of the polynomial, the better the approximation.

The reader is encouraged to verify that Eq. D.1.4 provides the desired values f_i for $k = 1, 2$, and 3.

Interpolation Error

In solving differential equations, interpolation on one set of data points needs to be transformed to interpolation on another set that is obtained by dropping one point and adding a new point, to march ahead on a time grid. The form of Taylor and Newton backward difference polynomials suggests that it may be possible to take advantage of the fact that most of the data remain the same. To go from approximating f near t_0 using $f_0, \ldots, f_0^{(k-1)}$, to using $f_0, \ldots, f_0^{(k)}$, by Taylor polynomials, Eq. D.1.2 yields

$$T_{k+1}(t) = T_k(t) + \frac{f_0^{(k)}}{(k)!}(t - t_0)^k$$

To achieve something equally simple in the case of increasing the data set from k to $k + 1$ points for Newton backward difference interpolation requires that

$$P_{k+1}(t) = P_k(t) + R_{k+1}(t) \tag{D.1.5}$$

where $R_{k+1}(t)$ must be of degree k. That is, adding a data point amounts to adding a correction term to the current interpolating polynomial. Substituting Eq. D.1.4 into Eq. D.1.5 and solving for $R_{k+1(x)}$ yields

$$R_{k+1}(t) = \frac{\nabla^k f_n}{(k)!\, h^k} \prod_{j=0}^{k-1} (t - t_{n-j}) \tag{D.1.6}$$

TABLE D.1.2 NEWTON BACKWARD-DIFFERENCE APPROXIMATION OF $e^{1.1}$

k	$P_k(1.1)$	E_k
1	2.718281828	−0.285884195
2	2.976960546	−0.027205478
3	3.001577080	−0.002588944
4	3.003919653	−0.000246371
5	3.004142579	−0.000023445
6	3.004163793	−0.000002231
7	3.004165812	−0.000000212
8	3.004166004	−0.000000020
9	3.004166022	−0.000000002
10	3.004166024	−0.000000000

It is shown in Ref. 16 that

$$\frac{\nabla^k f_k}{(k)!\, h^k} = \frac{f^{(k)}(\xi)}{(k)!} \tag{D.1.7}$$

for some ξ in $[t_{n-k}, t_n]$. The error in interpolating $f(t)$ by $P_k(t)$ is estimated, assuming that $f(t) \approx P_{k+1}(t)$, by combining Eqs. D.1.6 and D.1.7,

$$f(t) - P_k(t) \approx R_{k+1}(t) = \frac{f^{(k)}(\xi)}{(k)!} \prod_{j=0}^{k-1} (t - t_{n-j}) \tag{D.1.8}$$

for some ξ in $[t_{n-k}, t_n]$. How good the estimate is depends on how much $f^{(k)}$ varies over this interval.

An important application of Newton backward difference polynomials is use of past function values to extrapolate ahead in time, as was illustrated in Example D.1.2. Equation D.1.8 indicates that if the kth derivative of the function being approximated grows rapidly, approximation error increases. Even though the kth derivative of the function is not known, the error estimate on the right of Eq. D.1.8 may be approximated by the backward difference term on the left of Eq. D.1.7, which can be monitored during computation. If the error estimate indicates a problem, the time step h may be reduced to decrease the magnitude of error. This process is akin to driving a vehicle backward and attempting to steer using only the road path previously traversed, as shown in Fig. D.1.1. If the road suddenly changes direction (i.e., if curvature rapidly increases), the prudent driver will decrease speed of the vehicle to avoid disaster. This is equivalent to decreasing the distance traveled in a unit of time, the time step, to reduce error in following the road path.

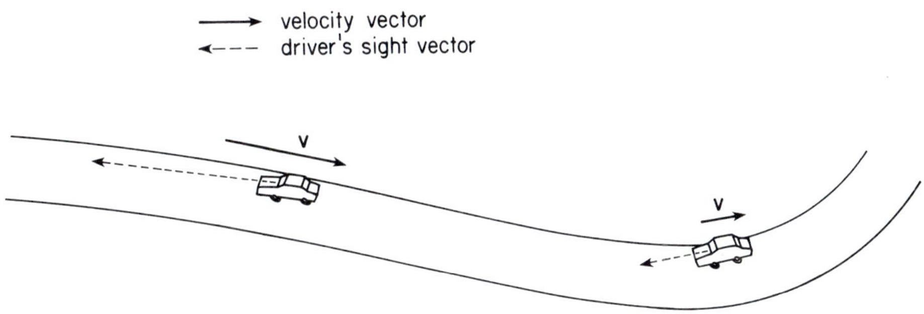

Figure D.1.1 Backward driving analogy.

D.2 NUMERICAL INTEGRATION OF INITIAL-VALUE PROBLEMS

Interpolation methods for approximating the solution of

$$\begin{aligned} \dot{\mathbf{x}} \equiv \mathbf{x}^{(1)}(t) &= \mathbf{f}(\mathbf{x}, t) \\ \mathbf{x}(t_0) &= \mathbf{x}^0 \end{aligned} \tag{D.2.1}$$

on an interval $[a, b]$ employ a mesh of grid points $\{t_0, t_1, \ldots\}$ in $[a, b]$, with $t_0 = a$. It is presumed that these grid points are equally spaced, in which case

$$t_n = t_0 + nh, \qquad n = 0, 1, \ldots$$

The notation $\mathbf{x}_n$ means an approximation of the solution $\mathbf{x}(t)$ of Eq. D.2.1 at grid point t_n; that is,

$$\mathbf{x}_n \approx \mathbf{x}(t_n)$$

Because $\mathbf{x}(t)$ satisfies Eq. D.2.1, an approximation of $\mathbf{x}(t_n)$ leads to an approximation of $\mathbf{x}_n^{(1)}(t_n)$; that is,

$$\mathbf{f}_n = \mathbf{f}(\mathbf{x}_n, t_n) \approx \mathbf{x}^{(1)}(t_n) = \mathbf{f}(\mathbf{x}(t_n), t_n)$$

The basic computational task in numerical integration is to advance the numerical solution to t_{n+1}, after having computed $\mathbf{x}_0, \mathbf{x}_1, \ldots, \mathbf{x}_n$. The solution of Eq. D.2.1 can be written as

$$\begin{aligned}\mathbf{x}(t_{n+1}) &= \mathbf{x}(t_n) + \int_{t_n}^{t_{n+1}} \mathbf{x}^{(1)}(t)\, dt \\ &= x(t_n) + \int_{t_n}^{t_{n+1}} \mathbf{f}(\mathbf{x}(t), t)\, dt\end{aligned}$$

The *Adams family of numerical integration methods* approximate this solution by replacing $\mathbf{f}(\mathbf{x}(t), t)$ with a polynomial that interpolates previously computed values $\mathbf{f}_i$, which is then integrated.

Adams–Bashforth Prediction

A convenient notation for the polynomial to be employed in the Adams–Bashforth formula of order k at t_n, which interpolates the computed values of $\mathbf{f}_i$ at k preceding points, is

$$\mathbf{P}_{k,n}(t_{n+1-j}) = \mathbf{f}_{n+1-j}, \qquad j = 1, 2, \ldots, k$$

Values of $\mathbf{f}_i$ and $\mathbf{x}_i$ must be stored from preceding steps. An approximation, or prediction, of the solution at t_{n+1} is obtained as

$$\mathbf{x}_{n+1}^p = \mathbf{x}_n + \int_{t_n}^{t_{n+1}} \mathbf{P}_{k,n}(t)\, dt \tag{D.2.2}$$

Newton's backward difference polynomial for $\mathbf{P}_{k,n}$, from Eq. D.1.4, is

$$\mathbf{P}_{k,n}(t) = \mathbf{f}_n + \sum_{i+1}^{k-1} \frac{\nabla^i \mathbf{f}_n}{(i)!\, h^i} \prod_{j=0}^{i-1} (t - t_{n-j})$$

This may be substituted into Eq. D.2.2 to obtain the *Adams–Bashforth predictor formula* of order k,

$$\mathbf{x}_{n+1}^p = \mathbf{x}_n + h \sum_{i=1}^{k} \gamma_{i-1} \nabla^{i-1} \mathbf{f}_n \tag{D.2.3}$$

where $\nabla^0 \mathbf{f}_n = \mathbf{f}_n$ and

$$\gamma_0 = 1 \qquad \text{(D.2.4)}$$

$$\gamma_i = \frac{1}{(i)!\,h} \int_{t_n}^{t_{n+1}} \prod_{j=0}^{i-1} \frac{t - t_{n-j}}{h}\, dt$$

To see that the values γ_i are independent of h, introduce the change of variable $s = (t - t_n)/h$, to obtain

$$\gamma_i = \frac{1}{(i)!} \int_0^1 \prod_{j=0}^{i-1} (s + j)\, ds$$

since $(t - t_{n-j})/h = [t - t_n + (t_n - t_{n-j})]/h = s + j$. Numerical values of the first six *Adams–Bashforth coefficients* are given in Table D.2.1.

The Adams–Bashforth formula of Eq. D.2.3 is called an *explicit method*, because the term $\mathbf{x}_{n+1}$ appears only on the left side of the equation and is determined explicitly by prior data.

The approximation $\mathbf{x}_{n+1}^p$ in Eq. D.2.3 has two sources of error. One is due to approximating $\mathbf{x}^{(1)}(t)$ by an interpolating polynomial, known as *local truncation error*. The other is due to errors that are present in the previously calculated values $\mathbf{x}_n, \mathbf{x}_{n-1}, \ldots$. It has been rigorously shown [16] that the latter contribution is insignificant, compared to local truncation error.

It is shown in Ref. 16 that the local truncation error is

$$\boldsymbol{\tau}_{n+1}^p = \gamma_k h^{k+1} \mathbf{x}^{(k+1)}(\xi) \approx h\gamma_k \nabla^k \mathbf{f}_n$$

for constant step size h and ξ in $[t_{n+1-k}, t_{n+1}]$. If enough differences are kept and if $h < 1$, it is possible to adjust order and/or step size to make the estimated local truncation error less than any given *error tolerance*.

Example D.2.1: Consider the initial-value problem

$$\dot{x} = f(x, t) = x$$
$$x(0) = 1 \qquad \text{(D.2.5)}$$

The solution is $x(t) = e^t$. A numerical approximation of the solution is to be constructed using Eq. D.2.3, with $h = 0.1$. Notice that at $t = 0$, the only available information is $x(0) = 1$, as shown in Table D.2.2; hence k must be 1. From Eq. D.2.3,

$$x_1 = x_0 + h(\gamma_0 \nabla^0 f_0)$$
$$= 1 + (0.1)(1)(1) = 1.1$$

TABLE D.2.1 ADAMS–BASHFORTH COEFFICIENTS

γ_0	γ_1	γ_2	γ_3	γ_4	γ_5
1	$\frac{1}{2}$	$\frac{5}{12}$	$\frac{3}{8}$	$\frac{251}{720}$	$\frac{95}{298}$

TABLE D.2.2 NEWTON BACKWARD DIFFERENCES

k	∇	f_0		f_1		f_2		f_3	f_4
1	$\nabla^0 f$	1.0							
2	$\nabla^0 f$	1.0		1.1					
	$\nabla^1 f$		0.1						
3	$\nabla^0 f$	1.0		1.1		1.215			
	$\nabla^1 f$		0.1		0.115				
	$\nabla^2 f$			0.015					
4	$\nabla^0 f$	1.0		1.1		1.215		1.342875	
	$\nabla^1 f$		0.1		0.115		0.127875		
	$\nabla^2 f$			0.015		0.012875			
	$\nabla^3 f$				−0.002125				
5	$\nabla^0 f$	1.0		1.1		1.215		1.342875	1.484013
	$\nabla^1 f$		0.1		0.115		0.127875	0.141138	
	$\nabla^2 f$			0.015		0.012875		0.132630	
	$\nabla^3 f$				−0.002125		0.000385		
	$\nabla^4 f$					0.002510			

At $t = 0.1$, x_1 and x_0 are available, hence $\nabla^1 f_1$ can be evaluated, as shown in Table D.2.2. Therefore, $k = 2$ may be used and

$$
\begin{aligned}
x_2 &= x_1 + h(\gamma_0 \nabla^0 f_1 + \gamma_1 \nabla^1 f_1) \\
&= 1.1 + (0.1)\{(1)(1.1) + (\tfrac{1}{2})(0.1)\} \\
&= 1.215
\end{aligned}
$$

Moving on to $t_3, t_4, \ldots$, with increasing values of k, yields

$$
\begin{aligned}
x_3 &= x_2 + h(\gamma_0 \nabla^0 f_2 + \gamma_1 \nabla^1 f_2 + \gamma_2 \nabla^2 f_2) \\
&= 1.215 + (0.1)\{(1)(1.215) + (\tfrac{1}{2})(0.115) + (\tfrac{5}{12})(0.015)\} \\
&= 1.342875 \\
x_4 &= x_3 + h(\gamma_0 \nabla^0 f_3 + \gamma_1 \nabla^1 f_3 + \gamma_2 \nabla^2 f_3 + \gamma_3 \nabla^3 f_3) \\
&= 1.342875 + (0.1)\{(1)(1.342875) + (\tfrac{1}{2})(0.127875) + (\tfrac{5}{12})(0.012875) \\
&\quad + (\tfrac{3}{8})(-0.002125)\} \\
&= 1.484013
\end{aligned}
$$

In Table D.2.3, a summary of approximate solutions of Eq. D.2.5 at $t = 10$, with $k = 1, 2, \ldots, 5$ is provided. In each case, the order of the algorithm starts at 1 and is increased up to k and then held fixed. Note that the error E_k decreases as k is increased. Note also that beyond $k = 4$, the error does not decrease significantly. This is because the error introduced in the first three steps at low order is larger than the error in all following steps at high order. The calculations above are repeated with $h = 0.01$ and summarized in Table D.2.4. As expected, with a smaller step size, more accurate approximations are obtained.

TABLE D.2.3 ADAMS–BASHFORTH PREDICTOR VALUES OF e^{10} FOR SEVERAL VALUES OF k, $h = 0.1$

k	$x^p(10)$	E_k
1	13780.612339815	−8245.853515625
2	21090.171400251	−936.294433594
3	21841.518610582	−184.947204590
4	21904.347159013	−122.118637085
5	21911.635344949	−114.830459595

TABLE D.2.4 ADAMS–BASHFORTH PREDICTOR VALUES OF e^{10} FOR SEVERAL VALUES OF k, $h = 0.01$

k	$x^p(10)$	E_k
1	20959.1556337710	−1067.310302734
2	22016.255843831	−10.209951401
3	22025.280168435	−1.185626507
4	22025.356977975	−1.108816862
5	22025.361257073	−1.104537964

Adams–Moulton Correction

The value $\mathbf{x}_{n+1}^p$ obtained by the Adams–Bashforth method may not be the best approximation to $\mathbf{x}(t_{n+1})$, because $\mathbf{f}(\mathbf{x}(t_{n+1}), t_{n+1})$ was not involved in determination of $\mathbf{x}_{n+1}$. Indeed, if $\mathbf{f}(\mathbf{x}(t_{n+1}), t_{n+1})$ should unexpectedly change in going from t_n to t_{n+1}, the difference between $\mathbf{x}_{n+1}$ and $\mathbf{x}(t_{n+1})$ could be greater than $\boldsymbol{\tau}_{n+1}^p$. This is because the factor $\mathbf{x}^{(k+1)}(\xi)$, which previously was smooth, has suddenly changed in magnitude in the interval t_n to t_{n+1}.

It seems plausible that a better approximation to $\mathbf{x}(t_{n+1})$ could be obtained if the Adams–Bashforth predicted value $\mathbf{x}_{n+1}^p$ of Eq. D.2.3 were improved by incorporating it into an interpolating polynomial for correction. The *Adams–Moulton corrector formula* of order $k + 1$ at t_n uses a polynomial $\mathbf{P}_{k+1,n}^*(t)$ that also interpolates the previous k values of $\mathbf{f}_i$,

$$\mathbf{P}_{k+1,n}^*(t_{n+1-j}) = \mathbf{f}_{n+1-j}, \qquad j = 1, \ldots, k$$

and the approximation for $\mathbf{f}$ at t_{n+1} obtained from the Adams–Bashforth predictor,

$$\mathbf{P}_{k+1,n}^*(t_{n+1}) = \mathbf{f}(\mathbf{x}_{n+1}^p, t_{n+1}) = \mathbf{f}_{n+1}^p$$

The corrected approximate solution $\mathbf{x}_{n+1}^c$ is then given by

$$\mathbf{x}_{n+1}^c = \mathbf{x}_n + \int_{t_n}^{t_{n+1}} \mathbf{P}_{k+1,n}^*(t)\, dt \tag{D.2.6}$$

The backward difference form of Eq. D.2.6, derived in the same way as Eq. D.2.3, is [16]

$$\mathbf{x}_{n+1}^{c} = \mathbf{x}_n + h \sum_{i=1}^{k+1} \gamma_{i-1}^{*} \nabla^{i-1} \mathbf{f}_{n+1}^{p} \tag{D.2.7}$$

where, with $s = (t - t_n)/h$,

$$\gamma_0^* = 1$$

$$\gamma_i^* = \frac{1}{i!} \int_0^1 (s-1)(s) \cdots (s+i-2)\, ds, \qquad i \geqslant 1$$

and

$$\nabla^0 \mathbf{f}_{n+1}^p = \mathbf{f}_{n+1}^p$$

$$\nabla^1 \mathbf{f}_{n+1}^p = \nabla \mathbf{f}_{n+1}^p = \mathbf{f}_{n+1}^p - \mathbf{f}_n$$

$$\vdots$$

$$\nabla^i \mathbf{f}_{n+1}^p = \nabla^{i-1} \mathbf{f}_{n+1}^p - \nabla^{i-1} \mathbf{f}_n$$

The first six *Adams–Moulton coefficients* are given in Table D.2.5.

Use of the Adams–Bashforth predictor or order k and the Adams–Moulton corrector of order $k + 1$ is called a *predictor–corrector algorithm* of order $K + 1$.

➧ **Example D.2.2:** The initial-value problem of Eq. D.2.5 is revisited for application of the Adams–Moulton corrector formula. The first predicted x_1 is, from Eq. D.2.3,

$$x_1^p = x_0 + h(\gamma_0 \nabla^0 f_0) = 1 + (0.1)(1)(1) = 1.1$$

This prediction is corrected by one higher-order formula; that is, from Eq. C.2.7,

$$\begin{aligned} x_1^c &= x_0 + h(\gamma_0^* \nabla^0 f_1 + \gamma_1^* \nabla^1 f_1) \\ &= 1 + (0.1)\{(1)(1.1) + (-\tfrac{1}{2})(0.1)\} = 1.105 \end{aligned}$$

The value at the next step is predicted as

$$x_2^p = x_1 + h(\gamma_0 \nabla^0 f_1) = 1.105 + (0.1)(1)(1.105) = 1.2155$$

and is corrected as

$$\begin{aligned} x_2^c &= x_1 + h(\gamma_0^* \nabla^0 f_2 + \gamma_1^* \nabla f_2) \\ &= 1.105 + (0.1)\{(1)(1.2155) + (-\tfrac{1}{2})(0.1105)\} = 1.221025 \end{aligned}$$

TABLE D.2.5 ADAMS–MOULTON COEFFICIENTS

γ_0^*	γ_1^*	γ_2^*	γ_3^*	γ_4^*	γ_5^*
1	$-\frac{1}{2}$	$-\frac{1}{12}$	$-\frac{1}{24}$	$-\frac{19}{720}$	$-\frac{3}{160}$

Similarly,

$$
\begin{aligned}
x_3^p &= x_2 + h(\gamma_0 \nabla^0 f_2) \\
&= 1.221025 + (0.1)(1)(1.221025) = 1.3431275 \\
x_3^c &= x_2 + h(\gamma_0^* \nabla^0 f_3 + \gamma_1^* \nabla^1 f_3) \\
&= 1.221025 + (0.1)\{(1)(1.3431275) + (-\tfrac{1}{2})(0.1221025)\} \\
&= 1.3492326 \\
x_4^p &= 1.3492326 + (0.1)(1)(1.3492326) = 1.4841559 \\
x_4^c &= 1.3492326 + (0.1)\{(1)(1.4841559) + (-\tfrac{1}{2})(0.1349233)\} \\
&= 1.4909205
\end{aligned}
$$

In Table D.2.6, a summary of the predictor–corrector solution of Eq. D.2.5 at $t = 10$, predicted by Adams–Bashforth formulas and corrected by one higher order of Adams–Moulton formulas is provided. Calculations are repeated with $h = 0.01$, and results are summarized in Table D.2.7. Note that solutions are substantially more accurate than those obtained in Example D.2.1 using only Adams–Bashforth predictors. Note also that for $k > 3$, the error that accumulated in starting the calculation with lower order formulas cannot be overcome.

TABLE D.2.6 ADAMS–MOULTON CORRECTOR VALUES OF e^{10} FOR SEVERAL VALUES OF k, $h = 0.1$

$k + 1$	$x^c(10)$	E_k
2	21688.414370387	−338.051452637
3	21980.499079835	−45.966720581
4	21994.743244024	−31.722553253
5	21990.708589885	−35.757209778
6	21986.702800628	−39.763000488

TABLE D.2.7 ADAMS–MOULTON CORRECTOR VALUES OF e^{10} FOR SEVERAL VALUES OF k, $h = 0.01$

k	$x^c(10)$	E_k
2	22022.822441367	−3.643353462
3	22026.251549431	−0.214245379
4	22026.188149769	−0.277645051
5	22026.130130982	−0.335663855
6	22026.088790048	−0.377004802

Similar to the previous development, the local truncation error $\boldsymbol{\tau}_{n+1}^c$ of the corrected value is defined by

$$\mathbf{x}^c(t_{n+1}) = \mathbf{x}(t_n) + \sum_{i=1}^{k+1} \gamma_{i-1}^* \nabla^{i-1}\mathbf{f}_{n+1}^p + \boldsymbol{\tau}_{n+1}^c$$

It is shown in Ref. 16 that for constant step size h, this truncation error is given by

$$\boldsymbol{\tau}_{n+1}^c = \gamma_{k+1}^* h^{k+2}\mathbf{x}^{(k+2)}(\xi) \approx h\gamma_{k+1}^* \nabla^{k+1}\mathbf{f}_{n+1}^p$$

with ξ in $[t_{n-1}, t_{n+1}]$.

The work involved in doing numerical integration on a computer is measured by the number of evaluations of $\mathbf{f}(\mathbf{x}, t)$ that are required. For complicated functions, such as those encountered in mechanical system dynamics, the remaining computation is relatively small, so this practice is a convenient way to measure work that is also machine independent. The predictor–corrector procedure costs two function evaluations at each step, so it is natural to ask: "Why not use the Adams–Bashforth method, which costs only one function evaluation at each step?" The predictor–corrector approach is much more accurate, as seen in Examples D.2.1 and D.2.2. Furthermore, it is so much better with respect to propagation of error that it can use steps that may be more than twice as large. Finally, error estimates are more reliable, which leads to a more effective selection of step size h.

Consider finally a *second order differential equation* and associated initial conditions,

$$\begin{aligned} \ddot{\mathbf{q}} &= \mathbf{f}(\mathbf{q}, \dot{\mathbf{q}}, t) \\ \mathbf{q}(t_0) &= \mathbf{q}^0 \\ \dot{\mathbf{q}}(t_0) &= \dot{\mathbf{q}}^0 \end{aligned} \tag{D.2.8}$$

By defining new variables $\mathbf{s} = \mathbf{q}$ and $\mathbf{r} = \dot{\mathbf{s}} = \dot{\mathbf{q}}$, a first order system that is equivalent to Eq. D.2.8 may be written as

$$\begin{aligned} \dot{\mathbf{s}} &= \mathbf{r} \\ \dot{\mathbf{r}} &= \mathbf{f}(\mathbf{s}, \mathbf{r}, t) \\ \mathbf{s}(t_0) &= \mathbf{q}^0 \\ \mathbf{r}(t_0) &= \dot{\mathbf{q}}^0 \end{aligned} \tag{D.2.9}$$

This first order initial-value problem may now be integrated for $\mathbf{s}$ and $\mathbf{r}$, equivalently $\mathbf{q}$ and $\dot{\mathbf{q}}$.

D.3 ALGORITHM FOR SOLVING DIFFERENTIAL-ALGEBRAIC EQUATIONS

Numerous efficient numerical integration algorithms are available for computing solutions of ordinary differential equations with initial conditions given. The purpose of this section is to show that *differential-algebraic equations* can be treated using differential equation methods.

It was demonstrated in Appendix B that if the constraint Jacobian has full row rank, it is theoretically possible to reduce the differential-algebraic equations of Eqs. C.3.1 to C.3.3 of Appendix C to a set of second order differential equations in independent generalized coordinates. The generalized coordinate partitioning method is used here to implement this method, employing a numerical reduction, rather than the explicit reduction outlined in Appendix C. Based on the knowledge that a theoretical reduction exists and has attractive mathematical properties such as a high degree of differentiability and nonsingularity of the coefficient matrix in Eq. C.3.14, an implicit computational implementation, called the *generalized coordinate partitioning algorithm* [12], may be carried out, as follows:

1. Begin with an assembled configuration that satisfies Eqs. C.3.2 to C.3.5 at the initial time t_0 (i.e., with $\mathbf{q}(t_0)$ and $\dot{\mathbf{q}}(t_0)$ that satisfy constraints).
2. Evaluate $\mathbf{\Phi_q}(t_0)$ and factor the matrix as in Eq. B.1.8 of Appendix B, to determine dependent and independent variables $\mathbf{u}$ and $\mathbf{v}$, respectively. This partitioning of the generalized coordinate vector $\mathbf{q}$ will be retained until computational tests indicate the need for refactorization and definition of a new partitioning.
3. At a typical time step t_i, solve Eq. C.3.1 for $\ddot{\mathbf{u}}$, $\ddot{\mathbf{v}}$, and $\boldsymbol{\lambda}$.
4. Integrate the first-order system

$$\begin{aligned}\dot{\mathbf{r}} &= \ddot{\mathbf{v}}\\ \dot{\mathbf{s}} &= \mathbf{r}\end{aligned} \tag{D.3.1}$$

   from t_i to t_{i+1}, using the Adams–Bashforth–Moulton predictor–corrector integration algorithm, to obtain $\mathbf{v}(t_{i+1})$ and $\dot{\mathbf{v}}(t_{i+1})$. Use the same integration algorithm and the first order system

$$\begin{aligned}\dot{\mathbf{x}} &= \ddot{\mathbf{u}}\\ \dot{\mathbf{y}} &= \mathbf{x}\end{aligned} \tag{D.3.2}$$

   to approximate $\mathbf{u}(t_{i+1})$.
5. Solve Eq. C.3.2 for an exact value of $\mathbf{u}(t_{i+1})$, using the Newton–Raphson method with the initial approximation of step 4. Solve Eq. C.3.10 for $\dot{\mathbf{u}}(t_{i+1})$.
6. If t_{i+1} exceeds the final time, stop. Otherwise, update i to $i + 1$ and continue.
7. If small pivot elements occur in Gaussian elimination with $\mathbf{\Phi_u}$, return to step 2, with t_0 replaced by t_{i+1}. Otherwise, return to step 3.

This implementation of the generalized coordinate partitioning algorithm has been used extensively and has been found to be reliable and accurate. It satisfies constraints to the desired degree of precision specified by the user and in general maintains good error control. For a more detailed presentation of methods for numerical solution of differential-algebraic equations, see Refs. 12 and 23.

REFERENCES

1. H. F. Davis, and A. D. Snider, *Introduction to Vector Analysis*, 5th Edition, Allyn and Bacon, Boston, 1987.
2. G. Strang, *Linear Algebra and Its Applications*, 2nd Edition, Academic Press, New York, 1980.
3. A. E. Taylor, and W. R. Mann, *Advanced Calculus*, 3rd Edition, Wiley, New York, 1983.
4. I. N. Sneddon, *Elements of Partial Differential Equations*, McGraw-Hill, New York, 1957.
5. E. A. Coddington and N. Levinson, *Theory of Ordinary Differential Equations*, McGraw-Hill, New York, 1955.
6. R. M. Rosenberg, *Analytical Dynamics of Discrete Systems*, Plenum Press, New York, 1977.
7. G. Hamel, *Theoretische Mechanics*, Springer-Verlag, Berlin, 1949.
8. L. A. Pars, *A Treatise on Analytical Dynamics*, Ox Bow Press, Woodbridge, Conn., 1979.
9. H. Goldstein, *Classical Mechanics*, 2nd Edition, Addison-Wesley, Reading, Mass., 1980.
10. J. Wittenburg, *Dynamics of Systems of Rigid Bodies*, B. G. Teubner, Stuttgart, 1977.
11. C. A. Truesdell, *Essays in the History of Mechanics*, Springer-Verlag, New York, 1968.
12. E. J. Haug, *Computer Aided Kinematics and Dynamics*, Vol. I: *Basic Methods*, Allyn and Bacon, Boston, 1989.
13. E. J. Haug, S. S. Kim, and F. F. Tsai, *Computer Aided Kinematics and Dynamics*, Vol. II: *Advanced Methods*, Prentice Hall, Englewood Cliffs, N.J., 1992.
14. I. S. Duff, A. M. Erisman, and J. K. Reid, *Direct Methods for Sparse Matrices*, Clarendon Press, Oxford, 1986.
15. M. A. Hussain, and B. Noble, "Application of Symbolic Computation to the Analysis of Mechanical Systems, Including Robot Arms," *Computer Aided Analysis and Optimization of Mechanical System Dynamics* (ed. E. J. Haug), Springer-Verlag, Heidelberg, 1984, pp. 283–306.
16. K. E. Atkinson, *An Introduction to Numerical Analysis*, Wiley, New York, 1988.
17. L. D. Petzold, "Differential/Algebraic Equations Are Not ODE's," *SIAM Journal of Scientific and Statistical Computing*, Vol. 3, No. 3, 1982, pp. 367–384.
18. W. H. Enright, "Numerical Methods for Systems of Initial Value Problems: The State

of the Art," *Computer Aided Analysis and Optimization of Mechanical System Dynamics* (ed. E. J. Haug), Springer-Verlag, Heidelberg, 1984, pp. 309–322.

19. R. E. Roberson, and R. Schwertassek, *Dynamics of Multibody Systems*, Springer-Verlag, Berlin, 1988.
20. D. S. Bae, and E. J. Haug, "A Recursive Formulation for Constrained Mechanical System Dynamics: Part I, Open Loop Systems," *Mechanics of Structures and Machines*, Vol. 15, No. 3, 1987, pp. 359–382.
21. D. S. Bae, and E. J. Haug, "A Recursive Formulation for Constrained Mechanical System Dynamics: Part II, Closed Loop Systems," *Mechanics of Structures and Machines*, Vol. 15, No. 4, 1987–88, pp. 481–506.
22. D. S. Bae, J. G. Kuhl, and E. J. Haug, "A Recursive Formulation for Constrained Mechanical System Dynamics: Part III, Parallel Processor Implementation," *Mechanics of Structures and Machines*, Vol. 16, No. 2, 1988, pp. 249–269.
23. E. J. Haug, and R. Deyo, *Real-Time Integration Methods for Mechanical System Simulation*, Springer-Verlag, Heidelberg, 1990.

Index